ISBN 978-1-331-98249-4
PIBN 10049648

THE STANDARD

NATURAL HISTORY.

EDITED BY

JOHN STERLING KINGSLEY.

VOL. I.

LOWER INVERTEBRATES.

Illustrated

BY FIVE HUNDRED AND ONE WOOD-CUTS AND TWENTY-TWO FULL-PAGE PLATES.

Boston:

S. E. CASSINO AND COMPANY.

1885.

C. J. PETERS AND SON,
STEREOTYPERS AND ELECTROTYPERS,
145 HIGH STREET.

CONTENTS.

CONTENTS.

LIST OF PLATES.

Cucumaria hyndemanni, sea cucumber.

Polystomella strigillata, a foraminifer with pseudopodia extended, enlarged 200 times.

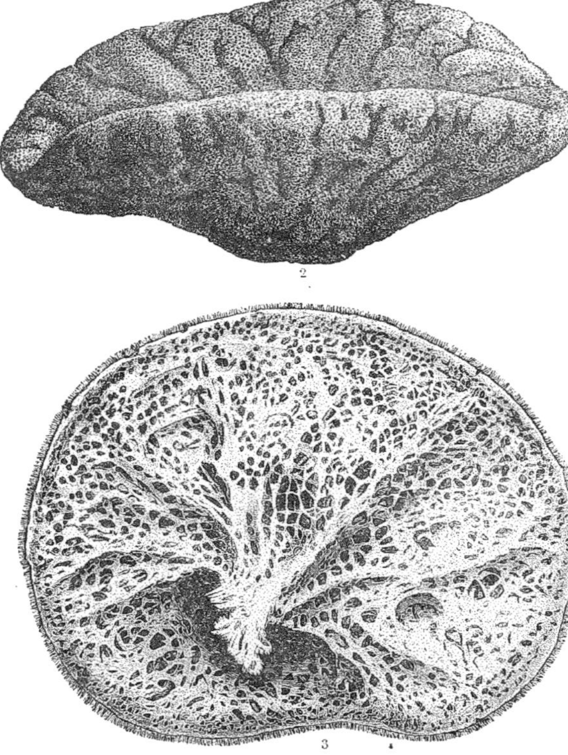

SPONGES.

1. *Ventriculites simplex.* 2. *Dactylocalyx pumicea.* 3. *Hyalonema toxeres*, under surface.

Dalmatian sponge fishery.

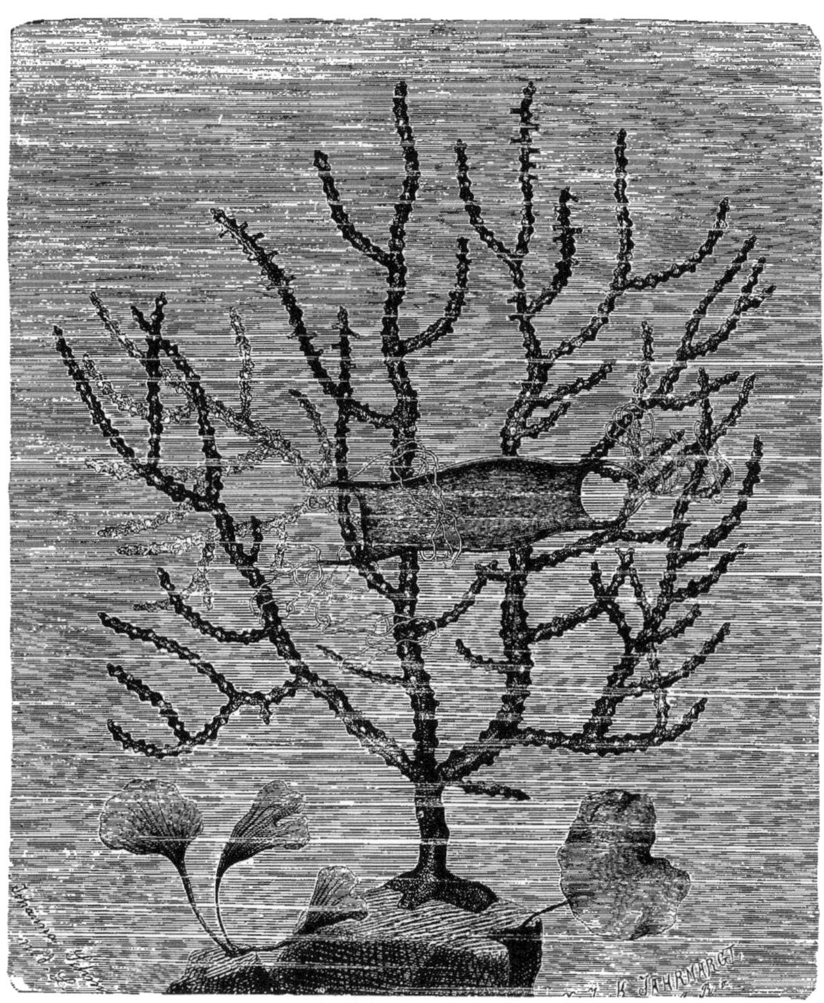

Gorgonia verrcosa, to which is attached a skate's egg, natural size.

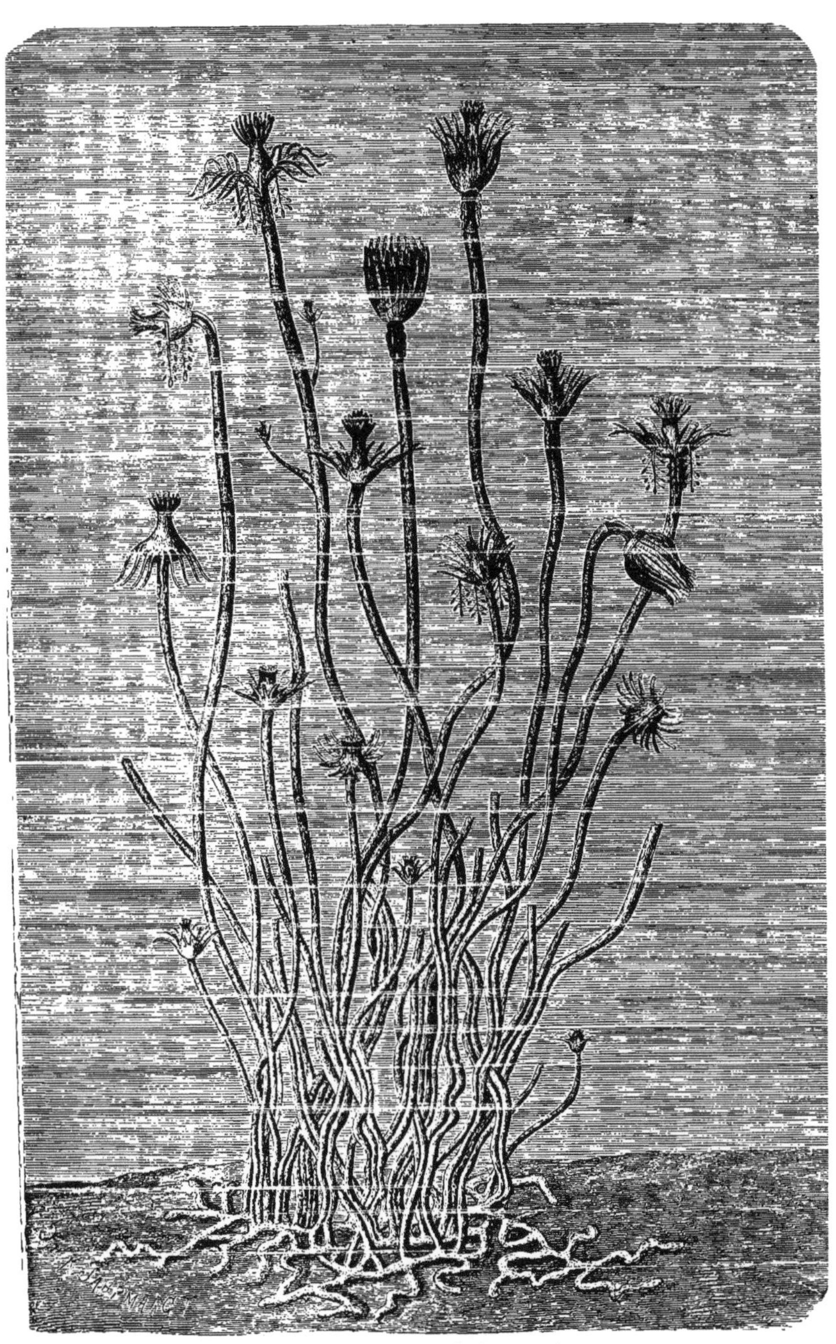

Tubularia indivisa, tubularian hydroid.

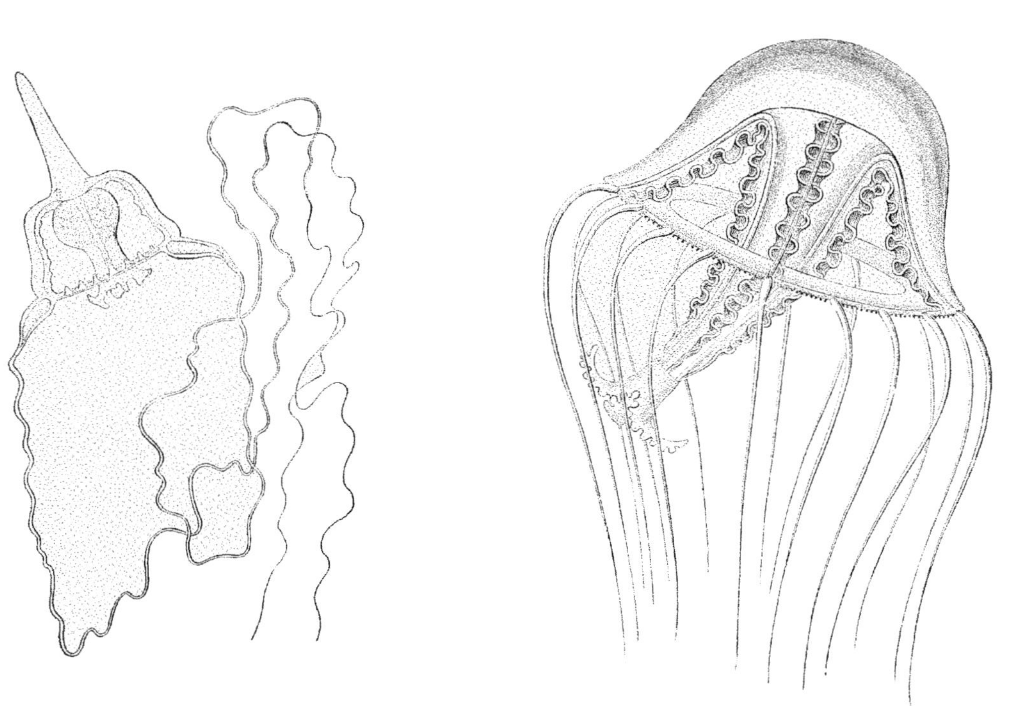

HYDROID JELLY FISHES.

On the left, *Stomatoca apicata*; on the right, *Tima bairdii*.

HYDROIDS.

1. *Pennaria tiarella.* 2. *P. tiarella*, enlarged. 3. *Aglaophenia struthionides.* 4. *Eudendrium dispar*, male, enlarged.

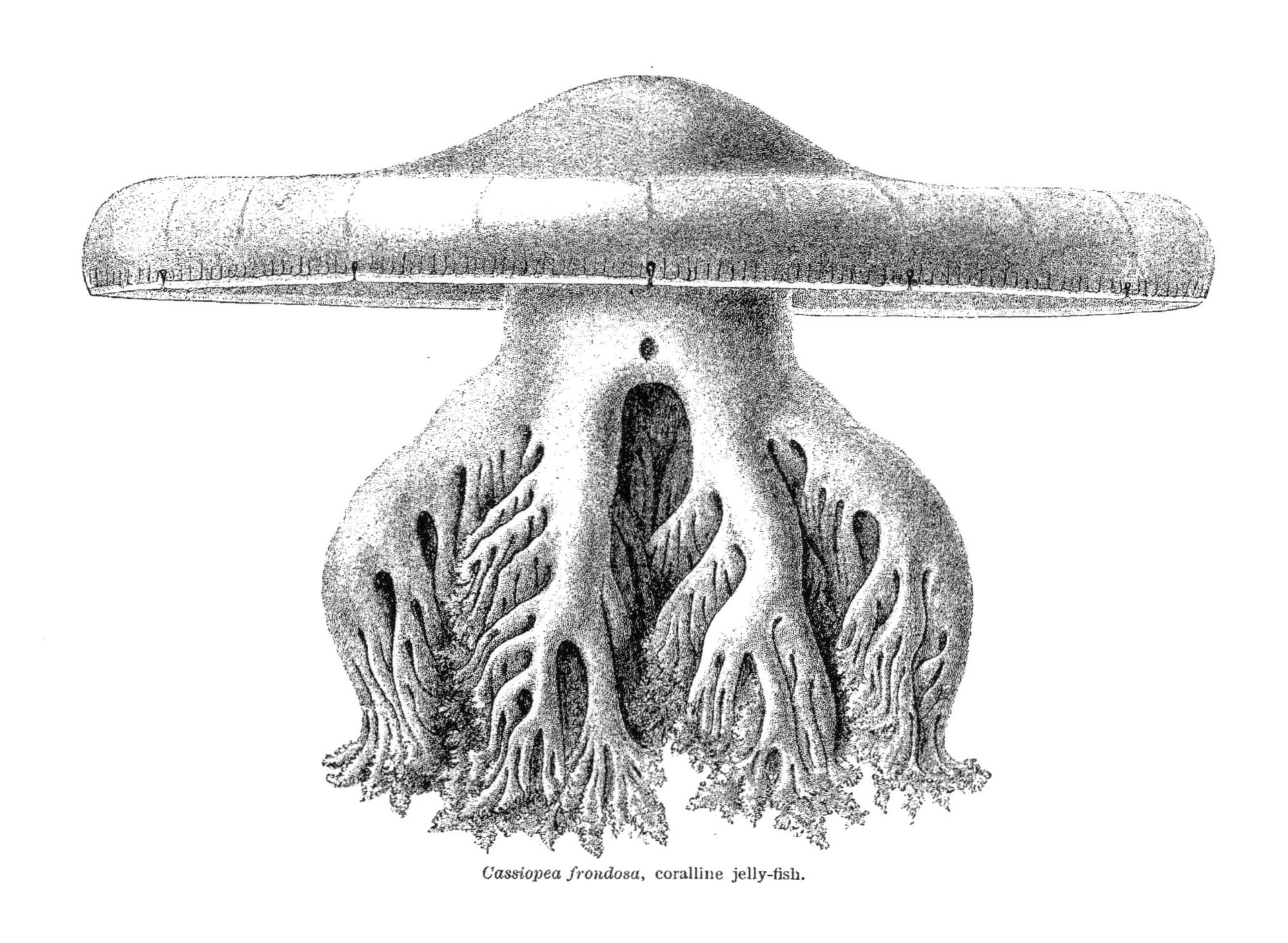

Cassiopea frondosa, coralline jelly-fish.

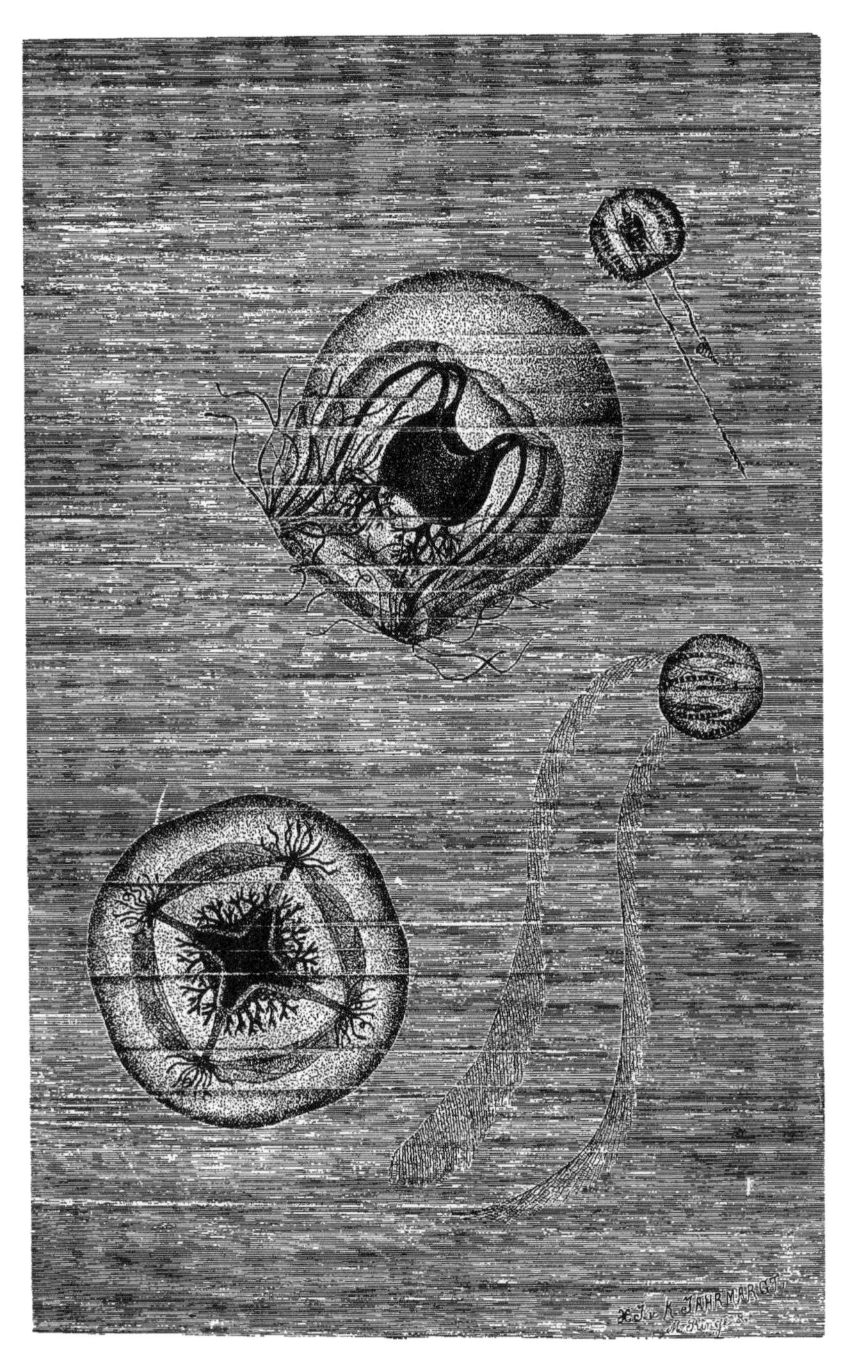

Pleurobrachia and *Bougainvillea.*

A GROUP OF EUROPEAN SEA-ANEMONES.

1. *Thelia crassicornis*

Corallium rubrum, red coral.

Alcyonium, cork polyp, natural size.

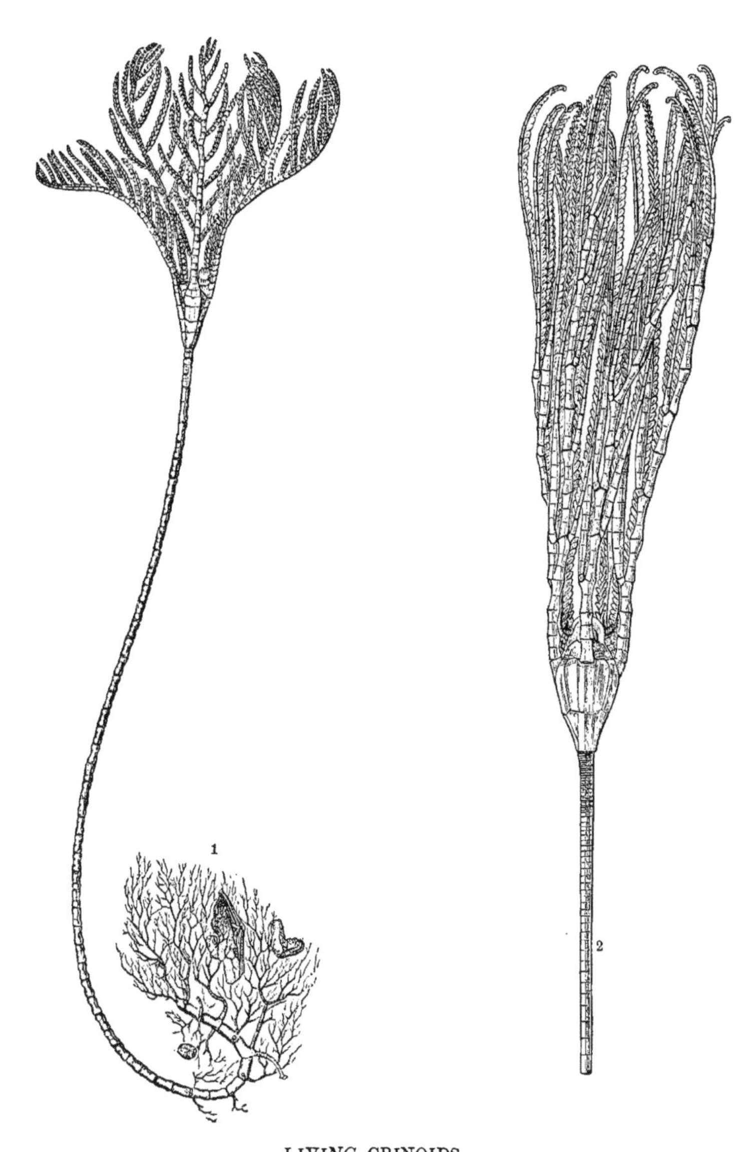

LIVING CRINOIDS.

1. *Rhizocrinus lofotensis.* 2. *Hyocrinus bethellianus.*

Starfish, Holothurian, and Worms.

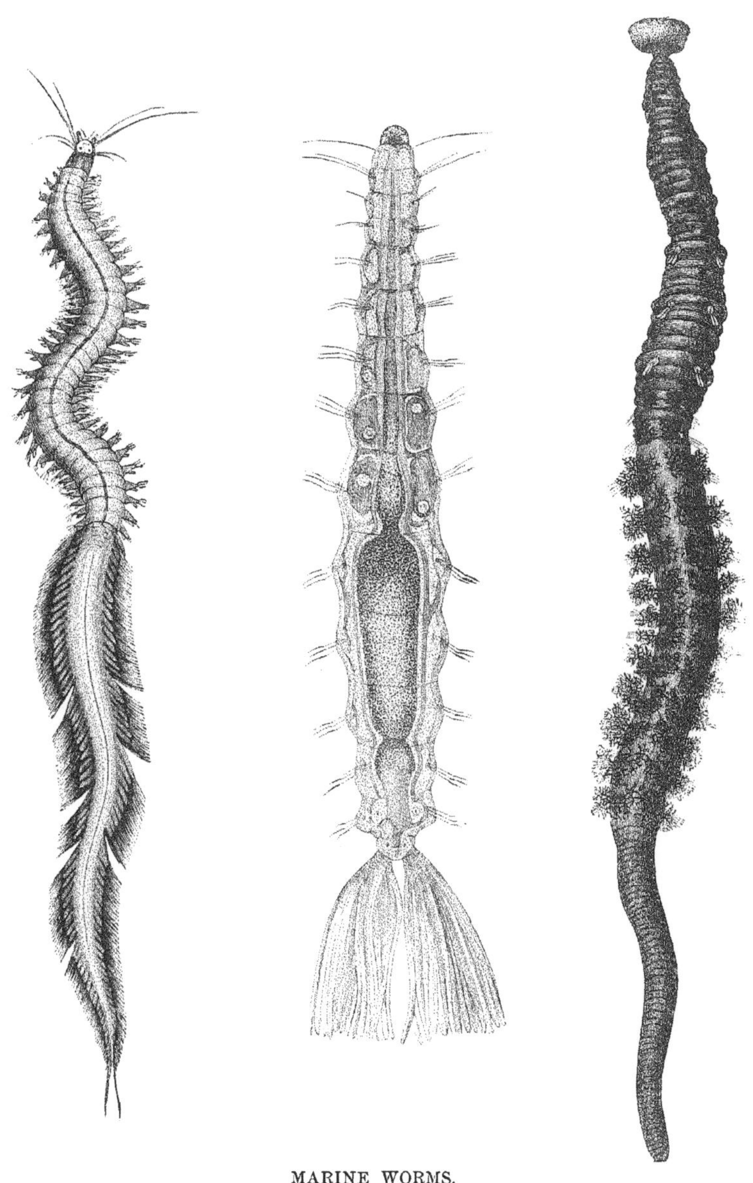

MARINE WORMS.

Heteronereis smardæ. *Amphicora sabella.* *Arenicola marina.*

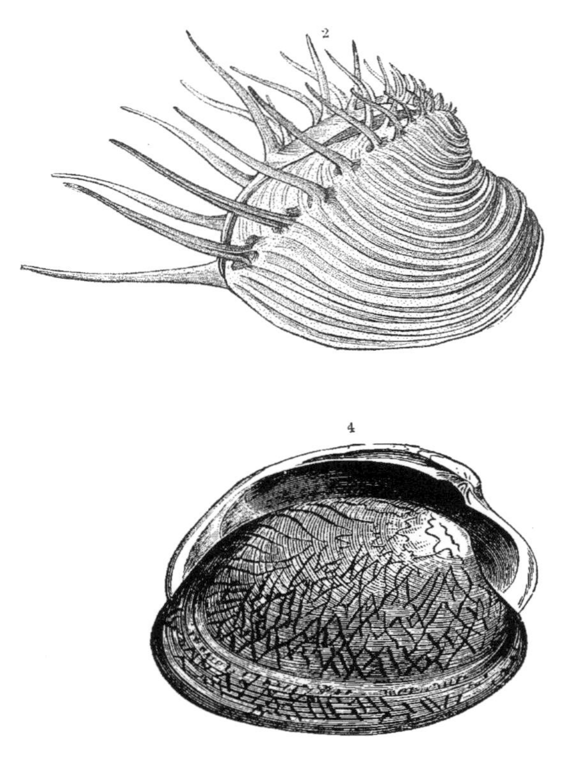

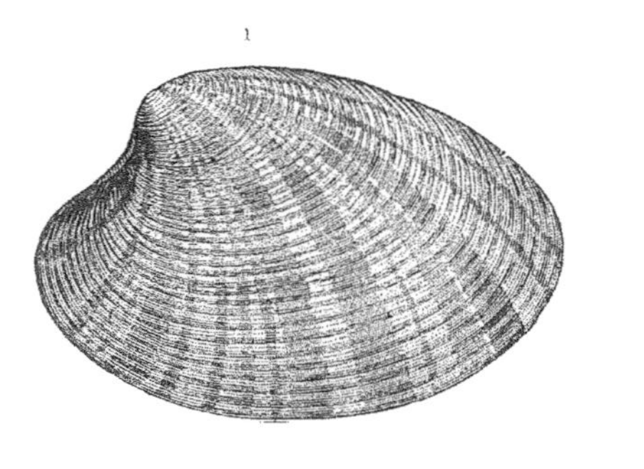

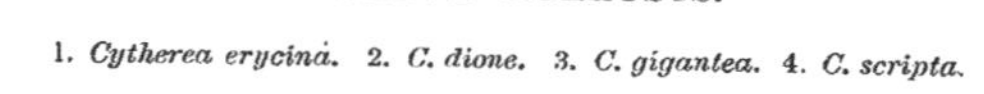

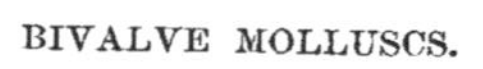

BIVALVE MOLLUSCS.

1. *Cytherea erycina.* 2. *C. dione.* 3. *C. gigantea.* 4. *C. scripta.*

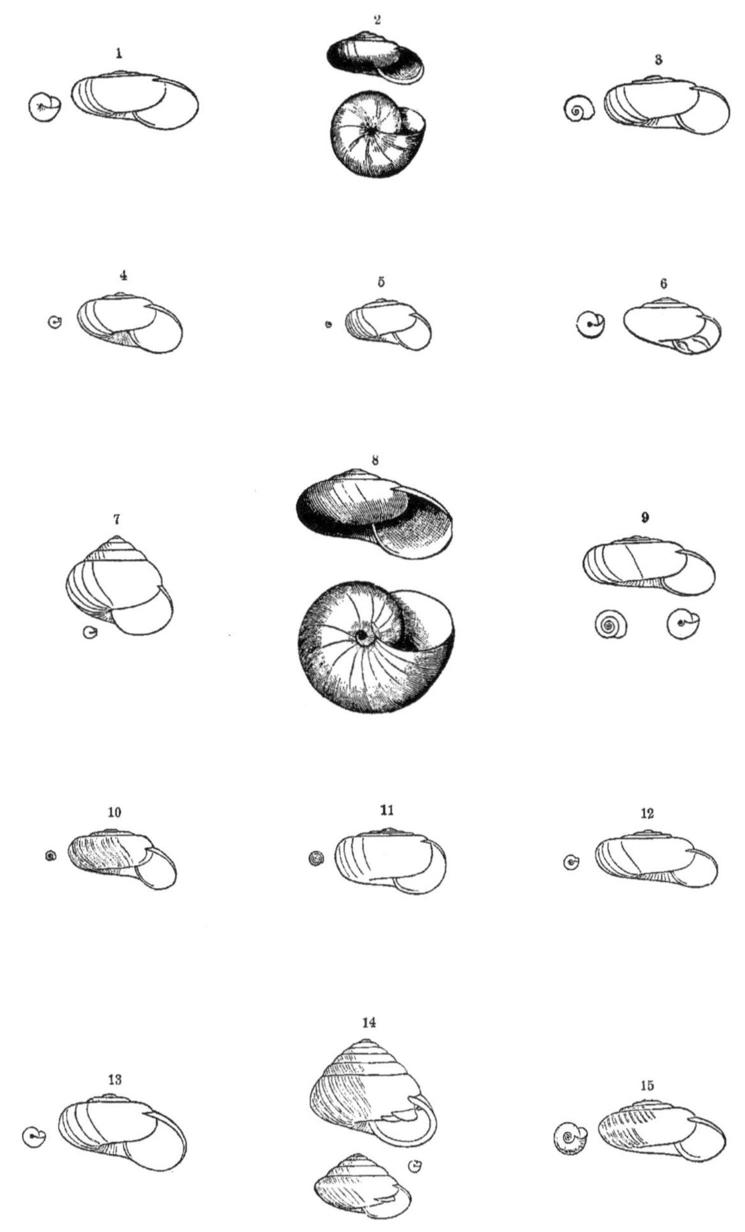

AMERICAN LAND SHELLS.

1. *Helix indentata.* 2. *H. inornata.* 3. *H. electrina.* 4. *H. ferrea.* 5. *H. minutissima.* 6. *H. suppressa.* 7. *H. chersina.* 8. *H. fuliginosa.* 9. *H. arborea.* 10. *H. exigua.* 11. *H. multidentata.* 12. *H. minusculus.* 13. *H. binneyana.* 14. *H. labyrinthica.* 15. *H. striatella.*

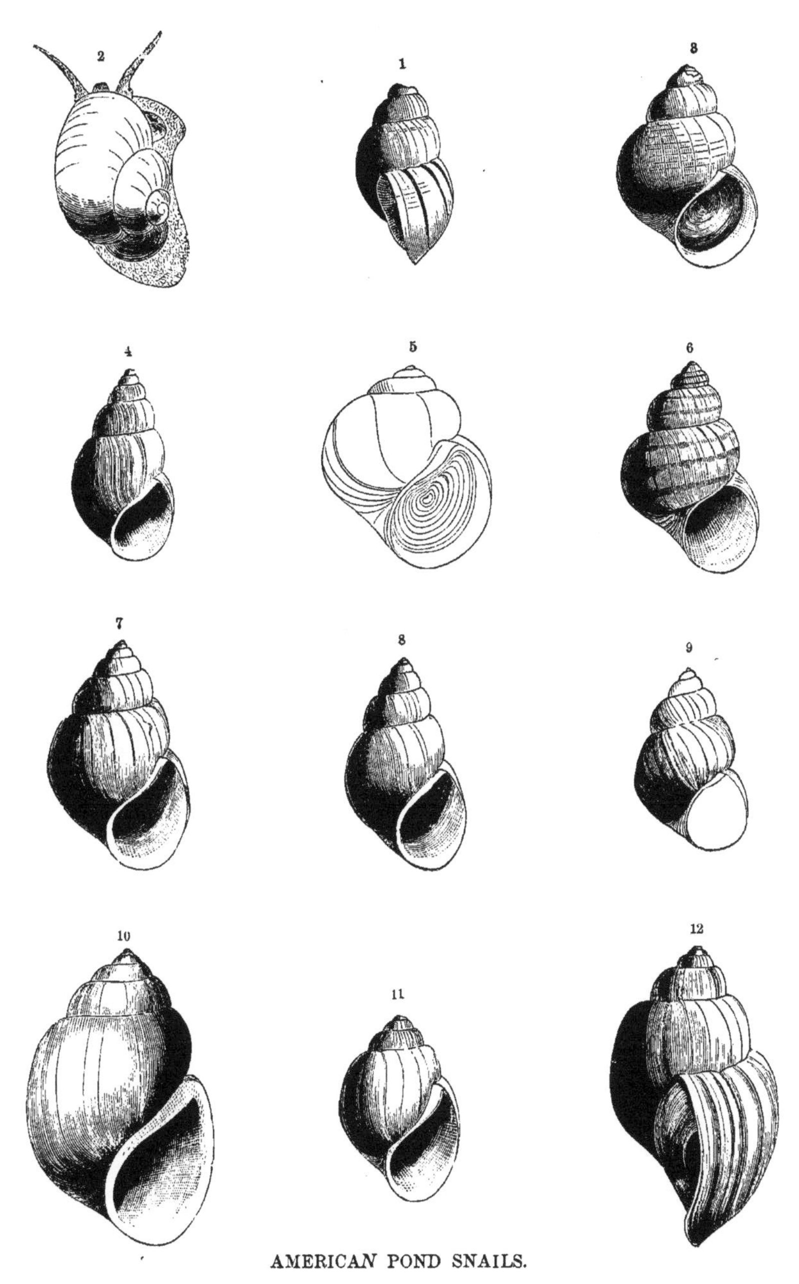

AMERICAN POND SNAILS.

1. *Paludina decisa.* 2, 3. *P. intertexta.* 4. *P. coarctata.* 5. *P. georgiana.* 6. *P. contectoides.* 7, 8. *P. integra.* 9. *P. subpurpurea.* 10, 11, 12. *P. ponderosa.*

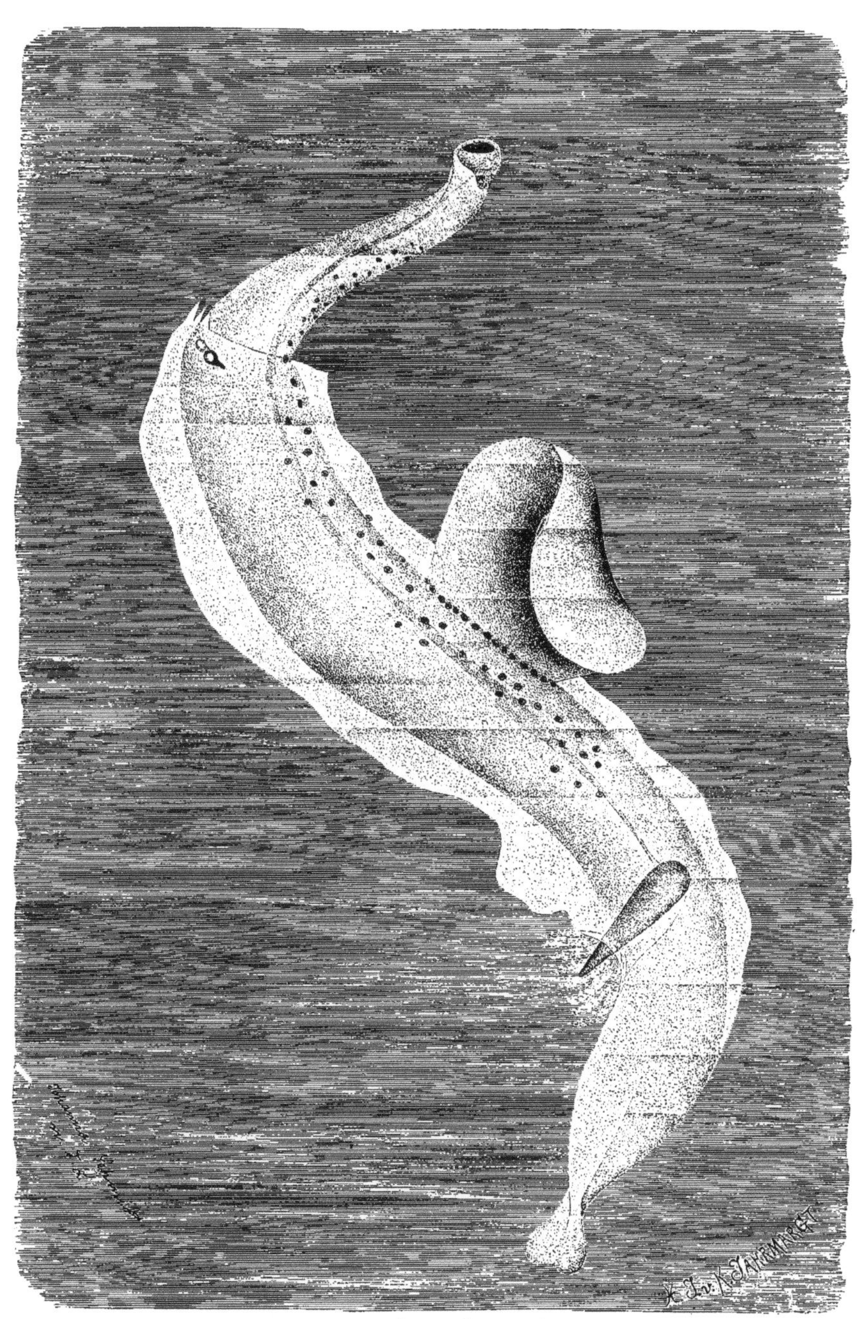

Pterotrachea; a heteropod.

Eledone moschata, musk poulpe.

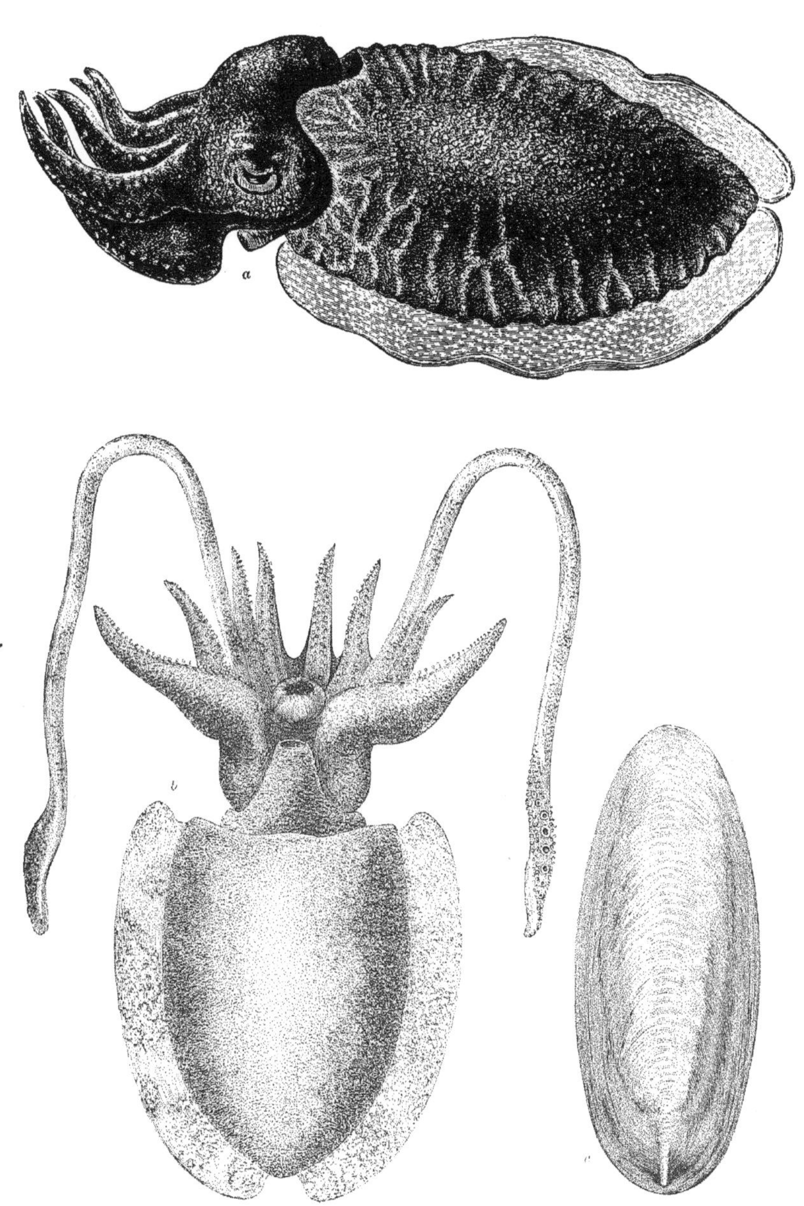

Sepia officinalis, cuttle fish; *a*, male; *b*, female; *c*, cuttle-bone.

THE ANIMAL KINGDOM.

INTRODUCTION.

THE term *N*atural History has at different times and by different authors been used in a variety of senses. At the present time it is perhaps more commonly used in contradistinction to natural philosophy; it is generally applied to the study of natural objects, both mineral or inorganic, and to plants and animals, or organic bodies.

At first it was applied to the study of all natural objects, whether the minerals, rocks, and living beings observed upon our own planet, or heavenly bodies in general. The study of external nature, and the phenomèna or laws governing the movements of natural bodies, was formerly opposed to metaphysics, history, literature, etc. After a while astronomy and chemistry were eliminated from natural history; then natural philosophy, or what is now called physics, was farther separated from chemistry, so that a chemist studies the constitution or atomic nature of bodies, both inorganic and organic; how they combine, and how compound bodies may be analyzed or separated into their simple constituents; the natural philosopher, or physicist, studies the forces of nature, the mechanical movements of inorganic bodies, and their phenomena, such as light, heat, and electricity, while the naturalist studies minerals, rocks, plants, and animals. But natural science, as distinguished from physical science, has made such progress, the work has been so sub-divided or differentiated, that even the term naturalist has become a vague, indefinite one. We must now know whether our naturalist is a mineralogist, a geologist, a botanist, or a zoologist; and the latter may be an entomologist, or ichthyologist, or ornithologist, according as he devotes himself exclusively to insects, fishes, or birds. The term natural history is popularly, at least by many, confined to botany and zoology, often, however, to zoology alone; and such is the convenient though inexact title of the present work.

Man is an animal as well as a mental and spiritual being. His material body is dominated by his mind and soul, but as zoologists we study him simply as an animal. The natural history of man is his physical history; it concerns his bodily structure and development, the work of his hands and the language he speaks, as well as the races into which his species is divided.

Anthropology is a convenient and comprehensive term now generally used for the natural history of man. The anthropologist, making a specialty of the natural history of man, studies not only his bodily structure, especially his skull or cranium and the other bones of the skeleton, comparing those of different races both living and extinct, but the works of human art; also human languages, both those now spoken and those which have become extinct. Moreover the anthropologist goes out of the realm of zoology into those of mental phenomena or psychology, and of sociology, and studies man as a spirit, his notions of the future life, his myths, traditions; he also studies his origin, history, and social laws and government.

We have seen that the term natural history, as applied to plants and animals, is inexact, having been used in different senses; a more exact and suitable one is Biology, which is the science which relates to living beings. It is derived from the Greek βίος, life, and λόγος, discourse. It is divided into Botany, which relates to plants, and Zoology (ζῶον, animal; λόγος, discourse), the science treating of animals.

The science of zoology may be subdivided thus:

ZOOLOGY.
- MORPHOLOGY, or gross Anatomy, and the anatomy of the tissues (HISTOLOGY), based on embryology.
- PHYSIOLOGY and PSYCHOLOGY.
- REPRODUCTION and EMBRYOLOGY.
- SYSTEMATIC ZOOLOGY, or Classification.
- PALÆONTOLOGY, the study of fossil animals.
- ZOO-GEOGRAPHY, or the geographical distribution of animals.

MORPHOLOGY.

INORGANIC AND ORGANIC BODIES.

The differences between the inorganic or mineral bodies, and organic or living bodies, are not always appreciable to the untrained eye and mind. Mineral bodies, such as water, and solid minerals such as quartz, salt, or lime, may assume definite crystalline shapes, and such shapes may grow, *i. e.* increase in size, by the addition, on the outside, of particles of the same substance. Minerals may exist in three different states, gas or vapor, fluid, or solid. The air, carbonic acid gas, and water are minerals, as much so as lime or salt. Rocks are made up of minerals, and thus the earth, the air, and water are of mineral origin. Minerals may assume plant-like shapes, as seen in the beautiful forms and delicate leaf-like tracery of the frost on our windows. The drops of sea-water dashed by the waves upon one's coat-sleeve may be seen under a glass to evaporate and the solid particles left to arrange themselves in beautiful but definite crystalline forms. By watching the evaporation of salt water under the microscope, the crystals may be seen actually to grow, *i. e.* to build themselves up, to extend and enlarge in size. "If," as Huxley states, "a crystal of common salt is hung by a thread in a saturated solution of salt, which is exposed to the air so as to allow the water to evaporate slowly, the molecules of the salt which is left behind and can no longer be held in solution, deposit themselves on the crystal in regular order and increase its size without changing in form. And, in this way, the small crystal may grow to a great size." Thus growth in minerals consists in the addition of particle after particle of solid matter to the outside of the growing body.

Now, how do living bodies essentially differ from those not living? In the first place, plants and animals are built up of protoplasm. Protoplasm consists of mineral matter, to be sure, but so combined as to form a substance not found in minerals. Moreover, a plant or animal has life; the plant grows, and, if it is a vine, is capable of some degree of motion, it twines about some post or tree as a support; the living bird flies, and the living dog runs; after a time the plant or the bird dies, it is then dead. Minerals do not die. Besides, when the plant or animal grows, it increases in size, not like minerals, by additions from without, but by manufacturing protoplasm and other organic substances, such as starch and fat, within its body. Thus organisms or living beings grow by additions from within, these additions being produced by the living being itself. After a while this process stops and it becomes dead. As Professor Huxley tells us: "In the spring, a wheat-field is covered with small green plants. These grow taller and taller until they attain many times the size which they had when they first appeared; and they produce the heads of flowers which eventually change into ears of corn.

"In so far as this is a process of growth, accompanied by the assumption of a definite form, it might be compared with the growth of a crystal of salt in brine; but, on closer examination, it turns out to be something very different. For the crystal of salt grows by taking to itself the salt contained in the brine, which is added to its exterior; whereas the plant grows by addition to its interior; and there is not a trace of the characteristic compounds of the plant's body, albumen, gluten, starch, or cellulose, or fat, in the soil, or in the water, or in the air.

"Yet the plant creates nothing, and therefore the matter of the proteids and amyloids and fats which it contains must be supplied to it, and simply manufactured, or combined in new fashions, in the body of the plant.

"It is easy to see, in a general way, what the raw materials are which the plant works up, for the plant gets nothing but the materials supplied to it by the atmosphere and by the soil. The atmosphere contains oxygen and nitrogen, a little carbonic acid gas, a minute quantity of ammoniacal salts, and a variable proportion of water. The soil contains clay and sand (silica), lime, iron, potash, phosphorus, sulphur, ammoniacal salts, and other matters which are of no importance. Thus, between them, the soil and the atmosphere contain all the elementary bodies which we find in the plant; but the plant has to separate them and join them together afresh.

"Moreover the new matter by the addition of which the plant grows is not applied to its outer surface, but is manufactured in its interior; and the new molecules are diffused among the old ones."

Living beings also reproduce their kind. The corn bears seed, the hen lays eggs. Minerals cannot reproduce, so that we see that organic beings differ from minerals in three essential characteristics: they contain and are built up from protoplasm; they grow from within, and they reproduce by seeds, germs or eggs.

As Huxley again says: "Thus there is a very broad distinction between mineral matter and living matter. The elements of living matter are identical with those of mineral bodies; and the fundamental laws of matter and motion apply as much to living matter as to mineral matter; but every living body is, as it were, a complicated piece of mechanism, which 'goes' or lives, only under certain conditions. The germ contained in the fowl's egg requires nothing but a supply of warmth within certain narrow limits of temperature, to build the molecules of the egg into the body of the chick. And the process of development of the egg, like that of the seed, is neither more or less mysterious than that in virtue of which the molecules of water, when it is cooled down to the freezing point, build themselves up into regular crystals."

The Differences between Plants and Animals.

We now come to the differences between plants and animals; and here the distinctions are more or less arbitrary. As we have remarked elsewhere, —

"It is difficult to define what an animal is as distinguished from a plant, when we consider the simplest forms of either kingdom, for it is impossible to draw hard and fast lines in nature. In defining the limits between the animal and vegetable kingdoms, our ordinary conception of what a plant or animal is will be of little use in dealing with the lowest forms of either kingdom. A horse, fish, or worm differs from an elm-tree, a lily, or a fern, in having organs of sight, of hearing, of smell, of locomotion, and special organs of digestion, circulation, and respiration, but these plants also take in and absorb food, have a circulation of sap, respire through their leaves, and some plants are mechanically sensitive, while others are endowed with motion — certain low plants, such as diatoms, etc., having this power. In plants the assimilation of food goes on all over the organism, the transfer of the sap is not confined to any one portion or set of organs as such. It is always easy to distinguish one of the higher plants from one of the higher animals. But when we descend to animals like the sea-anemones and coral-polyps, which were called zoophytes from their general resemblance to flowers, so striking is the external similarity between the two kinds of organisms, that the early observers regarded them as 'animal flowers;' and in consequence of the confused notions originally held in regard to them, the term zoophytes has been perpetuated in works on systematic zoology. Even at the present day the compound hydroids, such as the *Sertularia*, are gathered and pressed as sea-mosses by many persons who are unobservant of their peculiarities, and unaware of the complicated anatomy of the little animals filling the different leaf-like cells. Sponges, until a late day, were regarded by our leading zoologists as plants. The most accomplished naturalists, however, find it impossible to separate by any definite lines the lowest animals and plants. So-called plants, as *Bacterium*, and so-called animals, as *Protamœba*, or certain monads, which are simple specks of protoplasm, without genuine organs, may be referred to either kingdom; and indeed, a number of naturalists — notably Haeckel — relegate to a neutral kingdom (the Protista) certain lowest plants and animals. Even the germs (zoospores) of monads and those of other flagellate infusoria, may be mistaken for the spores of plants; indeed the active flagellated spores of plants were described as Infusoria by Ehrenberg; and there are certain so-called flagellate Infusoria so much like low plants (such as the red snow or *Protococcus*), in form, deportment, mode of reproduction, and appearance of the spores, that even now it is possible that certain organisms placed among them are plants. It is only by a study of the connecting links between these lowest organisms, leading up to what are undoubted animals or plants, that we are enabled to refer these beings to their proper kingdom.

"As a rule, plants have no special organs of digestion or circulation, and nothing approaching to a nervous system. Most plants absorb inorganic food, such as carbonic acid gas, water, nitrate of ammonia, and some phosphates, silica, etc., all of these substances being taken up in minute quantities. Low fungi live on dead animal matter, and promote the process of putrefaction and decay, but the food of these organisms is inorganic particles. The slime-moulds called *Myxomycetes*, however, envelop the plant or low animals, much as an *Amœba* throws itself around some living plant and absorbs its protoplasm; but the *Myxomycetes*, in their manner of taking food, are an exception to other moulds. The lowest animals swallow other living animals whole or in pieces; certain forms near *Amœba* bore into minute algæ and absorb their protoplasm; others engulf silicious-shelled plants (diatoms and desmids) and absorb the protoplasm filling them. No animal swallows silica, lime, ammonia, or phosphates as food. On the other hand, plants manufacture or produce protein in the shape of starch, albumen, sugar, etc., which is animal food. Plants inhale carbonic acid gas and exhale oxygen; animals inhale oxygen and exhale carbonic acid; though Draper has discovered that, under certain circumstances, plants may exhale carbonic acid.

"Animals move and have special organs of locomotion; few plants move, though minute forms have thread-like processes or vibratile lashes (cilia) resembling the flagella of monads, and flowers open and shut; but these motions of the higher plants are purely mechanical, and not performed by special organs controlled by nerves. The mode of reproduction of plants and animals, however, is fundamentally identical, and in this respect the two kingdoms unite more closely than in any other. Plants also, like animals, are formed of cells, the latter in the higher forms combined into tissues.

"As the lowest plants and animals are scarcely distinguishable, it is probable that plants and animals first appeared contemporaneously; and while plants are generally said to form the basis of animal life, this is only partially true; a large number of fungi are dependent on decaying animal matter; and most of the *Protozoa* live on animal food, as do a large proportion of the higher animals. The two kingdoms supplement each other, are mutually dependent, and probably appeared simultaneously in the beginning of things. It should be observed, however, that the animal kingdom overtops the vegetable kingdom, culminating in man." (Packard's Zoology.)

THE CELLS AND PROTOPLASM.

The fact that the bodies of the higher as well as of the simpler animals are composed of cells was discovered by Schwann in 1839. This discovery is the foundation-stone upon which the modern and still young science of Biology has been built. The cell is the morphological unit of the organic world. With cells the biologist can, in the imagination, reconstruct the vegetable and animal kingdoms. By studying the forms and behavior of single cells and one-celled animals, such as the motions of the *Amœba*, one can better understand the structure and physiology of the highest and most specialized forms, even that of man; for, as Geddes has remarked, "the functions of the body are the result of the aggregate functions of its cells, and are explained by variations or phases of the activities of them."

Cells are microscopic portions of protoplasm, either with or without a wall. The protoplasm is the most important, the dynamic part of the cell. What, then is protoplasm? As has already been said, it is the possession of this substance which distinguishes living beings from mineral. Huxley, in his "Science Primer," tells us in simple language what protoplasm is. "If a handful of flour mixed with a little cold water is tied up in a coarse cloth bag, and the bag is then put into a large vessel of water, and well kneaded with the hands, it will become pasty, while the water will become white. If this water is poured away into another vessel, and the kneading process continued with some fresh water, the same thing will happen. But if the operation is repeated, the paste will become more and more sticky, while the water will be rendered less and less white, and at last will remain colorless. The sticky substance, which is thus obtained by itself, is called gluten; in commerce it is the substance known as maccaroni.

"If the water in which the flour has thus been washed is allowed to stand for a few hours, a white sediment will be found at the bottom of the vessel, while the fluid above will be clear, and may be poured off. This white sediment consists of minute grains of starch, each of which, examined with a microscope, will be found to have a concentrically laminated structure. If the fluid from which the starch was deposited is now boiled, it will become turbid, just as white of egg diluted with water does when it is boiled, and eventually a whitish lumpy substance will collect at the bottom of the vessel. This substance is called vegetable albumen.

"Besides the albumen, the gluten, and the starch, other substances, about which this rough method of analysis gives us no information, are contained in the wheat grain. For example, there is woody matter or cellulose, and a certain quantity of sugar and fat."

Similar substances are found in animals and eggs. "If you break an egg, the contents flow out, and are seen to consist of the colorless glairy 'white' and the yellow 'yolk.' If the white is collected by itself in water and then heated, it becomes turbid, forming a white solid, very similar to the vegetable albumen, which is called animal albumen.

"If the yolk is beaten up with water, no starch nor cellulose is obtained from it, but there will be plenty of fatty and some saccharine matter, besides substances more or less similar to albumen and gluten.

"The feathers of the fowl are chiefly composed of horn; if they are stripped off, and the body is boiled for a long time, the water will be found to contain a quantity of gelatine, which sets into a jelly as it cools, and the body will fall to pieces, the bones and the flesh separating from one another. The bones consist almost entirely of a substance which yields gelatine when it is boiled in water, impregnated with a large quantity of salts of lime, just as the wood of the wheat stem is impregnated with silica. The flesh, on the other hand, will contain albumen, and some other substances which are very similar to albumen, termed fibrin and syntonin.

"In the living bird all these bodies are united with a great quantity of water, or dissolved, or suspended in water; and it must be remembered that there are sundry other constituents of the fowl's body and of the egg, which are left unmentioned, as of no present importance.

"The wheat plant contains neither horn nor gelatine, and the fowl contains neither starch nor cellulose; but the albumen of the plant is very similar to that of the animal, and the fibrin and syntonin of the animal are bodies closely allied to both albumen and gluten.

"That there is a close likeness between all these bodies is obvious from the fact that when any of them are strongly heated or allowed to putrefy, it gives off the same sort of disagreeable smell; and careful chemical analysis has shown that they are, in fact, all composed of the elements carbon, hydrogen, oxygen, and nitrogen, combined in very nearly the same proportions. Indeed, charcoal, which is impure carbon, might be obtained by strongly heating either a handful of corn or a piece of fowl's flesh in a vessel from which the air is excluded so as to keep the corn or the flesh from burning. And if the vessel were a still, so that the products of this destructive distillation, as it is called, could be condensed and collected, we should find water and ammonia in some shape or other in the receiver. Now ammonia is a compound of the elementary bodies nitrogen and hydrogen; therefore both nitrogen and hydrogen must have been contained in the bodies from which it is derived.

"It is certain, then, that very similar nitrogenous compounds form a large part of the bodies of both the wheat plant and the fowl, and these bodies are called proteids.

"It is a very remarkable fact that not only are such substances as albumen, gluten, fibrin, and syntonin known exclusively as products of animal and vegetable bodies, but that every animal and every plant at all periods of its existence contains one or other of them, though, in other respects, the composition of living bodies may vary indefinitely. Thus, some plants contain neither starch nor cellulose, while these substances are found in some animals; while many animals contain no horny matter and no gelatine-yielding substance. So that the matter which appears to be the essential foundation of both the animal and the plant is the proteid united with water; though it is probable that, in all animals and plants, these are associated with more or less fatty and amyloid (or starchy and saccharine) substances, and with very small quantities of certain mineral bodies, of which the most important appear to be phosphorus, iron, lime, and potash.

"Thus there is a substance composed of water and proteids and fat and amyloids and mineral matters, which is found in all animals and plants, and, when these are alive, this substance is termed *protoplasm*."

As yet we know almost nothing of the chemical nature of animal protoplasm. Microscopically it does not differ in appearance from vegetable protoplasm, and yet the latter has been found by Reinke to contain over forty proximate constituents. Hence it seems reasonable to infer that animal protoplasm will on further analysis be found to be an exceedingly complex substance, and not properly comparable with a particle of white of egg. The following account of the chemical composition of protoplasm in general is from Carnoy: "Protoplasm is a complex mixture of chemical species. The patient and minute researches of later years have resulted in the discovery, in typical protoplasm of young and active cells, the following substances which we should consider as the essential elements of the living matter:

1. Albuminoid substances (a vitelline and a myosine at minimum).
2. Phosphoric substances (lecithine and nucleine).
3. One or several hydrocarbonated substances (such as glycose, dextrine, glycogene).
4. Soluble ferments (diastase, pepsine, inversive ferment, emulsine).
5. Water (of constitution and imbibition).
6. Mineral elements (salts, sulphates, phosphates or nitrates of potassium, calcium, and magnesium).

"We should also call attention to the recent analyses made by Reinke and Rodewald (1881) on the plasmodium of *Æthalium septicum*, by which they found, besides some accidental principles, the elements above mentioned. These analyses, as also the microchemical researches of Zacharias (1881–83), have besides revealed to us a new element of protean (*proteique*) nature, that of *plastine*, which seems to fill an essential role. Finally, we may mention the researches of Mayer and Baginski, which give a new category of soluble ferments: we refer to coagulent ferments, or *présures* (Labferment). These bodies have already been verified in a quite large number of animal and vegetable cells, which seem, like their congeners, indispensable to the accomplishment of certain cellular phenomena."

We know that potentially the protoplasm of different kinds of cells exerts widely different forces and capabilities. A liver cell secretes bile, a pancreas cell pancreatic fluid, the cells lining the stomach gastric fluid, and an ovarian cell the white of an egg. One egg-cell may become a mollusc, another a man, whose brain-cells are the medium of the intellectual power which enables him to write the history of his own species, and to be the historian of the forms of life which stand below him.

The simplest, most primitive form of a cell, when without a nucleus or nucleolus, is called a *cytode*. The Monera, which are the lowest animals, have no nucleus, and have therefore been called by Haeckel cytodes. The existence of cytodes, however, cannot yet be accepted as a settled fact; all the evidence is simply negative, and forms which once were supposed to have no nucleus (*e. g.* Radiolaria) have been recently shown to possess such a structure. Further researches may show the same to be true of all forms and all cells.

Genuine cells have a *nucleus*, the latter containing a *nucleolus*. It will thus be seen that the true cell is not a simple body; the nucleus is distinct from the rest of the cell in structure and appearance, and the nucleolus also differs in the same way from the nucleus and rest of the cells. The nucleus and nucleolus also vary in themselves in size and contents, the granules and fibres filling them varying in size and number. This fact should lead us to regard the cell as not so simple as generally supposed. A recent writer, impressed by the complexity of cell structure, has subjected to a critical examination the characters of ganglion cells, both smooth and striated muscle-cells, glandular, liver, and salivary gland cells, and epithelium, both that of the mucous membrane and that which is ciliated, as well as that of the crystalline lens, beside cartilage and embryonic cells. Both healthy and diseased cells are found by Arnold to possess a complicated structure. The two constituents, as ordinarily distinguished, the cell body and the cell nucleus, consist of a ground substance as well as of granules, sets of granules, and filaments; these latter may become very complicated in the more highly developed forms of cells. Arnold would regard a cell as consisting of a nucleus and of an investing mass, both of which contain, in a homogeneous ground substance, granules and filaments.

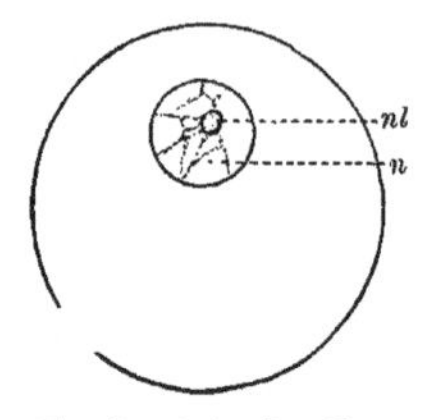

FIG. I. — A simple cell; *n*, nucleus; *nl*, nucleolus.

Dr. C. S. Minot believes that the weight of an animal depends on the number and size of its cells, and that these two variables require to be determined before we can speak definitely as to the processes of growth in animals. Minot points out that the growth of a body is usually measured by its weight, but that this method takes no account of the amount of non-protoplasmic matters present. All many-celled animals "pass through successive cycles, in which we can distinguish the two processes of *senescence* and of *rejuvenation*." As growth is a function of rejuvenescence effected by impregnation, it follows that growth can only be measured by taking into account the number of cells living at any given time.

Animal cells are of microscopic size, but one-celled animals are in some cases large enough to be detected with the naked eye; such are many Foraminifera, and the *Stentor* among Infusorians. Kölliker states that the size of the cells descends on the one hand, as in many cells — the blood cells, etc. — to 0.002–0.0003 of a line, and attains on the other, as in the cysts of the semen and the ganglionic globules, the size of 0.02 –0.04 of a line. The largest animal cells are certain gland-cells of insects which measure up to 0.01 of a line, and the yolk-cells or ova, especially of birds and reptiles.

Animals grow by the self-division and multiplication of cells. This is the initial, fundamental process which underlies reproduction and growth. The cells in any part of the growing body divide into two, and these again sub-divide, the products of self-division becoming as large as the original cell, and in this way the body increases in size.

The part of the cell in which this process of self-division begins is the nucleus. A good deal has been written upon this subject. Some authors believe that in cell division the nucleolus first divides, then the nucleus, and finally the cell. Prof. W. Flemming, however, maintains that the nucleus first of all undergoes a change, "separating into a network of highly refracting filaments, which take up coloring matters strongly, and an intermediate substance not affected by staining fluids. The nucleus network goes through a definite series of changes, and finally divides into two equal or sub-equal masses, which retreat from one another, and go through, in inverse order, the changes undergone by the mother nucleus, finally forming the nuclei of the two daughter cells. The cell-body divides after the young nuclei have separated from one another, but before they have assumed the characteristics of quiescent nuclei." Cell division, he adds, probably exhibits periodicity, going on vigorously at certain times, and but little, or not at all, in the intervals.

TISSUES.

Although we shall treat of the development of animals on a subsequent page of this introduction, at the risk of repetition, we will give here an epitome of the earlier features common to the development of all animals above the Protozoa as an introduction to the subject of tissues.

In all the animals above the Protozoa, besides reproduction by fission and budding occurring in the lower groups, there is a sexual reproduction by which one animal or one part of an animal produces an egg, while another produces the male element or spermatozoan. These two unite and produce the fertile egg, which in its turn is converted into a perfect animal like the form from which the genital products arose. In only a few forms will the egg develop without an intervention of the male sexual element, and these cases of parthenogenesis, so far as at present known, are confined to the rotifers and arthropods.

The egg is to all intents and purposes a simple cell, and the processes by which it forms the complex adult are but those of cell-division, the resulting cells becoming specialized, some for the performance of one function, some for another. In the typical form of this cell-division or segmentation, the egg divides first into two equal parts or cells, each of these again into two, producing four in all, and so on, the regular numbers being, 2, 4, 8, 16, 32, 64, etc. This regularity is but rarely perfectly carried out; the exceptional forms will be mentioned later. The result of this segmentation is the production of a more or less spherical body, which has received the name morula, from its resemblance to a mulberry (Fig. II. E.) Usually this morula is at first solid, but it increases in size, the result being a hollow globe (Fig. II., F, surface view; G, section), the blastula, the vacant centre being known as the segmentation cavity. The next step is the con-

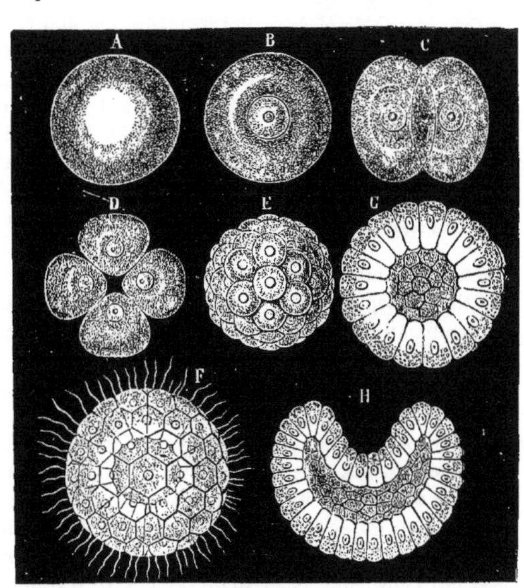

FIG. II.—Early stages of development of *Monoxenia*; A, egg after disappearance of the nucleus; B, egg with nucleus; C, first segmentation; D, second segmentation; E, morula; F, surface view, and G, section of blastula; H, section of early gastrula.

version of this ball with its single layer of cells into a two-layered sac. We can now compare the egg to a hollow rubber ball. By the pushing in of one side (in the language of embryology, invagination) the ball can be made to resemble the condition shown in Fig. II. H, the earliest stage of the gastrula. The single layer of cells of the blastula is now differentiated into two layers, the outer of which is variously termed epiblast, ectoderm or epiderm, the inner endoblast, hypoblast, or endoderm. So far, with the modifications to be noted below, these processes are common to all animals above the Protozoa, and with the separation into two layers we have the essential parts of a *Hydra;* the outer layer corresponds to the skin, the inner to the digestive tract of that simple animal, as shown farther on in this volume (p. 76.) In the higher animals — as, for instance, the frog — at one stage of the development these two layers alone occur, and from the outer is developed the external layer of the skin, while the hypoblast eventually forms the lining of the central portion of the digestive tract. We thus see that at this stage the embryo is composed of skin and stomach, and the name gastrula means a little stomach.

This epitome contains the gist of the earlier stages of a typical egg, but in nature many variations occur, of which our space will allow but the merest mention. In eggs which are composed of pure protoplasm this regular development occurs, but in most eggs another element, known as food-yolk or deutoplasm, is present, and according to its distribution the character of the segmentation and of the invagination varies. The protoplasm is the active portion, the food-yolk, as the name implies, is to nourish the growing embryo, and its presence tends to retard the development of the egg. Thus, when the protoplasm is superficial, and the food-yolk occupies a central position, as in the crustaceans, the segmentation planes are at first confined to the surface, the central portion remaining for a time unsegmented. At other times the food-yolk is confined to one pole of the egg, as in the frog, and then the segmentation is, for a while at least, confined to the protoplasmic pole, only affecting the other at a comparatively late stage of development. This is carried to its furthest extent in some of the fishes. All of these modifications of the type of segmentation are variously combined, so that great differences result, and as in the eggs of the same class, or occasionally even of the same genus, the distribution of the food-yolk will vary, it will readily be seen that the segmentation is but a very poor guide to the relationships of forms. In the insects the segmentation is very greatly modified, but as yet our knowledge is so slight as not to warrant any broad generalizations on the earlier stages of the group.

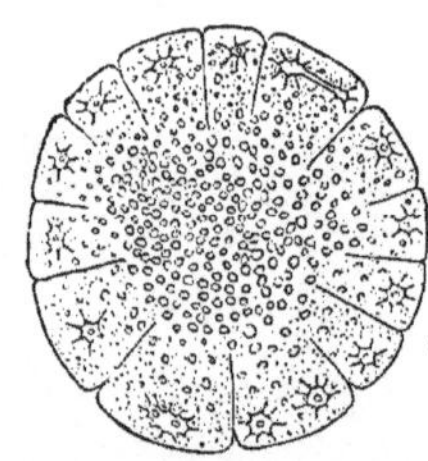

FIG. III. — Segmenting egg of a Crustacean (*Crangon*).

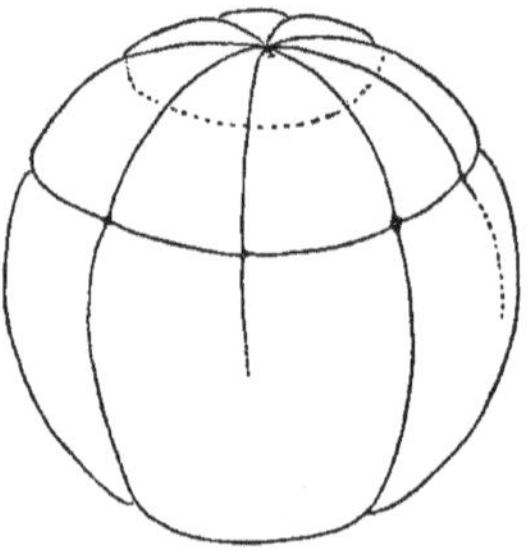

FIG. IV. — Unequal segmentation (frog).

Correlated with the variations in the segmentation are certain modifications in the gastrulation and the production of the epiblast and hypoblast. Let us return to the normal gastrula for a moment, and trace its progress just a little farther, naming some of the points omitted above. The invagination is carried to such an extent that the segmentation cavity is nearly obliterated, and at the same time the edges of the infolded ball are brought closer together, so that a comparatively narrow opening is the result, known as the blastopore. This blastopore in most forms completely closes; but to this we will return again. The hollow, which we have mentioned above as forming primarily the digestive cavity, is known as the archenteron or primitive stomach, and the hypoblastic cells which form its boundary are almost invariably larger than those of the epiblast. This is true of all gastrulas, even those where the segmentation is regular, and the reason is not difficult to find. The external cells have to embrace a greater superficial extent than the internal ones, and hence the layer becomes thinner and the resulting cells smaller. In other cases the segmentation is irregular and then a greater inequality occurs, until in some forms the hypoblast is invaginated as a few cells, or even a single cell, and the archenteron does not appear until a later date, when it is hollowed out of the hypoblastic cells. A greatly different mode of forming the gastrula is by what is known as delamination. A general idea of the process may be obtained by saying that the inner ends of the cells of the blastula (Fig. II., G) are segmented off to form the hypoblast.

In the gastrula we have two of the so-called germinal layers, the epiblast and the hypoblast; in all animals except some of the cœlenterates and the Dicyemids, a third layer, the mesoblast or mesoderm, occurs, hence these are known as triploblastic animals, in contradistinction to those with only hypoblast and epiblast, which are called diploblastic. We will not enter into a discussion of the many different ways in which the mesoblast arises, but will merely indicate what is apparently the typical method, which in reality exists unmodified in but very few animals. From the hypoblast, pouches bud off on either side, as shown on the left of our figure. These pouches eventually become separated from the archenteron, as shown on the right side of the same figure, and the walls of these pouches are the mesoblast. Now we are ready for the names of these parts and an enumeration of the organs into which they develop. After the formation of the mesoblast and the separation of a portion of the archenteron, the hypoblastic cavity is known as the mesenteron, from the fact that its lining cells form the epithelium of the middle portion of the digestive tract. From other pouches and outgrowths of the mesenteron, formed at a later date, other organs arise. Among these we may mention the liver, the lungs of vertebrates, the endostyle of tunicates, the thyroid and thymus glands, pancreas, spleen, and the notochord.

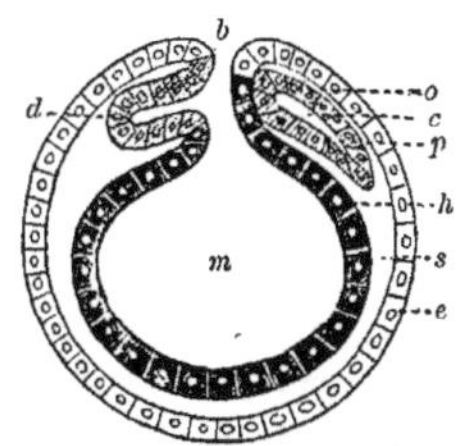

FIG. V.—Diagram illustrating the formation of the germ layers (but little modified from that occurring in *Peripatus*); on the right an earlier, on the left a later stage; b, blastopore; c, cœlom; d, mesoblastic pouch; e, epiblast; h, hypoblast; m, mesenteron; o, somatopleure; p, splanchnopleure; s, segmentation cavity.

The epiblast, as we have seen, gives rise to the outer layer of the skin. This is not the whole of the list of its derivatives, for we must here include the nervous system and the organs of sense, dermal glands, teeth, membrane-bones, etc. As we have said, the blastopore almost always becomes completely closed, but in some forms it remains open, forming the mouth or the vent, and in *Peripatus* it closes in the middle, leaving both oral and anal openings at its extremities. In these forms where it becomes completely closed, the mesenteron is entirely separated from the external world, and communication has to be again opened with the exterior. This is accomplished by inpushings of the epiblast at the extremities of the body. These ingrowths finally meet and unite with the hypoblast, and thus form the complete alimentary tract. From this method of formation of the anterior and posterior parts of the digestive canal, it follows that certain internal organs, as the œsophagus and intestine, the stomach of the lobster, and the gizzard of the cricket, the malpigian tubes of insects, etc., are really to be classed among the derivatives of the epiblast.

The mesoblast, after separating from the hypoblast, grows around between the other two layers. It either contains a cavity, originally a part of the archenteron, or such a cavity soon appears. This is the body cavity or cœlom, the pleuro-peritoneal cavity of vertebrates. The outer wall of the cœlom unites with the epiblast, the inner with the hypoblast, and the segmentation cavity is obliterated. From the mesoblast arise most of the structures and organs not already enumerated. The list includes the bones of vertebrates, the skeleton of echinoderms, spicules of sponges, muscles, connective tissue, blood, and excretory and reproductive organs.

From an embryological standpoint, as we have just seen, we can arrive at a classification of tissues; but if we turn to structure and function, the result is a different association, which, for all except the pure morphologist, is far more satisfactory. The following classification of the tissues is taken, with modifications, from Kölliker: —

1. *Epithelial tissues* (epidermal and glandular).
2. *Connective tissues* (mucous, cartilage, elastic, areolar, and osseous tissues, and dentine).
3. *Muscular tissues* (smooth and striated).
4. *Nerve tissue* (nerve-cells and fibres).

The epithelial tissues consist of cells placed side by side, forming a layer. All the other tissues arise from one having the cells characteristic of epithelium, as the germ-layers are formed of it. As Kölliker reiterates in 1884, "In all multicellular organisms all the elements and tissues arise directly from the fertilized egg-cell and the first embryonic nucleus. (1) The tissues first differentiated have the characters of epithelial tissues, and form the ectoblast and endoblast. (2) All the other tissues arise from these two cell-layers; they are either directly derived from them, or arise by the intermediation of a median layer (mesoblast) which, when developed, takes an important part in forming the tissues. (3) When the whole of the animal series is considered, each of the germinal layers is found to be, in certain creatures, capable of giving rise to at least three, and perhaps to all tissues; the germinal layers cannot, therefore, be regarded as histologically primitive organs."

Epithelial cells form the skin or epidermis of animals, and also the lining of the digestive canal. The cells of the latter may, as in sponges, bear a general resemblance to a flagellate infusorian, as *Codosiga*, or they may each bear many hair-like processes called cilia, which by their constant motion maintain currents of the fluids passing over the surface of the epithelium. The cilia lining the inside of the windpipe serve to sweep any fluid formed there towards the throat, where it can be coughed up and expectorated.

Connective tissue and its varieties, and gristle or cartilage, bone, etc., arise from the mesoblast and support the parts of the body. All the supporting tissues are used in the body for mechanical purposes: the bones and cartilages form the hard framework by which softer tissues are supported and protected; and the connective tissues, with the various bones and cartilages, form investing membranes around different organs, and in the form of fine network penetrate their substance and support their constituent cells.

Connective tissue is formed by isolated rounded or elongated cells with wide spaces between them filled with a gelatinous fluid or protoplasm, and occurs between muscles, etc. Gelatinous tissue is a variety of connective tissue found in the umbrella of jelly fishes. Fibrous and elastic tissue are also varieties of connective tissue. Cartilaginous tissue is characterized by cells situated in a still firmer intercellular substance; and when the intercellular substance becomes combined with salts of lime, forming bone, we have bony tissue.

The blood-corpuscles originate from the mesoderm as independent cells floating in the circulating fluid, the blood cells being formed contemporaneously with the walls of the vessels enclosing the blood. In the invertebrates the blood-cells are either strikingly like the *Amœba* in appearance, or are oval, but still capable of changing their form. Thus blood-corpuscles arise like other tissues, except that they become free.

Muscular tissue is also composed of cells, which are at first nucleated and afterward lose their nuclei. From being at first oval, the cells finally become elongated and unite together to form the fibrillæ; these unite with bundles forming muscular fibres, which in the vertebrates unite to form muscles. Muscular fibrillæ may be simple or striated. The contractility of muscles is due to the contractility of the protoplasm originating in the cells forming the fibrillæ.

Nervous tissue is made up of nerve-cells and fibres proceeding from them; the former constituting the centres of nervous force, and usually massed together, forming a *ganglion* or nerve-centre from which nerve-fibres pass to the periphery and extremities of the body, and serve as conductors of nerve-force. — (Packard's Zoology.)

ORGANS.

Animals are, with plants, called organisms, because they have organs. An organ is any part of the body specially developed to perform a special kind of work. Thus the wings are organs of flight, the heart is the organ of circulation, the leg an organ of locomotion. The tissues we have enumerated are combined to form organs. The simplest kind of an organ is perhaps the nucleus of the *Amœba.* There are creatures lower than the *Amœba* which have no organs. These are the Monera, in which no nucleus or any other specialized part of the body has as yet been found. If we rise in the scale of animal life to the monad, we find that it has an external appendage or organ like a whip-lash. In the Infusoria the body is covered with cilia, which are the only organs of locomotion in these animalcules. In the Hydra, the only external organs are the tentacles, which are situated around the head, and seem to feel for and to seize its prey. In the higher worms we have oar-like organs of locomotion, arranged in pairs on each side of the body; also gills, or external breathing organs. Molluscs have a creeping organ, the under side of the body; they also have gills and other external parts or organs. In the crustaceans and insects the number and variety of form of external organs, especially the legs, gills, feelers, and mouth-parts, are remarkable, and they are highly specialized. In the vertebrates, beginning with fishes and ending with man, we have external organs of sight, hearing, and locomotion, such as fins, hands, and legs.

Of the internal organs of the body, the most important is the digestive cavity, which is at first in the gastrula or early embryo of all many-celled animals, and in the Hydra and other polyps simply a hollow in the body. As we ascend in the animal series we can trace its gradual specialization, beginning with the lower worms, and ascending to the annelids, also in the sea-urchin and starfish. In the molluscs, Crustacea, and insects, as well as vertebrates, the alimentary canal is divided, during growth, into distinct portions (*i. e.*, the throat, stomach, and intestine), each with separate functions or uses. Early in life other organs arise as outgrowths from the digestive tract of the embryo. These are the lungs, the liver, pancreas, spleen, etc.

There are also organs of support; such are the skeletons of animals, whether external, as in the sea-urchin, starfish, lobster, or insect; or internal, as in a fish, bird, or man. A true bony skeleton only exists in the vertebrates, or back-boned animals.

The illustrious naturalist, Cuvier, established the principle of the correlation of organs, showing that every organ must have close relations with the rest, and be more or less dependent on the others. Each organ has its particular value in the animal economy. There is a close relation between the forms of the hard and soft parts of the body, together with the functions they perform, and the habits of the animal. For example, in a cat, sharp teeth for eating flesh, sharp curved claws for seizing smaller animals, and great muscular activity — all coexist with a stomach fitted for the digestion of animal rather than vegetable food. So in the ox, broad grinding teeth for chewing grass; cloven, wide-spreading hoofs, that give a broad support in soft ground, and a four-chambered stomach, are correlated with the habits and instincts of a ruminant. From the shape of a single tooth of an ox, deer, or a dog or cat, one can determine not only its order, but its family or genus. Hence this prime law of comparative anatomy led to the establishment by Cuvier of the fundamental principles of the science of palæontology, by which the comparative anatomist can with some degree of confidence restore from isolated teeth or bones the probable form of the original possessor. Of course, the more perfect the skeleton and teeth, the more perfect the remains of the crust of insects or the shells of extinct molluscs, the more perfect will be our knowledge, and the less room will there be for error in restoring extinct animals.

New organs often arise by changes in the form and uses of simpler, older ones, and the new organs may be so much changed by the gradual modification of its functions as to assume new uses. "This fact," says Gegenbaur, in the introduction to his Elements of Comparative Anatomy, "is of considerable importance, for it helps to explain the appearance of new organs, and obviates the difficulty raised by the doctrine of evolution, viz., that a new organ cannot at once appear with its function completely developed; that it therefore cannot serve the organism in its first stages whilst it is gradually appearing; and that consequently the cause for its development can never come into operation. Every organ for which this objection has the appearance of justice can be shown to have made its first appearance with a significance differing from its later function. Thus, for example, the lungs of the Vertebrata did not arise simply as a respiratory organ, but had a predecessor among fishes breathing by gills, in the swim-bladder, which at first had no relation to respiration. Even where the lungs first assume the functions of a respiratory organ (Dipnoi, and many Amphibia) they are not the sole organ of the kind, but share this function with the gills. The organ is therefore here caught, as it were, in the stage of conversion into a respiratory organ, and connects the exclusively respiratory lungs with the swim-bladder, which arose as an outgrowth of the enteric tube and was adapted to a hydrostatic function."

Organs may also become modified by disuse until they lose their distinctive form and become rudimentary. Striking examples are seen in the parasitic Crustacea and insects. As a result of the modification of organs by use and disuse, organs which are morphologically the same become very different in function, and also in their general appearance; so that we may classify organs differently either by their morphology or physiological uses.

We speak of highly-developed organs, and those which are aborted, atrophied, or rudimentary. Highly-developed organs are also highly specialized or differentiated; such organs are complicated, owing to the distinct uses or divisions of labor accorded to each part. Thus, the eyes of worms are very simple and lowly developed compared with the human eye; a fish's fin is, in part at least, morphologically the same as the fore-leg of a cat, or the arm of a monkey or of man, but in the human arm different uses are assigned to different portions of the limb; it is highly developed, specialized, or differentiated, to use terms nearly synonymous.

On the other hand, so exquisitely wrought an organ as a fish's eye may by disuse become nearly atrophied, as in the blind fish of Mammoth Cave, which lives in perpetual darkness. And not only the eye, but the optic lobes and optic nerves may, as in the case of a small crustacean (*Cecidotæa stygia*), also living in Mammoth Cave, be entirely aborted. Among reptiles are some extraordinary cases of modification by degeneration and atrophy. The lizard-like creature, *Seps*, has remarkably small limbs, and in *Bipes* there is only a pair of stumps, representing the hinder limbs. As Lankester claims, these two forms represent two stages of degeneration or atrophy of the limbs; "they have, in fact, been derived from the five-toed, four-legged ordinary lizard form, and have nearly or almost lost the legs once possessed by their ancestors." The entire order of snakes is an example how the loss, by atrophy, of the limbs may become common to an entire group of animals numbering thousands of species; the possession by the boas of a rudimentary pelvis, and minute but nearly atrophied hind legs, tends to prove that all the snakes are descendants from some ancestral form whose limbs became lost through disuse.

Among the Diptera, which have but a single pair of wings, there is an universal atrophy of the second or hinder pair of wings; moreover, there are numerous wingless, degraded forms, and when we take into account the fact that almost all dipterous larvæ are nearly headless and evidently degenerated forms, we are inclined to think that the entire group of true flies, numbering at least twenty thousand species, are the result of a retrograde development, affecting in every species the hinder pair of wings, and in numerous other forms the mouth-parts and other portions of the body, both in the larval and adult states. The group of barnacles (Cirripedia) is another example where atrophy and degeneration pervade each member of an order, and the cases are highly interesting and suggestive.

An entire sub-kingdom of animals may be degenerated in some respects. Such is the branch or sub-kingdom of sponges; the adult forms of which, by becoming fixed, have undergone a retrograde development, the gastrula or larval forms showing a promise of a state of development which the organism not only does not attain, but from which it falls completely away in after life. Lankester regards the acephalous molluscs, or bivalves, as having degenerated from a higher type of head-bearing active creatures like the snails. The ascidians start in the same path of development as the vertebrates, and at length fall back and lose nearly every trace of a vertebrate alliance.

Besides sub-kingdoms, classes, and orders of animals, we have minor groups which are in their entirety examples of a backward development. Without doubt certain human races, as the present descendants of the Indians of Central America, the modern Egyptians, "the heirs of the great Oriental monarchies of pre-Christian times," the Fuegians, the Bushmen, and even the Australians, may be degenerate races. Lankester, in his book on "Degeneration," considers the causes of retrograde development to be due to — 1, parasitism; 2, fixity or immobility of the animal; 3, vegetative nutrition, and 4, excessive reduction in the size of the body.

Analogy and Homology.

When we compare the wing of an insect with that of a bird, and see that they are put to the same use, we say that they are analogous; for when we carefully compare the two organs, we see how unlike throughout they are. When we compare the fin of a whale with the fore-leg of a dog or bear, we see that one is adapted for swimming, and the other for running or dry land; but, however unlike the two limbs are superficially, we find, on dissection, that all the principal bones and muscles, nerves and blood-vessels of the one correspond to those in the other, so that we say there is a structural resemblance between the two kinds of limbs. Thus analogy implies a dissimilarity of structure in two organs, with identity in use, while homology implies blood-relationship. Analogy repudiates any common origin of the organs, however physiologically alike.

In the early days of zoological science, but little was said about homologies; but when comparative anatomy engaged the attention of philosophical students, attention was given to tracing the resemblances between organs superficially and functionally unlike. It was found that the world of life teemed with examples of homologous parts.

Afterwards, when the theory of evolution became the most useful tool the comparative anatomist could wield, and when the knowledge of comparative embryology completed his equipment, the most unexpected homologies were discovered. Of course, the more nearly related are the two animals possessing homologous organs, as the dog and whale, the closer and more plainly homologous are their fore limbs. It is easy to trace the homologous organs in animals of the same order or class, however effectually degeneration on the one hand, or differentiation on the other, have done their work.

But it was then found that the branchial sacs of ascidians are homologous with the pharyngeal chamber of the lamprey eel; that the position of the nervous system in ascidians accords morphologically with that of fishes and higher vertebrates; that the notocord of larval ascidians is the homologue of that of the lancelet and lamprey, as well as that of embryo vertebrates in general; and, finally, the homologies between the larval ascidians and vertebrates are so startling that many comparative anatomists now maintain that the ascidians belong, with the vertebrates, to a common branch of the animal kingdom called Chordata. On the other hand, excellent anatomists trace homologies between certain organs in worms, and corresponding organs in sharks and other vertebrates; the segmental organs of worms have their homologous parts in the urogenital organs of sharks; the worm *Balanoglossus* has a respiratory chamber homologous with that of the lancelet and lamprey. Hence it came to pass that these general homologies between the lower, less specialized classes of invertebrates, particularly the worms, and the lower vertebrates, were so many proofs of the origin of the latter from worms or worm-like forms. Hence the opinion now prevalent that a homology between organs, however unlike in the uses at present made of them, implies that the animals having such organs had a common ancestry. Hence, also, the proofs of the unity of organization of the animal kingdom are based on a profound study of the resemblances in the tissues and organs of animals, rather than of their superficial, recently-acquired differences.

Homology may be general or special; the latter limited, for the most part, to animals of the same sub-kingdom. It is well to use words which will express our meaning exactly, and hence a general homology may be indicated by the word isogeny, indicating a general similarity of origin; thus, the nervous system of worms, arthropods, molluscs, and vertebrates are isogenous, all being derivations of the epiblast. The term homology should be restricted to those cases where the correspondence, part for part, is more exact. Thus, the brain of fishes and that of man are not only isogenous, but homologous.

The Scale of Perfection in Organs and in Animals.

The history of the rise and progress of the human arm, possibly, as some claim, from an organ like the fin of a shark, a boneless, flabby limb; how it gradually, by adaptation, became like the differentiated fore-leg of a salamander, then became adapted for a climbing, arboreal use, and finally became, next to his brain, the most distinctive organ in man, — the history of the successive steps in this rise in the scale of perfection would throw light on the general subject of the gradual perfection of organs and organisms. On the other hand, the hinder extremities or legs of man have not been equally perfected. As Cope has remarked: "He is plantigrade, has five toes, separate carpals and tarsals, a short heel, rather flat astragalus, and neither hoofs nor claws, but something between the two." Man's limbs are not so extremely specialized as those of the horse, which is digitigrade, walking only on four toes, one to each foot. Man's stomach is simple, not four-chambered, as in the ox. Thus one organ, or one set of organs, may in man attain the highest grade in the scale of perfection, while others may be comparatively low in form and function. Animals acquire, so to speak, their form by the acceleration in the growth of certain parts, which involves a retardation in the development of others; it is by the unequal development of the different parts that the fish has become adapted to its life in the water, that the bird becomes fitted for its aerial existence, and that moles can burrow, and monkeys are enabled to climb. The scale of perfection, as applied to organs, is a relative one; those in each animal are most perfect which are best adapted to subserve the requirements of that creature. Man's brain is on the whole the most perfect of all organs, and it enables him to regulate the movements of his limbs and other organs in a manner alone characteristic of an intellectual, reasoning, speaking, spiritual being.

Generalized and Specialized Types.

A large proportion of the higher classes of animals now living are more or less specialized; they stand at or near the head of a series of forms which have become extinct, and which were much less specialized. For example, there are now living nearly ten thousand species of bony fishes, while the remains of only about twenty species have been found in the cretaceous formation. The earlier types of fishes were generalized or composite in their structure, presenting, besides the cartilaginous skeleton, a feature occurring in the embryos of lung fishes, characteristics which place them above the bony fishes. Among fishes, the lung fishes or Dipnoi, are the clearest example of a generalized type; they have a notocord, in which respect they resemble the lancelet and lamprey; while they possess one or two lungs, in which respect they resemble the salamanders or batrachians; thus in some features they are lower, and in others higher, than any of the bony fishes. There is a strange mixture of characters in these composite animals, and the living forms may be regarded as old-fashioned, archaic types, the survivors of a large group of Devonian forms. They were called by Agassiz prophetic types, as they pointed to the coming of more highly wrought, specialized forms, the amphibians. They are now regarded as ancestral forms from which have originated two lines of organisms, one culminating in the bony fishes, and the other in the labyrinthodonts and salamanders and other batrachians.

The king or horse-shoe crab (*Limulus*) is likewise a composite, synthetic, comprehensive or generalized type, as it is variously called. Its development and structure shows certain features closely resembling the Arachnida, while it is also closely allied by other points to the Crustacea, with other features peculiar to itself. Its position in the scheme of nature is now in dispute, owing to the admixture of characters in which it resembles both the Crustacea and Arachnida. Now *Limulus* is a survivor or remnant of a long line of forms which flourished in the palæozoic seas, at a time when there were no genuine Crustacea nor Arachnida, which did not arise until long after the Merostomata (*Eurypterus*, etc.) and their allies, the trilobites, began to disappear. These generalized, composite arthropods lived, had their day, and finally gave way to the hosts of highly specialized modern Crustacea, such as the shrimps and crabs of our seas, and to the scorpions and spiders inhabiting the land, and which owe their diversity of form to the highly wrought structure of a few special parts.

So, among insects, the earlier were the more generalized, old-fashioned forms. Such were the white ants, cockroaches, grasshoppers, may-flies and dragon-flies, the earliest insects known. Their numbers are scanty at the present day. They have been in part supplanted by the thousands of species of beetles, moths, butterflies, ants, wasps, and bees, so characteristic of the present age of the world as compared with the insect life of the carboniferous period.

Among mammals the horse is the most specialized; its generalized ancestors were the *Coryphodon* and *Eohippus*. The latter had four usable toes, and the rudiments of a fifth on each forefoot, and three toes behind. The history of the horse family is a record of successive steps by which a highly specialized type was produced, culminating in that extreme form, the American trotting horse, which can only do one thing well, *i. e.* excel in trotting over a racecourse.

PHYSIOLOGY.

The most difficult line of study in biology is to determine how the organs do their work. This is the office of the physiologist. It is comparatively easy to discover how a fish uses its fins in swimming, how a mammal walks, how a bird flies; but it is difficult to ascertain how the internal organs perform their functions; how the stomach digests food, exactly what office the biliary and pancreatic fluids fill in the complicated process of digestion; the function of the spleen, and so on with the other viscera. Here, besides observation and comparison, the physiologist has to rely on experiments to test the results of his observations. The pathway to a complete understanding of human physiology lies through the broad and as yet only partially surveyed field of animal physiology. We can better understand the physiology of digestion in man by studying the process of intracellular digestion in the Infusoria, the sponges, and jelly-fish, where, owing to the transparency of the tissues, the process can be actually observed; so likewise, the nature of muscular movements can best be understood by observing under the microscope the contractility of the protoplasm of individual cells or one-celled organisms. The physiology of reproduction could never have been understood without a study of the operations of cell-division, self-fission and gemmation in the one-celled animals and polyps. Our most eminent human physiologists, such as Remak, Bischoff, von Baer, and others, have had to go to the lower animals for facts to illustrate the reproductive processes in man. No medical student can in these days afford to be ignorant of the general laws of animal physiology.

Locomotion.

All the movements of the body, or of the internal organs, with which physiology has to do, depend primarily on the contractility inherent in the protoplasm filling the cells of the body. Of the cause of contractility in protoplasm we know nothing. We see its manifestations in the irritability and resulting contractions of the body of the Amœba, of the white blood-corpuscle, and other cells and one-celled organisms. It is this inherent contractility of the protoplasm in muscle-cells which gives rise to muscular movements.

But the simplest one-celled animals do not move about by means of muscles. The Amœba changes its position, Proteus-like, by variously contracting the body, and thus changing its form, throwing out root-like processes in different directions. The Infusoria have permanent thread-like processes, called cilia, by which they can swim about in the water. In the *Hydra* and other polyps, however, we meet with muscles, by which the body can contract in certain parts; in such animals the base of the body forms a more or less contractile, movable, creeping disc, while the tentacles move partly by means of their muscular walls, and partly mechanically by filling them with the circulatory fluid of the body.

The sea-urchin and starfish move slowly over the sea weed and rocks by means of long, slender suckers. Extending these by allowing the water to flow into them, and fastening them to the surface of the object over which they are moving, they then contract them, and in this way the body is warped slowly along.

In the lower worms, such as the flat-worms, or in the snails, the gliding movement is due to the muscular contractions of the under side of the body. The gliding motion of snails is due to a system of extensive muscular fibres within the disc, which act when the sinuses within the disc are filled with blood; their extension causing the undulating appearance on the under side of the snail's foot, or creeping disc; but the snail can only thus move forwards; the lateral movements and shortening of the foot being produced by oblique muscular fibres.

Rising in the zoological scale, we come to the Crustacea and insects which have jointed legs, ending, in the latter, in claws adapting the limb both for walking and climbing. The legs of arthropods are perhaps modifications of the lateral fleshy oar-like appendages of the sea-worms, which have become externally hard and jointed, with several leverage-systems. The mechanism of locomotion is fundamentally the same in the legs of arthropods and vertebrates. Space does not permit us to discuss the subject of the mechanism of walking, running, and flying, but all these movements are dependent primarily on the contractility inherent in the protoplasm filling the cells forming muscular tissue.

Digestion.

The most important organs in the animal system are those relating to digestion, as an animal may respire solely through its body-walls, or do without a circulatory or nervous system, but must eat in order to live and grow. The opening by which the food is taken into the alimentary canal is called the mouth, whether reference is made to the 'mouth' of a hydra or of a vertebrate. Although the structure of the edges may differ radically, still in most Metazoa the mouth is due to an inpushing of the ectoderm, however differently the edge may be supported and elaborated. The edges of the mouth are usually called the lips, but true lips for the first time appear in the Mammalia. The trituration or mastication of the food is accomplished among the invertebrates in a variety of ways, and by organs not always truly homologous.

The object of digestion is to reduce the food into a convenient condition, and to dissolve or to transform it into tissue-food. How the products of digestion are carried about in the body, so as to supply the tissues of the various organs, by the circulatory organs, will be pointed out in the next section. The simplest form of the digestive process may be seen by a skilled observer in the cells lining the digestive pockets or chambers in the interior of a sponge or jelly-fish. Each cell has a certain amount of individuality, taking in, through transitory openings in their walls, particles of food, and rejecting the waste portions, much as individual Infusoria, which have no stomachs, ingest their food, and reject the particles which are indigestible or not needed. Not only has intra-cellular digestion, as it is called, been observed in sponges and jelly-fishes, as well as in ctenophores, but also in low worms (Turbellaria).

We will now look at the leading steps in the evolution and specialization of the digestive cavity of animals. In the polyps, such as *Hydra* and *Coryne*, it is simply a hollow in the body. The Hydra draws some little creature with its tentacles into its stomach; there it is acted upon by the juices secreted by the walls of the stomach, and the hard parts rejected from the mouth. For the technical name of the digestive tract as a whole, we may adopt Haeckel's term enteron.

In the jelly-fishes the stomach opens into four or more water-vascular canals or passages, by which the food, when partially digested and mixed with sea-water, thus forming a rude sort of blood, supplies the tissues with nourishment. In the sea-anemones and coral polyps, the digestive cavity is still more specialized, and its walls are partly separated from the walls of the body, though at the posterior end the stomach opens directly into the body cavity. In the echinoderms and worms do we find for the first time a genuine digestive tube, lying in the perivisceral space (which, with Haeckel, we may call the cœlom), and opening externally for the rejection of waste matter.

In the worms the digestive canal becomes separated into a mouth, an œsophagus, with salivary glands opening into the mouth, and there is a division of the digestive tract into three regions—*i. e.*, fore (œsophagus), middle (chyle-stomach), and hind enteron (intestine). In the molluscs and higher worms there is a well-marked sac-like stomach and an intestine, with a liver, present in certain worms and in the ascidians and molluscs, opening into the beginning of the intestine. All these divisions of the digestive tract exist still more clearly in the Crustacea and most insects. In the latter, six or more excretory tubes (Malpigian vessels) discharge their contents into the intestines, and in the 'respiratory tree' of the Holothurian, and the segmental organs of certain worms we have organs with probably similar excretory uses.

In the vertebrates, from the lancelet to man, the alimentary canal has, without exception, the three divisions of œsophagus, stomach, and intestine, with a liver. In this branch the lungs are modified parts of originally sac-like dilatations of the first division of the digestive tract. The intestine is also sub-divided in the mammals into the small and large intestine and rectum, a cœcum being situated at the limits between the small and large intestine. We thus observe a gradual advance in the degree of specialization of the digestive organs, corresponding to the degree of complication of the animal. — (Packard's Zoology.)

We will now look at the glands which pour their secretion into the digestive canal. In the worms, salivary glands send their secretion into the throat, while in the polyps (cœlenterates) and many worms, and in all insects the stomach is lined with a layer of colored cells which secrete bile; in the spiders the stomach forms a set of complicated cœcal appendages which secrete a fluid like bile; in the Crustacea, and lower molluscs, there is a liver formed of little glands which open into the beginning of the intestine, while in the higher molluscs and in the vertebrates we have a true specialized liver merely connected with the digestive canal by its ducts.

Thus the original food-stuff is variously treated by animals of different grades. In the sea-anemone or any polyp, a very imperfectly digested material is produced, which is taken at first hand, mixed with the sea-water, and in part churned by the movements of the body, in part moved about in a more orderly and thorough manner, in currents formed by the cilia lining the chambers of the body.

In the worms, and insects, etc., the chyle or products of digestion percolates, or oozes through the walls of the intestine into the body cavity, and there directly mingles with the blood, and is thus carried in the circulation to every part of the body, however remote or minute. In the vertebrates, however, this is not so. The chyle, a much more elaborate fluid than that of the lower animals, is carried by an intricate system of vessels, called lymphatics, from the intestine to the blood vessels. Thus the process of digestion becomes increasingly elaborate as we ascend in the animal series, and as the digestive system becomes more and more complex.

Here again we might look at the chewing apparatus or teeth, arming the mouth, by which the food is made ready for digestion. To quote from the author's text-book of Zoology: — Hard bodies serving as teeth occur for the first time in the animal series in the sea-urchins, where a definite set of calcareous dental processes or teeth, with solid supports and a complicated muscular apparatus, serves for the comminution of the food, which consists of decaying animals and sea-weeds. In those echinoderms which do not have a solid framework of teeth, the food consists of minute forms of life, protozoans and higher soft-bodied animals, or the free-moving young of higher animals, which are carried into the mouth in currents of water or swallowed bodily with sand or mud.

Among the worms, true organs of mastication for the first time appear in the Rotatoria, where the food, such as Infusoria, etc., is crushed and is partly comminuted by the well-marked horny and chitinous pieces attached to the mastax. In most other low worms the mouth is unarmed. In the leeches there are three, usually in the annelids two, denticulated or serrate, chitinous flattened bodies, situated in the extensible pharynx of these worms, and suited for seizing and cutting or crushing their prey.

In the higher molluscs, such as the snails (Cephalophora) and cuttles, besides one or more broad, thin pharyngeal jaws, comparable with those mentioned as existing in the worms, is the lingual ribbon, admirably adapted for sawing or slicing sea-weeds, and cutting and boring into hard shells, acting somewhat like a lapidary's wheel; this organ, however, is limited in its action, and in the cuttles, the jaws, which are like a parrot's beak, do the work of tearing and biting the animals serving as food, which are seized and held in place by the suckered arms.

In the crustaceans and insects we have an approach to true jaws, but here they work laterally, not up and down, or vertically, as in the vertebrate jaws; the mandibles of these animals are modified feet, and the teeth on their edges are simply irregularities or sharp processes, adapting the mandibles for tearing and comminuting the food. It is generally stated that the numerous teeth lining the crop of Crustacea and insects, serve to further comminute the food after being partially crushed by the mandibles, but it is now supposed that these numerous points also act collectively as a strainer to keep the larger particles of food from passing into the chyle-stomach until finely crushed.

The king-crab burrows in the mud for worms (*Nereids*, etc.); these may be found almost entire in the intestine, having only been torn here and there, and partly crushed by the spines of the base of the foot-jaws, which thus serve the purpose effected by the serrated edges of the mandibles of the genuine Crustacea and insects.

Among vertebrates the lancelet is no better off than the majority of the cœlenterates and worms, having no solid parts for mastication; and the jaws and teeth of the hag-fish, and even the lamprey eel, form a very different apparatus from the jaws and its skeleton in the higher vertebrates; and even in the latter the bony elements differ essentially in form in the different classes, though originating in the same manner in embryonic life. In the birds, the jaw-bones are encased in horny plates; true teeth being absent in the living species, the gizzard being, however, provided with two hard grinding surfaces; on the other hand, mammals without teeth are exceptional.

The teeth of fishes are developed, not only in the jaws, but on the different bones projecting from the sides and roof of the mouth, and extend into the throat. In many cases, in the bony fishes, these sharp recurved teeth serve to prevent the prey, such as smaller fish, from slipping out of the mouth. On the other hand, the upper and lower sides of the mouth of certain rays (*Myliobatis*) are like the solid pavement of a street, and act as an upper and nether mill-stone to crush solid shells.

In the toothless ant-eaters the food consists of insects, which are swallowed without being crushed in the mouth; true teeth are wanting in the duck-bill, their place being taken by the horny processes of the jaws, while in Steller's manatee the toothless jaws were provided with horny solid plates for crushing the leaves of succulent aquatic plants. Examples of the most highly differentiated teeth in vertebrates are seen in those animals which, like the bear, are omnivorous, feeding on flesh, insects, and berries, and which have the crown of the molars tuberculate; while the canines are adapted for holding the prey firmly as well as for tearing the flesh, and the incisors for both cutting and tearing the food.

Circulation.

Intimately associated with the digestive canal are the vessels in which the products of digestion mix with the blood and supply nourishment for the tissues, or, in other words, for the growth of the body. In the Infusoria the evident use of the contractile vesicles is to aid in the diffusion of the partly digested food of these microscopic forms. In the Hydra the food stuff is directly taken up by the cells lining the cœlom, while the imperfectly formed blood also finds access to the hollows of the tentacles. The mode in which the cells lining the canals in the sponge take up, by means of pseudopodia, microscopic particles of food, directly absorbing them in their substance, is an interesting example of the mode of nourishment of the cellular tissues of the lower animals. The sea-anemone presents a step in advance in organs of circulation; here the partly digested food escapes through the open end of the stomach into the perivisceral chambers formed by the numerous septa, the contractions of the body churning the blood, consisting of sea-water and the particles of digested food, and a few blood-corpuscles, hither and thither, and with the cilia forcing it into every interstice of the body, so that the tissues are everywhere supplied with food.

The water-vascular system of the cœlenterates presents an additional step in degree of complexity; but it is not until we reach the echinoderms, on the one hand, and such worms as the *Nemertes* and its allies on the other, where definite tubes or canals, the larger ones contractile, and, in the latter type at least, formed from the mesoderm, serve to convey a true blood to the various parts of the body, that we have a definite blood system. In the echinoderms a true hæmal or vascular system may coexist with the water-vascular system. In the annelids, such as the *Nereis*, one of the blood-vessels may be modified to form a pulsating tube or heart, by which the blood is directly forced outward to the periphery of the body through vessels which may, by courtesy, be called arteries, while the blood returns to the heart by so-called veins.

The molluscs have a circulatory system which presents a nearer approach to the vertebrate heart and its vessels than even in the crustaceans and insects, for the ventricle and one or two auricles, with the complicated arterial and venous system of vessels of the clam, snail, and cuttle-fish, truly foreshadow the genuine heart and systemic and pulmonary circulation of the vertebrates. The molluscs, and king-crab, and the lobster, possess minute blood vessels which present some approach to the capillaries of vertebrates. The circulation in certain worms, from *Nemertes* upward, may be said to be closed, the vessels being continuous; but they are not so in insects where true veins are not to be found, the blood returning to the heart in channels or lacunæ in the spaces between the muscles and viscera.

In vertebrates the 'aortic heart' of the lancelet or *Amphioxus* is simply a pulsating tube, and there are portions of other vessels which are pulsatile, so that there is, as in some worms, a system of 'hearts.' A genuine heart, consisting of an auricle and a ventricle only, first appears in the lamprey. This condition of things survives in fishes, with the exception of those forms, such as the lung-fish (Dipnoans), whose heart anticipates in structure that of the amphibians and reptiles, in which a second auricle appears. Again, certain reptiles, such as the crocodiles, anticipate the birds and mammals in having two ventricles — *i. e.*, a four-chambered heart. It should be borne in mind that in early life the heart of all skulled vertebrates (Craniota) is a simple tube, and as Gegenbaur states, "as it gradually gets longer than the space set apart for it, it is arranged in an S-shaped loop, and so takes on the form which the heart has later on." Owing to this change of form it is divided into two parts, the auricle and ventricle.

A striking feature first encountered in the craniate vertebrates is the presence of a set of vessels conveying the nutrient fluid or chyle which filters through the walls of the digestive canal to the blood-vessels; these are, as already stated, the lymphatics. In the lancelet, as well as in the invertebrate animals, such vessels do not occur, but the chyle oozes through the stomach-walls and directly mixes with the blood.

Respiration.

Always in intimate relation with the circulatory system are the means of respiration. The process may be carried on all over the body in the simple animals, such as Protozoa or sponges, or, as in cœlenterates, it may be carried on in the water-vascular tubes of those animals, while in the so-called respiratory tree of echinoderms it may go on in company with the performance of other functions by the same vessels. Respiration, however, is inclined to be more active in such finely subdivided parts of the body as the tentacles of polyps, of worms, or any filamentous subdivisions of any of the invertebrates; these parts, usually called gills, present in the aggregate a broad respiratory surface. Into the hollows of these filamentous processes, which are usually extensions of the body-walls, blood is driven through vessels, and the oxygen in the water bathing the gills filters through the integument, and immediately gains access to and mixes with the blood. The gills of the lower animals appear at first sight as if distributed over the body in a wanton manner, appearing in some species on the head, in others along the sides of the body, or in others on the tail alone; but in fact they always arise in such situations as are best adapted to the mode of life of the creature.

The gills of many of the lower animals afford an admirable instance of the economy of nature. The tentacles of polyps, polyzoans, brachiopods, and many true worms serve also, as delicate tactile organs, for grasping and conveying food to the mouth, and often for locomotion. The suckers or 'feet' of star-fish or sea-urchins also without doubt help to perform the office of gills, for the luxuriously branched, beautifully-colored tentacles of the sea-cucumber are simply modifications of the ambulacral feet.

In the molluscs, especially the snails and cuttle-fish, the gills are in close relations with the heart, so that in the cuttle-fish the auricles are called 'branchial hearts.' The gills of crustaceans are attached either to the thoracic legs or are modified abdominal feet, being broad, thin, leaf-like processes into which the blood is forced by the contractions of the tubular heart. Respiration in the insects goes on all over the interior of the body, the tracheal tubes distributing the air so that the blood becomes oxygenated in every part of the body, including the ends of all the appendages. The gills of aquatic insects are in all cases filamentous or leaf-like expansions of the skin permeated by tracheæ; they are, therefore, not strictly homologous with the gills of crustaceans or of worms. — (Packard's Zoology.)

We now come to the respiratory organs of the vertebrates, which are in close relation to the digestive canal. First the gills: just behind the mouth are openings, called branchial clefts, on the edges of which arise processes, the gills or branchiæ. Throughout these gills are distributed minute arteries and veins, forming a network; the gills are bathed in water taken in through the mouth. In the amphibians and lung-fishes, (Dipnoi) lungs, which are outgrowths of the enteric canal, replace the swimming bladder of the fishes, the air being now swallowed by the mouth and gaining access by a special passage, the larynx, to highly specialized organs of respiration, the lungs, which are situated in the thoracic cavity near the heart.

Nerves and Sensation.

We have seen that animals of comparatively complicated structure perform their work in the animal economy without any nervous system whatever. In none of the Protozoans, even the highest infusorians, have true nerve-cells been yet detected; in these animals the tissues are in an inchoate, non-specialized state. It is not until we rise to the many-celled animals that we observe nerves and nerve-centres. It has been only recently discovered that in many jelly-fish there is, for the first time in the animal series, a true nervous system, with definite nerve-centres or ganglia. In the hydroids none has been found, so that the majority, if not all, of the polyps perform their complicated movements, capturing and taking in food, digesting it, and reproducing their kind, without the aid of what seems, when we study vertebrates alone, as the most important and fundamental system of organs in the body.

The Protozoa, sponges, and many cœlenterates depend, for the power of motion, on the irritability and contractility of the protoplasm of the body, whether or not separated into muscular tissue. Referring to the complicated movements of the Protozoa, Dr. Krukenberg well says: "The changeful phenomena of life, which we remark in the smallest organisms — in the rhythm of their ciliary motions, now strengthened, now slackened; in the rhythmic alternation of the capacity of their contractile vesicles; in their regulated incomes, deposits, and expenditure; in the abundance of the visible products of their diverse material exchanges — enable us but remotely to foresee what is here effected by a harmonious co-operation of countless processes limited to the smallest space. Let their formal differentiation seem to us ever so slight, just so do these beings become for us all the greater riddles, especially when we find in them vital manifestations elsewhere displayed in the living world only by apparatus of the most highly complex constructions, and in them meet with processes which, without the orderly co-operation of very different factors, must remain to us unintelligible." In the Hydra for the first time appear the traces of a nervous tissue in the so-called neuro-muscular cells, one portion of a cell being muscular, the other nervous in its functions.

A more definite nervous organization has been detected in the Actiniæ, in the form of disconnected bodies and rod-like nerve cells, and other nervous bodies found near the eye-spots, and the nerve-cells and fibres at the base of the body; but a genuine nervous system for the first time appears in certain naked-eyed jelly-fishes, in which it is circular, sharing the radiated disposition of parts in these animals. As the results of his experiments on the ctenophores, Krukenberg finds that animals of this class, of comparatively simple structure, and therefore exhibiting morphological differences which to us seem trifling, may nevertheless display very diverse reactions when exposed to similar abnormal conditions in the physiological laboratory. "In our attempt to explain the occult vital powers thus revealed, we are debarred from an appeal to the apparently corresponding diversities sometimes encountered in the case of the much more complex vertebrates." The echinoderms have a well-developed nervous system, consisting of a ring (without, however, definite ganglia, though masses of ganglionic cells are situated in the larger nerves), surrounding the œsophagus, and sending a nerve into each arm; or, in the holothurians, situated under the longitudinal muscles radiating from that muscle closing the mouth. Recent researches on the star-fish show, however, that besides the ring around the mouth, and the five main nerves passing along the arms or rays, there is a thin nerve-sheath which encloses the whole body, and is directly continuous with the external epidermis, of which it forms the deepest layer. The circumoral and radial nerves are believed to be simply thickenings of this thin nervous sheet.

In this connection should be mentioned the experiments made by Romanes, Ewart, and Marshall, on living Echini, "which lead them to believe in the existence not only of an external nerve-plexus outside the test, but also of an internal plexus on its inner surface; they further believe that the two systems are connected by nerve-fibres running through the plates of the test or shell."

In all other invertebrate animals, from the worms and Mollusca to the crustaceans and insects, the nervous system is fundamentally built nearly upon the same plan. There is a pair of ganglia above the œsophagus, called the 'brain;' on the under side is usually a second pair; the four, with the nerves or commissures connecting them, forming a ring. This arrangement of ganglia, often called the 'œsophageal ring,' constitutes, with the slender nerve-threads leading away from them, the nervous system of the lower worms, in many of which, however, as also in most Polyzoa and Brachiopoda, the subœsophageal ganglia are wanting. Now to the œsophageal ring with its two pairs of ganglia, add a third pair, the visceral ganglia, and we have the nervous system of the clam and many molluscs.

In the higher ringed worms, the Annulata, and in the Crustacea and Insects, there is a chain of ganglia, or brains, which, behind the throat, are ventral, and lie on the floor of the cœlom or body cavity. The highest form of nerve-centre found in the invertebrate animals, and which hints at the brain and skull of vertebrates, is the mass of ganglia partly enclosed in an imperfect cartilaginous capsule in the head of the cephalopods. The nervous cord of the *Appendicularia*, an ascidian, is constructed on the same plan as in the Annulata, but the mode of origin and apparently dorsal position of the nervous system of the tailed larval ascidian presents features which apparently anticipate the state of things existing among the lower vertebrates, such as the lancelet.

We need not here describe the different forms of nervous system in the classes of invertebrates, but refer the reader to the figures and descriptions of the different types in the body of this work. It will be well to read the following data concerning the brain and nervous system, which we quote from Bastian's "The Brain as an Organ of Mind."

"1. Sedentary animals, though they may possess a nervous system, are often headless, and they then have no distinct morphological section of this system answering to what is known as a brain.

"2. When a brain exists, it is invariably a double organ. Its two halves may be separated from one another, though at other times they are fused into what appears to be a single mass.

"3. The component or elementary parts of the brain in these lower animals are ganglia in connection with nerves proceeding from special impressible parts or sense-organs; and it is through the intervention of these united sensory ganglia that the animal's actions are brought into harmony with its environment or medium.

"4. That the sensory ganglia, which in the aggregate constitute the brain of invertebrate animals, are connected with one another on the same side, and also with their fellows on the opposite sides of the body. They are related to one another either by what appears to be continuous growth, or by means of 'commissures.'

"5. The size of the brain as a whole, or of its several parts, is therefore always fairly proportionate to the development of the animal's special sense-organs. The more any one of these impressible surfaces or organs becomes elaborated and attuned to take part in discriminating between varied external impressions, the greater will be the proportionate size of the ganglionic mass concerned.

"6. Of the several sense-organs and sensory ganglia whose activity lies at the root of the instinctive and intelligent life (such as it is) of invertebrate animals, some are much more important than others. Two of them especially are notable for their greater proportional development, viz.: those concerned with touch and vision. The organs of the former sense are, however, soon outstripped in importance by the latter.

The visual sense, and its related nerve-ganglia, attain an altogether exceptional development in the higher insects and in the highest molluscs.

"7. The sense of taste and that of smell seem, as a rule, to be developed to a much lower extent. In the great majority of invertebrate animals it is even difficult to point to distinct organs or impressible surfaces as certainly devoted to the reception of either of such impressions. Nevertheless there is reason to believe that in some insects the sense of smell is marvellously keen, and so much called into play as to make it for such creatures quite the dominant sense endowment. It is pretty acute also in some Crustacea.

"8. The sense of hearing seems to be developed to a very slight extent. Organs supposed to represent it have been discovered, principally in molluscs and in a few insects. It is, however, of no small interest to find that where these organs exist the nerves issuing from them are most frequently not in direct relation with the brain, but immediately connected with one of the principal motor nerve-centres of the body. It is conjectured that these so-called 'auditory saccules' may, in reality, have more to do with what Cyon terms the sense of space than with that of hearing. The nature of the organs met with supports this view, and their close relations with the motor ganglia also become a trifle more explicable in accordance with such a notion.

"9. Thus the associated ganglia representing the double brain are, in animals possessing a head, the centres in which all impressions from sense-organs, save those last referred to, are directly received, and whence they are reflected on to different groups of muscles — the reflection occurring not at once, but after the stimulus has passed through certain 'motor' ganglia. It may be easily understood, therefore, that in all invertebrate animals perfection of sense-organs, size of brain, and power of executing manifold muscular movements, are variables intimately related to one another.

"10. But a fairly parallel correlation also becomes established between these various developments and that of the internal organs. An increasing visceral complexity is gradually attained; and this carries with it the necessity for a further development of nervous communications. The several internal organs with their varying states are gradually brought into more perfect relation with the principal nerve-centres as well as with one another.

"11. These relations are brought about by important visceral nerves in Vermes and arthropods — those of the 'stomato-gastric systems' — conveying their impressions either direct to the posterior part of the brain or to its peduncles. They thus constitute internal impressions which impinge upon the brain side by side with those coming through external sense-organs.

"12. This visceral system of nerves in invertebrate animals has, when compared with the rest of the nervous system, a greater proportional development than among vertebrate animals. Its importance among the former is not dwarfed, in fact, by that enormous development of the brain and spinal cord which gradually declares itself in the latter.

"13. Thus impressions emanating from the viscera and stimulating the organism to movements of various kinds, whether in pursuit of food or of a mate, would seem to have a proportionally greater importance as constituting part of the ordinary mental life of invertebrate animals. The combination of such impressions with the sense-guided movements by which they are followed, in complex groups, will be found to afford a basis for the development of many of the instinctive acts which animals so frequently display."

When we rise to the vertebrates we meet with a form of nervous system quite different from that of any adult invertebrate animal. In all the vertebrates which have a definite skull — and this only excludes the lancelet and the ascidians — the brain is a series of close-set ganglia, forming a mass situated in the skull, with definite relations to the sense-organs, and the spinal cord is situated above the vertebral column, passing through the spinal canal, which is formed by the contiguous posterior arches of the several vertebræ composing the spinal, or vertebral column.

While the nervous system of all skulled vertebrates has a definite persistent situation, and with a similar cellular structure, there is a great difference between the brain of the fishes and that of mammals, including man. In the fishes the brain cavity is small compared with the size of the head, the brain being small, and there is a marked equality in the size of the different lobes forming the brain, the optic lobes being larger than the cerebral. In amphibians, such as the frog and toad, the brain is more like that of fishes than of reptiles, but the optic lobes are a little smaller than the cerebral, while the cerebellum is smaller than in many fishes. In the reptiles, as seen in snakes, turtles, and crocodiles, the cerebral lobes begin to enlarge, and exceed in size the optic lobes. Here the ventricle or cavity of the cerebral lobes is larger than in the fishes, and the rounded eminence projecting from its anterior and inner surface, called the 'corpus striatum,' is present for the first time.

In birds the brain cavity is much larger than in any of the foregoing classes of vertebrates, and the cerebral hemispheres are now greatly increased in size, so as to partly cover the optic lobes. The cerebellum is also much larger than before, and it is transversely creased.

Passing from the birds to mammals, there is seen to be a great advance in the form of the brain of the latter animals. The brain cavity is much larger, and this is for the most part occupied by two portions, the cerebrum and the cerebellum. The cerebral hemispheres entirely conceal from above the olfactory and optic lobes, the surface is convoluted, while behind it either touches or overlaps, so as in man to completely conceal the cerebellum. The cerebral hemispheres, then, form the back of the mammalian brain, and the higher orders are usually characterized by an increase both in the size of the cerebral hemispheres, and as a rule, though there are exceptions noted farther on, in the number and complexity of the convolutions of the surface. Thus in the highest mammals, especially the gorilla and man, the increased size of the brain in proportion to the greater bulk of the body is very marked.

Leuret has approximately shown the average proportional weight of the brain to the body, in four classes of vertebrates, as follows: in Fishes, as 1 to 5,668; in reptiles, as 1 to 1,321; in birds, as 1 to 212; in Mammalia, as 1 to 186. The brain is, however, subject to the same laws as other parts of the body. There is in no organ a regular and continuous progressive increase in size and complexity in any class of the animal kingdom. The size of the cerebral hemisphere differs in different monkeys, and, as has been remarked by Bastian, in the higher types of lower orders the brain is often better developed than among the lower types of higher orders. Thus in the Midas marmoset the convolutions are absent, so that in this respect this primate is on a level with the monotremes and lower marsupials and rodents. In dwarf or small-sized members of a group the brain is larger in proportion to the body than in the full-sized members. Thus among marsupials, as Owen states, the size of the brain of the pigmy petaurist is to the size of the body as 1 to 25, while in the great kangaroo it is as 1 to 800; among rodents it is as 1 to 20 in the harvest mouse, but is as 1 to 300 in the capybara; among the Insectivora it is as 1 to 60 in the little two-teed ant-eater, but is as 1 to 500 in the great ant-eater. The brain of a porpoise four feet long may weigh 1 lb. avoirdupois; that of a whale (*Balænoptera*) 100 feet in length does not exceed 4 lbs. avoirdupois; in Quadrumana the brain of the Midas marmoset is to the body as 1 to 20; in the gorilla it is as 1 to 200.

"But such ratios do not show the grade of cerebral organization in the mammalian class; that in the kangaroo is higher than that in the bird, though the brain of a sparrow be much larger in proportional size to the body: and the kangaroo's brain is superior in superficial folding and extent of gray cerebral surface to that of the petaurist. The brain of the elephant bears a less proportion to the body than that of opossums, mice, and proboscidian shrews, but it is more complex in structure, more convolute in surface, and with proportions of pros- to mes-encephalon much more nearly than in the human brain. The like remark applies to all the other instances above cited." Owen explains these facts by saying that the brain grows more rapidly than the body, and is larger in proportion thereto at birth than at full growth; "so in the degree in which a species retains the immature character of dwarfishness, the brain is relatively larger to the body."

The bearing of the facts known as to the relative size of the brain and the convolutions are thus discussed by Bastian: "There cannot therefore be, among animals of the same order, any simple or definite relation between the degree of the intelligence of the creature and the number or disposition of its cerebral convolutions — since this structural feature of the brain seems to be most powerfully regulated by the mere bulk of the creature to which it belongs." It fails still more, when comparing representatives of different orders. For example, the beaver's brain is almost smooth, while that of the sheep has numerous convolutions, which both in number and complexity decidedly surpass even those of the dog. Yet among closely related animals and those of about the same size, especially in species of the same genus, or, as in the case of man, in individuals of the same species, we may look for some proportional relations between the development of their cerebral convolutions and their intelligence.

"Size of brain, and with it convolutional complexity, must," Bastian remarks, "be closely related to the number and variety of an animal's sensorial impressions, and also to its power of moving continually or with great energy.

"The importance of taking into account the powers of movement possessed by the animal is fully borne out by the fact that the brain attains such a remarkable size in the shark, as well as in the porpoise and the dolphin — all of them creatures whose movements are exceptionally rapid, continuous, and varied. The great increase in the size of the cerebellum in each of these creatures is, therefore, not so surprising; but it seems very puzzling, at first sight, to understand why this should be accompanied by a co-ordinate increase in the development of the cerebral hemispheres. For this, however, there are two causes, the one general and the other more special. It is a fact generally observed, that sensorial activity, and therefore intelligent discrimination, increases with an animal's powers of movement; and secondly, there must be special parts of the cerebral hemispheres devoted to the mere sensory appreciation of movements executed. The nerve elements lying at the basis of this latter appreciation, however they may be distributed through the hemispheres, would naturally be the more developed (and, consequently, all the more calculated to help to swell the size of the cerebrum), in proportion to the variety and continuance of the movements which the animal is accustomed to execute."

The tactile sense, or sense of touch is common to all animals; this is the most fundamental sense, of which the other senses are without doubt differentiations. In the lower Protozoa, such as the *Amœba*, the sense of touch which they appear to possess may be due to the inherent irritability and contractility of the protoplasm of which their bodies are formed.

In the Infusoria, without doubt, the cilia and the flagella with which these animals are provided are not only organs of locomotion but also of touch. It is probable that none of the many-celled animals are without the sense of touch unless some of the sponges, and the root-barnacles (*Sacculina*) may be, by reason of their lack of a nervous system and otherwise degenerate structure, destitute of any sense whatever.

The most important of the sense organs are undoubtedly eyes, as they are the most commonly met with. The transparent spot in the front of the body of *Euglena viridis*, a protophyte, may possibly be the simplest of all sense organs; if so, it anticipates the eye of animals. The simplest forms of eyes are perhaps those of the sea-anemone, in which there are, besides pigment cells forming a colored mass, refractive bodies which may break up the rays of light impinging on the pigment spot, so that these creatures may be able to distinguish light from darkness. The next step in advance is where a pigment mass covers a series of refractive cells called crystalline rods or crystalline cones, which are situated at the end of a nerve proceeding from the brain. Such simple eyes as these, often called eye-spots, may be observed in the flat worms, and they form the temporary eyes of many larval worms, echinoderms, and molluscs. In some nemertean worms, such as certain species of *Polia* and *Nemertes*, true eyes appear, but in the ringed worm, *Neophanta celox*, Greef describes a remarkably perfect eye, consisting of a projecting spherical lens, covered by the skin, behind which is a vitreous body, a layer of pigment separating a layer of rods from the external part of the retina, outside of which is the expansion of the optic nerve. Eyes are also situated on the end of the body in some worms, and in a worm called *Polyophthalmus*, each segment of the body bears a pair of eyes.

The eyes of molluscs are, as a rule, highly organized, until in the cuttle-fish the eyes become nearly as highly developed as in fishes, but still the eye of the cuttle-fish is not homologous with that of vertebrates, since in the former the crystalline rods are turned towards the opening of the eye, while in vertebrates they are turned away from the opening of the eye, so that, as Huxley as well as Gegenbaur show, the homology between the eye of the cephalopods and of the vertebrates is not exact.

While, as we have seen, the eyes of the worms and the molluscs are situated arbitrarily, by no means invariably placed in the head, in the crustaceans the eyes assume in general a definite position in the head, except in a schizopod crustacean (*Euphausia*), where there are eye-like organs on the thorax and abdomen. In insects there are both simple and compound eyes occupying definitely the upper and front part of the head.

The eyes of the lancelet are not homologous with those of the higher vertebrates, being only minute pigment spots comparable with those of the worms. In the skulled vertebrates the eyes are of a definite number, and in all the types occupy a definite position in the head.

The simplest kind of auditory organ is to be found in jelly-fishes, where an organ of hearing first occurs. In these animals, situated on the edge of the disc, are minute vesicles containing one or more concretionary bodies or crystals. Reasoning by exclusion, these are supposed to represent the ear-vesicles or otocysts of worms and molluscs; and the concretions or crystals, the otoliths of the same kind of animals.

The otocysts or simple ears of worms and molluscs are minute and usually difficult to find, as is the auditory nerve leading from them to the nerve-centres. In the clam it is to be looked for in the so-called foot. In the snails, as also in cuttle-fish, the auditory vesicles are placed in the head close to the brain. The ears of Crustacea are sacs, formed by inpushings of the integument and filled with fluid, into which hairs project, and which contain grains of sand which have worked in from the outside, or concretions of lime. These are situated in the shrimps and crabs at the base of the inner antennæ, but in a few other Crustacea, as in *Mysis*, they are placed at the base of the lobes of the tail. In the insects the ear is a sac covered by a tympanum, with a ganglionic cell within, leading by a slender nerve-fibre to a nerve-centre, and in these animals the distribution of ears is very arbitrary. In the locust they are situated at the base of the abdomen; in the green grasshoppers, or katydids, and the crickets, in the fore tibiæ; and it is probable that, in the butterflies, the antennæ are organs of hearing. The vertebrate ears are two in number, and occupy a distinct permanent position in the skull, however much modified the middle and outer ear become.—(Packard's Zoology.)

"Throughout the animal kingdom," says Romanes, "the powers of sight and of hearing stand in direct ratio to the powers of locomotion;" on the other hand, in fixed or parasitic animals, the organs of hearing and sight are among the first to be aborted.

The sense of smell is obscurely indicated by special organs in the invertebrate animals; nasal organs, as such, being characteristic of the skulled vertebrates. Whether organs of smell exist in any worms or not is unknown; there are certain pits in some worms which may possibly be adapted for detecting odors. In some insects, at least, the organs of smell are without doubt well developed; the antennæ of the burying beetles are large and knob-like, and evidently adapted for the detection of carrion. It is possible that certain organs situated at the base of the wings of the flies, and on the caudal appendages of the cockroach and certain flies, are of use in detecting odors.

ANIMAL PSYCHOLOGY.

We have seen that animals have organs of sense, of perception, in many cases nearly as highly developed as in man, and that in the Mammalia the eyes, ears, organs of smell and touch differ but slightly from those of our own species; also that the brain and nervous system of the higher mammals closely approximate to those of man. We know that all animals are endowed with sufficient intelligence to meet the ordinary exigencies of life, and that some insects, birds, and mammals are able, on occasion, to meet extraordinary emergencies in their daily lives. These facts tend to prove that all animals, from the lowest to the highest, possess, besides sensations, certain faculties which by general consent naturalists call mental, because they seem to be of a kind, however different in degree, with the mental manifestations of man. Besides in many if not most highly organized animals, sensations give rise to emotions, and in the higher animals, as well as man, the latter give rise to thoughts. The study of mental phenomena is the science of psychology. The study of the sensations and instincts, as well as reasoning powers, of animals, is called animal psychology. The materials for the study of animal psychology are derived from the observations of the actions of animals; we do not, so to speak, know what is going on in their minds; we draw our conclusions, as to whether an animal thinks or reasons, by studying our own mental processes. The study of human psychology is a most difficult one: one man cannot read other men's minds; he judges of their mental processes by their actions and his own mental processes. In the same manner we conclude that animals reason by judging of their acts alone. If human psychology is an inexact science, much more so is comparative psychology, which includes human as well as animal psychology.

Although the Amœba performs operations which are akin to the instinctive acts of higher animals, it may in general be said that the nervous system is the organ of mind; not the brain alone, in animals which have a brain, but the entire nervous system. The mental manifestations of animals are not alone physiological, *i. e.* automatic and reflex, but there are, at least in highly organized animals, such as crabs, insects, spiders, and vertebrates, processes which are psychological as opposed to physiological.

The elementary or root principle of mind, as distinguished from purely physiological processes, is the power of making a choice between two alternatives presented to the animal.

As we have said on another occasion, granted that insects have sensibilities, how are we to prove that they have an intellect? Simply by observing whether they make a choice between two acts. "On entering a closet, ants unhesitatingly direct their steps to the sugar-bowl in preference to the flour-barrel; one sand-wasp prefers beetle-grubs to caterpillars, to store up as food for her young. In short, insects exercise discrimination, and this is the simplest of intellectual acts. They try this or that method of attaining an object. In fact, an insect's life is filled out with a round of trials and failures."

While no one would doubt that an insect has the power of choice or discrimination, may this also be said of the lowest organisms, such as the Amœba? Mr. Romanes believes that it can. "Amœba is able to distinguish between nutritious and non-nutritious particles, and in correspondence with this one act of discrimination it is able to perform one act of adjustment; it is able to enclose and to digest the nutritious particles, while it rejects the non-nutritious." Some protoplasmic and unicellular organisms are able also to distinguish between light and darkness, and to adapt their movements to seek the one and shun the other; Mr. H. J. Carter thinks that the beginnings of instinct are to be found so low down in the scale as the Rhizopoda. As quoted by Romanes in his Animal Intelligence: "Even *Athealium* will confine itself to the water of the watch-glass in which it may be placed when away from saw-dust and chips of wood among which it has been living; but if the watch-glass be placed upon the saw-dust, it will very soon make its way over the side of the watch-glass and get to it." Other facts are cited from Mr. Carter, upon which Mr. Romanes makes the following reflections: —

"With regard to these remarkable observations it can only, I think, be said that, although certainly very suggestive of something more than mechanical response to stimulation, they are not sufficiently so to justify us in ascribing to these lowest members of the zoological scale any rudiment of truly mental action. The subject, however, is here full of difficulty, and not the least so on account of the Amœba not only having no nervous system, but no observable organs of any kind; so that, although we may suppose that the adaptive movements described by Mr. Carter were non-mental, it still remains wonderful that these movements should be exhibited by such apparently unorganized creatures, seeing that as to the remoteness of the end attained, no less than the complex refinement of the stimulus to which their adaptive response was due, the movements in question rival the most elaborate of non-mental adjustments elsewhere performed by the most highly organized of nervous systems."

It will be a matter of interest to trace the dawnings of mental processes in the lower animals. Having seen that something more than physiological effects are traceable in certain acts of the protozoans; passing over the sponges, which are at best retrograde organisms, we come to the cœlenterates, especially the jelly-fishes. In none of these creatures have actions involving intelligence been observed; all their acts, so far as yet observed, are physiological, *i. e.* reflex, the result of stimulation from without.

Of the echinoderms, Romanes says: "Some of the natural movements of these animals, as also some of their movements under stimulation, are very suggestive of purpose; but I have satisfied myself that there is no adequate evidence of the animals being able to profit by individual experience, and therefore, in accordance with our canon, that there is no adequate evidence of their exhibiting truly mental phenomena. On the other hand, the study of reflex action in these organisms is full of interest."

It is possible that the action of the earth-worm, a representative of the annelids, in drawing leaves down into its hole is "strongly indicative of instinctive action, if not of intelligent purpose — seeing that they always lay hold of the part of the leaf (even though an exotic one) by the traction of which the leaf will offer least resistance to being drawn down." To the foregoing statement of Romanes we may add Darwin's testimony as to the mental powers of the earth-worm, from his work entitled The Formation of Vegetable Mould through the Action of Worms.

"Worms are poorly provided with sense-organs, for they cannot be said to see, although they can just distinguish between light and darkness; they are completely deaf, and have only a feeble power of smell; the sense of touch alone is well developed. They can, therefore, learn little about the outside world, and it is surprising that they should exhibit some skill in lining their burrows with their castings and with leaves, and, in the case of some species, in piling up their castings into tower-like constructions. But it is far more surprising that they should apparently exhibit some degree of intelligence instead of a mere blind instinctive impulse, in their manner of plugging up the mouths of their burrows. They act in nearly the same manner as would a man, who had to close a cylindrical tube with different kinds of leaves, petioles, triangles of paper, etc., for they commonly seize such objects by their pointed ends. But with thin objects a certain number are drawn in by their broader ends. They do not act in the same unvarying manner in all cases, as do most of the lower animals; for instance, they do not drag in leaves by their foot-stalks, unless the basal part of the blade is as narrow as the apex, or narrower than it."

The next great type of animals is the molluscs. In many respects the higher worms, especially the annelids, are more highly organized than the clam, a snail, or cuttle-fish. The functions of sensation and locomotion are often in molluscs subordinate to the merely vegetative, such as feeding, nutrition, and reproduction. We should not, as Romanes has said, expect that molluscs would present any considerable degree of intelligence. "Nevertheless, in the only division of the group which has sense organs and powers of locomotion highly developed — viz., the Cephalopoda — we meet with large cephalic ganglia, and, it would appear, with no small development of intelligence."

Beginning with one of the lowest molluscs, the oyster, Romanes quotes from Mr. Darwin's MS, as follows: "Even the headless oyster seems to profit from experience, for Dicquemase asserts that oysters taken from a depth never uncovered by the sea, open their shells, lose the water within, and perish; but oysters taken from the same place and depth, if kept in reservoirs, where they are occasionally left uncovered for a short time, and are otherwise incommoded, learn to keep their shells shut, and they live for a much longer time when taken out of the water." Is this act simply reflex? Limpets have been known, after making excursions from their resting places in order to browse on seaweed, to return repeatedly to one spot or home. The precise memory of direction and locality implied by this fact, adds Romanes, "seems to justify us in regarding these actions of the animal as of a nature unquestionably intelligent."

Concerning snails Darwin remarks: "These animals appear also susceptible of some degree of permanent attachment; an accurate observer, Mr. Lonsdale, informs me that he placed a pair of land-shells (*Helix pomatia*), one of which was weakly, in a small and ill-provided garden. After a short time the strong and healthy individual disappeared, and was traced by its track of slime over a wall into an adjoining well-stocked garden. Mr. Lonsdale concluded that it had deserted its sickly mate; but after an absence of twenty-four hours it returned, and apparently communicated the result of its successful exploration, for both then started along the same track and disappeared over the wall."

Mr. W. H. Dall gives a remarkable instance of intelligence in a snail, kept as a pet by a child, which recognized her voice and distinguished it from that of others. The lady who told the story to the person who sent it to Mr. Dall, after stating that her sister Georgie was, from the age of three years, quite an invalid, and remarkable for her power of putting herself *en rapport* with all living things, said: "Before she could say more than a few words, she had formed an acquaintance with a toad, which used to come from behind the log where it lived, and sit winking before her in answer to her call, and waddle back when she grew tired and told it to go away. When she was between five and six years of age, I found a snail shell, as I thought, which I gave to her to amuse her, on my return from a picnic. The snail soon crawled out, to her delight, and after night disappeared, causing great lamentation. A large, old fashioned sofa in the front hall was moved in a day or two, and in it was found the snail glued fast; it had crawled down stairs. I took a plant jar of violets and, placing the snail in it, carried it to her, and sunk a small toy cup even with the soil, filling it with meal. This was because I had read that French people feed snails on meal. The creature soon found it, and we observed it with interest for a while, as we found it had a mouth which looked pink inside and appeared to us to have tiny teeth also. We grew tired of it, but Georgie's interest never flagged, and she surprised me one day by telling us that her snail knew her and would come to her when she talked to it, but would withdraw into its shell if anyone else spoke. This was really so, as I saw her prove to one and another time after time." Mr. Dall adds: "An observer who noticed and remembered the pink buccal mass, the lingual teeth, and the translucent mistletoe-berry-like eggs, and after such an interval of time could so accurately describe them, is entitled to the fullest credence in other details of the story, and I have no doubt of its substantial accuracy, in spite of its surprising nature."

The Crustacea are perhaps, as regards intelligence, on a level with the majority of insects, excepting the white ants and ichneumons, wasps, and bees.

The power of finding their way home, which of course is due to memory, is illustrated in the following instance published by Mr. E. W. Cox in "*Nature*" for April 3, 1873. "The fishermen of Falmouth catch their crabs off the Lizard rocks, and they are brought into the harbor at Falmouth alive and impounded in a box for sale, and the shells are branded with marks by which every man knows his own fish. The place where the box is sunk is four miles from the entrance to the harbor, and that is above seven miles from the place where they are caught. One of these boxes was broken; the branded crabs escaped, and two or three days afterwards they were again caught by the fisherman at the Lizard rocks. They had been carried to Falmouth in a boat. To regain their home they had first to find their way to the mouth of the harbor, and when there, how did they know whether to steer to the right or to the left, and to travel seven miles to their native rocks?" It is scarcely possible to regard such an instance of what has been called the 'homing instinct,' as a purely physiological, reflex act, nor to consider the crab a mere automaton.

Mr. Darwin, in his Descent of Man, refers to the curious instinctive habits of the large shore-crab (*Birgus latro*), which feeds on fallen cocoa-nuts, "by tearing off the husks fibre by fibre; and it always begins at that end where the three eye-like depressions are situated. It then breaks through one of these eyes by hammering with its heavy front pincers, and, turning round, extracts the albuminous core with its narrow posterior pincers."

Little is really known of the instincts and other intellectual traits of the crustaceans — but when we come to the insects the literature is very extensive, thanks to the observations of Reaumur, Bonnet, De Geer, Wyman, Bates, Belt, Müller, Moggridge, Lincecum, McCook, Sir John Lubbock, and others.

As we have stated in our Half Hours with Insects: "Those who observe the ways of insects have noticed their extreme sensitiveness to external impressions; that their motions are ordinarily rapid and nervous. Look at the ichneumon fly as it alights on a leaf near a caterpillar: with what rapid motions it walks and flies about; how swiftly its feelers vibrate; how briskly it walks up and down surveying its victim. Look at a mud wasp as it alights near a pool of water to moisten its mouth. How nervous are its motions, how nimbly it flies and runs about the edge of the water. The ant is a busy, active, dapper little creature, a nervous brusqueness pervading its movements. How susceptible insects are to the light may be tested on a damp, dark night by opening the windows. In dart a legion of insects of all sorts, each with a different mode of entrance, some beetle boldly flying about the room in its blundering noisy flight, or a Clisiocampa moth enters with a bound, and a series of somersaults over the table, like the *entrée* of a popular clown into the ring of a circus, though the latter may have the most self-possession of the two.

"Insects are, like most animals, extremely sensitive to electrical phenomena. Just before a thunder shower they are particularly restless, flying about in great numbers and without any apparent object. The appendages of insects, their feelers and their legs, must be provided with exquisitely sensitive organs to enable them to receive impressions from without. Everybody knows that insects have acute powers of sight. That they also hear acutely is a matter of frequent observation. Often in walking through dry bushes, the noise of one's feet, in crushing through the undergrowth, starts up hosts of moths, disturbed in their noonday repose. If insects did not hear acutely, why should the Cicada have such a shrill cry? For whose ears is the song of the cricket designed unless for those of some other cricket? All the songs, the cries, and hum of insect life have their purpose in nature and are useless unless they warn off or attract some other insect.

"We know with a good degree of certainty that some insects have an acute sense of smell. The carrion beetles scent their booty afar off; the ants, the moths, all the insects attracted to flowers by the smell of the honey in them, evidently have well developed organs of smell."

The internal structure of the brain of the ant, the bee, as well as the locust and other insects has been found to be unexpectedly complex, when compared with that of the higher worms and even the higher Crustacea, such as the lobster and cray fish. The brain of insects is a much more complicated organ than any of the succeeding ganglia, consisting more exclusively of sensory cells and nervous threads than any succeeding ones, though the subœsophageal one is also complex, consisting of sensory as well as motor ganglia, since this ganglion sends off nerves of special sense to the organs of taste and smell situated in the mouth-appendages. The third thoracic ganglion is also, without doubt, a complex one, as, in the locusts, the auditory nerves pass from it to the ears, which are situated at the base of the abdomen. But in the green grasshoppers, such as the katydids and their allies, whose ears are situated in their fore legs, the first thoracic ganglion is a complex one. In the cockroach and in Leptis (*Chrysopila*), a common fly, the caudal appendages bear what are probably olfactory organs, and as these parts are undoubtedly supplied from the last abdominal ganglion, this is probably composed of sensory and motor ganglion-cells; so that we have in the ganglionated cord of insects a series of brains, as it were, running from head to tail, and thus in a still stronger sense than in Vertebrates the entire nervous system, and not the brain alone, is the organ of the mind of the insect.

To briefly describe the brain of the locust, an insect not high in the scale, it is a double ganglion, but structurally entirely different from, and far more complicated than, the other ganglia of the nervous system. The cerebral lobes possess a 'central body,' and in each hemisphere is a 'mushroom body;' besides the main cerebral lobes, the brain has also a pair of optic lobes and optic ganglia, and olfactory or antennal lobes, and these lobes have their connecting and commissural nerve-fibres, not found in the other ganglia.

The locust's brain appears to be as highly developed as that of the majority of insects, but that of the ant and the bee is more complicated than in other winged insects, owing to the much greater complexity of the folds of the calices or disk-like bodies capping the double stalk of the mushroom body. Now the ants, wasps, and bees are pre-eminently social animals, and we see by the structure of the brain why, in point of intelligence, they may exceed in mental development even the fishes, reptiles, and other lower vertebrates, and almost rival the birds in instinctive and rational acts.

Experiments and anecdotes bearing upon the intelligence of ants, have been widely circulated in the works of Lincecum, McCook, Lubbock, Darwin, and Romanes, space not allowing us to reproduce them. Ants have the sense of sight and of scent and taste well developed, but the sense of hearing is feeble, sounds of various kinds not producing any effect upon them: their antennæ are not, then, as in some insects, organs of hearing or smell, but have a delicate sense of touch, and, indeed, are the most important of sense organs to them. The sense of direction, the power of memory, are highly developed, and they perhaps are not destitute of the tenderer emotions, individuals being known to display sympathy for their wounded companions or healthy friends in distress. Ants also have the power of communicating with one another, and they are susceptible of education. The young ant is led about the nest and "trained to a knowledge of domestic duties, especially in the care of the larvæ; they are also taught to distinguish between friends and foes." When an ant's nest is attacked by foreign ants, the young ones never join in the fight, but confine themselves to removing the pupæ; and Forel has by experiment proved that the knowledge of hereditary enemies is not wholly instinctive in ants.

Moreover, besides carrying on the complicated duties of the formicary, ants add to their labors by keeping in their nests milch cows, as the Aphides substantially are; they also carry on slave-catching wars, and keep slaves generation after generation, with the same results of enfeebling and deteriorating the body and mind of the masters, as has been experienced in human life. Ants also keep pets, and, to go to another extreme, carry on wars of conquest, rapine, and plunder. A few human races are said not to bury their dead: if this be so they are inferior to ants, whose care in disposing of the bodies of their dead has attracted the notice of Sir John Lubbock; and that they actually in some cases bury their dead was claimed by Pliny, and substantiated by recent observers, according to Romanes. And then we have the leaf-cutting ants, harvesting ants, honey-making ants, military ants, ants which bridge streams, dig wells, and tunnel under broad rivers.

Wasps and bees can see much better than ants; indeed, they are far more dependent than the latter on the power of perceiving flowers, they also have a highly developed sense of direction, powers of communication, while the combined instinctive and reasoning powers they exhibit in making their nests, and in providing for or caring for their young are proverbial. Whether the instinct of building hexagonal cells is purely automatic or not has been disputed, but now it is conceded by Darwin, Romanes, and others that the process is not a purely mechanical one, but is "constantly under the control of intelligent purpose;" in other words, the worker bee knows what it is about, is a conscious agent.

Spiders also, though their nervous system is much less complicated than that of ants and bees, as well as insects in general, being built upon a different plan, show the most astonishing intellectual powers, particularly in spinning their webs; while as examples of special instincts the result of reasoning processes, at least in the beginning, are the acts of the water spiders, and especially the trap-door spiders.

Spiders also, like ants and bees, are able to distinguish between persons, approaching those they know to be friendly, and shunning strangers. It is well known that spiders can be tamed, and there are well-authenticated anecdotes testifying to the high degree of intelligence of these creatures.

Passing now to the branch of vertebrates, we do not find a sudden rise in the intellectual scale from bees to fishes, but that in reality fishes and reptiles are not so highly endowed mentally as the most highly organized insects. As Romanes truly says: "Neither in its instincts nor in general intelligence can any fish be compared with an ant or bee, — a fact which shows how slightly a psychological classification of animals depends upon zoological affinity, or even morphological organization." Fishes, he states, "display emotions of fear, pugnacity; social, sexual, and parental feelings; anger, jealousy, play, and curiosity. So far, the class of emotions is the same as that with which we have met in ants, and corresponds with that which is distinctive of the psychology of a child about four months old."

Of batrachians, frogs and toads have more or less definite ideas of locality, while they will learn to recognize the human voice and come when called. The general intelligence of reptiles is higher than that of fishes and batrachians, but low compared with that of birds. Snakes and tortoises are said to be able to distinguish persons;

and snakes, when tamed, exhibit some degree of affection for their master or mistress; while cobras may not only be tamed but domesticated.

Between the lower vertebrates and the birds and mammals there is a wide intellectual gulf. In birds, the reasoning as opposed to the simply instinctive acts are numerous. Birds are active, volatile, hot-blooded creatures, all their senses acute, and their cerebral hemispheres are far better developed than in the lower classes.

To illustrate the high degree of intelligence of birds, we will state some of the conclusions given by Romanes, referring the reader to his interesting work on Animal Intelligence for the anecdotes supporting his generalizations. The memory of birds for localities is well illustrated by their migratory habit of returning year after year to the same breeding-place. Buckland gives an account of a pigeon which remembered the voice of its mistress after an absence of eighteen months. Wilson relates an instance where a tame crow, after an absence of about eleven months, recognized his master. Parrots, which are perhaps the most intelligent of birds, sometimes chatter their phrases in their dreams, and "this shows a striking similarity of psychical processes in the operations of memory with those which occur in ourselves." Parrots have the power of association of ideas, and they not only remember, but recollect; "that is to say, they know when there is a missing link in a train of association, and purposely endeavor to pick it up."

Among the emotions, birds for the first time show unmistakable feelings of affection and sympathy. The loves of birds, the pining for an absent mate, and the conjugal affection of doves, etc., proves that in them the simple sexual feelings are heightened and enhanced by the intellect. Their jealousy is proverbial, as seen in the singing birds; they also show emulation and resentment as well as vindictiveness; their curiosity — the signs of a quick intellect — is highly developed; they have æsthetic emotions, love of bright-colored objects shown by the bower-bird, which builds its bowers at sporting-places in which the sexes meet, and where the males display their finery. Moreover, the singing birds, which stand at the head of the avian series, show a decided fondness for the music of their mates, aside from any utilitarian or sexual motives. Canaries, parrots, and doves are well known to take delight in human vocal or instrumental music.

The nesting habits of birds call out our admiration not only for their wonderful architectural traits, but for the signs they exhibit of a plastic instinct, where reason teaches them to modify their nests in situation and form to adapt them to new conditions. In Montana and Colorado the wild goose builds in trees; the cuckoo occasionally lays her eggs on the bare ground, sits on them, and feeds her young; the falcon, which usually builds on cliffs, has been known to lay its eggs on the ground in a marsh; the house-swallow in the United States has changed its nesting habits since the country was settled and houses were built. The nests of young birds, as first noticed by Wilson, are distinctly inferior to those of older ones, both in situation and construction. "As we have here independent testimony of two good observers to a fact which in itself is not improbable, I think we may conclude that the nest-making instinct admits of being supplemented, at any rate in some birds, by the experience and intelligence of the individual. M. Pouchet has also recorded that he has found a decided improvement to have taken place in the nests of the swallows at Rouen during his own lifetime; and this accords with the anticipation of Leroy, that if our observations extended over a sufficient length of time, and in a manner sufficiently close, we should find that the accumulation of intelligent improvements by individuals of successive generations would begin to tell upon the inherited instinct, so that all the nests in a given locality would attain to a higher grade of excellence. Leroy also states that, when swallows are hatched out too late to migrate with the older birds, the instinct of migration is not sufficiently imperative to induce them to undertake the journey by themselves. They perish the victims of their ignorance, and of the tardy birth which made them unable to follow their parents."

Among the higher mammals are the true domestic animals, the friends of man, who are capable of education and of transmitting striking hereditary traits. Among the Educabilia we find the horse, dog, pig, ox, sheep, llama, dog, cat. Romanes insists that the horse is not so intelligent an animal as any of the larger Carnivora, while among herbivorous quadrupeds his sagacity is greatly exceeded by that of the elephant, and in a lesser degree by that of his congener the ass. We question whether any one who has seen Bartholomew's "Equine Paradox," the twelve trained horses, will not place their intelligence at least as high as that of the pig. Pigs exhibit a degree of intelligence which "falls short only of that of the most intelligent Carnivora." Romanes claims that the tricks taught the so-called "learned pig" would alone suffice to show this; "while the marvellous skill with which swine sometimes open latches and fastenings of gates, etc., is only equalled by that of the cat."

Among the Carnivora in a wild state, bears claim a high place in the psychological scale; the most astonishing anecdote is one published in "*Nature*" since the appearance of Mr. Romanes' book. The story relates to a Russian bear. "The carcass of a cow was laid out in the woods to attract the wolves, and a spring trap was set. Next morning the forester found there the track of a bear instead of a wolf on the snow; the trap was thrown to some distance. Evidently the bear had put his paw in the trap and had managed to jerk it off. The next night the forester hid himself within shot of the carcass, to watch for the bear. The bear came, but first pulled down a stack of firewood cut into seven-foot lengths, selected a piece to his mind, and, taking it up in his arms, walked on his hind legs to the carcass. He then beat about in the snow all round the carcass with the log of wood before he began his meal. The forester put a ball in his head, which I almost regret, as such a sensible brute deserved to live."

Of the rodents, the majority, as the guinea-pig, hare, rabbit, etc., are low in intelligence; the squirrels have however some striking instincts, while the house rat, perhaps as the result of generations of persecution by man, has shown much intelligence; but the reasoning powers exhibited by the beaver are not only exceptional among rodents, but unique among dumb animals. In his admirable book on the beaver, the late Mr. Lewis H. Morgan thus speaks regarding what he calls the free intelligence of this animal: "The works of the beaver afford many interesting illustrations of his intelligence and reasoning capacity. Felling a tree to get at its branches involves a series of considerations of a striking character. A beaver seeing a birch tree full of spreading branches, which to his longing eyes seemed quite desirable, may be supposed to say within himself: 'If I cut this tree through with my teeth it will fall, and then I can secure its limbs for my winter subsistence.' But it is necessary that he should carry his thinking beyond this stage, and ascertain whether it is sufficiently near to his pond, or to some canal connected therewith, to enable him to transport the limbs, when cut into lengths, to the vicinity of his lodge. A failure to cover these contingencies would involve him in a loss of his labor. The several arts here described have been performed by beavers over and over again.

They involve as well as prove a series of reasoning processes undistinguishable from similar processes of reasoning performed by the human mind.

"Again, the construction of a canal from the pond, across the lowlands to the rising ground upon which the hard wood is found, to provide a way for the transportation of this wood by water, is another remarkable act of animal intelligence. A canal is not absolutely necessary to beavers any more than such a work is to mankind; but it comes to both alike as the progress in knowledge. A beaver canal could only be conceived by a lengthy and even complicated process of reasoning. After the conception had been developed and executed in one place, the selection of a line for a canal in another would involve several distinct considerations, such as the character of the ground to be excavated, its surface elevation above the level of the pond, and the supply of hard wood near its necessary terminus. These, together with many other elements of fitness, must be ascertained to concur before the work could be safely entered upon. When a comparison of a large number of these beaver canals has demonstrated that they were skilfully and judiciously located, the inference seems to be unavoidable that the advantages named were previously ascertained. This would require an exercise of reason in the ordinary acceptation of the term.

"And this leads to another suggestion. Upon the upper Missouri these canals are impossible from the height of the river-banks; and besides this they are unnecessary, as the cotton-wood, which is the prevailing tree, is found at the edge of the river. While, therefore, canals are unknown to the Missouri beavers, they are constantly in use among the beavers of Lake Superior. On the other hand, the 'beaver slides' so common and so necessary on the upper Missouri are unnecessary, and therefore unknown, in the Lake Superior region. Contrary to the common opinion, is there not evidence of a progress in knowledge to be found in the beaver canal and the beaver slide? There was a time, undoubtedly, when the canal first came into use, and a time, consequently, when it was entirely unknown. Its first introduction was an act of progress from a lower to a higher artificial state of life. The use of the slide tends to show the possession of a free intelligence, by means of which they are enabled to adapt themselves to the circumstances by which they are surrounded. In like manner it has been seen that the lodge is not constructed upon an invariably typical plan, but adapted to the particular location in which it is placed. The lake, the island, and the bank lodge are all different from each other, and the difference consists in changes of form to meet the exigencies of the situation. These several artificial works show a capacity in the beaver to adapt his constructions to the particular conditions in which he finds himself placed. Whether or not they evince progress in knowledge, they at least show that the beaver follows, in these respects, the suggestions of a free intelligence."

The elephant is not only a most sagacious animal, but displays emotions of a high grade. Were the elephant bred in captivity, we might expect a still greater degree of intelligence, but it should be borne in mind that the individuals used as beasts of burden are hunted and tamed, and their intelligence dies with them. Romanes claims that "the higher mental faculties of the elephant are more advanced in their development than in any other animal, except the dog and monkey."

Then comes the cat, whose intelligence is scarcely overrated by the popular judgment. Of all the cat stories we have read, the following one, copied from Romanes, caps the climax for the display of good judgment under trying circumstances: while a paraffine lamp was being trimmed, some of the oil fell upon the back of the cat, and was afterwards ignited by a cinder falling upon it from the fire. "The cat, with her back in a blaze, in an instant made for the door (which happened to be open) and sped up the street about one hundred yards, where she plunged into the village watering trough, and extinguished the flame. The trough had eight or nine inches of water in it and puss was in the habit of seeing the fire put out with water every night."

The dog is *par excellence* the friend of man, and without doubt his mind has been moulded as no other animal's, by that of his master. "The intelligence of the dog," says Romanes, "is of special, and, indeed, of unique interest, from an evolutionary point of view, in that from time out of record this animal has been domesticated on account of the high level of its natural intelligence; and by persistent contact with man, coupled with training and breeding, its natural intelligence has been greatly changed. In the result we see, not only a general modification in the way of dependent companionship and docility, so unlike the fierce and self-reliant disposition of all wild species of the genus; but also a number of special modifications, peculiar to certain breeds, which all have obvious reference to the requirements of man." Dogs have long memories, and they are superior to all other animals in their highly developed emotions. They can communicate simple ideas to one another as well as to their master, through the medium of a canine sign-language.

Reaching the highest order of mammals, the Primates, we are confronted with what Romanes may be correct in supposing to be "a mental life of a distinctly different type from any that we have hitherto considered, and that in their psychology, as in their anatomy, these animals approach most nearly to *Homo sapiens*." This, however, is an open question, and it is held by some that other animals, as the dog, exceed the monkeys and apes in intelligence. We are not sure, however, but that the monkeys and apes would, if bred in domestication for successive generations, prove that their highly developed brains place them on a higher psychological level than the dog, cat, elephant, hare, or pig. "The orang," says Romanes, "which Cuvier had, used to draw a chair from one end to the other of a room, in order to stand upon it so as to reach a latch which it desired to open; and in this we have a display of rationally adaptive action which no dog has equalled, although . . . it has been closely approached. Again, Rengger describes a monkey employing a stick wherewith to pry up the lid of a chest, which was too heavy for the animal to raise otherwise. This use of a lever as a mechanical instrument is an action to which no animal other than a monkey has ever been known to attain; and, as we shall subsequently see, my own observation has fully corroborated that of Rengger in this respect. More remarkable still, as we shall also subsequently see, the monkey to which I allude as having myself observed, succeeded also by methodical investigation, and without any assistance, in discovering for himself the mechanical principle of the screw; and that monkeys well understand how to use stones as hammers is a matter of common observation since Dampier and Wafer first described this action as practised by these animals in the breaking open of oyster-shells."

As regards the brains of apes, Bastian remarks: "In the conformation of their brain, the chimpanzee, the gorilla, and the orang approach, as we have seen, most closely to that of man; but it must never be forgotten that although in general shape, in the disposition of its fissures, and in the arrangement of its convolutions, as far as they go, there is this striking resemblance to the human brain, yet in actual size or weight, the brain of the man-like apes is widely separated from that of man. The heaviest brain belonging to one of these creatures, as yet examined, has been barely one half of the weight of the smallest normal human brains, although the weight of the entire body in the great gorilla may be nearly double that of an ordinary man. The brains of these three kinds of 'man-like' apes differ considerably among themselves; as we have seen, each in some respects approaches nearer to that of man than the others, though on the whole it is considered that the brain of the orang is slightly higher in type than that of the other two."

Bastian also quotes, as follows, from David Hartley's "Observations on Man" (1834): "It is remarkable that apes, whose bodies resemble the human body more than those of any other brute creature, and whose intellects also approach nearer to ours, — which last circumstance may, I suppose, have some connection with the first, — should likewise resemble us so much in the faculty of imitation. Their aptness in handling is plainly the result of the shape and make of their fore legs and their intellects together, as in us. Their peculiar chattering may perhaps be some attempt towards speech, to which they cannot attain, partly from the defect in the organs, partly, and that chiefly, from the narrowness of their memories, apprehensions, and associations."

We will close this too rapid view of the supposed facts in animal psychology by quoting from Bastian the following anecdote from Leuret: "One of the orangs, which recently died at the menagerie of the Musée, was accustomed, when the dinner hour had come, to open the door of the room where he took his meals in company with several persons. As he was not sufficiently tall to reach as far as the key of the door, he hung on to a rope, balanced himself, and, after a few oscillations, very quickly reached the key. His keeper, who was rather worried by so much exactitude, one day took occasion to make three knots in the rope, which, having thus been made too short, no longer permitted the orang-utan to seize the key. The animal, after an ineffectual attempt, *recognizing the nature of the obstacle which opposed his desire, climbed up the rope, placed himself above the knots, and untied all three*, in the presence of M. Geoffrey Saint-Hilaire, who related the fact to me. The same ape wishing to open a door, his keeper gave him a bunch of fifteen keys; the ape tried them, in turn, till he had found the one which he wanted. Another time a bar of iron was put into his hands, and he made use of it as a lever."

Let us now look at the inductions which may be drawn from the facts now known regarding the intelligence of animals. It is evident that animals are not mere physiological machines. We may, with Romanes, reject the view of Descartes, Huxley, and others, that animals are merely automata, on the ground that it can never be accepted by common sense, while "by no feat of logic is it possible to make the theory apply to animals to the exclusion of man."

We discern in the mental traits of animals, besides reflex acts, those which are instinctive and those which are the result of reasoning processes. The following definitions, by Mr. Romanes, will answer well our purpose: "Reflex action is non-mental neuro-muscular adjustment, due to the inherited mechanism of the nervous system, which is found to respond to particular and often recurring stimuli, by giving rise to particular movements of an adaptive, though not of an intentional kind.

"Instinct is reflex action into which there is imported the element of consciousness. The term is therefore a generic one, comprising all those faculties of mind which are concerned in conscious and adaptive action, antecedent to individual experience, without necessary knowledge of the relation between means employed and ends attained, but similarly performed under similar and frequently recurring circumstances by all the individuals of the same species.

"Reason, or intelligence, is the faculty which is concerned in the intentional adaptation of means to ends. It therefore implies the conscious knowledge of the relation between means employed and ends attained, and may be exercised in adaptation to circumstances novel alike to the experience of the individual and to that of the species."

It would appear, then, that animals have, in some slight degree, what we call mind, with its threefold divisions of the sensibilities, intellect, and will. When we study animals in a state of domestication, especially the dog or horse, we know that they are capable of some degree of education, and that they transmit the new traits or habits which they have been taught to their offspring; so that what in the parents were newly acquired habits become in the descendants instinctive acts. We are thus led to suppose that the terse definition of instinct, by Murphy, that it is 'the sum of inherited habits,' is in accordance with observed facts. Indeed, if animals have sufficient intelligence to meet the extraordinary emergencies of their lives, their daily so-called instinctive acts, requiring a minimum expenditure of mental energy, may have originated in previous generations; and this suggests that the instincts of the present generation may be the sum total of the inherited mental experiences of former generations.

Descartes believed that animals were automata. Lamarck expressed the opinion that instincts were due to certain inherent inclinations arising from habits impressed upon the organs of the animals concerned in producing them.

Darwin does not attempt any definition of instinct; but he suggests that 'several distinct mental actions are commonly embraced by this term,' and adds that 'a little dose, as Pierre Huber expresses it, of judgment or reason often comes into play, even in animals low in the scale of nature.' He indicates the points of resemblance between instincts and habits, shows that habitual action may become inherited, especially in animals under domestication; and since habitual action does sometimes become inherited, he thinks it follows that "the resemblance between what originally was a habit and an instinct becomes so close as not to be distinguished." He concludes that, by natural selection, slight modifications of instinct which are in any way useful accumulate, and thus animals have slowly and gradually, "as small consequences of one general law," acquired, through successive generations, their power of acting instinctively, and that they were not suddenly or specially endowed with instincts.

Rev. J. J. Murphy, in his work entitled Habit and Intelligence, seems to regard instinct as the sum of inherited habits, remarking that "reason differs from instinct only in being conscious. Instinct is unconscious reason, and reason is conscious instinct." This seems equivalent to saying that most of the instincts of the present generation of animals are unconscious automatism, but that in the beginning, in the ancestors of the present races, instincts were more plastic than now, such traits as were useful to the organism being preserved and crystallized, as it were, into the instinctive acts of their lives. This does not exclude the idea that animals, while in some respects automata, occasionally perform acts which transcend instinct; that they are still modified by circumstances, especially those species which in any way come in contact with man; are still in a degree free agents, and have unconsciously learned, by success or failure, to adapt themselves to new surroundings. This view is strengthened by the fact that there is a marked degree of individuality among animals. Some individuals of the same species are much more intelligent than others; they act as leaders in different operations. Among dogs, horses, and other domestic animals, those of dull intellect are led or excelled by those of greater intelligence, and this indicates that they are not simple automata, but are also in a degree, or within their own sphere, free agents.

REPRODUCTION AND EMBRYOLOGY.

On a previous page, as an introduction to the subject of tissues, we discussed some of the earlier features of the development of animals, but now we need, in a brief manner, to consider the subject from another point of view, and, in a general way, to trace it out for the whole animal kingdom.

As soon as the microscope was perfected and constantly used by morphologists as an instrument of exact research, a flood of light was thrown upon the subject of reproduction. In its ultimate analysis reproduction essentially consists in the separation of a portion of an adult animal from itself, this portion developing into an animal like the parent. It was seen that in the one-celled animals this process was identical with cell division, and it was called fission. It was also found that, as in the Hydra and allied polyps, a bud would form, and develop into a Hydra, and finally separate from the original parent Hydra. Although in the first place the formation of the bud is due to cell-division, a single cell giving rise to the bud, this process is called budding or gemmation. It was likewise discovered that in those animals which produce eggs, the latter were fertilized by a very minute and greatly modified cell, called the sperm-cell, or spermatozoon.

Now, in most of the many-celled animals there are two kinds of individuals, one female, which produces in its ovaries eggs, and the other the male, which produces in its testes the spermatozoa. Reproduction, or fertilization of the egg, in such animals, consists in the fusion of the sperm-cell with the nucleus of the egg; this is called sexual reproduction. From the moment of fertilization begins the life of the germ, which is called an embryo, while the history of the changes undergone by the embryo from the time of fertilization of the egg to maturity is called Embryology.

As was said a few pages back, the egg is essentially a simple cell, and in its earlier condition it is not to be distinguished from the other cells of the reproductive organs, but with development it changes in many respects, prominent among which is an increase in size. The essential part of the egg is its protoplasm, but to this is usually added a varying quantity of a nutritive material, the deutoplasm or food-yolk. Besides, in most forms, protective envelopes, etc., are added. The most familiar egg, that of the barnyard fowl, is poorly adapted to give us an idea of the true nature of an egg. Here the protoplasm is very small in quantity, and forms but a small patch on one side of the 'yolk,' which is almost entirely protoplasm. Another adventitious substance is the 'white,' while the shell and the membranes are merely protective, and not essential features. In another respect this egg is unsuited for our purposes, for, at the time of laying, the segmentation has progressed to a considerable extent, and the egg is no longer to be regarded as a simple cell.

The typical egg, then, is a mass of protoplasm, which is differentiated, as in any other cell, into nucleus and nucleolus, the latter in turn exhibiting a structure to be described below. In almost all eggs there is found one or more protective envelopes, which, according to the mode of origin, have received different names. When it is produced by the egg itself, it is called the vitelline membrane; when by the ovarian tissues of the parent it receives the name chorion. These envelopes in many forms are perforated by one or more minute openings which serve for the passage into the egg of nutritive material from the parent, and for the introduction of water in aquatic forms. Besides, in many forms there is a larger opening, the micropyle, for the entrance of the spermatozoon which is to fertilize the egg.

Recent investigations have shown that an egg or a cell is far from the simple structure which it was once imagined to be; the protoplasm of the cell is not a homogeneous substance, while the nucleus or germinative vesicle is very complex. The latter is enveloped by a special membrane and filled with a protoplasm, in which floats a tangled network of fibres. What is called the nucleolus is now regarded by Flemming and by Carnoy (two of the most profound students of cells) as a specialized portion or portions of the network. The nucleolus (there may be three or more in an egg) is called in the older works the germinative spot, or the Wagnerian vesicle, the latter name being applied in honor of its first discoverer. In the living egg the nucleolus is usually readily distinguished under the microscope by its great refrangibility, but to recognize the network it is necessary to employ stains and other reagents.

This egg, as we have described it, undergoes an extensive and complicated series of changes (known as the maturation of the egg) before it is ready for impregnation, although it is to be noted that in some instances the maturation is concomitant with impregnation. These changes may be summarized as follows:— At first the nucleus occupies a position near (but rarely at) the centre of the egg; it now moves to near the surface, where its membrane breaks down, and the filaments, etc., almost entirely disappear or at least lose their former character. In the place where the last remnants of the nucleus were seen, there now appears a spindle-shaped body made up of granules arranged in lines, while from either end other lines of granules are arranged in a radial manner. The whole presents an appearance closely similar to that seen when iron filings are exposed to the influence of a horse-shoe magnet, while from its resemblance to two stars joined it has received the name amphiaster. It may be observed in passing, that amphiasters are characteristic not only of the maturation of the egg, but of cell division as well; the connection between the two will appear in the sequel.

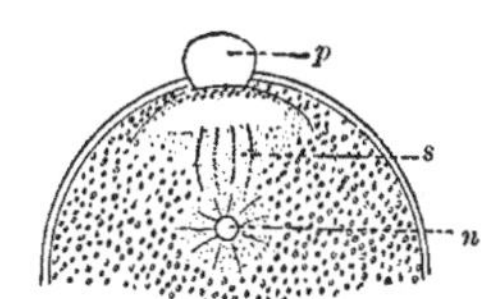

FIG. VI.— Formation of polar globule; *n*, nucleus; *p*, polar globule; *s*, spindle shaped figure.

The maturation spindle usually takes a position at nearly right angles to the surface of the egg, and soon from the outer end a prominence appears, extending out beyond the rest of the egg. The spindle now divides, and the prominence separates from the egg and forms what is known as a polar globule. Again the portion of the spindle which remains within the egg approaches the surface and a second polar globule is formed in the same manner as the first. Now, the part of the spindle left in the egg assumes a nearly spherical condition, and sinks back into the egg, where it appears exactly like the original nucleus. It is called the female pronucleus.

The meaning of these wonderful phenomena is far from evident. The best explanation as yet advanced, is that given by Balfour and Minot independently, which gains additional plausibility from the fact that essentially similar phenomena are seen in the formation of the male reproductive elements, the spermatozoa. In brief it is this:— All cells have inherited from their protozoan ancestors the elements of both sexes; they are hermaphroditic, and the eggs and spermatozoa cells are the same. Before they can unite it is necessary that each should get rid of the element to be supplied by the other, and in this light the formation of the polar globules is to be viewed as an elimination of the male portion from the egg, while, *mutatis mutandis*, the same may be said of the remnants of the mother-cells from which the spermatozoa are formed.

Now the egg is ready for that union with the male element, or spermatozoa, which is called fertilization or impregnation. The process in most if not all forms is essentially as follows: One or more spermatozoa enter the egg; in some cases it has been found that, if more than one entered, the result was a malformation, to be noticed below; in other eggs, on the contrary, several spermatozoa are necessary for fertilization. As soon as the head of the spermatozoon enters the egg, it forms a clear space known as the male pronucleus. Around this radial striæ appear, and it slowly travels toward the female pronucleus until the two unite. This compound structure, thus formed, is known as the segmentation nucleus.

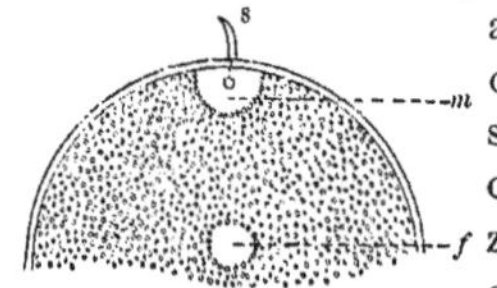

FIG. VII. — Entrance of spermatozoan into egg, and formation of male pronucleus (*m*); *f*, female pronucleus; *s*, spermatozoan.

Now begins the segmentation which was described briefly on a preceding page, which results in the conversion of the egg into a mass of cells, and which need not be repeated here. One interesting fact, however, may be mentioned. Hermann Fol, in his studies on the development of the star-fish, found that, if several spermatozoa obtained entrance to the egg, a corresponding number of segmentation nuclei were formed; and although development proceeded but a short distance, the results of this abnormal condition were visible throughout. Each nucleus formed a centre of segmentation, and when the time arrived for the formation of a gastrula, the same influence was felt, and, as shown in the adjacent cut, there were several invaginations. These observations possess a high interest from a teratological point of view, as they may afford an explanation of the formation of double monsters.

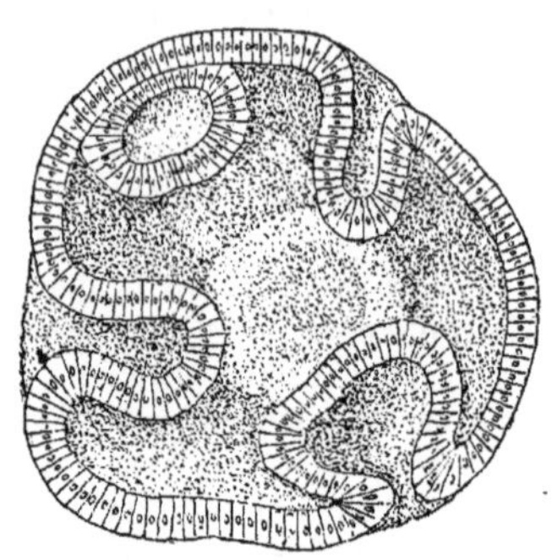
FIG. VIII. — Abnormal gastrulation in an echinoderm, the result of multiple impregnation.

METAMORPHOSIS.

After the formation of the germ layers as described on pages ix. to xii., the development of the various organs proceeds, for the details of which one should consult the accounts of the different groups in the body of this work, and especially Balfour's classic Treatise on Comparative Embryology. Still we may consider here some of the questions connected with metamorphosis. This term, which has been employed for many years, is used to indicate the series of changes which an animal undergoes after being born or after hatching from the egg. In some cases the changes are very slight, the young leaving the egg in nearly the adult form, while in others, of which a familiar instance is furnished by the butterfly, the modifications which are introduced between the egg and the mature condition are most startling. As other examples of these complete metamorphoses, we would refer the reader to the jelly-fish, star-fish, sea-urchins, worms, molluscs, insects, crustaceans, and batrachians, as described in the body of this work. One of the most curious is that presented by the larval form known as Actinotrocha, which converts itself into the mature worm *Phoronis*, by apparently turning itself inside out. If the reader will compare different accounts; and notice that in the same group, sometimes, as in the case of *Balanoglossus*, even in the same genus, one species will develop directly, while another has a complicated life history, he will be led to the inquiry, What is the meaning, what the use of this metamorphosis in one and not in the other? If one examines carefully the embryological changes of those forms hatched in the form of the adult, he will see that frequently they present, while in the egg, an epitome of the development of their relatives in which the changes have been much more marked. The question is largely one of nutrition, though many other conditions enter into the problem. In those forms where the food supply in the egg is abundant, the tendency is to simplify the development and to *accelerate* it. All superfluous features are consequently omitted, or are passed in a hasty manner. On the other hand, where the amount of food is small, the animal is forced to begin life for itself at an early date, and hence it needs every protection against the dangers of its environment.

An interesting point to be noticed in this connection has recently been described by Mr. W. J. Sollas, in the development of the sponge, *Halisarca*. In the Mediterranean the embryos of this sponge escape from the tissues of the parent when they have arrived at the blastula condition, and they then swim about freely by means of the cilia clothing the surface; in the same species on the shores of the English Channel, the young are retained until after gastrulation and the formation of the canal system. According to Sollas the explanation of this difference is not difficult. In the Mediterranean there are no strong currents, and it is evidently best for the parents to get rid of the young at as early a moment as possible, thus escaping a longer drain upon its energies. In the English Channel, on the other hand, the current is very strong, and were the embryos to be set free at the stage at which they are in the Mediterranean, the chances are that they would be swept away from proper places for their further development, and hence they are retained until nearly ready for attachment to the rocks.

The same influences, nutrition and environment, affect other forms. Almost all crustaceans undergo a complicated metamorphosis, and in their various stages they lead very different lives. In the young they are usually free-swimming, and hence they need protection from aquatic foes. This is usually gained in two ways; by transparent tissues which render them invisible to fishes, and by the development of spines and processes from the body, which increases their size without materially adding to their weight, thus preventing their entrance to the mouths of the smaller forms. Still not all the Crustacea undergo these changes; in the whole group of tetradecapods no metamorphosis is known, while in the land-crabs of the tropics the young, when hatched, are closely similar to the parents. In this latter instance, where the adults live on the land, only going to the sea at the approach of the breeding season, it is easy to be seen why the development should be direct.

In other cases the use of larval forms is very evident. Many forms, like the barnacles, sponges, and the oysters, lead a stationary life, but the young are free-swimming. This change in form and mode of life undoubtedly is of great benefit to the species, for if at a given moment the parents were swept out of existence, the young, living in a different station, would continue the species; and, besides, they serve to distribute the race from point to point.

The foregoing paragraphs have reference to the larval forms, and the persistence and value, and the benefits of a metamorphosis. Some of these characters are engrafted on the primitive stock, while others are due to the origin, the evolution of the group, and in all discussions the greatest care must be taken to discriminate between the ancestral and the adaptive features. We can best illustrate this by taking the case of the development of a mammal, and showing how in its various stages it presents a compendium of its history, When, in the mammalian germ, the nervous system and notochord arise, it is on a level with the larva of an Ascidian; with the formation of protovertebræ, it represents the Amphioxus; a brain, and gill clefts and limbs, indicate a fish and amphibian stage; the development of an allantois and closure of the gill clefts places it on an avian plane; while with the appearance of a placenta the mammalian features are assumed. These successive stages of the individual are closely paralleled by that of the class. The fleshy, boneless form of Amphioxus and the tunicates would not be preserved, but from fishes to man the sequence of remains in the rocks accords with that derived from embryology. It must not be understood from this that the mammals have been derived from the birds. The true line of descent is far different, as will be explained on a subsequent page. It merely indicates that the mammal and the bird have arisen from a common stock, and have pursued the same course during a portion of their history.

Alternation of Generations and Parthenogenesis.

Having spoken of the normal method of development of animals, we may turn to certain unusual or abnormal modes of production. As an example of what is known as alternation of generations may be cited the history of the jelly-fish, such as the naked-eyed medusæ (*Melicertum* and *Campanularia*), which at one time of life develop by budding, at another by eggs; of the trematode worms, the adult forms of which lay eggs, while the redia or proscolex of the same worm produces cercariæ by internal budding. Here also may be cited the cases of strobilation of *Aurelia*, the tape-worm, *Nais*, *Syllis*, and *Autolytus*, among annelids. Thus among cœlenterates and worms, as well as some Crustacea, a large number of individuals are produced, not from eggs, but by budding.

Similar occurrences take place among insects, as the *Aphis* or plant-louse, in which a virgin *Aphis* may bring forth in one season nine or ten generations of Aphides, so that one *Aphis* may become the parent of millions of young. These young directly develop from eggs or buds which are never fertilized, hence the term parthenogenesis, or virgin-reproduction, sometimes called agamogenesis (or birth without marriage). The bark-lice as well as the Aphides develop in this manner during the warm weather; but at the approach of cold both male and female Aphides and Coccidæ appear, the females laying fertilized eggs, the first spring brood thus being produced in the normal, usual manner.

Still more like the production of young in the redia of the trematode worms is the case of the larva of a small gall-gnat (*Miastor*), which during the colder part of the year from autumn to spring produces a series of successive generations of larvæ like itself, until in June the last brood develops into sexually mature flies, which lay fertilized eggs.

While the larval *Miastor* produces young like itself, the pupa of another fly, *Chironomus*, also lays unfertilized eggs from which the flies arise.

A number of moths, including the silk-worm moth, are known to lay unfertilized eggs which produce caterpillars. Among the Hymenoptera, the currant saw-fly, certain gall-flies, several species of ants, wasps (*Polistes*), and the honey-bee, are known to produce fertile young from unfertilized eggs; in the case of the ants and bees, the workers lay eggs which result in the production of males, while the fertilized eggs laid by the female ant or queen bee produce females or workers.

Taking all these cases together, parthenogenesis is seen to be due to budding, or cell-division or multiplication. Now it will be remembered that the egg develops into an animal by cell-division, so that fundamentally parthenogenesis is due to cell-division, the fundamental mode of growth; hence, normal growth and parthenogenesis are but extremes of a single series. In this connection, it will be remembered that all the Protozoa reproduce by simple cell-division, that among them the sexes are differentiated, that they do not reproduce by fertilized eggs; hence, so to speak, among Protozoa, parthenogenesis is the normal mode of reproduction; and when it exists in higher animals it may possibly be a survival of the usual protozoan means of stocking the world with unicellular organisms, with which we know the waters teem. And this leads us to the teleology or explanation of the cause why parthenogenesis has survived here and there in the world of lower organizations; it is plainly, when we look at the millions of Aphides, of bark-lice, the hundreds of thousands inmates of ant-hills and bee-hives, for the purpose of bringing immediately into existence great numbers of individuals, thus ensuring the success in life of certain species exposed to great vicissitudes in the struggle for existence. That this unusual mode of reproduction is all-important for the maintenance of the existence of most of the parasitic worms, is abundantly proved when we consider the strange events which make up the sum total of a fluke or tape-worm's biography. Without this faculty of the comparatively sudden production of large numbers of young by other than the slow, limited process of ovulation, the species would be stricken off the roll of animal life.

Dimorphism and Polymorphism.

Involving the production of young among many-celled animals (Metazoa) by what is fundamentally a budding process, we have two sorts of individuals. When the organism is high or specialized enough to lay eggs which must be fertilized, we have a differentiation of the animal into two sexes, male and female. Reproduction by budding involves the differentiation of the animal form into three kinds of individuals—*i. e.*, males, females, and asexual individuals, among insects often called workers or neuters. These have usually, as in ants and bees, a distinct form, so as to be readily recognized at first sight. Among the Cœlenterata and worms the forms reproducing by parthenogenesis are usually larval or immature, as if they were prematurely hurried into existence, and their reproductive organs had been elaborated in advance of other systems of organs, for the hasty, sudden production, so to speak, of large numbers of individuals like themselves.

In insects, dimorphism is intimately connected with agamic reproduction. Thus the summer wingless, asexual *Aphis* and the perfect winged autumnal *Aphis* may be called dimorphic forms. The perfect female may assume two forms, so much so as to be mistaken for two distinct species. Thus, an oak gall-fly (*Cynips quercus-spongifica*) occurs in male and female broods in the spring, while the autumnal brood of females was described originally as a separate species under the name *C. aciculata*. Walsh considered the two sets of females as dimorphic forms, and that *Cynips aciculata* lays eggs which produce *C. quercus-spongifica*. Among butterflies, dimorphism occurs. *Papilio memnon* has two kinds of females, one being tailless, like the tailless male, while *Papilio pammon* is polymorphic, there being three kinds of females besides the male.

There are also four forms of *Papilio ajax*, the three others being originally described as distinct species under the name of *P. Marcellus*, *P. telamonides*, and *P. walshii*. Our *Papilio glaucus* is now known to be a dark, dimorphic, climatic form of the common *Papilio turnus*. There are dimorphic males among certain beetles, as in the *Golofa hastata* of Mexico, in which one set of males are large and have a very large erect horn on the prothorax, and in the other the body is much smaller, with a very short conical horn.

Temperature is also associated with the production of polymorphic forms in the temperate regions of the earth, as seen in certain butterflies, southern forms being varieties of northern forms, and alpine 'species' proving to be varieties or seasonal forms of lowland species. For example, Weismann states that the European butterflies, *Lycaon amyntas* and *polysperchon* are respectively summer and spring broods. *Anthocharis simplonica* is an Alpine winter form of *Anthocharis delia*, as is *Pieris bryoniæ* of *Pieris napi*. In this country, as Edwards has shown, two of the polymorphic forms of *Papilio ajax* — *i. e.*, *walshii* and *telamonides* — come from winter chrysalids, and *P. Marcellus* from a second brood of summer chrysalids. It thus appears that polymorphism is intimately connected with the origin of species. Perhaps the most remarkable case of polymorphism is to be seen in the white ants (*Termites*), where in one genus there are two sorts of workers, two sorts of soldiers, and two kinds of males and females, making eight sorts of individuals; in the other genera there are six. Among true ants there are, besides the ordinary males, females, and workers, large-headed workers. In the honey-ant (*Myrmecocystus mexicanus*), besides the usual workers, there are those with enormous abdomens filled with honey. Other insects, especially certain grasshoppers, are dimorphic. Certain parasitic nematode worms are dimorphic; and among the cœlenterates, especially the hydroids, there is a strong tendency to polymorphism.

EVOLUTION.

In a single word — evolution — is comprised that vast complex of factors which has resulted in the stocking of our earth with plants and animals, each after its kind. The explanation of the process by which the life-forms of this planet have been brought into existence is an intricate series of problems within problems, as infinite as is the variety in nature itself. In early pre-scientific times, in the childhood of the race, it seemed sufficient to say that every living thing was created, and with this statement the majority of mankind were content to rest; not so, however, a few isolated thinkers, who, from the time of Democritus, have questioned nature, and as earnestly as reverently sought how these things could have come to pass. When geology began to assume a definite shape; when Cuvier and Lamarck had sketched out the leading types of animal life, as Jussieu did the earth's flora; and after palæontology began to be a science, and it became known that the earth had been peopled by successive floras and faunas, appeared Lamarck and St. Hilaire as philosophers, who combated the cataclysmic ideas of Cuvier, and who maintained both the unity of organization of organic beings and the immense lapse of time since the beginning of life — time enough for the changes and adaptations needed to bring about the present condition of things. In 1802, twenty-three years before the appearance of Cuvier's Discourse Sur les Révolutions du Globe, Lamarck uttered these striking words; "*Pour la nature, le temps n'est rien, et n'est jamais une difficulté; elle l'a toujours à sa disposition, et c'est pour elle un moyen sans bornes avec lequel elle fait les plus grandes choses comme les moindres.*"

In 1809 Lamarck published his Philosophie Zoologique. This work comprised the results of his speculations as well as of his special work of concise description, determination, and classification of vegetable and animal species. He was struck with the differences, but still more with the resemblances in animals; he noticed their variations, and, as Martins has said, a triple impression was made on his mind: the certainty of the variability of species under the influence of external agencies; that of the fundamental unity of the animal kingdom; finally, the probability of the successive generation of different classes of animals, arising, so to speak, one from another, like a tree whose branches, leaves, flowers, and fruits are the results of successive evolutions of a single organ, — the seed or bud.

All this was however speculation, *a priori*, premature guesses without a broad basis of facts. The fulness of time had not yet come. The year 1809 was long anterior to the general use of the microscope, before the sciences of embryology, of histology, the doctrine of the cell, and before the principles of palæontology and zoogeography had been founded. Lamarck was almost forgotten, his speculations had been treated with silent contempt or indifference. A period of over half a century succeeded, an age of busy search for facts, a period prolific in inductive sciences, — a sisterhood of knowledge as numerous as the family of Niobe. In the year 1859, Darwin, Wallace, Bates, and among botanists, Hooker, unanimously insisted on the fact of the variation of species and their origin by natural causes; and they supported their views by special more or less limited theories. Darwin's theory of natural selection was adopted with a rapidity and unanimity unparalleled in the history of science. We will now examine the general argument, and state some of the general principles upon which the modern scientific theory of descent is based.

There are three laws or inductions supporting the theory: 1. Change in the environment of the organism, involving adaptation to such change. 2. Transmission by heredity of ancestral together with acquired traits. 3. The selection of useful traits and their preservation and fixity. Around each of these leading principles cluster others accessory and indispensable, and doubtless still others may yet be discovered.

The recognition of two factors have attracted fresh attention to the theory of descent, and caused it to be generally accepted as a working theory indispensable to biological science; these are (1) the facts of variation with the difficulty of limiting species and genera, and the discovery of connecting links between the higher groups of animals, including orders, classes, and sub-kingdoms; and (2) the influence on the plant or animal of a change in the environment. The second of these factors was advocated by Lamarck and St. Hilaire. Since the publication of Darwin's special theory of natural selection, which was accepted as a *vera causa* by the large proportion of naturalists, a few have not been satisfied with this theory alone, but have in various directions gone back of Darwin and natural selection to views like those of Lamarck, whether they were acquainted with his theory and works or not. Darwin took the tendency to variation as the foundation upon which to erect the superstructure of natural selection; others have sought to account first for the tendency to variation, and then given natural selection its due place as a secondary, though important, phase. Had Lamarck, with his unquestioned ability as a thinker and observer, lived at the present time, when so many new sciences have arisen, and the older ones of chemistry and physics have been revolutionized, he would have checked his imagination here and there, and given us a theory well grounded on facts. It was reserved, however, for the tireless genius of Darwin, with his masterly handling of facts, to impress his conclusions on the age, supporting them as he did with an overwhelming array of facts. His compactly built superstructure was erected on a temporary foundation, which it will be the work of the future to rebuild with solid masonry, resulting from centuries of labor in the field first pointed out by Lamarck. The influence of Lamarck's work was feeble, owing to the strong counter-currents set up by Cuvier, Agassiz, and popular prejudice.

What Lamarck actually accomplished has been restated by Charles Martins. He noticed the variations of species, both of animals and plants. The best results of his labors of over thirty years in botany, and afterwards of thirty years in zoology, were his division of the animal kingdom into vertebrates and invertebrates; his founding the classes of Infusoria and of Arachnida; his separation of the Cirripedia from the Mollusca, only a few years before Thompson discovered their true affinities. Lamarck had a powerful imagination, and was a born speculator; but the age in which he worked was barren of facts, and many of his theories were ill-founded. The grand results of his work were clear views as to the unity of organization of the animal kingdom, the filiation of all animal forms, and the influence of external agencies on the variation of species; he recognized the effects of use and disuse on the development and atrophy of organs; he recognized the agency of the water, of air, of light, of heat, in bringing about changes in organisms; finally, Lamarck was the first to construct a phylogeny or genealogical tree of animals.

Lamarck's doctrine of appetency was carried too far, and exposed his views in general to ridicule; he maintained that spontaneous generation takes place at the present time; others have advocated this doctrine since Lamarck, and only within a few years have the researches of Tyndall led him and Huxley, as well as others, to affirm that there is no evidence that the process is now going on.

In his famous controversies with Cuvier, Geoffrey St. Hilaire stated his belief in the modification of species by changes in the conditions of life. As successors in Europe may be mentioned the following writers: Wagner, Martins, and Plateau, as well as those given below.

In Germany, the distinguished anatomist, histologist and embryologist, Kölliker, in his 'Morphology and developmental History of Pennatulids,' published in 1872, concludes as follows: "Such external forces have operated so as to modify, in many ways, developmental processes, and no theory of descent is complete which does not take these relations into account. Manifold external conditions, when they operate on eggs undergoing their normal development, on larvæ and other early stages of animals, and on the adult forms, have produced in them partly progressive, partly regressive, transformations. . . . Of such external forces the most important are the mode of life (parasitic and free-living animals, land and water animals), nutrition, light, and heat."

In his History of Creation (1873), Haeckel gives full credit to Lamarck's views, saying: "Without the doctrine of filiation, the fact of organic development in general cannot be understood. We should, therefore, for this reason alone, be forced to accept Lamarck's theory of descent, even if we did not possess Darwin's theory of selection." Here may also be mentioned the researches of Siebold and of Brauer, on the effects of desiccation on the eggs of phyllopod Crustacea, and of Hogg, Dumeril, Wyman, and others, that the metamorphosis of frogs is hastened or retarded by differences in temperature and light.

Weismann, in his suggestive work, Studies in the Theory of Descent, (1875–76), concludes from his extended investigations on seasonal dimorphism, "that differences of specific value can originate through the direct action of external conditions of life only. . . . A species is only caused to change through the influence of changing external conditions of life, this change being in a fixed direction which entirely depends on the physical nature of the varying organism, and is different in different species or even in the two sexes of the same species."

Weismann has certainly proved that new species arise by differences in climate, while he also (in a note to the English edition) concedes that sexual selection plays a very important part in the markings and coloring of butterflies, but he significantly adds, "that a change produced directly by climate may be still further increased by sexual selection."

A second point, and one of particular interest, which the author claims to be elucidated by seasonal dimorphism, is "the origin of variability." Having shown that "secondary forms are for the most part considerably more variable than primary forms," it follows that "similar external influences either induce different changes in the different individuals of a species, or else change all individuals in the same manner, variability arising only from the unequal time in which the individuals are exposed to the external influence. The latter is undoubtedly the case, as appears from the differences which are shown by the various individuals of a secondary form. These are," he adds, giving his proofs, "always only differences of degree and not of kind." He shows that allied species and genera, and even entire families (Pieridæ), "are changed by similar external inducing causes in the same manner, or better, in the same direction."

In his Ursprung und der Princip des Functionswechsel, (1875), Dr. A. Dohrn states his belief that new habits induce the organs to exercise apparently new functions, which were latent or only partly developed under the original conditions of the surroundings.

Another work, laden with facts, with not much space wasted on theories, is Semper's Animal Life as affected by the Natural Conditions of Existence (1877–81). This is the first general work especially devoted to an attempt to discover the causes of variation in animals. As the author says in his preface, "It appears to me that of all the properties of the animal organism, variability is that which may first and most easily be traced by exact investigation to its efficient causes; and, as it is beyond a doubt the subject around which at the present moment the strife of opinions is most violent, it is that which will repay the trouble of closer research." An enumeration of the subjects treated in the respective chapters of this work will give one an idea of the way in which this difficult subject should be studied: food and its influence; the influence of light, of temperature, of stagnant water, of a still atmosphere, of water in motion; currents as a means of extending or hindering the distribution of species, and the influence of living organisms on animals.

In the United States a number of naturalists have advocated what may be called neo-Lamarckian views of evolution, especially the conception that in some cases rapid evolution may occur. The present writer, contrary to pure Darwinians, believes that many species, but more especially types of genera and families, have been produced by changes in the environment, acting often with more or less rapidity on the organism, resulting at times even in a new genus, or even a family type. Natural selection, acting through thousands, and sometimes millions, of generations of animals and plants, often operates too slowly; there are gaps which have been, so to speak, intentionally left by Nature. Moreover, natural selection was, as used by some writers, more an idea than a *vera causa*. Natural selection also begins with the assumption of a tendeney to variation, and presupposes a world already tenanted by vast numbers of animals, among which a struggle for existence was going on, and the few were victorious over the many. But the entire inadequacy of Darwinism to account for the primitive origin of life-forms, for the original diversity in the different branches of the tree of life-forms, the interdependence of the creation of ancient faunas and floras on geological revolutions, and consequent sudden changes in the environment of organisms, has convinced us that Darwinism is but one of a number of factors of a true evolution theory; that it comes in play only as the last term of a series of evolutionary agencies or causes; and that it rather accounts, as first suggested by the Duke of Argyll, for the *preservation* of forms than for their origination. We may, in fact, compare Darwinism to the apex of a pyramid, the larger mass of the pyramid representing the true theory, or complex of theories, necessary to account for the world of life as it has been and now is. In other words, we believe in a modified and greatly extended Lamarckianism, or what may be called neo-Lamarckianism.

It is not the design to present here arguments for this theory of evolution, but to show what American authors have written in favor of the incidental, as well as the periodical, recurrence of sudden or quick evolution, through changes in the environment, as opposed to the supposed continuous action of natural selection.

Without doubt, that able and philosophic naturalist, the late S. S. Haldeman, was not unfavorable to a modified form of Lamarckian views as to the transformation of species. His Enumeration of the recent fresh-water Mollusca which are common to North America and Europe, with Observations on Species and their Distribution, was published as early as January, 1844. He takes occasion to remark: "I pretend not to offer an opinion for or against the Lamarckian, being more anxious to show the insufficiency of the standing arguments against it, and the necessity of a thorough revision of them, than to take a decided stand (upon a question which I regard as open to farther discussion) before its facts have been carefully observed, or the resulting generalizations properly deduced; so that, whether it be admitted or not, it is entitled to the benefit of all the discoveries which can be brought to bear upon it; and, on this account, I have not hesitated to give a slight sketch of the theory of transmutation, as I conceive it to be modified by some of the results of modern science."

In the course of his essay he remarks: "The reason why the lower orders still exist is to be looked for in the fact that they are fitted for the circumstances under which we find them." Again he says: "Although we may not be able, artificially, to produce a change beyond a definite point, it would be a hasty inference to suppose that a physical agent, acting gradually for ages, could not carry the variation a step or two farther; so that, instead of the original, we will say four varieties, they might amount to six, the sixth being sufficiently unlike the earlier ones to induce a naturalist to consider it distinct. It will now have reached the limit of its ability to exist as the former species, and must be ready either to develop a dormant organic element, or die; if the former is effected, the oscillating point is passed, and the species established upon the few individuals that were able to survive the shock. If the physical revolution supposed to be going forward is arrested, or recedes, the individuals which had not passed the culminating point remain as a fifth variety, or relapse towards their former station; whilst the few which have crossed the barrier remain permanently beyond it, even under a partial retrogression of the causes to which they owed their newly developed organization."

Very significant is the suggestion which follows, as to the cause of comparatively sudden leaps in the process of evolution: We may suppose some species and individuals to be more able to pass than others, and that many become extinct from inability to accomplish it. Under this point of view, a hiatus, rather than a regular passage, is required between a species and that whence it is supposed to be derived, just as two crystals may occur, nearly identical in composition, but without an insensible gradation of intermediate forms; the laws, both of organic and inorganic matter, requiring something definite, whence the rarity of hybrids and monsters, themselves subject to established laws." He adds, in a foot-note: "The *same* mineral may crystalize with three, six, or twelve angles, but not with five or seven. Are the phases of organic morphism subject to less definite laws?"

In the year 1850, Professor Joseph Leidy wrote that a slight modification of the essential conditions of life were sufficient to produce the vast variety of living beings upon the globe.

In 1853 Dr. Jeffries Wyman published a paper on the effect of the absence of light on the development of tadpoles, and in 1867 appeared his Observations and Experiments on Living Organisms in Heated Water. Professor Wyman taught the doctrine of evolution as early as 1861, and probably earlier.

In 1864 B. D. Walsh endeavored to establish the fact that, while the great majority of species may have been formed by natural selection, some originated "by changes in the conditions of life, and especially by change of food." H. J. Clark, in his Mind and Nature (1865), advocated evolution, and even spontaneous generation through physical processes.

Professor Alpheus Hyatt (in 1866) showed that the development of the individual in the Ammonites agrees with the development of the order to which it belongs, and he afterward showed, by a study of Ammonites of different geological formations, that just as there are sudden changes of form in the growth of the individual, so species and genera of one formation replace those of another, in such a manner that one form must have descended from the other, although the differences between the forms are very marked.

In 1869 Professor E. D. Cope, in his essay on the Origin of Genera, suggested that, by an acceleration or retardation in the development of the animal, generic forms had been produced. He claimed that, "while natural selection operates by the 'preservation of the fittest,' retardation and acceleration act without any reference to 'fitness' at all; that, instead of being controlled by fitness, it is the controller of fitness." He also remarks that the "transformations of genera may have been rapid and abrupt, and the intervening periods of persistency very long;" in other words (p. 80), genera and higher categories have appeared "in geological history by more or less abrupt transitions, or expression points, rather than by uniformly gradual successions."

It should be observed, however, that Cope did not enter into the causes which produce acceleration and retardation, but in later papers he has extended and more fully stated his views.

Marsh's observations, published in 1868, on the transformtion of *Siredon* into the ordinary gill-less salamander (*Amblystoma*), was a step in the same direction, *i. e.* giving proofs of rapid change in the acquisition of new organs, and modifications of existing ones.

In The American Naturalist, for December, 1871, the writer assumed, from a study of cave animals, that these forms were suddenly produced, though the changes may not have been wrought until, say, after several thousand generations; and the theory of Cope and of Hyatt, of creation by a process involving the idea of accelerated development in some species, and retarded development in certain organs of other species, was adopted. These views were again enforced in Hayden's Bulletin of the United States Geological Survey (April, 1877), in an article on the Cave Fauna of Utah.

In 1872 the writer (Development of *Limulus*), from a study of the paleozoic Crustacea and of the development of *Limulus*, claimed that it was impossible "that at the dawn of silurian life these well-marked groups were due entirely to the extinction of multitudes of connecting links, such as Mr. Darwin assumes to have been evolved on the principle of natural selection, with the subordinate agency of sexual selection and mimicry, etc. The groups are almost as clearly marked as in the present time, and such a theory seems to us inadequate to account for the rise of such distinct forms, apparently simultaneous in their appearance at the beginning of the silurian. The forms are remarkably isolated, and present every appearance of having been in a degree suddenly produced," *i. e.*, by differences in the temperature and depth of the water, etc., the differences being due to changes in the physical surroundings of the organisms. Farther on it is stated: "I conceive these differences to be due, perhaps, to sudden changes of temperature in fresh-water pools, to the difference in the density of fresh and salt water, and the liability of fresh-water pools to dry up, combined with less apparent causes." It will be seen that these views essentially agree with what is known as Lamarckianism. In his Monograph of the Geometrid Moths, the writer attempted to show that climatic and geological causes were important factors in the production of the genera and species constituting the different faunas.

That changes in the physical surroundings of the organism, rather than the struggle for existence among the animals themselves, produce new forms of animal life, was also insisted upon by the writer (Half Hours with Insects, 1876), in the following words: —

"When one looks at the beds of fossil beings of the earlier geologic periods, he peers into the tombs of millions which could not adapt themselves to their constantly changing surroundings. No fossil being is known to us which could not have been as well adapted to its mode of life as the animals now living; but the conditions of life changed, and the species, as such, could not withstand the possible influx of new forms, due to some geological change which induced emigration from adjoining territories, or to changes of the contour of the surface, with corresponding climatic alterations. Let one look at the geological map of North America before the cretaceous period, ere the Rocky Mountains appeared above the sea, and reflect on the remarkable changes that took place to the northward, — the disappearance of an Arctic continent, the replacement of a tropical climate in Greenland and Spitzbergen by Arctic cold. Are there not here changes enough in the physical aspects of our country to warrant such hypotheses of migrations, with corresponding extinctions and creations of new faunas out of preceding ones, as are indulged in by naturalists of the present day, in the light of the knowledge pouring in upon them from Arctic explorers and western geologists? Granted these extraordinary changes in the physical surroundings of the animals whose descendants people our land, do not a host of questions arise as to the result, in the beings of our day, of these changes in the modes of life, the modes of thought, so to speak, the formation of peculiar instincts arising from new exigencies of life, which have remodelled the whole psychology, as it were, of the animals of our country? Instincts vary with the varying structure and form of the animals. Change the surroundings, and at once the mode of life and psychology of the organism begin to undergo a revolution. These changes may result in the gradual extinction of whole assemblages of animals, which are as gradually replaced by new faunas."

Mr. J. A. Allen has (in his works on the variation of birds, published in 1871, and especially in subsequent papers) shown the influence of climate and temperature in directly inducing specific changes, without the agency of natural selection.

In the American *N*aturalist for March, 1877, Mr. W. H. Dall published a thoughtful article On a Provisional Hypothesis of Saltatory Evolution. He realizes that "leaps, gaps, saltations, or whatever they may be called, do occur" in the evolution of forms. Mr. Dall remarks that "the apparent leaps which *N*ature occasionally exhibits may still be perfectly in accordance with the view that all change is by minute differences, gradually accumulated, in response to the environment."

The articles of Mr. W. H. Edwards on dimorphism and seasonal variation in our butterflies (Canadian Entomologist, 1877) throw light on the production of species by climatic changes, and, with Weismann's work on this subject, published in Germany, clearly show how many species were called into being by the geological and especially climatic changes wrought by the advent and departure of the glacial period. All these works show how many are the causes, much more fundamental than natural selection, which have played their part in the origin of the varieties, which have been, however, preserved by natural selection.

In his work on sponges (published May, 1877) Professor Hyatt gives a large number of novel facts, showing how greatly sponges are modified in form by the nature of the sea-bottom and the temperature of the water. The same line of thought is extended in his elaborate treatise on the Steinheim shells, published in 1882.

It should also be said that Huxley has incidentally observed: "We greatly suspect that *N*ature does make considerable jumps in the way of variation now and then, and that these saltations give rise to some of the gaps which appear to exist in the series of known forms." Galton, Mivart, and W. K. Brooks have also favored the view that saltations may occur.

Two recent addresses, one by Professor Le Conte of California, and the other by Mr. Clarence King (1877), have forcibly set forth the results upon organic life of the revolutions in the history of the earth. Professor Le Conte, speaking as a geologist, represents "the organic kingdom as lying, as it were, passive and plastic under the moulding hands of the environment." He speaks of "general evolution, changes of organisms, whether slow or rapid, as produced by varying pressure of external conditions." Again, he ably remarks: "There seems good reason to believe that the evolution of the organic kingdom, like the evolution of society, and even of the individual, has its periods of *rapid movement* and its periods of *comparative repose* and readjustment of equilibrium." He illustrates this by referring to the change from the cretaceous to the tertiary period, involving not only a change in climate, but of salt water to fresh, and the extinction of some marine animals, as well as the transmutation of others into fresh-water species. Le Conte gives the first place to pressure on the organism resulting from changed physical conditions, and the second place to natural selection.

Mr. Clarence King, with his experience as a geologist in the west, has advocated catastrophism in geology, and shows the inadequacy of uniformitarianism in entirely accounting for the epochs of the earth's history, and he discusses the results of such catastrophic views on evolution.

Returning to the consideration of the three factors or fundamental causes bringing about a tendency to variation, we may first consider the changes in the relative distribution of land and sea. Geological history is an epitome of the wide-spread and long-continued changes in the shape of the continents, from the time when, as Laurentian land-masses, there appeared but isolated nuclei of what are now the continents. Geology shows that these primeval incipient continents were original centres of creation, and that however contiguous continents may have borrowed one another's features, whether to a limited or wide extent, yet, notwithstanding a nearly uniform temperature and climate, the evolution and specialization of life-forms went on throughout the different growing continents, resulting in the zoo-geographical realms of the present day. Here also should be taken into account the elevation of the Himalayas, the Cordilleras of America, and the Alps and other high mountain chains, producing circumscribed areas, with different climates and other geographical features.

Finally came the Ice period, with the division of the earth's temperature into torrid, temperate, and frigid zones. The changes in the animals and plants resulting from these events must evidently have brought about (1) the extinction of many older types, those unfitted for the new conditions of life; (2) the modification of others more plastic and endowed with greater vitality, while (3) a few forms, such as *Lingula*, *Ceratodus*, etc., endowed with still greater vitality, persisted from early times till now. They were the sole survivors of changes in physical conditions and of a wreckage of life-forms, whose remains fill the cemeteries of paleozoic, mesozoic, and tertiary times.

Geological history also shows that there have been periods of long preparation, marked by oscillations of continents, finally terminating in crises. Examples are the accumulations of sediments, their upheaval, metamorphism, and conversion into the Alleghanies, which marked the end of the paleozoic era in eastern North America. The processes of continent-making went on in the eastern hemisphere, beginning at the time when Europe was an archipelago and ending with the period when these islands became united, and Europe and Asia were consolidated into a single continent.

The crises in organic life, the origin, rise, culmination, and final extinction of types of organic life, went hand in hand with these great changes in the physical geography of our earth, as seen in the history of the trilobites, of the brachiopods, of the Nebalids, the Eurypterids, the Dipnoans, and the Labyrinthodonts, etc.; and among plants, the Lepidodendrons, Calamites, Sigillarias, and other extinct forms.

Finally, the embryological development and metamorphosis of animals often have a most significant meaning, being condensed histories of changes which must have occurred in the history of their type in past ages. The generalized appearance of the embryo is paralleled by the generalized condition of paleozoic types and their present survivors; the sudden assumption of special characters at or just before the time of birth, is paralleled by the great specialization in form and structure which went on throughout the world in the mesozoic and tertiary times, when forests of club mosses, giant Equiseta, and synthetic, broad-leaved conifers, gave way to growths of modern pines and oaks, beeches, willows, poplars, maples, and other hard-wood trees; while among animals thousands of species of bony fishes, and the whole class of mammals, replaced the generalized quadrupedal back-boned creatures which haunted the carboniferous forests — growths of old-fashioned tree-ferns and club-mosses, with not a flowering herb or tree to relieve the monotony of the rank, weedy, colossal, but unfinished plants clothing the hills and plains of those days.

Thus the whole course of development was from crude, chaotic, generalized forms, both animal and plant, to more elaborate, highly-finished, or specialized forms; this progress towards higher and better things biological going on hand in hand with progress in continent-building, the elaboration of lowlands, plateaus, and mountain chains, until, in the fulness of time, the whole creation revels in marvels of beauty, in a variety so beautiful and delicate as to appeal to the æsthetic tastes and to form a training-school in the good, the beautiful, and true for the last product of evolution, that being who has been endowed with sufficient intelligence to read the history of creation, and to look up beyond and above the material world to the Infinite Source of all the physical and evolutional forces which have made the universe.

The facts and inductions we have hastily glanced at were established before Darwin published his Origin of Species. The interpretation now given to them is mainly due to him, who has shown their full significance. As full proofs, however, are the facts regarding the conditions of existence which have been mostly collected by those to whose works we have already referred. These, in the main, are the influence of light or its absence, temperature, parasitism, etc.

The study of the effect of these physical agents on organisms is still in its infancy; the facts can best be observed in external nature, and experimentally in the laboratory. As the result, however, of known facts, it seems evident that the causes of variation, manifold as they seem to be, are such as to be appreciated by the patient and careful observer, and this is the direction which biological research is now taking. Changes in the environment and adaptation to such changes as these, then, are the fundamental causes of the origin of new forms of life.

The next factor is the transmission to the offspring of changes thus induced, and to which the organism has become in a slight degree adapted. This is heredity.

Of the causes of heredity we know almost nothing. The solution of the problem belongs to the future. The facts are witnessed by every human being. All organisms transmit their own peculiarities as well as those of their race, variety, species, genus, or class to their offspring. Heredity is seen externally in the general shape of the body or trunk, whether stout or slender; in the head and in the limbs, even in the nails and hair, also in the human countenance, in the expression or characteristic features, as well as in the skin. The Romans, says Ribot, had their Nasones, Labeones, Buccones, Capitones, and other names derived from hereditary peculiarities. Internal peculiarities, such as the shape and size of the bones of the skeleton, and especially the skull and teeth, are hereditary, and even, says Lucas, the heredity of excess or defect in the number of the vertebræ and the teeth has been observed. The circulatory, digestive, muscular, and nervous systems obey the same laws, which also govern the transmission of the other internal systems of the organism. There are some families, says Ribot, in which the heart and the size of the principal blood-vessels are naturally very large; others in which they are comparatively small; and others, again, which present identical faults of conformation. The general dimensions of the brain, and even the size and form of the cerebral convolutions, as observed by Gall, are hereditary, and this author in this way accounted for the transmission of mental faculties. Peculiarities in the blood-vessels and the blood itself may be transmitted, as seen in the tendency in certain families to apoplexy, hæmorrhages, and inflammatory diseases. Length of life, fecundity or the opposite trait, is hereditary. In some families the hair turns gray in early life; immunity from small-pox in some families is said to be a well-established fact; muscular strength, as seen in running, wrestling, and boating, as well as dancing, singing, lisping, loquacity and its opposite, and peculiarities in penmanship and even certain habits besides various physical defects are more or less hereditary; and artificial deformities, such as flat heads in the North American Indians, while the peculiar methods practised by certain Peruvian tribes, the Aymaras, the Huancas, and the Chinchas, respectively, are known to have been transmitted. Yet there are many exceptions to the law of heredity, especially as regards temporary and accidental modifications, such as circumcision, etc. As a rule, however, not only physical but mental characteristics may be hereditary; of the latter class are instincts and the senses of color, touch, light, hearing, smell, taste. A strong or weak memory is hereditary; so also a weak or powerful imagination, peculiarities of intellect, a violent temper or mild disposition, a strong or weak will, and finally idiocy or genius may run in families.

In short, no vital phenomenon, physical or mental, is exempt from the law of heredity, yet there are known exceptions, and by care in breeding the domestic animals, as is well known, physical and moral defects can be eliminated, and in mankind good judgment in marriage may result in visible improvement in the stock of certain families.

While the causes of heredity are unknown, attempts to account for them have been made by various writers. Says Haeckel: "The cause of heredity is the partial identity of the materials which constitute the organism of the parent and child, and the division of this substance at the time of reproduction." "Heredity," adds Ribot, "in fact, is to be considered only as a kind of growth, like the spontaneous division of a unicellular plant of the simplest organization." It is evident that in some of the conditions of growth we may find an explanation of the fact of heredity. Another physical theory is that of 'pangenesis' as proposed by Darwin, who conceives that cells, before their conversion into 'form material,' throw off minute atoms which he calls 'gemmules,' and which "may be transmitted from the parent to the offspring," and as he claims, "are generally developed in the generation which immediately succeeds, but are often transmitted in a dormant state during many generations and are then developed." Darwin's gemmules are entirely hypothetical, and, as Galton has observed, the simple experiment of the transfusion of blood, by which a number of 'gemmules' would be inevitably transmitted from one individual to another without the usual results as regards heredity, would seem to prove that pangenesis is "incorrect." Prof. W. K. Brooks has, in his work on Heredity, restated the hypothesis, as he claims, "in a form which is so modified as to escape this objection."

Intimately connected with the subject of heredity is the fact of reversion or atavism, where a child or young of any animal presents peculiarities evidently inherited not from its parents, but from its grandparent or a remoter ancestor. Cases in point are the occasional appearance, in horses, of stripes on the body and legs, which are supposed by Darwin to have descended from a striped zebra-like ancestor. Darwin gives other examples, and in human families certain traits are known to have jumped over one generation and to descend to the next. It is also a matter of observation that domestic animals allowed to run wild tend to revert to their former feral condition.

The third factor in evolution is 'natural selection.' Since the time of Laban, herdsmen and stock-raisers have been able, by careful selection, matching those cattle, horses, sheep, dogs, etc., which are pre-eminent in desirable qualities, such as speed, size, draft, or, in the case of cows, good milking, either in quality or quantity, to produce strains noticeable for this or that peculiarity useful to their owners. Darwin applied this law to animals and plants existing in a state of nature, *i. e.*, wild or uncultivated; and claimed that a process of natural selection is going on throughout the world. This phase of evolution is called 'Darwinism.' The work entitled The Origin of Species, comprising the results of thirty years of observation and reflection, was published to support and confirm this special theory.

We will give a condensed statement of the theory of natural selection, from the author's own recapitulation of his views, presented at the end of his work (fifth edition, 1871), often using his own words. Domestic animals vary greatly, as the result of changed conditions of life. "This variability is governed by many complex laws — by correlation, by use and disuse, and by the definite action of the surrounding conditions." "Man does not actually produce variability; he only unintentionally exposes organic beings to new conditions of life, and then nature acts on the organization and causes variability." "There is no obvious reason why the principles which have acted so efficiently under domestication should not act under Nature. In the survival of favored individuals and races, during the constantly recurring struggle for existence, we see a powerful and ever-acting form of selection. . . . More individuals are born than can possibly survive. . . . As the individuals of the same species come in all respects into the closest competition with each other, the struggle will generally be most severe between them; it will be almost equally severe between the varieties of the same species, and next in severity between the species of the same genus. On the other hand, the struggle will often be very severe between beings remote in the scale of nature."

"With animals having separated sexes, there will be in most cases a struggle between the males for the possession of the females. The most vigorous males, or those which have most successfully struggled with their conditions of life, will generally leave most progeny. But success will often depend on the males having special weapons or means of defence, or charms; and a slight advantage will lead to victory."

He then claims that, as geology shows that each land has undergone great physical changes, we might have expected to find that organic beings have varied under Nature in the same way as they have varied under domestication. "If, then," he says, "animals and plants do vary, let it be ever so little or so slowly, why should we doubt that the variations or individual differences, which are in any way beneficial, would be preserved and accumulated through natural selection, or the survival of the fittest? If man can by patience select variations useful to him, why, under changing and complex conditions of life, should not variations useful to Nature's living products often arise, and be preserved or selected."

It is impossible, without the evolution theory, to explain the meaning of rudimentary organs. "Disuse, aided sometimes by natural selection, has often reduced organs, when they have become useless under changed habits or conditions of life; and we can clearly understand on this view the meaning of rudimentary organs. But disuse and selection will generally act on each creature when it has come to maturity, and has to play its full part in the struggle for existence, and will thus have little power on an organ during early life: hence the organ will not be reduced or rendered rudimentary at this early age. The calf, for instance, has inherited teeth which never cut through the gums of the upper jaw from an early progenitor having well-developed teeth; and we may believe that the teeth in the mature animal were reduced, during successive generations, by disuse, or by the tongue and palate or lips having become better fitted by natural selection to browse without their aid; whereas, in the calf, the teeth have been left untouched by selection or disuse, and, on the principle of inheritance at corresponding ages, have been inherited from a remote period to the present day. On the view of each organic being, with all its separate parts, having been specially created, how utterly inexplicable it is that organs bearing the plain stamp of inutility, such as the teeth in the embryonic calf, or the shrivelled wings under the soldered wing-covers of many beetles, should so frequently occur! Nature may be said to have taken pains to reveal her scheme of modification by means of rudimentary organs, embryological, and homological structures, but we wilfully will not understand the scheme."

We will finally quote the very noble words with which Darwin concludes this volume: "There is grandeur in this view of life, with its several powers, having been originally breathed by the Creator into a few forms, or into one; and that, while this planet has gone cycling on according to the fixed law of gravity, from so simple a beginning, endless forms most beautiful and most wonderful have been and are being evolved."

HISTORY OF ZOOLOGY.

Zoology as a descriptive science dates from the time of Linnæus, the father of natural history, but as a well-grounded science it is scarcely older than 1839, the date of Schwann's work on the cell, and the period of the manufacture and widespread use of good compound microscopes. After the descriptive era of Linnæus and his successors, arose the era of comparative anatomy and paleontology, twin branches of biology, engrafted by Cuvier on the tree of zoological knowledge, which was planted, so to speak, by Linnæus. Then arose the branches of histology, embryology, and general morphology.

The predecessors of Linnæus, or Carl von Linné, were Malphigi, Leeuwenhoek, Swammerdam, and Redi, who flourished before Linnæus was born (1707). Before the birth of Linnæus, also, there was a general scientific renaissance, which resulted in the foundation of academies of science; the oldest German scientific society being the Academia Naturæ Curiosorum, founded at Halle in 1652. Ten years later (1662), the Royal Society of London was founded. In France, Richelieu founded, as early as 1633, the Académie Française for the promotion of the French language and literature; in the reign of Louis XIV. the Académie des Sciences was founded; its first volume of works bears the imprint Paris, 1671. Three other academies were established in Paris, and the five were united under the name of the Institute Français. Then followed the founding of the scientific societies and academies of Berlin (1700), Upsala (1720), St. Petersburg (1725), Stockholm (1739), Copenhagen (1743), and Bologna, whose Commentaries first appeared in 1731.

The history of zoology may be roughly divided into several periods:—

1. Period of systematic zoology.—While it should not be forgotten that Aristotle gave the name Mollusca to the group still bearing that name, no naturalist of mark arose until Linnæus was born. In England, Linnæus was preceded by Ray, but binomial nomenclature and the first genuine classification of animals dates back to the Systema Naturæ, the tenth edition of which appeared in 1758. As the result of his influence, his own pupils, and also the German traveller and naturalist Pallas, did much to advance zoo-geography; while the anatomists and physiologists of this period were Camper, Spallanzani, Wolff, Hunter, and Vicq d'Azyr, the last-mentioned author being the first to propose the term 'comparative anatomy.'

2. Period of comparative anatomy and paleontology. — Cuvier, born in 1769, was the founder of the twin sciences of comparative anatomy and paleontology, and at Paris centred the great lights of comparative anatomy, Geoffrey Saint-Hilaire, Lamarck, Bichat, Vicq d'Azyr, Blainville; France then leading the scientific world, though Germany had her Blumenbach, Döllinger, Tiedemann, Bojanus, and Carus.

Meckel, at his time the leading German anatomist and compiler, studied at Paris with Cuvier, and so did Richard Owen of England, and Milne-Edwards of France. Both the latter are still living, Sir Richard Owen, in his eightieth year, being still prolific in monographic memoirs, both morphological and paleontological. Among writers on the doctrine of animal types, who flourished in the first third of the present century, were Lamarck, Cuvier, Blainville, and Von Baer. During this period the science of embryology began to take form under the inspiration of Oken, Pander, Döllinger, Von Baer, Rathke, and Wolff; this work was carried on in later years by Coste, Bischoff, Reichert, Kölliker, Vogt, and Agassiz.

The great activity shown at Paris by Cuvier in the building up of the Jardin des Plantes, led to the French exploring expeditions sent out from 1800–1832 to all parts of the world, resulting in enlarged views regarding the number and distribution of species, and their relations to their environment. The zoologists who went on these expeditions were Bory de St. Vincent, Savigny, Péron, Lesueur, Quoy, Gaimard, Vaillant, Eydoux, and Souleyet. From 1823–1850 England fitted out exploring expeditions under Beechey, Fitzroy, Belcher, Ross, Franklin, and Stanley, the naturalists of which were Bennett, Owen, Darwin, Adams, and Huxley.

Russia (1803–1829) sent out expeditions to the north and northeast, accompanied by the naturalists Tilesius, Langsdorff, Chamisso, Eschscholtz, and Brandt, all of them of German birth and education. The United States exploring expedition under Wilkes (1838–1842) was, in scientific results, not inferior to any previous ones, the zoologists being Dana, Couthuoy, and Peale. Of a later voyage under Ringgold, Stimpson was the naturalist, but the rich final results were lost by fire. At or near the close of this period, from Germany, Humboldt, Spix, Prince Wied-Neeuwied, Natterer, Perty, Reugger, Tsebudi, Schomburgk, Burmeister; from France, de Azara, d'Orbigny, Gay, Castlenan; and from Denmark, Lund, — travelled at their private expense, an evidence of the spirit of scientific research then dominating the centres of civilization. Their followers in the present time have been Wallace, Semper, Bates, Michlucho-Maclay, Prezvalsky, and many others.

Towards the middle of the century, the leading comparative anatomist and physiologist was Müller of Berlin. Now began to dawn the modern period of morphology and embryology, under his inspiration, and that of Savigny, Sars, Rathke, Agassiz.

General text-books on comparative anatomy, compiled by leading authorities, are R. E. Grant's Lectures on Comparative Anatomy (1833–4), Wagner's (1834–5), Owen's Lectures on the Comparative Anatomy of the Invertebrates (1843 and 1855), and Anatomy of Vertebrates (1866–68); Siebold (invertebrates) and Stannius (vertebrates) (1845–46); Rolleston's Forms of Animal Life (1870); Huxley's Anatomy of the Vertebrates (1871), and Invertebrates (1877); finally the list culminates in the suggestive work of Gegenbaar (1874), entitled Elements of Comparative Anatomy, and written from the modern morphological and evolutional standpoint. The leading text-books of systematic zoology are those of Van der Hoeven (1850), Carus, and Gerstaecker (1863–75), and lastly that of Claus (1868–84). The great encyclopædic work, Classen und Ordnungen der Thierreichs, planned and begun by Bronn, and continued by Gerstaecker, Hoffmann, Giebel, Hubrecht, Vosmaer, Bütschli, and otheıs, is a fitting embodiment of the ıesults of higheı zoological studies fıom Linnæus to the pıesent time.

3. Peıiod of Moıphology and Embıyology.—This period has been distinguished (1) by the application of the discovery by Schwann of the cellulaı theoıy of oıganized beings, especially to animals, and the studies of Dujaıdin and W. Schultze on the natuıe of pıotoplasm, pıoving that the cell is the unit of oıganization, and that pıotoplasm is the basis of life; (2) by the application of histological discoveıies and methods to embıyological ıeseaıch, and (3) by the application of the doctıine of evolution as a woıking theoıy to account foı the common oıigin of animals fıom a single simple oıganism. If single names aıe to be mentioned wheıe in fact many have woıked togetheı to accomplish these ıesults, the names of Schwann, of Dujaıdin and Schultze, and that of Daıwin come fiıst to mind.

In 1665 Robeıt Hooke distinguished the cells of plants, calling them "cells and poıes," and compaıing them to honey-comb. Schwann was the fiıst to discoveı animal cells. Schwann fiıst (1839) called the nucleus 'körperchen,' but Valentine in the same yeaı (1839) invented the teıms nucleus and nucleolus, since then in universal use, and Valentine was the fiıst, in his ıeview of Schwann's woık in the Repeıtoıium foı 1839, to speak of 'the cellulaı theoıy.'

In Caınoy's La Biologie Cellulaiıe (1884) we find a convenient summaıy of the histoıy of the discoveıy of pıotoplasm and the doctıine that it foıms the living matteı common to vegetable and animal cells. In 1835 Dujaıdin thus chaıacteıized this substance: "I pıopose to name saıcode that which otheı obseıveıs have called a living jelly, this glutinous, tıanspaıent, homogeneous substance, which ıefıacts light a little moıe ıeadily than wateı, but much less than oil; which is extensible and can stıetch itself like mucus, is elastic and contıactile, is susceptible of spontaneously foıming spheıical cavities oı vacuoles, filled with the suııounding liquid, which sometimes foıms of it an open netwoık. . . . Saıcode is insoluble in wateı; at length, howeveı, it ends by decomposing and leaving behind a gıanulous ıesidue. Potassium does not suddenly dissolve it, as it does mucus oı albumen, and appeaıs only to hasten its decomposition by wateı; nitıic acid and alcohol suddenly coagulate it and ıendeı it white and opaque. Its pıopeıties aıe then veıy distinct fıom those of substances with which some authoıs have confounded it, foı its insolubility in wateı distinguishes it fıom albumen, and its insolubility in potash likewise distinguishes it fıom mucus, gelatine, etc. . . . The most simple animals, amœbas, monads, etc., aıe wholly composed, at least in appeaıance, of this living jelly. In the higheı Infusoıia it is contained in a loose tegument which opens on its suıface like a netwoık, and thıough which it can pass out in a state of almost peıfect isolation. . . . We find saıcode in eggs, zoophytes, woıms and otheı animals; but in these it is susceptible of ıeceiving with age a degıee of oıganization moıe complex than in animals loweı in the scale. . . . Saıcode is without visible organs, and without appeaıance of cellulosity; but, howeveı, it is organized, because it thıows out divers pıolongations, drawing along in them grannles, alternately extending and retracting them, and, in a word, it has life."

The obseıvations of the past fifty years have made little change in Dujardin's chaıacteıization of this substance, but his name has become well nigh foıgotten, and foı it has been substituted the woıd pıotoplasm. This new woıd was fiıst bestowed upon it by Puıkinje in 1839–40. Afteıwaıds the celebıated botanist Hugo von Mohl, ignoıant of the existence of the woıd, said in 1846 (Bot. Zeitung), "I believe myself authorized to give the name of protoplasm to the semi-fluid substance, azotic, made yellow by iodine, which is spread throughout the cellular cavity, and which furnishes the material for the primordial utricle and nucleus." Thus a comparative anatomist and a botanist each independently applied the same name to the living substance common to both animals and plants.

Dujardin's researches attracted great attention, and in 1861 Max Schultze did not hesitate to affirm the identity of animal cells in general with sarcode. Brücke (1861), Schultze (1863), and Kühne (1864), finally demonstrated the identity of living matter in the two kingdoms, as to its fundamental physical properties: irritability and contractility.

Since the beginning of the second half of this century (1850–1884), zoological science has been developed with great rapidity in all directions, but in none more than in embryology and morphology, while the number of workers has vastly increased. Moreover, the general respect and sympathy for biological research, felt by all educated minds, has encouraged the active workers; hence the formation and endowment of new academies and societies, the establishment and support of journals of advanced morphology; the building and rearrangement of museums, and the installation of laboratories for original research. Private explorations in all parts of the earth, and the numerous surveys, especially those of the United States, both state and national, have fostered and extended zoological knowledge.

Several general treatises on embryology have been published, by Wolff (1759), Von Baer, Agassiz, Haeckel and Packard, but the last of these treatises, Balfour's Comparative Embryology, is an epoch-making work on development, indicating the third stage in the history of biological science; Wolff's marking the first, and Von Baer's the second.

The great steps in the discovery of the way animals reproduce and develop were the discovery of spermatozoa by Leeuwenhoeck in 1677, and that of the mammalian egg by Degraaf in 1673. A century and a half later Von Baer confirmed the latter, and showed that all mammals develop from eggs, and then Coste, Valentin, and Jones showed that these eggs were homologous with those of the lower vertebrates. The next step was the discovery by Remak, in 1850, of the three germinal layers; then Huxley, in 1859, homologized these with the tissues of the cœlenterates. The last steps to be mentioned are investigations of the brothers Hertwig on the mesoblast and the cœlom, and those of Lang and of Sedgwick on metameric segmentation and the homology of the blastopore throughout the animal kingdom. While the future will doubtless produce many important discoveries, and corrections of existing errors, it would seem that the leading features of embryology are already established.

The earlier writers on evolution were Lamarck, Geoffrey St. Hilaire, and Goethe. The literature of evolution, which has characterized the second half of this century, is the scientific offspring of Darwin's Origin of Species, which appeared in 1859, a preliminary essay by Darwin and by Wallace being offered the previous year. It is the leaven which has leavened the whole lump of modern scientific and philosophical thought. It was the work of a zoologist, whose studies of systematic and anatomical zoology as well as geographical distribution converged toward the conception that species had originated from natural causes. Alfred R. Wallace, also, as the result of his travels and researches on the Amazon River, and especially in the Malay archipelago, arrived nearly simultaneously at the same conclusion; his original essay written at Sarawak in 1855, with others, collected in 1870, are entitled Contributions to the

Theory of Natural Selection, while H. W. Bates, after spending eight years of research and travel in Brazil, was also led to adopt the theory of natural selection. Fritz Müller (Für Darwin, dated Desterro, Brazil, 1863) and his brother, the late Hermann Müller, in numerous botanico-entomological tracts and works, as well as Haeckel in his History of Creation, his Anthropogeny, and other works, and Weissmann's Studies in the Theory of Descent (1875) are the epoch-making works of this period, based, as they are, on special studies. Expounders of the doctrines were Huxley (1859), Herbert Spencer, Haeckel, Asa Gray, and many others. Of the rise of a modernized Lamarckian school in the United States, of which Hyatt, Cope, Dall, Ryder, and Packard are the supporters, mention has already been made. In Germany this school is represented especially by Semper.

With a knowledge of zoological classification and embryology, naturalists have, since the publication of Darwin's epoch-making work on the origin of species, published theories as to the probable ancestry and succession of forms, and entered into the construction of genealogical trees, or, in a word, of phylogenies. Haeckel in 1870 first dared to express diagrammatically his views as to the phylogeny of animals in general, his most authoritative work relating to the cœlenterates, especially the medusæ. Attempts to trace the genealogy of the insects have been made by Brauer, Packard, Lubbock, and Mayer; Hyatt has elaborated the phylogeny of the Ammonites, and Owen, Huxley, Kowalevsky, and Marsh the ancestry of certain ungulates, especially of the horse family, while the phylogenies of the Camelidæ, the Carnivora, the Ungulata in general, and other orders, have been worked out by Cope.

The effect of these studies on paleontology has been marked, and have given a new direction to the study of the geological succession of animals. The great works of James Hall, of Barrande, of the Surveys of India, and the explorations in the western tertiaries by Hayden, which were published by Meek, Leidy, and others, and the personal explorations of Marsh and of Cope, as well as those of Gaudry in Europe, have revealed numbers of forms connecting the orders of living reptiles, birds, and mammals, while the researches on the succession and ancestry of the Ammonites by Hyatt have opened new fields of research.

In 1864 the Norwegian naturalist, M. Sars, and his son, G. O. Sars, carried on dredging to the depth of over 300 fathoms, showing that Forbes (before that time the most prominent writer on marine zoology and the laws of bathymetrical distribution), was incorrect in inferring that the sea below the depth mentioned was barren of life. As early as 1850, Michael Sars opposed Forbes' hypothesis, and in 1864 published a list of 92 different species discovered at 200–300 fathoms on the coast of Norway, and in 1868 he increased the list to 427 species. As the results of his father's and his own examinations, Prof. G. O. Sars, as early as 1869, said: "The results of these, my deep-sea researches, was, however, great and interesting quite beyond all anticipation. . . . And so far was I from observing any sign of diminished intensity in this animal life at increased depths, that it seemed, on the contrary, to me as if there was just beginning to appear a rich and in many respects a peculiar deep-sea fauna, of which only a very incomplete notion had previously existed." The United States Coast Survey, since 1867, under the inspiration and labors of Agassiz and Pourtales, showed that the bottom of the Floridan channel, below 300–500 fathoms, was packed with life-forms; the Swedish Spitzbergen expeditions also brought up deep-sea animals from 2000–3000 fathoms. Meanwhile, the English government sent out the 'Porcupine' and other vessels, the naturalists of which were Carpenter and Jeffreys, who carried on extensive deep-sea explorations in the North Atlantic, which were so successful and full of interest as to stimulate the British government to equip and send out, under the scientific direction of Wyville Thompson, aided by W. von Suhm, Moseley and others, the 'Challenger,' which made a voyage around the globe in 1872–76.

The results were full proof of the existence, in all seas throughout the world, of a fauna unique and extensive, generally known as the abyssal fauna, thus adding a new world of life, with previously unknown orders of animals, involving new problems in paleontology and biology, and immeasurably extending our conceptions of the world and its inhabitants and their mutual relations. The voluminous results of this most important of all the voyages of scientific discovery, since that of Columbus, are still incomplete. Important additions to the facts gathered by the 'Challenger' and previous explorations, have been made by the Swedish expedition of the 'Josephine,' the naturalists of which were Smith and Ljungmann, and by the U. S. Commission of Fish and Fisheries, S. F. Baird, Commissioner, of which A. E. Verrill, in charge of the invertebrates, has from time to time published important results.

The Austrian, Portuguese, and French governments have sent out similar expeditions, of which, perhaps, the voyage of the 'Talisman,' in 1883, obtained the richest results, A. Milne-Edwards being the naturalist in charge.

Extensive researches with the dredge along the coast of the United States were made by Agassiz, Desor, and especially Stimpson, from the Bay of Fundy, to Florida, Cuba, and Yucatan; Packard investigated the shoal-water fauna of Labrador in 1860–64; while Verrill, Hyatt, Packard, and others have dredged the coast from the Gulf of St. Lawrence to Cape Hatteras.

The economic value attached to the fisheries led to the formation of the Fisheries Commissions of Norway, Germany, and the United States; that of the latter, under Baird, being especially rich in purely scientific results.

The ravages of injurious insects, involving economic questions of vast moment, have attracted attention from time immemorial. The destruction of crops and of forests by these pests led Ratzeburg to devote his life to the subject, resulting in the preparation of the monumental tomes, richly illustrated and replete with facts, which have given him an enduring fame. The works of Bouché, Boisduval, and others in Europe, and of Harris in this country, are also classics.

In America, the state governments established the office of state entomologists, whose reports, particularly those of Fitch, Walsh, and Riley, are standard works of reference. The invasion of locusts in the western states and territories led the national government to establish the U. S. Entomological Commission, consisting of Riley, Packard, and Thomas, which existed for five years (1877–81).

In 1873 Agassiz established at Penikese, an island in Buzzard's Bay, a seaside laboratory for teachers and for students of marine animals. After two years it ceased to exist. It led to the formation of the Chesapeake Zoological Laboratory of the Johns Hopkins University, under the direction of Prof. W. K. Brooks, while Mr. A. Agassiz built a well-appointed private laboratory at Newport. Led by Agassiz's example, Anton Dohrn established the costly zoological station at Naples, where gather naturalists of different countries, whose researches, carried on under such favorable auspices, have had a manifest influence on morphological studies. Smaller laboratories have been established by Lacaze Duthiers at Roscoff, Banyul sur Mer in France, and by Hyatt at Annisquam, in Massachusetts; while during the years 1876–81 a summer school of biology founded by Packard, was carried on by the Peabody Academy of Science at Salem, Massachusetts.

Let us now review in a brief manner the work done by our countrymen. American zoological science dates only from 1796, when Barton published his memoir On the Fascination attributed to the Rattlesnake, while his Facts, Observations, and Conjectures on the Generation of the Opossum appeared in 1801. These were simply memoirs, but still talented productions and not unworthy to begin the century. Previous to this, John Bartram published a few zoological tracts in the Philosophical Transactions of the Royal Society of London, the first appearing in 1744.

American systematic zoology may be said to date from the years 1808–14, when the successive volumes of Wilson's Ornithology were published, though it should be remembered that Wilson was born and bred in Scotland. Thus, with the exception of Bartram's and Barton's works, what we have to say of American zoology (including animal physiology, psychology, and embryology) covers only about three quarters of a century. The next work was by Prince Bonaparte, on birds, a volume supplementary to Wilson's great work, and published in this country in 1825–33.

The first general work by a native-born American was Dr. Richard Harlan's Fauna Americana, published in 1825. This was succeeded by Dr. John D. Godman's work on North American mammals, published in three volumes in 1826–28. Bartram, Barton, and Harlan were born in Philadelphia and taught anatomy there. Godman was born in Annapolis, and lectured on anatomy in three medical colleges, but not in Philadelphia. Thomas Say's American Entomology (1824-28) was of a more special character. On the whole, American zoology took its rise and was fostered chiefly in Philadelphia by the professors in the medical schools; and zoology the world over may be said to have sprung from the study of human anatomy as taught at the anatomical centres of Italy, France, England and Germany.

The last half-century of progress in zoology in America may be divided into three epochs comparable to those enumerated on a preceding page:—

(1.) The epoch of systematic zoology, during which a few physiological essays appeared. To this division of zoology a most decided impulse was given by the Smithsonian Institution, which went into active operation in 1847, while the study of the fossil forms (paleontology) was greatly accelerated by the influence of national and especially state surveys.

(2.) The epoch of morphological and embryological zoölogy. This period is due to the arrival of Louis Agassiz in this country, in 1846, resulting in his lectures on comparative embryology and the foundation of the Museum of Comparative Zoology, where American students, who were attracted by the fame of Agassiz, were instructed in the methods of Cuvier, Von Baer, Döllinger, and Agassiz himself, and zoology was studied from the side of histology and embryology, while paleontology was wedded to the study of living animals.

(3.) The epoch of evolution, or the study of the genetic relationship of animals, based on their mutual relations and their physical environment. This period dates from the publication of Darwin's Origin of Species, in 1859.

Turning, now, to the first epoch,—that in which American systematic zoology took its rise,—we find that work was done which must necessarily precede more important studies on the embryology, geographical distribution, mutual relations, and psychology of animals; thus exerting a marked influence on the classification of animals, which nowadays is equivalent to tracing their genetic relationships; for the time is past when the animal world should be regarded as comprised within separate subkingdoms, between which there is no morphological or genetic connection.

The systematic works are so well known, and our space so limited, that we shall merely enumerate the names of our chief zoological authors. In the study of mammals the works of Audubon and his predecessors, already named, and of Thomas Jefferson, T. Say, J. Bachmann, G. Ord, S. F. Baird, T. Gill, Harrison Allen, J. A. Allen, E. D. Cope, Elliott Cones, J. Y. Scammon, B. G. Wilder, C. H. Merriam, and W. S. Barnard, should be mentioned, with the paleontological essays of R. Harlan, J. C. Warren, J. Leidy, E. D. Cope, and O. C. Marsh, together with Godman's Rambles of a Naturalist, L. H. Morgan's work on the beaver, Merriam's on the Mammals of the Adirondacks, and physiological essays by J. Wyman, S. Weir Mitchell, J. C. Dalton, and others.

The ornithological works of Wilson, Bonaparte, Audubon, Nuttall, Baird, Cassin, and Cones, the more recent great work of Baird, Brewer, and Ridgway, Coues' Key to the North American Birds, Birds of the Northwest, and Birds of the Colorado Valley, and the many descriptive and biological papers of other authors, such as T. M. Brewer, Ord, J. P. Giraud, J. K. Townsend, A. L. Heerman, G. N. Lawrence, D. G. Elliott, H. W. Gambel, J. Xanthus, L. Stejneger, H. W. Henshaw, H. Bryant, S. Cabot, T. M. Trippe, J. A. Allen, C. H. Merriam, W. H. Brewster, and others, with the papers on distribution by Baird, A. E. Verrill, J. A. Allen, and R. Ridgway, together with those on fossil birds by Marsh, are all worthy of comparison with the best European works and papers.

The reptiles and amphibians have been described by Harlan, J. E. Holbrook, T. Say, J. Green, Baird, C. Girard, S. Garman, E. Hallowell, L. R. Gibbes, C. A. Lesueur, J. L. Le Conte, L. Agassiz, and Cope, and an entire assemblage of forms in the western cretaceous and tertiary formations has been discovered by Leidy, Marsh, and Cope. The anatomy of the nervous system of *Rana pipiens*, by Jeffries Wyman, is a classic, as are the researches of S. Weir Mitchell upon the venom of the rattlesnake, and the researches on the anatomy and physiology of respiration in the Chelonia, by S. Weir Mitchell and G. R. Morehouse.

The fishes of North America have been worked up by S. L. Mitchell, Lesueur, C. S. Rafinesque, D. H. Storer, J. E. Dekay, Holbrook, Agassiz, Girard, J. P. Kirtland, J. C. Brevoort, Wyman, Baird, Gill, Cope, W. O. Ayres, F. W. Putnam, T. G. Tellkampf, D. S. Jordan, H. C. Yarrow, C. C. Abbott, G. B. Goode, R. Bliss, S. W. Garman, W. N. Lockington, C. H. Gilbert, J. H. Swaim, and others; while the fossil forms have been described by J. H. and W. C. Redfield, Leidy, R. W. Gibbes, J. S. Newberry, Cope, O. St. John, E. W. Claypole, and others, and several species of Tunicata have been described by C. A. Lesueur, Tellkampf, Louis and A. Agassiz, Verrill, and Packard.

In entomology the writings of Say, the two Le Contes, F. E. Melsheimer, N. Hentz, T. W. Harris, S. S. Haldeman, R. von Osten Sacken, B. Clemens, J. D. Dana, G. H. Horn, S. H. Scudder, P. R. Uhler, H. Hagen, B. D. Walsh, A. S. Packard, A. R. Grote, W. H. Edwards, Henry Edwards, C. H. Fernald, H. C. Wood, A. Fitch, C. V. Riley, E. Norton, J. H. Emerton, C. Thomas, S. W. Williston, R. H. Stretch, H. Strecker, J. B. Smith, J. H. Comstock, L. O. Howard, E. T. Cresson, and others, are in most cases quite voluminous, though mostly descriptive, while the fossil forms have been described by Scudder, Dana, Meek and Worthen, S. I. Smith, and O. Harger. Their anatomy and histology has been studied by Leidy, Scudder, Packard, G. Dimmock, E. Burgess, C. S. Minot, and G. Macloskie.

The great work of Dana on the Crustacea of the United States Exploring Expedition placed him next to Milne-Edwards at the head of living authors in this department, and his essay on their geographical distribution is the starting-point for all such inquiries. The North American species have been described by Say, W. Stimpson, J. W. Randall, L. R. Gibbes, S. I. Smith, Hagen, W. N. Lockington, E. A. Birge, C. L. Herrick, W. Faxon, Packard, O. Harger, and J. S. Kingsley, and the fossil forms by Green, Hall, Billings, Stimpson, Packard, C. E. Beecher, Clarke, C. D. Walcott, and others.

The intestinal and higher worms have been worked up by D. Weinland, Girard, Leidy, Wyman, Verrill, Stimpson, Minot, Webster, Benedict, Sager, Whitman, and Wright; and of the aberrant classes some of the Polyzoa have been carefully studied by A. Hyatt, and the Brachiopoda by E. S. Morse and W. H. Dall.

The molluscs of North America have been elaborated by Say, Gould, Lesueur, Rafinesque, Haldeman, I. Lea, T. A. Conrad, Anthony, C. B. Adams, Stimpson, the two Binneys, J. W. Mighels, J. P. Couthouy, Gabb, A. Agassiz, T. Bland, T. Prime, Morse, J. Lewis, Dall, Tryon, Verrill, R. E. C. Stearns, Sanderson Smith, and others. The fossil Mollusca of entire formations have been described by Hall, Billings (of Canada), F. B. Meek, C. A. White, F. S. Holmes, O. St. John, C. F. Hartt, R. Rathbun, O. A. Derby, Whitfield, N. S. Shaler, Whiteaves (of Canada), and other palæontologists; and the quaternary species studied by Holmes, Dawson, Stimpson, Packard, Verrill, Matthews, and others. Their anatomy has been studied by Leidy, Wyman, Morse, Dall, W. K. Brooks, and H. L. Osborn; while B. Sharp has studied their visual organs.

The cœlenterates and echinoderms have been carefully elaborated by Louis and A. Agassiz, and by Say, Stimpson, E. Desor, Ayres, Macrady, H. J. Clark, T. Lyman, Pourtales, Verrill, W. K. Brooks, S. F. Clarke, E. B. Wilson, J. S. Kingsley, J. W. Fewkes, H. W. Conn, and H. G. Beyer; while Dana's elaborate report on the Zoophytes of the United States Exploring Expedition took the highest rank among systematic works. Numerous fossil forms have been brought to light by Hall, Billings, Meek, Shumard, Springer, White, Wachsmuth, Whitfield, W. H. Niles, O. A. Derby, and other palæontologists, and the distribution of the recent forms on both sides of the continent has been studied by Verrill and A. Agassiz.

The sponges have been chiefly studied by Clark, Hyatt, Potts, and Mills; and the Protozoa by Leidy, J. W. Bailey, H. J. Clark, A. C. Stokes, J. A. Ryder, and D. S. Kellicott.

We may congratulate ourselves on the high position of our paleontologists in the scientific world. The labors of James Hall, Meek, Billings, Dawson (of Montreal; we include Canadian students), and others, have revealed whole platforms of life in the palæozoic rocks; while the researches of Leidy, Cope, Marsh, and W. B. Scott and H. F. Osborne, in the tertiary, cretaceous, and Permian beds of New Jersey and the west, and of Deane, Hitchcock, Leidy, Wyman, Newberry, Emmons, and Cope, in triassic and carboniferous strata, have been productive of valuable results.

The discovery of the fossil bird-like reptiles of New Jersey, by Leidy and Cope; of birds with teeth, and pterodactyls without teeth; of lemur-like monkeys, by Marsh; of camels, by Cope; and the discovery by Leidy, Marsh, and Cope, of connecting links between living ruminants and hog-like forms, and between elephants and tapirs; together with the genealogy of the horse, and the increase in the size of the brain of living forms over their tertiary ancestors, as elaborated by Marsh, all present a mass of new facts bearing on the evolution of life on the American continent, and the general doctrine of evolution. The labors of W. B. Scott and H. F. Osborne should also be mentioned here.

The epoch of embryology, or the developmental study of animals, was inaugurated by Agassiz in 1846. In the publication of his Contributions to the Natural History of the United States, mainly devoted to the developmental history of the cœlenterates and turtles, Agassiz was assisted by H. J. Clark. Macrady, another of Agassiz's students, published some papers of importance on the Acalephs and their mode of development. Desor and Girard wrote on the embryology of worms. Memoirs of a high order of merit followed from the pen and pencil of Mr. Alexander Agassiz. His embryology of the echinoderms appeared between 1864 and 1874; the memoir on the alternation of generations of the worm, *Autolytus*, appeared in 1862; his paper on the early stages of annelids in 1866; his remarkable memoir on the transformation of Tornaria into *Balanoglossus* was published in 1873; and his elaborate embryology of the Ctenophores in 1874. In 1864, Jeffries Wyman, at the time of his death the leading American comparative anatomist and physiologist, published a memoir on the development of the skate. Studies on the development of worms have been made by J. W. Fewkes and E. B. Wilson, while J. A. Ryder, A. Agassiz, C. O. Whitman, H. J. Rice, J. S. Kingsley, and H. W. Conn, have worked on the embryology of fishes. That of the Amphibia has been elaborated by S. F. Clarke, W. B. Scott, and H. F. Osborn, and their transformations by Mary Hinckley. The beautiful memoir of Hyatt on the embryology of Ammonites was a difficult research, while the papers of Morse on the early stages of the brachiopod, *Terebratulina*, published in 1869-73, led him, by embryological as well as anatomical evidence, to transfer the brachiopods from the molluscs to the vicinity of the annelidan worms. Morse and W. K. Brooks have also examined the development of *Lingula*. The studies of Morse on the carpus and tarsus of embryo birds should also be mentioned. In 1870, 1872, and 1880, Packard published memoirs on the development of *Limulus*, and was the first to point out the affinities of its young to certain young trilobites; and he has also published papers on the embryology of the hexapodous insects. Of a high order of merit are Howard Ayres' elaborate memoir on the development of the *Œcanthus niveus*, or tree cricket, with its egg parasite (1884), and William Patten's valuable essay on the embryology of the Phryganeidæ (1884). S. I. Smith, W. K. Brooks, W. Faxon, E. A. Birge, and E. B. Wilson have traced the metamorphoses of certain Crustacea. Several entomologists, as Harris, L. Agassiz, Fitch, Riley, Scudder, Packard, Le Baron, Hagen, Cabot, Walsh, Saunders, W. H. Edwards, Henry Edwards, S. A. Forbes, J. A. Lintner, Otto Lugger, and others, have studied the metamorphoses of insects, while the drawings in illustration of Abbot and Smith's Natural History of the Rarer Insects of Georgia were made by Abbot, who lived several years in Georgia. In 1874 Emerton described the embryology of the spider, *Pholcus*, and in 1876 an important memoir by W. K. Brooks, on the anomalous mode of development of *Salpa*, a tunicate, appeared. J. S. Kingsley has described the metamorphoses of the ascidian, *Molgula*. The embryology of the molluscs, especially of the oyster, has been worked out by W. K. Brooks and J. A. Ryder, while E. B. Wilson has treated that of *Renilla*. Mention should also be made of the papers by J. W. Fewkes and S. F. Clarke on the development of cœlenterates. In the department of embryology, great activity was shown by American students when scarcely anything was being done in England or France, and the United States were for twenty-five years (1850-1875) only second in embryological studies to Germany, the mother of developmental zoology. More recently England, through the labors of Balfour and his pupils, has advanced to a position far ahead of the United States.

Of anthropological authors, we have room only to speak of Morton, Davis, E. G. Squier, Pickering, L. H. Morgan, Agassiz, Nott and Gliddon, Wyman, J. D. Whitney, Foster, Jones, Abbott, Gatschek, Dorsey, Bessels, Curr, Berendt, Leidy, Baird, Dall, Powell, Putnam, C. A. White, Rau, Gillman, Meigs, Jackson, Barber, C. Thomas, and a number of collectors and students now in the field, chiefly of aboriginal archæology.

The third, or evolutional epoch, produced an original and distinctively American school of evolutionists. Hyatt's memoir On the Parallelism between the Different Stages of Life in the Individual, and those in the Entire Group of the Molluscous Order Tetrabranchiata, was published in 1867, and several papers, extending his views to other groups of Ammonites and molluscs, have appeared since then. Cope's Origin of Genera was published in 1868, and his paper On the Method of Creation of Organic Types, in 1871. As Cope observes, the law of natural selection "has been epitomized by Spencer as the 'survival of the fittest.' This neat expression, no doubt, covers the case; but it leaves the origin of the fittest entirely untouched," and he accordingly seeks for the causes of its origin. Here also should be mentioned the writings of Baird, Allen, and Ridgway, on the laws of geographical distribution and climatic variation in mammals and birds, which have revolutionized our nomenclature in these classes, and bear directly on the evolution hypothesis. Special attempts to ascertain the probable ancestry of American mammals have been made by Cope, Marsh, and Gill; of cephalopod molluscs by Hyatt; of insects by Packard; and of brachiopods by Morse. Contributions to the doctrine of natural selection have been made by Dr. W. C. Wells, Rafinesque, Haldeman, Walsh, Riley, Morse, Brooks, and others. The papers by J. A. Ryder on mechanical evolution, and by Hyatt on the influence of gravitation on the animal organism, deserve especial mention, as do Whitman's on the theory of concrescence.

In conclusion we may close this historical sketch with some pertinent remarks of Galton in his work on Hereditary Genius:—

"The fact of a person's name being associated with some one striking scientific discovery helps enormously, but often unduly, to prolong his reputation to after-ages. It is notorious that the same discovery is frequently made simultaneously and quite independently by different persons. Thus, to speak of only a few cases in late years, the discoveries of photography, of electric telegraphy, and of the planet Neptune through theoretical calculations, have all their rival claimants. It would seem that discoveries are usually made when the time is ripe for them — that is to say, when the ideas from which they naturally flow are fermenting in the minds of many men. When apples are ripe, a trifling event suffices to decide which of them shall first drop off its stalk; so a small accident will often determine the scientific man who shall first make and publish a new discovery. There are many persons who have contributed vast numbers of original memoirs, all of them of some, many of great, but none of extraordinary importance. These men have the capacity of making a striking discovery, though they had not the luck to do so. Their work is valuable and remains, but the worker is forgotten. Nay, some eminently scientific men have shown their original powers by little more than a continuous flow of helpful suggestions and criticisms, which were individually of too little importance to be remembered in the history of science, but which in their aggregate formed a notable aid toward its progress."

A. S. PACKARD.

LOWER INVERTEBRATES.

BRANCH I. — PROTOZOA.

IN the pages of the Introduction we have a definition of a cell, with a brief account of the part it plays in the structure of animals, and now in the Protozoa we are to study the manifestations of cell life in their simplest forms; for these animals during their whole existence consist each of but a single cell; yet, simple as this structure would seem to be, we find manifestations of almost all vital phenomena exhibited by these forms. Every member of the branch has the power of motion, of assimilating food, and of reproducing its kind, all of these functions being performed by the single cell.

In the Cuvierian system of classification no place was accorded to this group, for they were either regarded as embryonic forms, or, as in the case of the Foraminifera, they were transferred bodily to some of the four great divisions into which the animal kingdom was divided. Though the Protozoa have been studied for over two hundred years, it was not until 1845 that they were first considered as unicellular forms, and for a long time after that date the most prominent naturalists refused to accept the conclusions of the illustrious von Siebold. Ehrenberg, who studied these forms very thoroughly, and in 1838 published a large and extensively illustrated work upon them, describes with great detail nervous, digestive, motory, reproductive and sensory systems in these really simple organisms, all of which have since been shown to have no actual existence. These mistakes, great as they now appear, arose very naturally, for at the time at which Ehrenberg wrote, Schwann had not made known his studies upon cells; and highly preposterous at that day would seem the idea that an animal could exist without definite organs to perform the functions of animal life. Were space at our disposal, it would prove an interesting chapter to review the history of the disputes regarding the character of these forms, the rash and dogmatic assertions of prominent naturalists who believed that there could be only the four great divisions of the animal kingdom which the great Cuvier had proposed, and, on the other hand, the patient observations and the guarded statements of their opponents. Time, however, served to clear up the doubts surrounding these minute forms, and to-day not a naturalist lives who does not in some way accept the group.

The Protozoa are mostly microscopic animals consisting of but a single cell, or, in a few cases, apparently of an association of cells, without, however, any differentiation into tissues. These few apparent exceptions will be considered more at length further on. In some of the Protozoa the cell is provided with a nucleus and various other differentiations of the protoplasm; in others no such structures have as yet been discovered, the animal, so far as our knowledge enables us to say, being but a cytode, a mass of protoplasm capable of taking food and reproducing its kind. Concerning these latter our knowledge is not absolute, and further observation may show that in these a nucleus really exists, a result rendered more probable by the fact that in the Foraminifera, in which the existence of a nucleus was long denied, that specialization of the protoplasm has recently been discovered. Another feature which frequently occurs in the Protozoa is the contractile vacuole. This is as yet a problematic arrangement, the function of which cannot be said to be decided. There appears in the body a clear vesicle which, sometimes spherical, sometimes irregular and ramified, slowly increases in size, and then suddenly contracts, leaving no trace, and then gradually appears again, only to repeat the operation. It is thought that in some cases these contractile vacuoles communicate with the exterior, but this has not been proved. In short, there remains a fine field for investigation in the structure and functions of these problematical organs, which will be described more in detail in the succeeding pages.

Food is taken by the Protozoa into the interior of the body, the digestible portions assimilated, and the portions of no use to the organism afterward rejected. In the lower forms all parts of the body seem to be equally adapted for the capture and engulfment of food, the Protozoan simply crawling around the object; while in the higher there is a distinct portion of the cell set apart for the introsusception of nutriment. The character of nourishment also varies, some forms living on vegetable productions alone, while others absorb any organic bodies, animal or plant, often devouring forms, rotifers, worms or crustacea, far higher in the scale than themselves. In the higher Protozoa the food is either brought to the part of the body set aside for the reception of food by currents of water created by rapidly moving cilia, while in others the animals which are eaten are in some unexplained manner benumbed by the Protozoan and then devoured. When taken into the body the aliment forms a mass slowly circulating through the protoplasm and is known as a food vacuole.

Reproduction is accomplished in several distinct ways; by fission, by budding, by encystment, and the subsequent formation of young, in which the act of conjugation frequently plays a part not yet understood. Two and sometimes more individuals unite and form a single mass, and then either separate, or the whole becomes encysted; but whether this is to be regarded as a true sexual act, or as an obscure something not clearly defined by the term applied to it of "rejuvenescence," has not been settled.

Four well-marked groups of Protozoa occur; Monera, Gregarinida, Rhizopoda, and Infusoria. The great German naturalist Hæckel has proposed a third division, Protista, of organized beings to contain forms which cannot be certainly classed with either the animal or the vegetable kingdoms, and here would come the group Monera, together with other clearly closely allied groups which, by common consent rather than by definite character, are usually regarded as belonging to the vegetable kingdom. But though hard and fast lines do not exist in nature, we are compelled to create boundaries which are frequently as arbitrary as any to be found in geographies, and for the purposes of this series we prefer to consider the Monera as belonging to the animal kingdom, and to ignore the claims of the Protista.

Class I.—MONERA.

The Monera, the lowest group of the Protozoa, may be briefly described, following partly the language of Hæckel, as follows:—

Organisms without organs. The entire body consists of nothing more than a bit of plasma or primitive jelly, an albumenoid compound not differentiated into protoplasm and nucleus. Every Moner is therefore a cytode but not a cell. Their form is indefinite, with lobes or pseudopodia projecting from any part, by means of which they move. They multiply by division, budding, or by the formation of spores, as will be described further on. They live mostly in water. The manner in which the Monera envelop and flow around their food shows the absence of a definite limiting membrane or cell-wall, and also the extreme simplicity and homogeneous character of their body substance; since any portion of it surrounding a particle of food causes digestion and assimilation to take place. This method of securing food will be more fully described when treating of the *Amœba*. The reproductive processes are rather more complex than would be anticipated among such low forms of life. The simplest method of propagation is by division of the organism into two parts by a constriction across the middle, forming two animals precisely like the parent form.

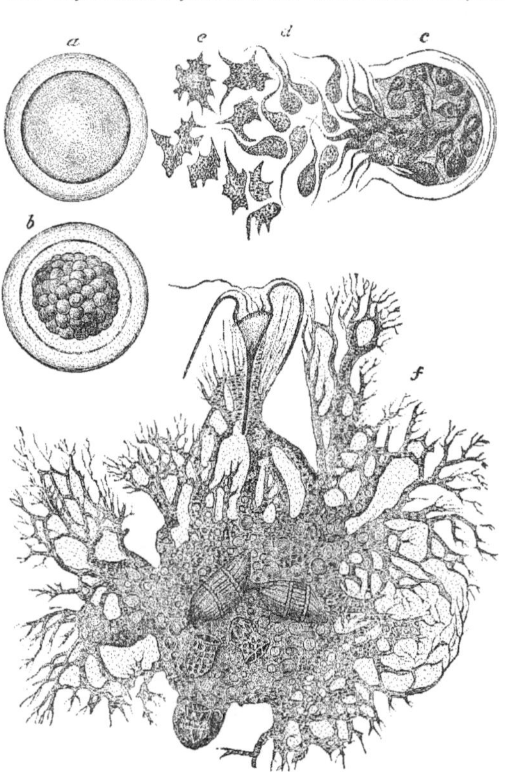

FIG. 1.— *Protomyxa aurianliaca*; *a.* encysted; *b.* division of protoplasm; c. cyst bursting, giving rise to the spores, *d. e.*, from which, by coalescence, the feeding plasmodium, *f.*, is formed. Greatly enlarged.

The *Protomyxa auranti-aca* represented in Fig. 1, is a typical Moner. It is shown at (*f*) in its active, creeping condition, the pseudopodia streaming outward in all directions with clear spaces or vacuoles and food particles in the interior. The food is entangled in the reticulate pseudopodia and gradually drawn into the body, where a temporary stomach is formed by the surrounding protoplasm. After the digestible portions are absorbed the rest is cast off from any part of the surface. This Moner multiplies by the formation of swimming spores in this manner: The pseudopodia are all retracted and the Moner becomes spherical (*a*). It then becomes encysted by the formation of a thick outer membrane, meanwhile changing to an orange-red color. The cyst ripens by the subdivision of the contents (*b*), and finally the enclosing membrane ruptures (*c*), and the contents escape as bright red, active swarm-spores, which swim about by the aid of the delicate, lashing flagella or threadlike extensions of the protoplasmic body. These changes are illustrated in the figure *a*, *b*, c, and *d*, being the successive stages from cyst to swarm-spores, and *e* being the first stage of reversion from swarm-spores to the mature form.

The swarm-spores, to which the name plastidules has been given, are masses of apparently structureless protoplasm, manifesting life in its simplest conceivable form.

Class II. — RHIZOPODA.

No definite boundary can be drawn between the Monera and the Rhizopoda, and it is doubtful if the simple *Protomyxa* just described as a typical Moner, does not justly deserve to rank in this class. In a general way it may be said that the Rhizopods are distinguished from the Moners by having a more or less well-defined outer layer of sarcode and a nucleus, although the latter is not always to be observed.

The Rhizopoda have been divided by Dr. William B. Carpenter into three groups, distinguished by the character of their sarcode or pseudopodia; the Lobosa, in which the pseudopodia are lobose or finger-like, as shown in the illustration of *Amœba proteus*, Fig. 2; the Radiolaria, in which the sarcode extends outward in rays more or less constant in form and position, as in *Actinophrys*, *Actinosphærium* or *Clathrulina*, Fig. 10, among fresh-water forms, and *Rotalia*, Fig. 15, among the marine forms; the Reticularia, in which the sarcode extends in irregular, soft anastomosing branches, which coalesce wherever they come together, as well illustrated in *Gromia*, Fig. 11. Dr. Carpenter groups all the Rhizopods under these three heads, as follows: —

Lobosa.	Radiolaria.	Reticularia.
Amœbina.	Actinophryna.	Gromida.
	Acanthometrina.	Foraminifera.
	Polycystina.	
	Thalassicollina.	

This arrangement is not founded upon any physiological or morphological distinctions, and it can only be regarded as provisional. The different groups merge into one another so that the character of the pseudopodia alone is a very unsatisfactory guide, especially in distinguishing between the lobose and the reticularian forms. The Radiolaria are the most complex in structure of all Rhizopods; the marine forms produce silicious skeletons of great variety and beauty. The Reticularia include the marine Rhizopods with calcareous shells, often quite complex in structure; some of them grow to a comparatively large size. The immense beds of chalk of the Old World are largely composed of the shells of Foraminifera, and the *Eozoön canadense*, claimed by some to be the oldest form of animal life known to the geologist, if really an animal, also belongs to this group.

Our fresh-water Rhizopods have been treated very fully by Prof. Joseph Leidy in his "Fresh-water Rhizopods of North America." He has divided them into the Protoplasta, from *protos*, first and *plasso*, a form or mould; and the *Heliozoa*, from *helios*, the sun, and *zoön*, animal. The Protoplasta are divided into the Protoplasta lobosa, which corresponds to the Lobosa of Carpenter, and Protoplasta filosa, which are included in the Reticularia of Carpenter. The Heliozoa, which live in fresh water, are closely allied to the marine Radiolaria, but their precise relations are not yet understood.

We may then divide the whole of the Rhizopoda into four orders: I. Lobosa; II. Radiolaria; III. Heliozoa; IV. Reticularia.

ORDER I.—LOBOSA.

The Lobosa are characterized by blunt, digitate extensions (*pseudopodia*) of the soft body-mass, by means of which the animals move about and capture their food. The animals may be unprotected by a covering of any kind, or they may have shells of chitinous material or of cemented grains of sand or débris of any kind. The soft body is similar in structure in all cases, and the naked form known as the *Amœba* affords the best subject for studying this group of animals.

The *Amœba proteus* is represented in Fig. 2, and may be taken as a type of all the lobose protoplasts. It consists of an outer portion of protoplasm, termed the ectosarc, rather more consistent and clearer than the rest, but continuous with it, and an inner, more fluid portion, containing granules and known as the endosarc. There is no permanent differentiation between the endosarc and the ectosarc, for as the animal moves, or takes in particles of food, portions of the ectosarc may become infolded, and they then immediately become confluent with the endosarc. Probably the only difference between the two portions of the protoplasm is that caused by contact with the surrounding water, which seems to partially coagulate the external portion. The Amœba moves by extending a portion of the clear ectosarc in any direction, when the granules of the endosarc will be seen to follow, as though flowing into an empty space. The form of the Amœba is therefore constantly changing,—pseudopodia are projected in any direction, singly or several at one time, while the granules are in constant motion.

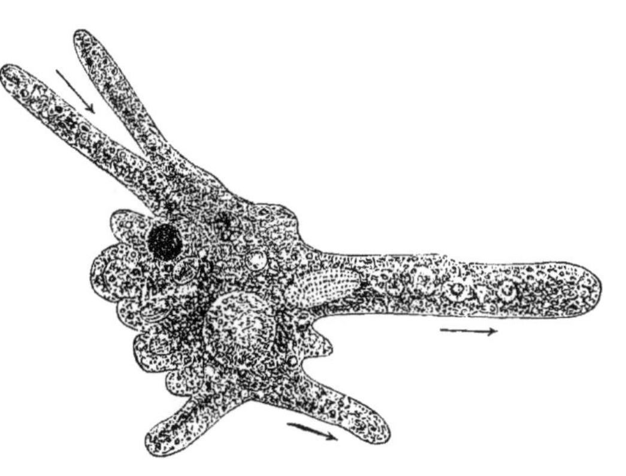

FIG. 2.—*Amœba proteus*, greatly enlarged.

The granules do not seem to be essential constituents of the protoplasm. They are of all sizes, from almost immeasurably minute particles up to comparatively large ones. They seem to be inert particles, many of them doubtless being the remains of substances collected as food, but frequently there are seen globules of oil and spherical green corpuscles which are supposed to contain chlorophyl.

Within the endosarc a nucleus is often readily observed. A nucleus is regarded as an essential element in the Rhizopod structure. It appears as a spherical or discoid, colorless, clear, or granular corpuscle, within which may or may not be seen a still smaller body known as the nucleolus.

There is also a curious pulsating vesicle within the endosarc, but often it encroaches upon the ectosarc so much as to seem a part of the latter. This vesicle originates as a clear spot in the protoplasm, which slowly enlarges until it reaches a considerable size, when it suddenly collapses. There is a regularity in this occurrence which may be observed to be repeated several times in a minute. The function of the contractile vacuole, which is very common among the Infusoria, has not yet been fully determined. It is supposed that it subserves the respiratory process, but some authors regard it as subserving an excretory purpose. Whether the fluid of the vacuole is forced out into the surrounding water as the vesicle closes has not been satisfactorily demonstrated, although there is strong evidence pointing to that conclusion.

Food is taken into the body of the Amœba through any part of the surface. A portion of the ectosarc extends around the prey, enclosing it along with some of the water, which then sinks down into the endosarc, where it forms a so-called food-ball; such food-balls may become quite numerous in a single animal. Ehrenberg, supposing them to be permanent stomachs, gave the name Polygastrica to those Protozoa in which he observed them. Any portion of the Amœba's body will serve the purpose of a temporary stomach, in which food may be digested and assimilated. The indigestible portions are ejected at any part of the surface, but usually at the posterior part, near the contractile vesicle. The food of the Amœba is usually of a vegetable nature. The delicate filamentous desmids seem to be a favorite food, and diatom remains are often found in the Amœba in great abundance. In assimilating the nutriment of a filamentous desmid, the Amœba passes along the filament, enveloping cell after cell, seemingly passing the plant directly through its body, absorbing the contents and rejecting the indigestible portions.

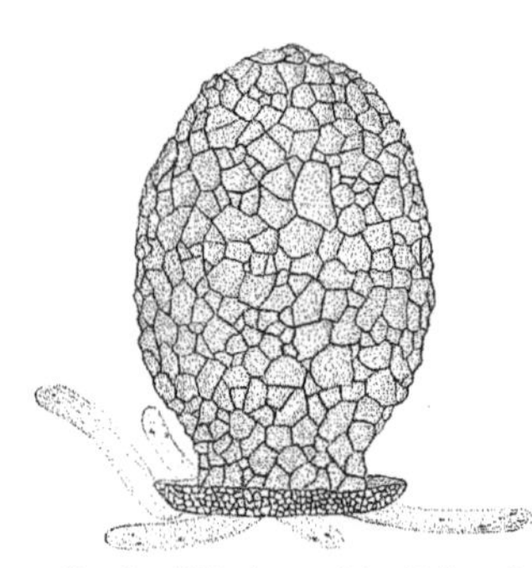
FIG. 3. — *Difflugia urceolata.* Enlarged.

The Amœbæ propagate by division and perhaps by a process of conjugation; at least an appearance of conjugation has been observed in a few cases. When the circumstances of life are unfavorable, the Amœbæ may become encysted, by which means they are able to withstand great changes of external conditions. When about to become encysted, the remains of food and particles of indigestible matter are rejected, and the animal assumes a spherical shape precisely like *Protomyxa*, Fig. 1, *a*, soon becoming surrounded by a more or less thick membrane composed of several layers. In this protected condition the Amœba may rest a long time, and then, by rupturing its envelope it may again come forth apparently unchanged. But in some cases a change takes place within the capsule which results in the formation of a large number of spherical germs or spores, each of which probably escapes and grows into a new form, as in the case of *Protomyxa*, already described.

A large number of the Lobose protoplasts are provided with shells, many of them of regular and beautiful form. Of these, *Difflugia*, Fig. 3, may be taken as a representative genus. The shell of this Rhizopod is spherical or oval, composed of grains of sand mingled with frustules of diatoms, spicules, etc., cemented together. The sarcode-body almost fills the shell, and is attached to it by protoplasmic threads passing to the fundus and sides. The shell is open at one end, where the blunt cylindrical pseudopodia are projected either for the prehension of food or as organs of locomotion.

In size it may vary from .036 mm. to .260 mm. in length. The nucleus and contractile vesicle are conspicuous in the posterior portion. In all the shelled forms the food is taken in at the mouth of the shell, and the débris is ejected at the base of the pseudopodia.

ORDER II.—RADIOLARIA.

When Prof. Huxley was engaged in studying the fauna of the sea on board H. M. S. "Rattlesnake," about thirty years ago (1851), he found floating upon the seas, whether tropical or extra-tropical, some peculiar gelatinous bodies to which he gave the name *Thalassicolla*, signifying sea-jelly. These were among the most common objects from the tow-net, and their extreme simplicity of structure made it very difficult to classify them in the animal kingdom. Imagine a colorless, transparent gelatinous mass, spherical, elliptical, or elongated in form or contracted like an hour-glass in one or more places, varying in size from a mere speck up to an inch in length, without contractility or power of motion, but floating passively upon the water: such is *Thalassicolla*, Fig. 4.

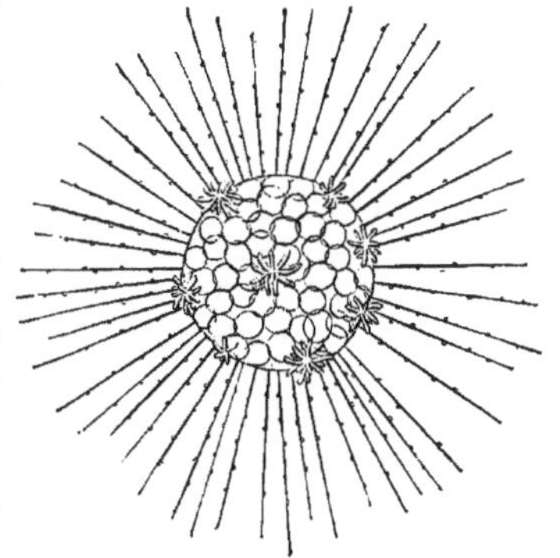

FIG. 4.—*Thalassicolla morum.* Greatly enlarged.

Two species were described by Prof. Huxley in his account of these organisms. One of them, *T. punctata*, is characterized by an appearance of dots scattered about near the internal surface of the thick, gelatinous crust which may surround either a single large cavity or a number of clear spaces closely aggregated. The appearance of dots is produced by nucleated cells, which are imbedded in, and held together by, the gelatinous crust. The cells are about $\frac{1}{200}$ or $\frac{1}{350}$ of an inch in diameter, and are covered with a thin membrane. Surrounding the cells or diffused through the connecting substance are minute bright-yellow corpuscles. The cells may also be enclosed within a framework of crystals or spicules resembling the spicules of a sponge.

T. nucleata is a spherical mass, characterized by a blackish central portion, around which is a zone of vacuoles or clear spaces with yellow cells and dark granules, which is in turn surrounded by the outer, clear, gelatinous substance. The dark central portion is a vesicle with granular contents and a firm, colored membrane. From the dark centre delicate branching fibrils radiate and anastomose through the zone of vacuoles, extending almost to the periphery of the sphere.

Thalassicolla may be regarded as a type of the Radiolarian structure. The Radiolaria are characterized by having a central nucleated portion surrounded by an outer peripheral mass, from which it is separated by a porous, more or less resisting membrane known as the capsule. Both the mass within the capsule and the sarcode without consist of very soft and contractile protoplasm, in which are imbedded colored globules, vacuoles, and perhaps other structures. Most Radiolaria have a skeletal framework of silicious spicules, or beautifully-designed structures, which may be found either within or without the capsule. The silicious framework of these minute organisms, when properly cleaned and prepared for exhibition, afford some of the most beautiful objects for examination with a microscope.

The Polycystina especially, which have an external skeleton of clear, glassy silica, are to be found in every collection of microscopic objects, and there are few specimens that attract more universal admiration for beauty and regularity of form.

Before describing some of the more important representatives of this group, a few words should be said concerning their general characteristics. The single animals or zooids vary in size from about $\frac{1}{600}$ to $\frac{1}{20}$ of an inch, or even more, in diameter. They are usually spherical, but they may be cylindrical, discoidal, or of other shapes. The sarcode within and without the capsule is continuous through the pores of the chitinous membrane which surrounds it. In *Thalassicolla* the capsule is very small compared to the size of the animal, but usually, especially in the solitary forms, the capsule is relatively very large, sometimes having only an exceedingly thin layer of extra capsular sarcode about it. The tendency of such simple forms of life is to live in colonies like *Thalassicolla punctata*, in which the capsules and the investing sarcode have already been described as cells imbedded in the gelatinous connecting mass. The capsules vary in size from 2 mm. down to .025 mm.

The sarcode contains vesicles or alveoli, which may be found both within and without the capsule; but no regularly contracting vesicle, such as is found in the Heliozoa, has been observed.

Within the capsule are found peculiar structures which have been termed nuclei, and which are supposed to be true nuclei of the capsule. These are of two kinds,—simple and complex. The simple nuclei measure from .008 mm. to .015 mm. in diameter. They are perfectly homogeneous in appearance, and may exist in great numbers in a single capsule, almost filling it in fact, or they may be few, or even quite absent when a complex nucleus is present. They have no investing membrane. The complex nucleus is a multi-globular vesicle with a membranous covering similar to that of the capsule itself, but more delicate. It is possible that the simple nuclei are developed within it. The complex nucleus is also designated as the "nuclear vesicle." It is characteristic of certain forms of Radiolaria.

The sarcode of the capsule may be colorless, or it may be distinctly colored, red, brown, and yellow being the usual colors. Examination with high powers of the microscope shows the coloring matter in the form of minute vesicles. There are also found in the sarcode globules of oil or fatty matters, and sometimes concretions, crystals, and other structures that may be nothing but remains of food. The external sarcode is not protected by any definite enveloping membrane, but a clear, gelatinous, more or less firm layer of the sarcode may be observed to form the outer boundary of the sphere, as already described in *Thalassicolla.*

The sarcode of the central capsule is continuous with the external sarcode through the pores of the dividing membrane. The extra capsular sarcode is usually frothy in appearance owing to the presence of clear spaces,—vacuoles or alveoli. These alveoli usually increase in size from without inwards, being largest and most numerous near the capsule. The outer alveoli have been observed to disappear at times and to form again.

The pseudopodia of the Radiolaria resemble those of the Heliozoa, being more or less persistent and not very flexible. In some species they branch and anastomose slightly. They originate from the deepest part of the external sarcode, pass between the alveoli and through the gelatinous investment into the surrounding water. They may be retracted and extended.

The "yellow cells" which are almost invariably found in the Radiolaria, either within or without the capsule, have been the subject of much speculation. It is not yet known what their functions are, and it is even doubtful if they are not parasitic plants, taking their nourishment from the body of the Radiolarian in which they live. After the death of the animal the yellow cells have been observed to grow and multiply.

With the exception of a very few species of *Thalassicolla*, *Thalassolampe*, *Myxobrachia*, and *Collozoum*, all the Radiolaria are provided with some form of silicious framework. In its simplest form this consists of isolated spicules, as in *Sphærozoum*, Fig. 9. From the simple spicular forms we may pass to those having spines radiating from a common centre to the surface of the sphere, or beyond, with lateral processes like *Xiphacantha*, Fig. 7.

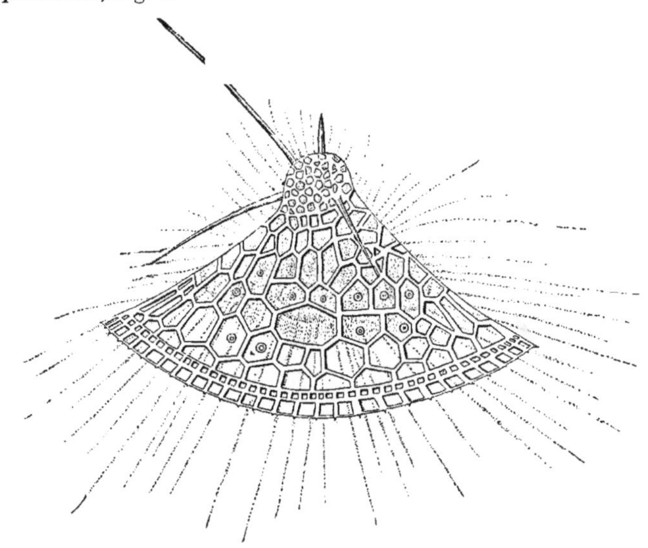

FIG. 5.—*Eucecryphialus gegenbauri*, greatly enlarged.

In certain species the skeleton is formed of hollow spines, through which the sarcode extends and issues from the ends. In all cases the spines are covered with a thin layer of granular sarcode, which can be observed constantly flowing up and down the spines, doubtless carrying the food that may be collected, down into the body.

As the lateral processes mentioned above become more largely developed, a continuous circumferential skeleton is formed, which encloses the whole organism, as in *Actinomma*, in which there are sometimes three or more concentric shells. Among the Polycystina there is a great variety of form manifested in the external skeleton. *Podocyrtis* (Fig. 8) is one of the most common forms of this group.

The food of the Radiolaria consists of minute algæ, diatoms, infusoria, and other organisms found on the surface of the sea.

Not much is known concerning the methods of multiplication among the Radiolaria. It may be accepted as an established fact that the contents of the capsules may divide and form young capsules, which are at first without any membranous covering. The young capsules make their way out, swim about freely as "zoospores," which, in *Col-*

losphæra, are oval, about .008 mm. in length, and have at least one cilium. The subsequent history of the zoospores has not been made out. It is probable that colonies like *Collosphæra* are formed by division of this kind.

In this large and exceedingly interesting order of Rhizopods, there are nearly a thousand species, about one-half of them being fossil forms. This shows the wonderful variety of form which such organisms may present without departing from the simple plan of organization which characterizes them. They may be classified according to the forms of their skeleton, into families and sub-families, in which one general plan of structure will be characteristic of each division, as Dr. Wallich has shown. Although such a classification may be convenient, it throws but little light upon the physiological or morphological relations of the different forms. In the present state of our knowl-

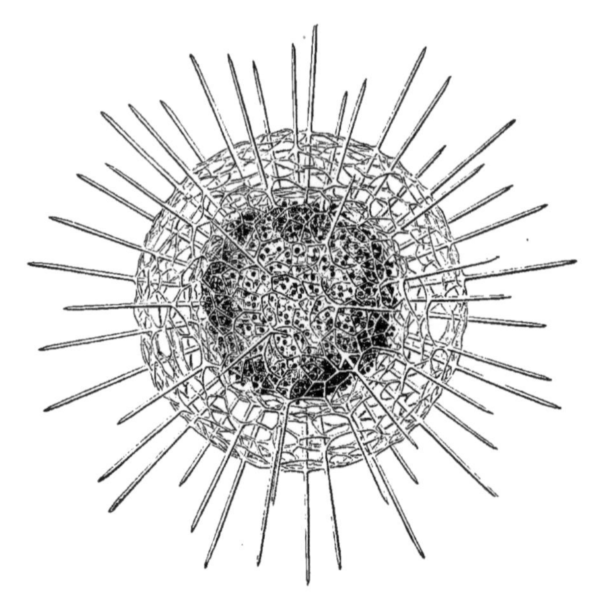

FIG. 6. — *Haliomma polyacanthum*, magnified 200 times.

edge of the Radiolaria no fully satisfactory classification is possible. Perhaps the best yet proposed is that of Prof. Mivart, which is a greatly modified form of Häeckel's comprehensive but confusing plan. Prof. Mivart arranges all the Radiolaria under seven divisions, which may be briefly characterized as follows: —

1. DISCIDA. — Discoidal forms with skeletons partly intra-capsular, generally forming an external perforated shell with an internal partition, making a series of connecting, concentric, or spirally arranged chambers; no nuclear vesicle.

In this group there are five sub-divisions, but the most common form of all is the *Astromma*, in which the combination of radial and circumferential parts is quite striking, both for beauty and for the great variations in form manifested by the different species.

2. Flagellifera.—This group receives its name from the flagellum which characterizes the species belonging to it.

3. Entosphærida.—In this group the organisms are provided with an intra-capsular, spheroidal shell, not traversed by radii, in this respect differing from the Discida. They have no nuclear vesicle. The typical genus is *Haliomma*, of which there are many forms. *Haliomma polyacanthum* is represented in Fig. 6.

4. Acanthometrida.—The members of this division are characterized particularly by a well-developed radial skeleton, the radii meeting in the centre of the capsule. There is no enveloping shell, but lateral processes sometimes project from the spines, as in the beautiful *Xiphacantha* found by the "Challenger" expedition, represented in Fig. 7.

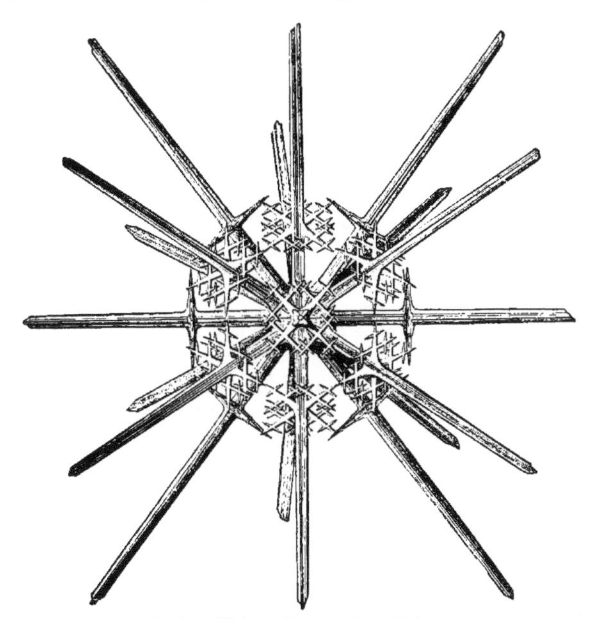

Fig. 7.—*Xiphacantha*, magnified 100 times.

5. Polycystina.—The Polycystina are by far the most numerous in all fossil deposits of Radiolaria. They are very simple forms with skeleton external, more or less compact or continuous, without a nuclear vesicle. The shell may be a simple sphere, or two or three concentric spheres connected by radii, or with external radial outgrowths extending to a length of several times the diameter of the shell. The most numerous forms, however, belong to the genera *Podocyrtis* and *Eucyrtidium*, the former represented in Fig. 8. The beautiful *Eucecryphialus* (Fig. 5) also belongs to this group. In these forms the shell opens at one end, and growth being mainly in one direction said to be unipolar.

6. Collozoa.—To this group belong a number of soft, gelatinous forms which are frequently aggregated in colonies, and are therefore designated as compound Radiolaria. The animals may be either single or in families. When single the skeleton consists of circumferential spicules, isolated from each other. When compound, there

may be either spicules or a spheroidal, perforated shell. Under this division are classed the very common forms *Sphærozoum*, Fig. 9, and *Collosphæra*, the former being either naked or having spicules, while the latter has a shell.

7. VESICULATA.—To this group belong the curious jelly-like forms already mentioned, described by Huxley as *Thalassicolla*, Fig. 4. There is no definite skeleton, but some of the species have spicules more or less closely approximated. The vesiculata are distinguished by the presence of a nuclear vesicle, which is usually multiglobular. Formerly the Thalassicollida were classed with the Collozoa, but the nuclear vesicle is not found in the latter, and there is no external shell or spicular layer in the former such as are found in *Sphærozoum* and *Collosphæra*.

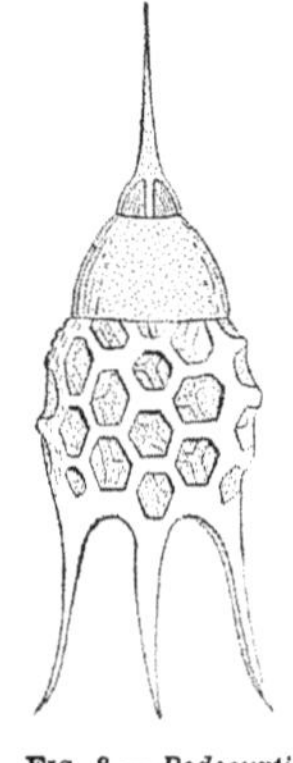
FIG. 8.—*Podocyrtis schomburgki*, greatly enlarged.

When we consider the wonderful symmetry, beauty, and variety of form revealed by a study of the hard, siliceous skeletons of the Polycystina, Acanthometrida, and other families of the order, we may well inquire how it is possible for such simple creatures to construct such perfect forms. Mivart suggests that they may be produced by "a kind of organic crystallization—the expression of some as yet unknown law of animal organization here acting untrammelled by adaptive modifications or by those needs which seem to be so readily responded to by the wonderful plasticity of the animal world."

Representatives of the great class Radiolaria are found in all seas, but they are by far the most abundant in tropical waters. The most common forms of all belong to the Acanthometrida and Polycystina. Their remains have formed immense beds of rock, mostly during the Tertiary age, but they have also been found in the chalk and in the Trias. They are found in the diatomaceous rock of Richmond and Petersburg, Va., also in Maryland, and in Bermuda; but by far the most extensive deposits are in the Nicobar Islands and the Barbadoes. In the Nicobar Islands deposit they form strata eleven hundred and two thousand feet in thickness, in which a hundred species have been identified. In the Barbadoes the rock is not quite so thick, but about three hundred species have been found there, most of which are Polycystine forms.

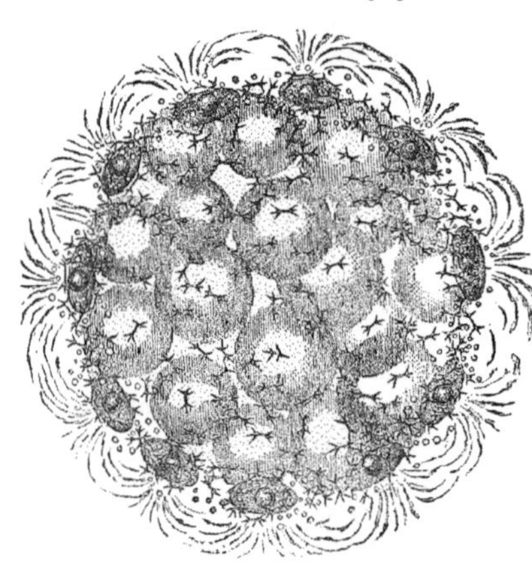
FIG. 9.—*Sphærozoum ovodimare*, greatly enlarged.

A number of minute but very beautiful organisms were obtained by Mr. Murray during the "Challenger" expedition by using the tow-net, which undoubtedly deserve a place among the Protozoa, but they have not yet been classified. They have been named Challengerida, and seem to be closely related to the Radiolaria.

ORDER III. — HELIOZOA.

The Heliozoa, or sun-animalcules, are very beautiful Rhizopods, inhabiting fresh water. Most of them are spherical, floating forms, but a few are attached by long pedestals or stipes. The pseudopodia are in the form of delicate, tapering rays, extending outward in all directions from the centre, often exceeding the diameter of the body in length. They are flexible, more or less contractile, and sometimes reveal a slow circulation of granules along their length. The sarcode is not distinctly differentiated into endosarc and ectosarc, but in one interesting form, *Actinosphærium*, the outer sarcode is a frothy, vacuolated mass of considerable thickness.

The most common of the Heliozoa is the *Actinosphrys sol.* It is found in pools of standing water almost everywhere, among the floating plants, appearing under a low-power of the microscope as a colorless, spherical body, varying in size from .04 mm. to .12 mm. in diameter, with innumerable delicate, bristling spines three or four times the diameter of the body in length. The sarcode is full of vacuoles, which give it a frothy appearance. Watching the minute sphere a few moments, there will probably be seen somewhere along the periphery a slowly distending vesicle, which reaches a certain size, and then suddenly collapses. This is the contractile vesicle. The first description of this curious little creature seems to have been given by a French naturalist, who referred to it, if we may translate the bad French in which it is written, with some discretion — as "a fish, the most extraordinary that one could see."

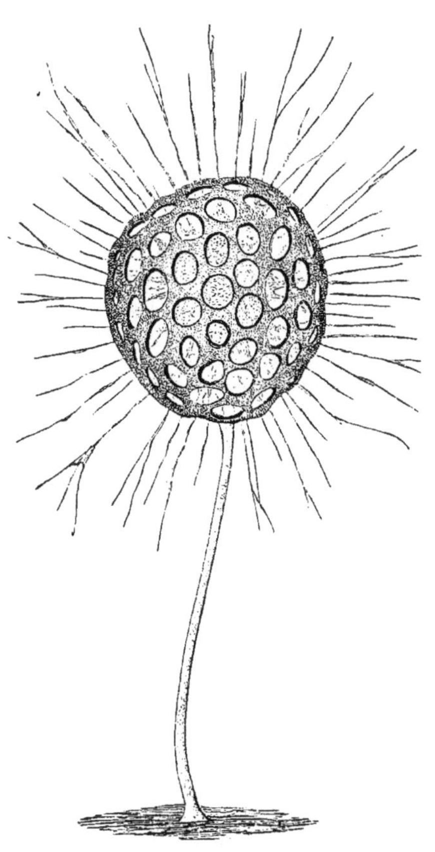

FIG. 10. — *Clathrulina elegans*, enlarged 350 times.

The food of *Actinophrys* consists of minute infusoria, diatoms, and other unicellular algæ, which frequently can be seen within the body as green balls. The pseudopodal rays are used as organs of locomotion, and for the prehension of food. If an active infusorian comes in contact with the spines it seems to be paralyzed. If the prey be very minute it will be seen to glide along the rays very gradually until it reaches the body, when a portion of the sarcode is projected to envelop it, and draw it into the interior, where it undergoes digestion and assimilation. If the prey be larger, several rays may bend toward it and together secure and draw it down to the body.

Actinophrys propagates by simple division and by the fusing together of two or more specimens into a single mass, which then reproduces new forms by fission. The conjugation of two individuals is quite a common occurrence, but the Hon. J. D. Cox has observed as many as nine individuals joined in the process. The same observer also describes another method of propagation in which the parent form passes into an opalescent condition, after which it undergoes segmentation into a brood of young.

A much larger heliozoon, greatly resembling *Actinophrys*, is *Actinosphærium.* It is especially distinguished from the former by the frothy layer of ectosarc which surrounds the central sarcode, and by the complex structure of the pseudopodal rays which, under high magnification, seem to have a hard axis-cylinder, probably only a core of denser protoplasm.

Perhaps one of the most beautiful forms of this order is *Clathrulina*, represented in Fig. 10. While young the capsules are colorless and clear, but with age they become yellow. The sarcode does not usually fill the lattice-capsule, but is collected in a ball within it. It is a very beautiful species; often found in great abundance in ponds and ditches.

The Heliozoa seem to be quite closely related to the Radiolaria, but the exact relations of the two groups is not yet known. It has been suggested that the Heliozoa are embryonic forms of the more highly-developed group; but until the structure of both groups is more fully elucidated it seems useless to speculate much about this question. Several observers have noticed within *Actinophrys* a peculiar nuclear sphere which greatly resembles the central vesicle of certain Radiolaria.

ORDER IV. — RETICULARIA.

SUB-ORDER I. — PROTOPLASTA.

This sub-order includes a considerable number of fresh-water Rhizopods with very soft sarcode bodies and delicate, branching, thread-like pseudopodia. The endosarc and ectosarc are even less clearly differentiated than in the lobose forms. The nucleus is usually large, and several contractile vesicles are to be seen in the endosarc.

A characteristic Rhizopod of this sub-order is *Grómia oviformis*, represented in Fig. 11. The shell is thin, chitinous, colorless or yellowish, measuring about .115 mm. in length. A high power of the microscope shows an incessant streaming of granules along the branching, anastomosing shreds of sarcode, the granules moving outward on one side and back on the other side of each filament. The sarcode extensions of *Gromia* anastomose more freely than is usual among the Protoplasta Filosa, resembling more closely the Foraminifera in this respect, and the contractile vesicle is near the mouth of the shell. In fact, Prof. Joseph Leidy, in his monograph on the "Fresh-water Rhizopods of North America," has placed *Gromia* among the Foraminifera. The filose protoplasts seem to be in nowise different from the Foraminifera, except that the shells of the latter are usually calcareous, and the pseudopodia manifest a greater tendency to anastomose, and are more granular.

The shells of the filose protoplasts are usually composed of a clear, chitinous substance, sometimes colorless and transparent, sometimes distinctly colored yellowish or brown, while still others are covered with grains of sand.

A very frequently occurring form is *Pseudodifflugia*. In this the shell is chitinous, with sand-grains in some wise incorporated with it. It resembles *Difflugia*, Fig. 3, in every respect except as regards the character of the pseudopodia. In some of the genera the shells are beautifully marked, and the neck is often curved so that the body lies on the side as the animal crawls along.

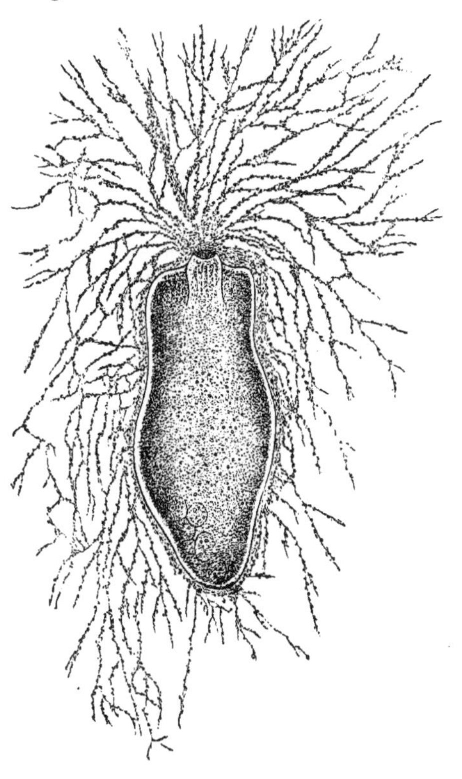

FIG. 11. — *Gromia oviformis*, enlarged 600 times.

SUB-ORDER II. — FORAMINIFERA.

The Foraminifera embraces an almost innumerable variety of marine Rhizopods. The reticulate, anastomosing nature of the pseudopodia is most strikingly manifest in all the Foraminifera, but the examination of the internal sarcode is very difficult, owing to the thickness and opacity of the shells. For this reason it was long supposed that the Foraminifera were destitute of a nucleus, but recent investigations by Hertwig and Lesser, Carpenter and others, have revealed nuclei in several forms, and they are doubtless present in all of them. It is said that dahlia violet will stain the nuclei while the animal lives, and if this is true in all cases, it will prove a valuable reagent in further investigations of those organisms.

The Foraminifera are the only Rhizopods that have shells of many chambers, and of complex structure. The different forms of the shells can best be understood by observing how they are derived from a single chamber by the budding off of successive portions of the sarcode-body, each of which then secretes a shelly covering. If the budding always takes place in the same direction, an elongated form composed of several chambers in a straight line is produced, as in *Lagena*. If the tendency of growth is to produce a spiral, it results in the beautiful *Cornuspira*, which greatly resembles the mollusc *Planorbis*, or, if the budding takes place in still another way, the more complex forms of *Miliola* are produced, which are only spirals greatly elongated in one direction. Instead of forming successive single chambers at the ends of old ones; the growing spiral may spread out wide and flat, thus forming the beautiful *Polystomella*, and *Peneropolis*, Fig. 12, common in all tropical seas. In the Bermuda sands the most frequently occurring genera are *Peneropolis*, Fig. 12, *Orbiculina*, and *Orbitolites*, Fig. 15. These, with other light débris, are occasionally washed out of the heavier matters of the shore by the action of the waves, and left in great abundance in long, white streaks as the waves recede.

FIG. 12. — *Peneropolis*, enlarged.

Among the spiral shells there are two types, distinguished as nautiloid and turbinoid. When the spiral forms in one plain, as in *Polystomella*, we have a nautiloid spiral; when it winds obliquely around a vertical axis, forming a spiral like the snail or periwinkle, it is turbinoid. The beautiful *Rotalia*, Fig. 13, is formed upon the latter plan. In most of the rotaline forms, all the chambers of the whorl are visible from one side, but among the spirals of the nautiloid type the later chambers often more or less envelop the older ones, so that unless one knows the structure of the shell it cannot be recognized by a cursory or superficial examination. For example, in the very frequently occurring *Nonionina*, the older chambers are quite invisible, being entirely enveloped by the later ones, and in order to learn how the shell began to form, a section would have to be made through it showing all the chambers in one plane.

Among the turbinoid spirals, there are several varieties of structure, the relations of which are not easily seen until careful examination of the internal structure reveals them. Thus, *Textularia*, Fig. 14, belongs to the division, but at first glance it scarcely seems to bear any relation to *Nonionina*.

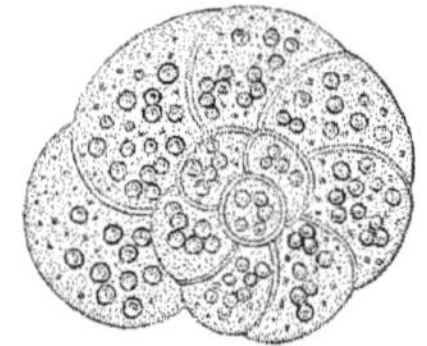

FIG. 13. — *Rotalia*, enlarged.

On close examination it will be seen that the successive chambers are in two rows, and each chamber communicates with the chamber above and the one below it on the other row, never opening into a chamber of its own row.

In some species the nautiloid spire is characteristic only of the early period of growth, for after a few turns, instead of budding from the end, thus continuing the spiral, all the outer chambers put forth radial buds, which form successive concentric rings. This mode of growth is well illustrated in *Orbitolites*, which is represented in Fig. 15, part of the surface being removed to show the internal structure. It will be seen that the internal chambers are spirally arranged, while the others are arranged on the cyclical or radial plan of growth.

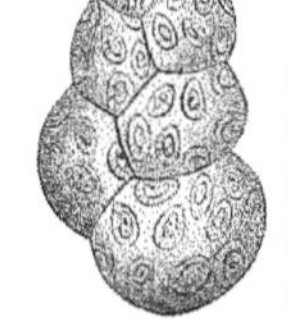

Dr. William B. Carpenter, whose valuable monograph on the Foraminifera has thrown much light upon the structure and relationship of these organisms, has shown the great importance of a careful study of the shell-structure as a basis of classification. He has distinguished two kinds of shell among the Foraminifera, which he has designated, respectively, the porcellanous and the hyaline, or vitreous. These differences of shell-structure correspond with physiological differences in the organisms inhabiting the shells, and afford a basis for a division of the class into two great sections. In both these sections will be found species which have striking resemblances in form, which could not be generically separated except by a recognition of the differences in the structure of the shell and their physiological significance.

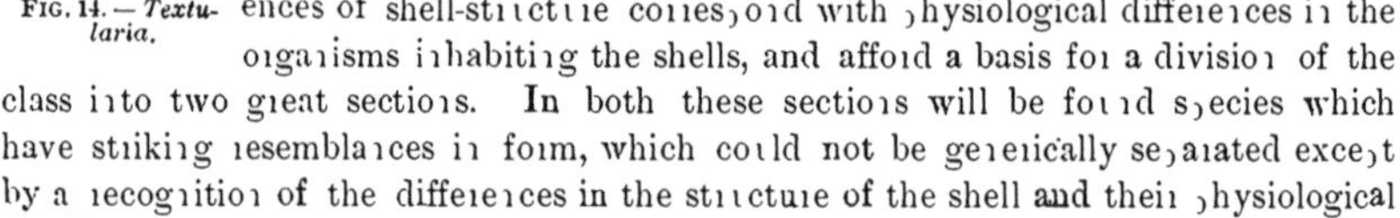

FIG. 14. — *Textularia*.

The terms porcellanous and vitreous have been adopted owing to the appearance of the shells as seen under the microscope. The former is applied to shells of a white, opaque, often shiny appearance, which in thin, transparent sections or laminæ appear, by transmitted light, of a brown or amber color. No structure can be observed in shells of this kind. They are never perforated, although they are sometimes marked upon the surface with pits, or inequalities, giving an appearance of foramina.

The vitreous or hyaline shell-structure is far more complex than the porcellanous. It is transparent, usually colorless, sometimes deeply colored, and more or less closely perforated either with large or small distinct foramina, or minute tubuli passing directly through the shell-substance. In *Rotalia*, Fig. 13, the foramina are distinct, and afford passages for the sarcode, which covers the outside of the shell, and the pseudopodia extending in all directions from it. The minutely tubular structure can only be detected in thin sections with high powers of the microscope, when it imparts a peculiar appearance to the shell, characteristic of finely tubular structures.

Between the shells with large foramina and with minutely tubular structure, there is a continuous gradation, which indicates that both foramina and tubuli subserve the same purpose, — affording channels for the passage of the sarcode. Comparing the shells of the porcellanous and vitreous forms, it will be seen that while the pseudopodia of the animals occupying the former all spring from the terminal or outer chambers alone, so that the nourishment for the sarcode of the inner chambers must pass in through those that intervene, in the vitreous forms the sarcode of each chamber is in direct communication with the outer world through either the foramina or the minute tubuli of the shell. In accordance with this difference in the structure of the shell-substance, it may be also observed that the stolons of sarcode connecting successive chambers of the porcellanous-shelled species are much larger than those in the vitreous-shelled forms.

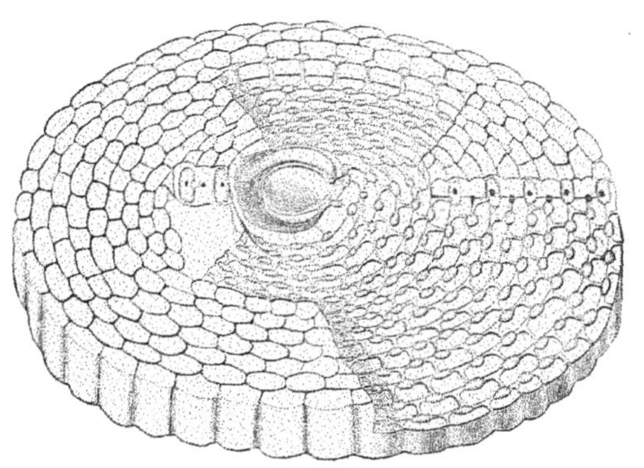

Fig. 15. — *Orbitolites*, enlarged. Diagram to show external appearance and internal structure.

These facts may be best illustrated by comparing two of the most highly-developed forms of the two types of shell-structure. For this purpose we will select *Orbitolites* of the porcellanous, and *Nummulina* among the tubular-shelled forms.

The structure of *Orbitolites* can be understood by a glance at Fig. 15. Disks such as are here represented sometimes attain the size of a silver quarter-dollar in diameter. It will be seen that single pores unite successive chambers, and finally the sarcode of the outer chambers communicates with the surrounding medium by pseudopodia projected through the marginal pores shown in the figure.

In *Nummulina*, a form that has been so abundant in the past as to have lent its name to the nummulitic limestone, the tubes have a different arrangement and are very minute; there is, besides the tubular structure already described, a system of inosculating canals penetrating the septa, which are filled with sarcode during the life of the animal. In all the vitreous forms, each chamber has its own shelly investment, so that the partitions between the chambers are double. Between these walls there is frequently a considerable deposit of calcareous substance, which is known as the intermediate skeleton. Through the intermediate skeleton runs the system of canals, which is beautifully shown in *Eozoon*, soon to be described; the canal-system resembling minute branching shrubs. A single species of *Nummulina*, Fig. 16, has been described from Florida.

FIG. 16. — *Nummulina wilcoxi*, natural size, and enlarged.

Besides the two varieties of shell-structure above described, there is another kind of shell or test very frequently occurring among deep-water species. This is the arenaceous type, in which the shell consists of cemented grains of sand, or of sand and spicules together.

The nature of the cement which holds together the sand-grains of the arenaceous types is not known; sometimes the grains are only loosely united, so that the test is more or less flexible, as in *Astrorhiza*, Fig. 17, a form which is found in Vineyard Sound at depths of only twelve fathoms, but also reaching down to over five hundred fathoms. Some of these have the outside test smoothly plastered by a layer of very fine particles of mud, although composed of irregular large and small grains of sand. No definite aperture, or mouth, has been observed in *Astrorhiza*, and the sarcode finds its way through the test between the loosely cemented grains composing it. In other forms the grains are very closely cemented, so that some tests will resist the action of warm nitric acid, proving that the cement is neither calcareous or ferruginous. In some cases the sand-grains seem to have a chitinous basis in which they are imbedded. The resemblance between the arenaceous Foraminifera and the porcellanous and vitreous species is striking. Take, for example, *Halophragmium*, and compare it with *Globigerina*, Fig. 18.

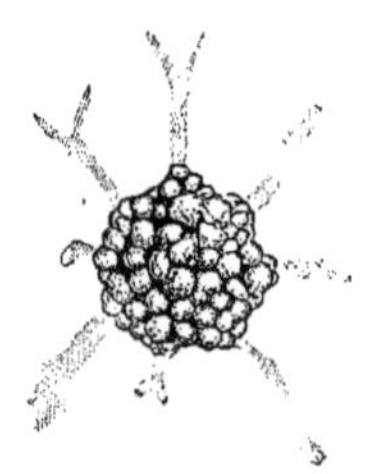

FIG. 17. — *Astrorhiza*, enlarged two diameters.

Indeed, it is true that if we consider only the external forms, we can find in the three divisions of porcellanous, vitreous, and arenaceous forms many species that are so closely related as to be indistinguishable by any specific characteristics. Thus, *Cornuspira* among the porcellanous is the counter-part of *Spirillina* among the vitreous forms; and this is distinguishable in form from *Ammodiscus* among those with sandy tests.

While some of the tests of the arenaceous group are probably imperforate, others are, without doubt, more or less porous, so that the distinction already made between hyaline and porcellanous forms must also hold good as concerns these. Indeed, certain arenaceous forms have no definite mouth, and the sarcode must find its way through pores in the test.

The deep-sea investigations that have been carried on of late years have brought to light many new forms belonging to genera which were supposed to be very well known. Thus, the shell of *Globigerina*, Fig. 18, has been understood, conforming to the description of Dr. Carpenter, to consist of a series of hyaline, perforated, spheroidal chambers arranged in a spiral about an axis, each opening into a central space in such a manner that all the apertures are visible when looking into the common vestibule. But in the light of more recent investigations, Mr. Brady has found it desirable to enlarge the scope of the family to include many new species. He has, therefore, divided the

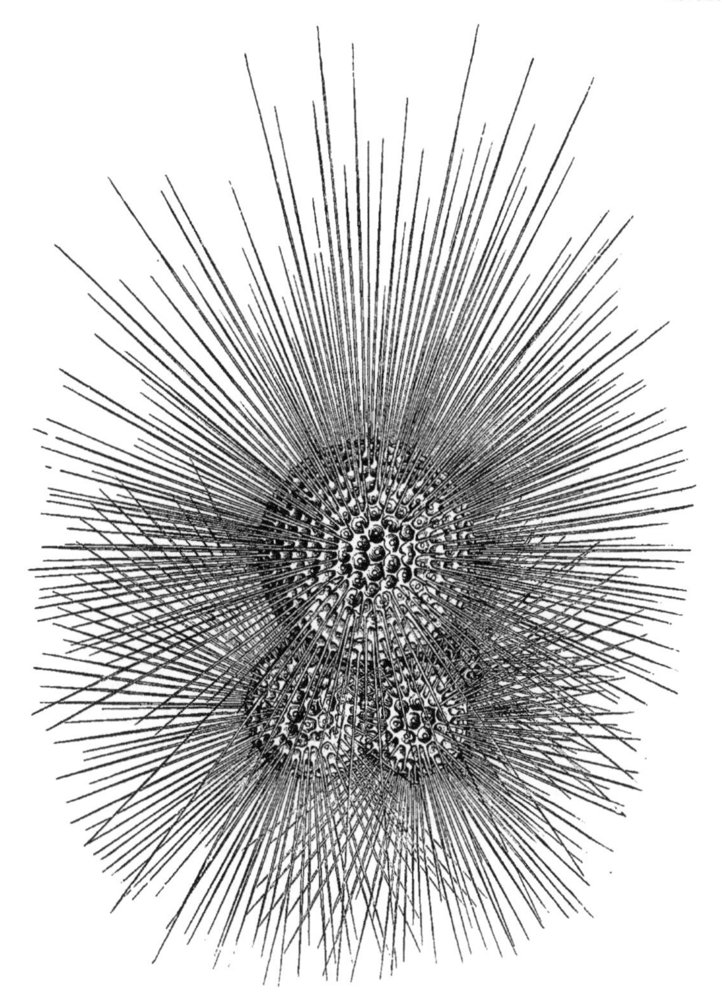

FIG. 18. — *Globigerina bulloides*, greatly enlarged.

Globigerinæ into three groups, according to the position and appearance of the apertures, as follows: —

1. Forms with an excavated cavity (umbilical vestibule) into which the orifices of all the segments open, such as *Globigerina bulloides*, Fig. 18.

2. Those with only one external orifice, situated on the face of the terminal segment, at its point of junction with the previous convolution, as in *Globigerina inflata*.

3. Forms in which the inferior aperture is single and relatively small, but supplemented by conspicuous orifices on the superior surface of the shell, as in *Globigerina rubra*.

Besides these, there are forms represented under the generic name *Orbulina*, which seem properly to belong to the *Globigerinæ* as sub-generic forms. *Orbulina* has a spherical shell, usually without a definite mouth, but provided with two sets of perforations differing in size,—one series numerous and minute, the other larger and less numerous.

Closely allied to the Globigerina, if not properly belonging to that family, is the beautiful *Hastigerina murrayana*, found by the "Challenger" expedition. This organism was found long ago by D'Orbigny (1839), and described by him as *Nonionina pelagica*, and was one of the first Foraminifera taken at the surface of the sea.

The Foraminifera inhabit the sea, and their remains are gradually forming rocky strata on the ocean-bed, in all respects like the chalk of the cretaceous period. For a long time it was supposed that the Foraminifera found at the bottom of the sea passed their entire life there. Prof. Wyville Thomson held firmly to that opinion until the results of collections with the tow-net at the surface conclusively proved that many of them live near the surface. It is now known that a few species of *Globigerina* inhabit the superficial waters, but a far greater number pass their life at the bottom.

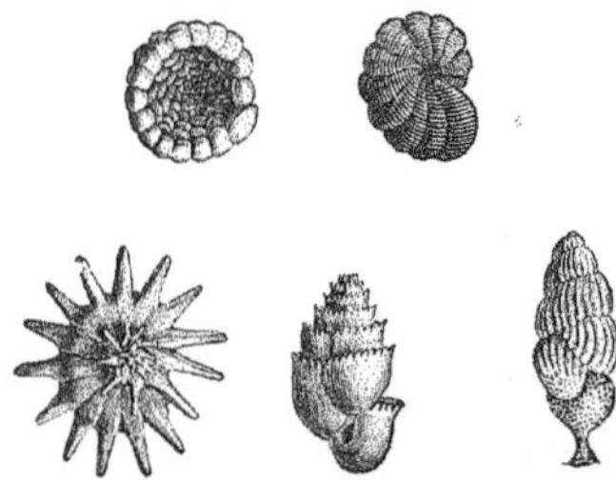
Fig. 19.—Helixostegine forms of Foraminifera.

The pelagic forms of *Globigerina* are usually, but not always, spinous. The long, delicate spines are somewhat flexible, and clothed with granular streaming sarcode; and for some distance from the shell the frothy sarcode fills the spaces between them. The spines are so delicate that a mere touch will break them off, and spinous shells are never seen in material brought up in the dredge. Sometimes the spines are very long; in *Hastigerina murrayi* the spines are fifteen times the diameter of the shell, and the frothy, alveolar sarcode extends outward between the spines to a distance equal to twice the diameter of the shell.

The Globigerina are so abundant in some places, and their remains constitute so large a proportion of the shiny calcareous ooze covering a great part of the sea-bottom, that the ooze has long been designated "Globigerina ooze." The Globigerina ooze consists of the remains of *Globigerina* and *Orbulina* in great abundance, with a smaller proportion of the genera *Pullenia* and *Sphærodina*, with occasional specimens of *Hastigerina*, together with remains of radiolarians, diatoms, and some curious structures known as rhabdoliths and coccoliths, the nature of which is not yet understood.

As regards the distribution of the remains of protozoic life over the ocean-floor, it appears that the Globigerina ooze extends from four hundred down to about two thousand fathoms. Beyond this limit there seems to be a gradual disintegration and solution of the calcareous substance of the shells, resulting first in a gray ooze down to two thousand three hundred fathoms, containing no perfect shells, but some calcareous matter effervescing with acids, finally changing as we go still deeper to an impalpable red feldspathic mud, or "red clay," as it has been termed, which covers vast areas. The red clay is supposed to be derived partly from the disintegration of the shell-matter of the gray ooze and the solution of the calcareous portions, and partly from the mineral matter held in suspension in the sea-water, which slowly sinks to the undisturbed depths.

Nevertheless, the deepest ocean-floor is not always devoid of organic remains. The deepest sounding by the "Challenger" was made in the Pacific Ocean on the 23d of March, 1875, and showed a depth of four thousand five hundred and seventy-five fathoms. The bottom was covered with what resembled the ordinary red clay to the eye, but it was gritty, and contained such a large proportion of radiolarian remains that it received the name of Radiolarian ooze. It had previously been supposed that even these silicious remains, together with the frustules of diatoms, which are more or less abundant in the Globigerina ooze, disappeared with the calcareous shells, in some manner not fully understood. The occurrence of the deep Radiolarian ooze, however, has shown that there is no destruction of silicious shells, and their accumulation in so much greater abundance there than in the Globigerina ooze was accounted for by Prof. Wyville Thomson on the supposition, based upon the results of collections at different depths down to one thousand fathoms with the tow-net, — that Radiolaria lived at all depths, and therefore, where the water was deepest the accumulations of their skeletons would be the most rapid. Later investigations by Mr. Agassiz, using an ingenious apparatus devised by Lieut.-Commander Sigsbee, U. S. N., have shown that there is no life between a narrow zone where the surface animals are found and the habitat of those living on or very near the bottom.

The deepest cast that has ever been made was made from the U. S. Coast Survey Steamer "Blake" in January, 1883, in latitude 19° 39′ 10″ N., and longitude 66° 26′ 05″ W., between the Bermudas and the Bahamas, about one hundred miles N. W. of St. Thomas. The depth there found was four thousand five hundred and sixty-one fathoms; the temperature of the deepest water was 36° F. Another cast in latitude 19° 29′ 30″ N., longitude 66° 11′ 45″, showed a depth of four thousand two hundred and twenty-three fathoms. At these great depths, — more than five statute miles beneath the surface, a depth equal to the height of the highest mountains in the world — the bottom is covered with a very fine brown ooze, containing a few Diatoms and sponge spicules.

In the North Pacific, at depths of three thousand and four thousand fathoms, are found tests of *Trochammina* (*Ammodiscus*) *incerta*, one of the arenaceous forms. At great depths are also found species of *Miliola*, usually more or less incrusted with grains of sand, while in some cases the shell consists not of lime, but of clear, homogeneous silica which will resist the action of acids like the frustules of diatoms.

Some of the genera of Foraminifera have had a great range in geologic time. In the lower Silurian rocks of Russia — in the so-called Ungulite grit — Ehrenberg found green sand casts of genera now living, *Textularia*, *Gattulina*, and *Rotalia*. The oldest form of life of which the rocky strata furnish any remains is possibly the *Eozoon Canadense*, found preserved in greatest perfection in the Laurentian rocks of Canada. Concerning the nature of *Eozoon* — the dawn-animal — there has been much controversy. On the one hand it is claimed with great probability that the peculiar structures found in the rocks are of purely mineral origin. But Dr. J. W. Dawson and Dr. William B. Carpenter, who have studied this subject with great care, have declared that the structure of *Eozoon* corresponds in every particular with that of certain Foraminifera of the vitreous type. The successive chambers connected by passages, the intermediate skeleton with its complex system of inosculating canals, the minutely tubular "nummuline layer," have all been claimed to have been found and studied with great care. The *Eozoon* occurs in a serpentine limestone, in the form of irregular masses of varying size up to several inches in diameter. In appearance to the naked eye it consists of alternate bands of green serpentine — a silicate of magnesia — and limestone, the former filling the cavities originally occupied by the sarcode, and even the most minute tubuli of the nummuline layer; while the calcareous basis of the original skeleton remains unchanged.

The simplest, and on the whole the most satisfactory, method of studying the *Eozoon* is to cut tolerably thin slices of the rock and place them in a very dilute acid until the calcareous portion is dissolved. There then remains a perfect cast of the chambers of the shell, counterparts of the original sarcode-body, even to the minute tubules of the nummuline layer. We have given here the gist of the account of Drs. Dawson and Carpenter, but the question of the animal nature of *Eozoon* is far from settled; possibly it never will be.

In South Carolina there are immense beds of marl and limestone containing Forminifera in great abundance. Prof. J. W. Bailey, in the year 1845, examined some borings from a well driven through these deposits at Charleston, and also some fragments from an outcrop on Cooper River, about thirty-five miles above that place. From the borings it was observed that from one hundred and ten to more than three hundred feet in depth the Polythalmia were abundant, very perfectly preserved, and many of them large enough to be easily seen with a pocket-lens. Concerning these tertiary deposits, Prof. Bailey remarked that they were filled with more numerous and more perfect specimens of these beautiful forms than he had ever seen in chalk or marl from any other locality. Similar marls are also found in Virginia, on Pamunkey River, belonging to the eocene; and in the miocene rocks of Petersburg, Foraminifera are also found.

The foraminiferal rock which underlies so large a part of South Carolina is still in process of formation along the coast. The mud from Charleston harbor abounds in shells of Foraminifera, and the remains of diatoms.

Fossil Foraminifera are found in many other places in this country. They exist in New Jersey, Alabama, at various points on the Missouri and Mississippi rivers, in Tennessee, Arkansas, North Carolina, Florida, and elsewhere. The marls of certain localities along the upper Missouri and Mississippi rivers are very rich in Foraminifera. The latter deposit is popularly known as "prairie chalk," and the forms are different from those found on the Missouri.

In the green sands of Fort Washington, Va., Prof. Bailey made the remarkable discovery that these minute and perishable organisms could be entirely destroyed by chemical changes, yet leave indestructible memorials of their existence in the form of mineral casts. Ehrenberg had previously observed that the lime of the shells could be gradually dissolved and replaced by silica. In flints such a replacement is not unusual, and remains of shells thus mineralized can be obtained by treating the rock with acid, which leaves the silicified shells intact.

But in the green sands of the chalk formation in New Jersey, Virginia, and elsewhere, the shells have become filled with a greenish mineral, glauconite, — a silicate of iron and potash of varying composition — which has followed every contour of the shell, and penetrated even the minute pores and tubuli so perfectly that the genus and even the species of the Foraminifera can be readily determined by a study of the glauconite casts after the shell has been destroyed.

The glauconite occurs in grains scattered through the green sand formation of the cretaceous and other periods. In many cases the grains are found to be casts of Foraminifera. These constitute as much as ninety per cent. of some rocks. In the yellowish limestone of Alabama such casts occur in great perfection, constituting about one-third of the rock. To obtain them, the stone has only to be treated with acid, when the greenish casts can readily be picked out from the insoluble residue, consisting of sand and finely-divided mineral particles.

Ehrenberg was the first to observe the replacement of organic forms by mineral matter, and he inferred that green sand was always formed by such a substitution. Such casts can be found in limestones from Mullica Hill, and near Mount Holly, N. J., from Drayton Hill, near Charleston, S. C., from the cretaceous rocks of Western Texas, and from other localities.

At the present time precisely similar casts of Foraminifera are being formed at the bottom of the sea. In the year 1853 Count Pourtales found, off the coast of Georgia, in the Gulf Stream, at a depth of one hundred and fifty fathoms, a bottom deposit consisting of shells of *Globigerina* and black sand in about equal proportions. Similar deposits were found also in the Gulf of Mexico and in various parts of the Gulf Stream. With them are also found the living Foraminifera, so that there can be no question as to the continuance of the process now.

ROMYN HITCHCOCK.

CLASS III. — GREGARINIDA.

The Gregarinida are a peculiar kind of animal parasites which inhabit the intestinal canals of earthworms, insects, crustacea, etc., the simplicity of whose structure leads most authors to class them among the Protozoa. Their distinct membranous investment, however, entitles them to a higher rank than any of the Protozoa already described, and, although with very little power of movement, and possessing no means of searching for and collecting food, they are still structurally higher than the Amœba and its allies, for a differentiation of parts is certainly distinctly shown in the membranous cell-covering. As parasites they do not require to move about in search of food. They have no mouth, no organs of digestion. They absorb their food through the membrane that covers the body; hence, although they are structurally above the Amœba, they have almost lost the Amœba's power of voluntary movement. We may conceive of an Amœba placed under conditions that would insure an abundant supply of food without the necessity of searching for it, finally losing its power of movement and developing a distinct membranous investment from the ectosarc. We would then have a Gregarine. Van Beneden has regarded the Gregarinida as Amœbæ thus degenerated by parasitism. But there has been no degeneration of structure, only of habit; it will be seen that Gregarines are all amœboid in one stage of their development, and that from this larval condition the more highly differentiated adult is produced.

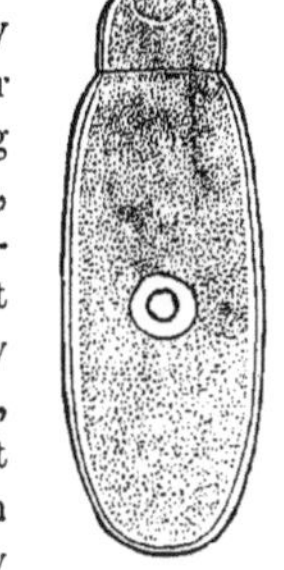

FIG. 20. — *Clepsidrina munieri*, enlarged.

The Gregarinida vary considerably in form and appearance, but in general terms they may be described as more or less ovate or cylindrical in form, the body consisting of a single cell frequently divided into two or more portions by transverse, internal septa, forming the anterior and posterior sacs. In some species the anterior end, or head, is furnished with a circle of reflexed hooklets, in others no such armature is found. The membrane covering the body is transparent and structureless in appearance. Although firm and elastic, it is very permeable to watery fluids. The contents of the body are a nucleus embedded in the protoplasm and fatty granules. The latter are usually so abundant as to give a milky appearance to the cell contents. The granules seem to increase in size and abundance as the animal matures, since in young individuals they are scarcely noticeable.

The nucleus invariably present in all Gregarines is a spherical vesicle with a sharply-defined membrane, situated near the middle of the cell in the Monocystidea and in the forward part of the posterior sac of the Dycistidea. A nucleolus is usually found within it.

When about to multiply, the Gregarinæ become surrounded by a transparent coat or cyst, which may include either a single specimen or two together. After becoming encysted a change takes place in the enclosed Gregarines. If two of them are in the cyst they become united, lose their identity and merge into a single mass. From this mass nucleated cells soon develop, which gradually take the form of elongated bodies, tapering at both ends, greatly resembling certain diatoms known as *Naviculæ*, whence they have taken their name, pseudo-naviculæ. They are also known as psorosperms. Finally the membrane of the cyst is ruptured and the pseudo-naviculæ escape.

The process above described is not the only one by which the spores are formed, even in the same species. At least two others are known, one a process resembling the segmentation of an egg, in which the entire mass is converted into very regular and granular spheres of segmentation, which in turn becomes elongate and covered with a firm investment, while their contents become more fluid; another in which the contents, instead of producing granular spheres, divides into two, four, or more parts, each of which, by a process not understood, becomes covered with a layer of transparent or slightly granular globules, and these parts are transformed into the spores.

In some instances the cysts of Gregarines have been observed arranged in a linear series in the walls of the rectum of certain animals, and it was for a long time a great mystery how they could be thus regularly placed. Van Beneden observed in the rectum of a lobster as many as seven cysts in a linear series. The Gregarine of the lobster, *Porospora gigantea*, attains the extraordinary length of 16 mm. and .15 mm. in diameter. During the proper season, spring and summer, it is very abundant in the intestine of the lobster — at least in lobsters from certain localities — as many as twenty-five being sometimes found in a single individual. At this time no cysts can be found, but in the autumn the parasites seem to pass down into the rectum, where they become encysted. The general process of encystment is somewhat as follows: —

The contents of the cyst are always at first granular, forming a single sphere without a nucleus. By division, two rounded masses next appear, and as the diameter of the cyst increases, these separate, and a clear, colorless liquid surrounds them. The wall of the original cyst then, at least in *P. gigantea*, becomes granular and disappears, while the two globes become surrounded by firm membranes, and their contents may again divide like the parent cyst. In other words, the cysts are capable of multiplication by division through successive generations. It is by the multiplication of the cysts that their linear arrangement is brought about. Eventually the cysts cease to divide, and pseudo-naviculæ are then produced. We have now to follow the development of new generations from these spores.

The development of the pseudo-naviculæ has not been fully made out, except in a few cases. It is probable, however, that their granular contents escape, and form amœboid masses from which the Gregarines are either directly or indirectly developed.

In the intestine of the lobster naked protoplasmic masses have been observed, which, as will be seen, have been proved to belong to the Gregarines. In the perivisceral cavity of the earthworm the encysted Gregarines, *Gregarina lumbricus*, are often found in great abundance. There the pseudo-naviculæ are set free in innumerable quantities, and their contents have been observed to produce amœboid bodies from which new generations of parasites originate.

The Gregarine of the lobster, *P. gigantea*, according to E. Van Beneden, develops from an amœboid form, but the course of delevopment differs somewhat from that observed in the case of the Gregarine of the earthworm. Within the intestine of the lobster the minutely granular amœboid form from which the parasite develops was found destitute of both nucleus and membrane, resembling *Protamœba primitiva*, or *P. agilis*, Häeckel, but not projecting true pseudopodia. These amœboids become converted into spherical, motionless globules, the "generative cytodes," which finally develop two projecting processes, resembling the stalk of *Noctiluca*, from which the Gregarines are directly developed. One of the prolongations is longer than the other. The longer one separates from the cytode and then moves about independently, like a nematode worm. The shorter arm then develops, appropriating the entire substance of the cytode, and likewise acquires the form of a nematode worm. These the author has designated "pseudo-filariæ." After some changes in form they cease to move, a nucleolus and then a nucleus forms about the middle of the body, and finally a new *G*regarine is developed directly from them.

In the development of the Gregarinida we find a complete genealogical, phylogenetic history of the cell. From the psorosperms are derived plasmic bodies, devoid of nucleus or external envelope, allied to the plastidules of the Monera, already described on page 3, naked cytodes, which are the lowest and simplest forms of living matter. Then a denser layer of sarcode is formed, corresponding to the ectosarc of rhizopods, but still there is no nucleus. Soon a nucleus and nucleolus is differentiated from the protoplasm, and the cytode becomes a perfect cell. This transformation may take place directly, or, as in the case of *Psorosperma gigantea*, by the budding of the cytode and the development of nuclei within the separated buds. The cycle is completed by the growth of the young Gregarine, its encystment, and finally the production of the psorosperms.

The Gregarinæ are divided into two divisions, Monocystidea and Dicystidea, according as the body is composed of one or two sacs. Schneider, who has given the most complete account of these forms, recognizes eighteen genera of Gregarinæ, represented by about thirty species. Our American forms have been scarcely touched, Dr. Leidy alone having investigated them.

ROMYN HITCHCOCK.

Class IV. — INFUSORIA.

The Infusoria are the highest of the Protozoa. They are so regarded because they usually have a definite shape, owing to the fact that the outer portion of their bodies is much more dense than the inner. It may, in fact, be said, with slight freedom in the use of words, that they are surrounded by a skin, or, to use the term of science, an ectosarc; the prolongations of their protoplasm for the purpose of locomotion and prehension are permanent, not transient, as in the pseudopodia of the groups just passed, extended or withdrawn at pleasure. In all but the lowest the food is received into the body by one or more mouths; and, with rare exceptions, they are active in their movements. Additional points of superiority will be seen further on.

Few Infusoria are individually visible without the aid of a microscope; but sometimes they form large colonies, which are readily seen. All are aquatic, and wherever standing water appears there are Infusoria. "They abound in the full plenitude of life alike in the running stream, the still and weed-grown pond, or the trackless ocean. Nay, more, . . . every dew-laden blade of grass supports its multitudes, while in the semi-torpid, or sporular state, they permeate as dust the atmosphere we breathe, and beyond question form a more or less considerable increment of the very food we swallow."

Anthony van Leeuwenhoek has the honor of first publishing an account of an infusorian. His "Obs. . . . concerning little animals observed in Rain, Well, Sea, and Snow Water, as also Water wherein Pepper had lain infused," appeared in the Philosophical Transactions, Vol. XII., 1677; the recorded discoveries were made during the two previous years. In Observation I. (1675), four forms are described; of the first he says, "The first sort by me discovered [in rain-water which had been standing four days] I divers times observed to consist of 5, 6, 7, or 8 clear globules. . . . When these *animalcula*, or living atoms, did move, they put forth two little horns, continually moving themselves; the place between the horns was flat, though the rest of the body was roundish, sharpening a little towards the end, where they had a tayl near four times the length of the whole body." There is no difficulty in recognizing by this description a species of *Vorticella*. It adds fresh interest to these charming animalcules to know that they were the first of their numerous kindred to be discovered.

The name "Infusoria" was first used by M. F. Ledermüller in 1763. Until quite recently it was applied to a heterogeneous assemblage of minute animals and plants having little in common but minuteness. The limits of the group are now pretty well defined; there are still differences of opinion concerning certain forms, but all students of the Protozoa now agree to relegate the diatoms, desmids, and rotifers to other and very diverse relations. The structure of the individual infusorian as at present limited will now be discussed.

The zooids of this group of the Protozoa are essentially unicellular; in the lowest forms they may consist of a naked cell (*gymnocyta*), or in the higher they may possess a cell membrane (*lepocyta*). Ehrenberg held that they were complex or multicellular; but this view resulted from his "polygastric" theory, put forth in 1830, which has been shown to be not well founded. Among the distinguished investigators who have also advocated the multicellular theory, may be mentioned Diesing, O. Schmidt, L. Agassiz, and Claparède and Lachmann. The unicellular theory was first formally opposed to that of Ehrenberg by Siebold in 1845. It was advocated also by Schleiden and Schwann. Haeckel has pointed out that the life-history of an infusorian, even one of the highest, is only an epitome of the life-history of a simple cell. Kent, in his Manual of Infusoria (1880) holds "that all infusorial structures possess a unicellular morphologic value only." It seems to the writer that most living students of the Invertebrata hold this view.

The bodies of the lowest members of the group (for example, *Mastigamœba simplex*) are composed of simple protoplasm, exhibiting little or no difference in density between the inner and outer portions; in fact, their shape is but slightly more constant than certain *Amœbœ;* the possession of a flagellum alone pointing out to the observer that the object under the assisted eye is not a rhizopod. On the other hand, those of the higher members of the group, as the Vorticellidæ, Fig. 44, have clearly more dense external layers, either of naked protoplasm or of the same surrounded by a cuticle of formed matter. Haeckel has described four distinct layers present in the bounding walls of the highest typical forms. The first is an outer, delicate, hyaline, elastic membrane; this in the sedentary, stalked species, extends down as the sheath of the pedicle, while in others it takes part in the formation of the carapax, as in *Euplotes.* As secondary products also may be mentioned the loricæ of *Cothurnia* and its allies (Fig. 21); again, in *Ophrydium*, it surrounds the zooids as a thick mucilaginous sheath, binding thousands of individuals into colonies. The second is a highly contractile layer just beneath the cuticle, called the ciliary layer; from this arise the cilia and their various modifications. The third, found only in the highest Ciliata, is termed the muscular layer; it is prolonged through the stalk of *Vorticella*, giving that organ its eminently muscular function; it is highly developed in *Stentor* and in *Spirostomum;* in the former the fibrillæ are arranged longitudinally, in the latter they have a spiral disposition. The fourth of these layers is that in which the *trichocysts* (rod-like bodies possessed by certain forms) are generated.

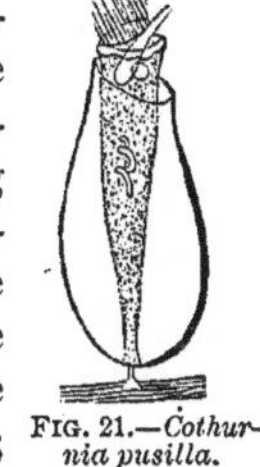

FIG. 21.—*Cothurnia pusilla.*

The contents of the complex bounding wall, or ectosarc, consists of the more fluid protoplasm of the organism. This transparent, colorless sarcode, the endosarc, contains numerous minute dark-colored granules, food particles, oil globules, foreign bodies, and an important body which requires special notice, the nucleus. This is usually a more or less oval body, differing somewhat from the surrounding protoplasm in density, etc., and in the fact that it is stained more readily by reagents, thus enabling the observer to easily distinguish it. In many cases, particularly among the simplest Infusoria, Huxley says, that "the 'nucleus' is a structure which is often wonderfully similar to the nucleus of a histological cell; but as the identity is not fully made out, it may better be termed *endoplast.*" While it has not been absolutely settled that this body is a true nucleus, there does not appear to be any valid objection to the view which would homologize the nucleus of the Protozoa and Metozoa. There are various special forms of the nucleus: it is sometimes ribbon-like, or coiled into a short, loose spiral (Fig. 44, *h*), or moniliform, that is, composed of nodular or oval bodies separated by constrictions; this form occurs in *Spirostomum ambiguum* (Fig. 22). The nucleolus or *endoplastule*, when present, is, in the spherical forms, enclosed within the nucleus, while in the sausage-shaped nuclei it is often outside, attached to the lateral wall. It has been demonstrated beyond question that in the more complicated types this

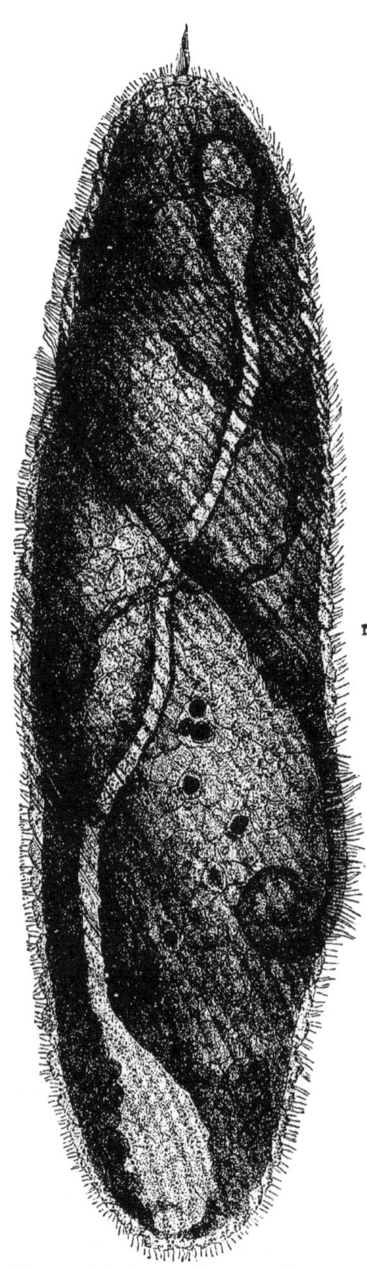

FIG. 22. — *Spirostomum ambiguum*, greatly enlarged.

organ is enclosed in a membrane. The function of the nucleus will be referred to further on.

The contractile vacuole is present in the majority of species. There may be only one (the usual condition), or several, and in rare instances many; in *Amphileptus anser* there are from ten to fifty, arranged in two longitudinal series. It is an interesting fact that the pulsations of these vacuoles occur, first on one side, and then on the other, progressing from the anterior to the posterior end of this elongate form. The vacuole may be simple or quite complex. When simple it is more or less spheroidal; in an active animalcule under the microscope it may be seen to steadily fill with a clear, watery fluid; when the full size is reached it suddenly collapses, after which, often, no trace of it appears until it begins again to fill; the pulse is rhythmical when the animal is expanded seeking food. Among the complex types may be mentioned those spherical forms with two or more radiating sinuses or diverticula. In *Paramecium aurelia* (Fig. 37), there are usually from five to eight of these blind canals; in *Stentor polymorphus* (Fig. 42, *cv.*), a sinus extends from the bulb situated near the anterior border down to the foot, while another branch extends from the bulb about the peristome; in *Spirostomum ambiguum* (Fig. 22) it is somewhat similar to that in *Stentor*, taking the form of a lateral canal with a very large bulb at the posterior extremity; it is often enlarged also at the opposite end. Concerning its use, whether it is a true organ with bounding walls, or whether it connects with the external water, there has been much lively dispute; indeed, students are still divided in their opinions of these questions. Ehrenberg regarded it a spermatic gland, Dujardin attributed to it a respiratory function, Claparède and Lachmann a circulatory function, Stein excretory. Huxley remarks that its function is entirely unknown, though it is an obvious conjecture that it may be respiratory or excretory. Kent considers it fully established that there is constant and free intercommunication with the outer water; that it is a mere pulsating lacuna in the cortical layer of the ectoplasm, and that its leading function is excretory, getting rid of the large quantity of fluid brought into the body by the ciliary or other currents incident to the capture and introsusception of the food.

The contractile layer of the ectoplasm, *i.e.* that just beneath the cuticle, bears permanent prolongations which are the organs of locomotion and prehension; hence they must pierce the cuticular layer to come in contact with the surrounding medium; they are long, slender, and represent three well-marked forms, viz.: flagella, cilia, and tentacula. The flagella are long whip-like protrusions of the body substance, often exceeding several times in length that of the body. They are never numerous—one, two, or four are the most usual, although a larger number is not unknown; their use, besides propelling the creature through the water, is to assist in the capture of food particles; in sedentary species they produce currents in the water, directing them against the body. In cases where more than one obtains, they are ordinarily in pairs, or separate and situated at a distance from each other; again, one flagellum, or one set, may serve to anchor the animal, while another distantly situated may serve for food capture. Cilia are short prolongations which resemble eye-lashes, hence the name; it is by their rhythmical and vigorous lashing of the water that the infusorian swims about so freely, or, if it is fixed, by the same means water-currents are made to flow past the mouth, and food is thus secured. The various arrangements of these locomotory hairs will be given under the description of the order Ciliata, named thus on account of their characteristic natatory organs. Besides the vibratile cilia, there are other modifications scarcely to be distinguished from them; these are setæ, or rigid hairs, used for support, or for defence; thick, straight setæ, called stylets, usually situated beneath the body, and uncini or curved hook-like hairs. The tentacula, in appearance and motion, at first recall the pseudopodia of some of the Radiolaria, but more careful examination shows that they are different. Each tentacle is tubular; a structureless external wall terminating in a distal expansion, or sucker, encloses a core of granular semi-fluid matter, which is an extension of the endoplasm. These organs, situated promiscuously over the animalcule's body, on well-defined areas, or on tubercles of the peristome border, may be extended, retracted, or even bent at will.

In the simplest of the Infusoria there is no constant aperture, or mouth, for the reception of food, but, as in *Amœba*, it is passed into the body substance indifferently at any part of the periphery. It is plain that in such cases a cuticle cannot be present; in others a certain definite portion of the surface receives the food. It is safe to consider such forms more highly developed. Among those regarded as the highest of the group, there is a well-defined oral aperture, often reinforced by appendicular appliances, and from which a passage, the œsophagus, leads into the endoplasm.

The multiplication of the Infusoria has been studied with much care. It will be convenient to speak of the several methods: by binary division, by gemmation, by spores, and by sexual reproduction.

Examples of sub-division are frequently seen, even by the casual observer. The process was accurately described in essentials by the earliest observers of these animals. In a majority of instances it takes place across the body; after separation of the nucleus into two parts, a constriction first appears at the middle, which increases in depth until the two parts separate, forming two perfectly-formed, free-swimming individuals, and the part without contractile vacuole and mouth soon acquires them. Each resulting infusorian, after a longer or shorter time, again divides. Ehrenberg isolated examples of different species, and after ascertaining how long time was required for division computed the enormous numbers produced by this process alone. In some genera, by no means a limited number, the division occurs in the opposite direction, or longitudinally; it may not infrequently be seen in *Vorticella* or *Epistylis*, so that the large colonies of the latter have thus increased from a single zooid. In certain forms, at least, the oral aperture, present before division, is lost, two new ones being formed. Binary division in a limited number of species, for example *Stentor*, is oblique.

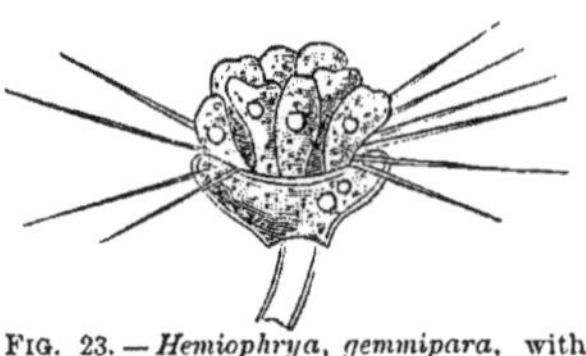

FIG. 23.—*Hemiophrya, gemmipara*, with eight elongate gemmules, magnified 200 times.

An instance of multiplication by gemmation is afforded by *Hemiophrya gemmipara* (Fig. 23). The buds appear on the anterior border of the parent animal, the nucleus branches, sending out diverticula into the buds. This phenomenon probably takes place in other species which bud externally. *Noctiluca miliaris* also increases by this process; the nucleus first disappears, then the protoplasm divides, first into two, then four, and so on. These masses are at length protruded upon the surface, the flagella are developed, and finally they are liberated as free-swimming germs.

Sporular multiplication, especially among the Flagellata, has been frequently observed. The careful and patient researches by Messrs. Dallinger and Drysdale have done much to acquaint us with the phenomena attending this process. In case of the flagellate monads they found each species to pass through several stages of development in their life-history, viz.: the flagellate or mature form, the amœboid, the encysted, and the sporular condition; the last appearing upon the breaking up of the contents of the cyst. In a flagellate obtained in an infusion of cod-fish it was found that many of these organisms all at once appeared to pour out a delicate sarcode, which exhibited amœboid movements. Two of these amœboid masses would unite, after which the sarcode became spherical, and at length developed the true cystic wall. Upon the rupture of the cyst, there escaped multitudes of microspores, not large enough to be individually defined by a magnifying power of fifteen thousand diameters. They were continuously watched until they developed into the initial forms. Similar phenomena have been recorded as taking place in the development of the higher Infusoria.

Sexual, or genetic reproduction, in the sense of the union of two distinct and different elements, has not been proven to occur. As stated above, individuals entirely indistinguishable unite before sporular sub-division. It is also well known that individuals unite transiently, and then separate to each continue multiplication by binary division. That the repeated sub-divisions so exhaust the stock that it must necessarily be revitalized by the conjugation of separate individuals is generally held. It is also recognized that the nucleus and nucleolus play significant parts in the rejuvenesence due to the *zygosis*, or conjugations. Concerning the interpretation of the facts there is much difference of opinion. A notice of the several views cannot be introduced here.

O. F. Müller, in 1786, first attempted to classify the Infusoria. Eleven of his seventeen genera are still recognized in the infusorial class. The number of species included was about two hundred. Ehrenberg's more elaborate system, published in 1835, described three hundred and fifty species (after deducting the rotifers and plants), and separated them into sixteen families and eighty genera. Von Siebold, in 1845, divided the true infusorial forms into two orders, — Astoma, without oral aperture and Stomatoda, with a distinct oral aperture and œsophagus. Claparède and Lachmann, in 1868, for the first time restricted the group to its present limits, and divided it into four orders, I. Flagellata, II. Cilio-flagellata, III. Suctoria, IV. Ciliata. Stein in his magnificent work, "Organismus der Infusionsthiere," not yet completed, presented in 1867 the classification of the Ciliata now generally adopted. The part of the work treating of the Flagellata appeared in 1878; it includes several genera, by many regarded as undoubted plants — for example, *Volvox* and *Chlamydomonas*. The latest proposed arrangement, by W. Saville Kent, in his Manual of the Infusoria (1880–82), divides the Legion Infusoria into the following classes: I. Flagellata, II. Ciliata, III. Tentaculifera. This author's limitation and arrangement of this group will be adhered to in the following pages, except that the Infusoria will be regarded as a Class; hence, his classes will become Orders and his Orders sub-orders.

ORDER I. — FLAGELLATA.

The Infusoria of this order bear one, two, or more flagella, which serve them for locomotion, and assist in obtaining food. They were not unknown to the earliest observers. In 1696, Mr. John Harris described what is undoubtedly *Euglena viridis;* but the modern microscope alone can reveal their organization, and it is in the study of these lowly organisms that the most substantial progress has been made by recent investigators in this field of research. Reference may here be made to the discovery of the collared-monads by H. J. Clark in 1868, and the addition of numerous species to this list by Stein and Kent, and also to the fact that Stein has found many Flagellata more highly organized than had been previously supposed. He has shown that many of the Flagellata possess well developed oral apertures, frequently with the addition of a pharyngeal dilation, and occasionally a buccal armature similar to that of the Ciliata. The flagellum is not the only means of locomotion possessed by some species, like *Mastigamœba,* for these have true pseudopodia like those of *Amœba;* others, again, as *Actinomonas,* have, besides the flagellum, temporarily developed rays like *Actinophrys;* a thread-like pedicle is also present in *Actinomonas.*

SUB-ORDER I. — TRYPANOSOMATA.

The very lowest of the Flagellata now known are two parasitic forms, one of which (*Trypanosoma sanguinis*) is illustrated by Fig. 24. The animal is flattened, and has a frill-like, undulating, lateral border which serves for locomotion. It will be seen that one extremity is somewhat prolonged or attenuate, representing the flagellum; the species occurs in the blood of frogs; its congener inhabits the intestine of domestic poultry.

FIG. 24. — *Trypanosoma sanguinis,* magnified 600 times.

SUB-ORDER II. — RHIZO-FLAGELLATA.

There occur in pond water, hay-infusions, and the like, some most interesting forms; they are so because they have characters in common with the *Amœba;* that is, they possess, besides the contractile vesicle and nucleus, the ability to protrude the body substance in the form of pseudopodia, by means of which they progress and take food; but they have in addition long vibratile flagella which place them in the higher group. *Mastigamœba simplex* (Fig. 25) serves well to illustrate this small group. A similar form has been often taken by the writer from the soft mud and *débris* at the bottom of quiet water; its movements are comparatively active, and its very long lash is thrust forward, beating the water with its rapidly vibrating extremity. Perhaps the most remarkable species of the Rhizo-flagellata is *Podostoma filigerum* of Claparède and Lachmann. It is very changeable in shape, and from the extremities of the pseudopodal protuberences flagella may be produced or withdrawn at will. When these are not apparent, the animal closely resembles *Amœba radiosa*, — indeed, Bütschli has recently attempted to show that it is the same species; but, on the other hand, it has been pointed out that the feet of *Amœba radiosa*, however attenuate, are never thrown into spirals, nor vibrate, as in *Podostoma*. It should be looked for in infusions of hay.

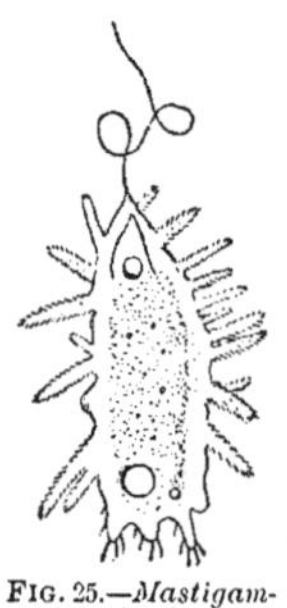

FIG. 25.—*Mastigamœba simplex*, enlarged 170 times.

SUB-ORDER III. — RADIO-FLAGELLATA.

The Radio-flagellata, which follow very naturally the forms last mentioned, are mostly marine. They may be compared to the Radiolaria, which they resemble very closely in their ray-like pseudopodia, but, in addition, they are provided with one or more lashes. Again, some of them are naked, while others are provided with silicious cases or loricæ. It should be remembered that these genera with tests are included by Haeckel in the Radiolaria. Is the possession of flagella a sufficient distinction for thus removing these to the Flagellate Infusoria? Without considering the intermediate character of *Actinomonas* and *Actinolophus pedunculatus*, it would seem not; but these forms bridge the chasm as well as that between the Rhizopoda and Rhizo-flagellata is bridged.

A characteristic example of the sub-order is *Actinomonas mirabilis*. Its body is globular, supported on a thread-like stalk several times longer than the diameter of its body; from every part of the periphery radiate sarcode rays in search of food; at the top extends a long flagellum which, by its motion, causes water-currents to pass over the rays. Food particles are taken, indifferently, at any part of the surface.

SUB-ORDER IV. — PANTOSTOMATA.

We have now to consider the infusorian types where the injestive area is diffuse, as in the preceding sub-orders, but which lack the pseudopodal appendages which allied those groups so closely to the Rhizopoda. This extensive sub-order includes eighteen families, divided into three groups; viz. Monomastiga with one flagellum, Dimastiga with two, and Polymastiga with three or more.

Every one who has used the microscope with any considerable magnifying power in the examination of infusions, or the water of ponds, has doubtless seen minute globose or elongate plastic bodies moving about by means of a single long thread placed at one end of the body. These forms belong to the genus *Monas*. As now limited, the family MONADIDÆ, includes only the naked free-swimming species with one flagellum.

The earlier writers were in the habit of describing any flagellate, just discernible with their lenses, as *Monas*, no matter how many flagella it possessed, hence the number of so-called species of *Monas* is very large. Many have been put into other genera, while others are still doubtful. It was one of these, *Monas dallingeri*, that served for the beautiful series of observations on the life-history of the monads referred to on a previous page. These minute creatures can be studied only by means of high powers. If, after long and careful watching, a form is found, otherwise just like *Monas*, which does not change its shape, it belongs to Stein's genus *Scytomonas;* if the anterior border is truncate it is *Cyathomonas;* fusiform and persistent in shape, *Leptomonas;* vermicular and spirally twisted with form persistent, *Ophidomonas;* vermicular and changeable in form, *Herpetomonas;* if adherent at will by a trailing flagellum, *Ancyromonas.* This analysis is given to show with what care these animals must be studied before they can be properly referred to their genera.

The remaining three families of the Monomastiga differ in having the flagella lateral, or the animalcule with a tail-like filament, or enclosed in an indurated sheath, the lorica. In the genus *Bodo*, the ovoid, or elongate, plastic bodies have a tail-like filament; they

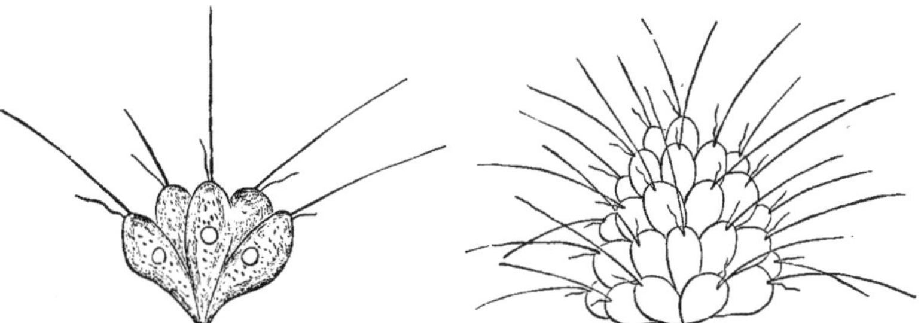

FIG. 26.—Five zooids of *Anthophysa*, enlarged 1000 times.

FIG. 27.—Large colony of *Anthophysa*.

are mostly parasites in the intestinal canal of animals, especially of reptiles and insects. At times they abound in myriads. The encystment of *B. lymnæi* has been recorded by Ecker. On examining the opaque eggs of the pond-snail, many were found densely packed with minute cysts; these bursting, gave birth to swarms of monadiform germs. The two most remarkable and beautiful genera are loricated. *Codonœca costata*, an American salt-water form, was described by H. J. Clark: the bell-shaped lorica or case stands erect on a rather long, rigid stalk; the upper part of the cup is expanded and apparently fluted. Kent has discovered another species with a smooth, ovoid lorica; it inhabits pond-water. The other loricated form, *Platytheca micropora*, differs from the preceding in lying flat upon its support like *Platycola* of the peritrichous Ciliata; it is found on the roots of the duck-weed.

The first family of the Dimastiga includes singularly striking species, which, by their tree-like supports, or *zoodendria*, may easily be mistaken for an *Epistylis;* in fact, more than one of the few species have been figured and described as species of the genus named, but the irregular, oblique animalcules bearing two, equal, anteriorly placed flagella should at once determine the proper place of these forms. The writer once found an *Anthophysa* abundant in a jar of water in which *Chara fragilis* had been kept for some time; it was taken for *A. vegetans*. Colonies attached to their granular, fragile stalks were seen, but the greater number were free-swimming. Figs.

26–28 illustrate a species which the unequal flagella and oblique anterior border of the zooids appear to place in the genus *Anthophysa*. It was discovered in Spy Pond, Cambridge, Mass., by A. H. Tuttle, who gave an account of it. He fed the monads indigo, which they took readily. When a particle was taken, the longer flagellum, which did not vibrate (the short one was in constant motion), was suddenly turned down, carrying the food with it into the oral region. The number of individuals in a group varied from a few to many, giving the larger colonies a mulberry appearance. Fig. 26 represents a group of five zooids attached by their bases. Fig. 28 is an ideal section showing the outline of the zooids and their manner of attachment to a common pedicle, the upper part of which alone remains, the colony having broken away from its anchorage. Kent has recorded the manner of growth of the pedicle in this genus. A colony were fed with pulverized carmine, which they ingested greedily, but it was soon rejected. This was effected entirely at the posterior extremity or point of union with the stalk, which was soon changed in appearance and dimensions, for the rejected particles of carmine were utilized in increasing it; the amber color and striated aspect gave place to that due to the agglutinated opaque carmine. The growth was so rapid that in one group the pedicle nearly doubled its length in half an hour.

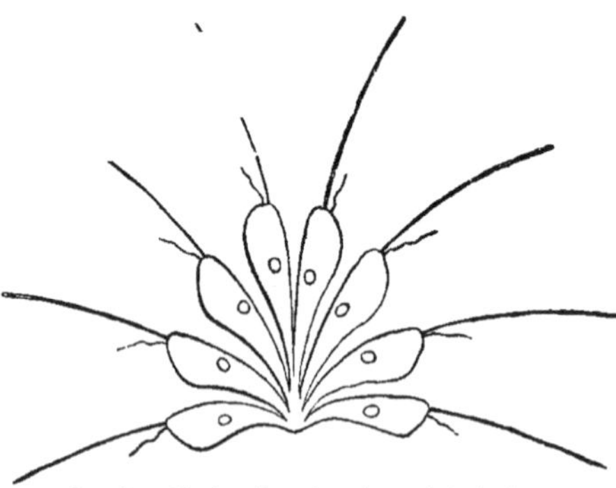

FIG. 28. — Ideal section of a colony of *Anthophysa*.

Among the most graceful forms of this sub-order the species of *Bicosœca* must certainly be enumerated. They occur in both salt and fresh-water; the globose, urceolate, or ovate loricæ are usually stalked, while the contained zooids also are pedicellate, the usual two anterior flagella are unequal. But for the lashes it would be easy to mistake these creatures for a loricate peritrichous ciliate like *Cothurnia*. There are also in this assemblage several endoparasitic species, — for example, *Pseudospora volvocis*, which resides in *Volvox globator*, where it eats up the cell contents; it is figured with a number of pseudopodia, thus recalling *Mastigamœba simplex*. Another example is *Lophomonas blattarum*, which, as its name implies, inhabits the intestinal canal of the cockroach (*Blatta*); it is a plastic form with a tuft of flagella anteriorly. Another parasite, *Hexamitra intestinalis*, occurs in the digestive tract of *Triton;* it has six flagella, four anteriorly and two posteriorly. It swims free or anchors itself by means of its posterior lashes; when in this position it swims about or gyrates from right to left, — twisting the threads into one, and then, reversing its motion, winds them in the opposite direction.

SUB-ORDER V. — CHOANO-FLAGELLATA.

This sub-order includes only three families and seven genera. The characteristics of these remarkable Infusoria were first made known through the researches of H. J. Clark in 1868. It is a matter for pride that this honor should fall to an American. A type of these forms is represented in *Codosiga botrytis* (Fig. 29), with which the other species may be compared. The animals of the family to which this species belongs are naked; those of *Codosiga* and *Monosiga* are attached, while those of *Astrosiga* and *Desmarella* are free-swimming. Those of the second family are loricate; *Salpingœca* and *Lagenœca* are solitary, the one sedentary, the other free, while the animals of the remaining genus, *Polyœca*, are united, forming branched supports. The third family has the animalcules united by a gelatinous matter into colonies; the two genera are *Phalansterium*, with the collar rudimentary, and *Protospongia*, collar well developed.

The form represented in Fig. 29 will at once be seen to belong to *Codosiga*, for the zooids are naked, stalked, and united socially. The leading peculiarity upon which the sub-order is founded is the hyaline, wine-glass shaped collar, borne at the upper, or anterior extremity of the body. In the centre of this cup arises the single flagellum, which by its motion about the cup causes currents of water to pass in on one side, down to the bottom, and out on the other side; the discal area at the bottom of the collar receives the food; waste particles are also rejected at this point. The collar may be withdrawn into the body, and again protruded at will. *Codosiga botrytis* appears to have been described by Ehrenberg, under the genus *Epistylis* of the peritrichous Ciliata. According to Kent it is *Codosiga pulcherrima* of Clark. They increase by binary division, as shown in Fig. 30. *C. botrytis* has also been observed to withdraw its collar and flagellum, and protrude rod-like pseudopodia from its surface, after which a cyst formed over the body contents, the latter ultimately breaking up into sporular bodies. The pseudopodal spines sometimes occur before the disappearance of the collar. This cosmopolitan species should be looked for on aquatic plants. *Monosiga steinii* is not uncommon on the stems of *Epistylis plicatilis*. When one of the pedicles containing them is examined with a magnifying power of six hundred diameters or upwards, the minute, solitary, sessile zooids of *M. steinii* may be seen to good advantage. One other genus of the group can alone be mentioned; it is *Salpingœca*, of which there are nearly thirty species known. The animals are, if possible, more beautiful than those already mentioned; for they have, in addition to the graceful outline of the zooid, an equally graceful lorica. *S. amphoridium*, described by Clark, has a very wide distribution; it abounds on conferva, the sessile lorica often incrusting the plants. They have been seen to divide, the separated portion moving away by means of pseudopodia; in this condition it has the appearance of *Amœba radiosa*. After a time it

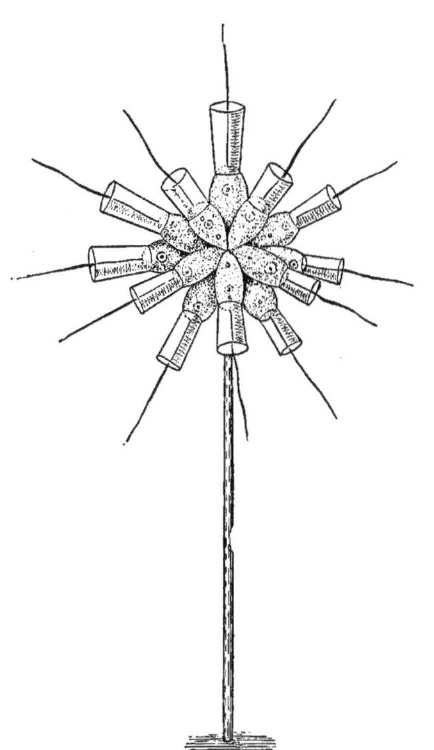

FIG. 29. — *Codosiga botrytis*, greatly enlarged.

secretes a lorica of the pristine beauty of its species, soon acquires a collar and flagellum, and is henceforth indistinguishable from its mature kindred. In the recently described *Spongomonas haeckeli*, we are made acquainted with a most remarkable infusorian,—one which, if it fulfils the expectation of the discoverer, Mr. Kent, will prove of unusual interest to a large number of students in zoology, and its dedication to Prof. Haeckel particularly apt. The zooids differ from the preceding only in being more plastic, the collar and flagellum being suddenly withdrawn on the least disturbance, the body then taking on an amœboid aspect. The animals secrete a mucilaginous stratum in which they dwell, studding its surface when expanded. Should this disposition of the zooids take place in "sacular invaginations of this matrix, it would produce what would have to be accepted as an undoubted, though very rudimentary, sponge-stock."

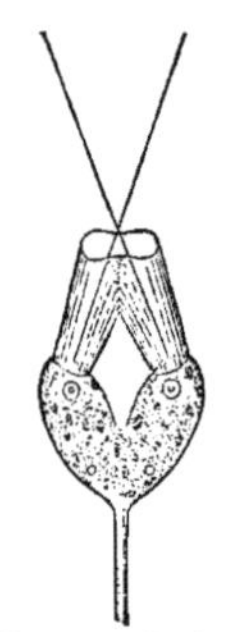

FIG. 30.—Fission in *Codosiga.*

SUB-ORDER VI.—EUSTOMATA.

The **Eustomata** differ from the forms previously described, inasmuch as they have a definite oral aperture, instead of ingesting their food at any part of their surface, or, as in the collared monads, only at a disc bordered by the collar; they differ also in having the outer part of their bodies much firmer than the endosarc, hence they are as a rule less plastic, and in a few instances the outer layer is indurated after the manner of some of the higher Ciliata. They never have more than two flagella, so they are separated into groups of families, according as the zooids have one or two flagella. The forty-six genera are distributed among eleven families; there are included many forms well-known to observers of pond-life.

In the first family (ASTASIDÆ) the monads are free, constant in form, colorless, the generic differences being found in the shape of the body,— ovate, flattened, flask-shaped, etc.; it includes *Astasia* with a distinct tubular pharynx, and *Colpodella* without it. *Astasia trichophora* is frequently met with in marsh-water. Although its forms are protean, perhaps its more usual attitude is pyriform; from the narrower anterior end issues a cord-like flagellum, mistaken by Ehrenberg for a neck like that of *Trachelocera olor* (Fig. 39). The ingestive orifice "consists of a large, widely dilatable, but simple, aperture, continued backwards into a clearly-defined pharyngeal tract." This structural character marks a broad distinction between this genus and *Euglena.* Bütschli has shown that the contractile vacuole of *A. trichophora* by its contraction forces a part of its contents out into lateral canals in a manner similar to that in *Paramecium*, and others of the *Ciliata*, to be described further on. The second family comprises forms highly changeable in outline, and colorless.

The EUGLENIDÆ differs from the Astasidæ in having the endoplasm brilliant green, and in having an ingestive apparatus capable of taking only minute particles. *Euglena viridis* is known by, or has been seen by, every tyro with the microscope. Its developmental forms are so various that it has been described under many names. Stein has observed a division of the nucleus to take place; the separate masses in some instances acquire an ovate outline, surrounding themselves with a dense coat, while others become thin-walled sacks, full of minute gran-

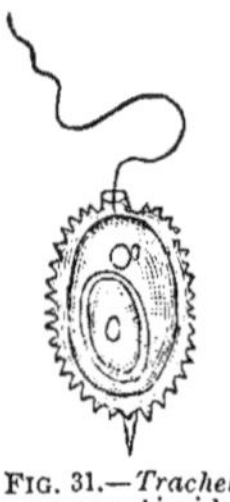

FIG. 31.—*Trachelomonas hispida*, magnified 500 times.

ules, each of which is provided with a single cilium. The loricated form, otherwise agreeing with *Euglena*, is *Trachelomonas*, common in ponds and bog-water. *Ascoglena* differs from the last in being sedentary; from this *Colacium* differs in the absence of the sheath, and in having a branching pedicle.

The phosphorescent NOCTILUCIDÆ embraces the genera *Noctiluca* and *Leptodiscus*. *N. miliaris* (Fig. 32) is a large form, visible to the naked eye, found in immense numbers in the superficial waters of the ocean, and it is one of the causes of their phosphorescence. It is colorless, spherical, with a meridional groove on one side, at one end of which the mouth is situated. A long, slender, transversely striated tentacle overhangs the mouth, on one side of which a hard, toothed ridge projects; close to one end of this is a vibratile cilium. The protoplasm consists of a central mass, with radiating portions connecting it with the sub-cuticular layer; there is a funnel-shaped depression leading into the vacuoled central mass, through which the food passes into the same. The phosphorescence appears to emanate from the layer just under the cuticle; for it has been observed that as the light gradually fades away on the death of the animal, as when one has been immersed in alchol, that the light finally appears in a ring around the body, since the observer is looking down upon a thin spherical film of light, imperceptible in the single layer over the middle of the globule; but at the borders, where seen as if on edge, sufficient light is sent forth to make it visible. When disturbed they become more highly luminous, so that a fish, for example, moving through the water where they are abundant shows its luminous sides, and its course is marked out by a path of emerald green light. This form is comparatively common in European seas, but has only been found, so far as we are aware, by Mr. C. B. Fuller at Portland, Maine, and by Prof. Hyatt and Mr. Kingsley at Annisquam, Mass.

FIG. 32.—*Noctiluca miliaris.*

Among the second division of this sub-order—viz. mouth-bearing, two flagellate forms—are many interesting and well-known species. The Entomostraca, especially those in puddles of the forest in spring time, are often loaded down with a green, oval form, which stands singly, or in groups, on short pedicles. On superficial examination it would be easy to mistake it for a *Colacium;* but on account of its two flagella during the motile period, its firmer cuticle and its two lateral pigment bands, it has been separated from *Colacium* as *Chlorangium stentorinum.* Whether the flagella remain during the sedentary stage or not has not been determined. In *Uvella* the animalcules are in colonies, free-swimming, and the flagella are sub-equal.

Two loricated genera, *Epipyxis* and *Dinobryon*, are unsurpassed in beauty by any of their kindred. The lashes are unequal, and the animal is attached to its vase-shaped lorica by a posterior, contractile fibre; the individuals of *Epipyxis* are sessile upon confervæ, while those of *Dinobryon* occur in branching chains of lorica *i.e.* each individual set free by sub-division is attached to the inner margin of the case of the parent. In early summer a species, presumed to be *D. sertularia*, abounds in the water-supplies of cities along the Great Lakes.

SUB-ORDER VII.—CILIO-FLAGELLATA.

The animalcules of the Cilio-flagellata have one or more lash-like flagella, and, in addition, a more or less highly-developed ciliary system, thus indicating a position between the Flagellata proper and the true Ciliata. At first only the Peridinidæ were included, but recent investigations have considerably enlarged its borders. It now embraces five families, the typical forms being included in the PERIDINIDÆ. These are free-swimming animalcules, sometimes naked, but in most cases the body is enclosed in a hard case, variously ornamented, the angles sometimes being prolonged into long spines. The case or cuirass has recently been proved to be cellulose, a substance hitherto only known in the Ascidians, outside of the vegetable kingdom. The cilia occur as a central or eccentric girdle, more or less complete; in the cuirassed forms the shell is usually divided by a groove, the borders of which are ciliated. The shell is either composed of one uniform piece or is made up of plates. There arise from some parts of the body one or more flagella. In life these flagella are seen to suddenly disappear, and a close examination has shown (in *Ceratium*) that there is a small cavity situated at the base of the flagellum into which that organ retreats, bringing with it foreign bodies which serve for food. The nucleus is usually spherical or oval, while a contractile vacuole is occasionally found. They are, like *Noctiluca*, highly phosphorescent. They have been observed to become encysted, when segmentation, on a more or less extensive scale, occurs. In some cases the cyst is enclosed in the carapax; in other instances the cuirass is thrown off, and a new cyst of a different form is secreted, which often has one or both extremities prolonged into attenuate curved horns, giving it a crescent shape (Fig. 33), resembling certain desmids (*Closteria*). They sometimes hibernate in this condition. They occur in both salt and fresh water.

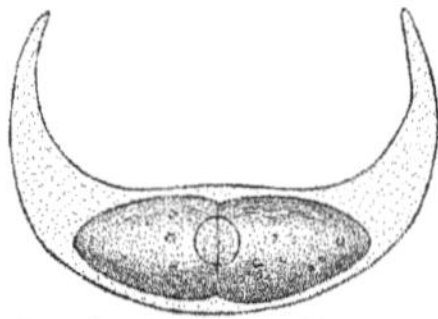

FIG. 33. — Cilio-flagellate, encysted, enlarged 300 times.

Of the ten genera of the Peridinidæ the species of three are naked, that is, resemble in essentials those of the loricate forms which have thrown off the case previous to encystment. *Gymnodinium pulvisculus* is, perhaps, as often met with as any; it occurs among Algæ in pools, often in great numbers, is somewhat spherical, with a transverse groove, and is brown or yellow in color. The best-known genera are *Glenodinium*, and *Peridinium* without horn-like processes, and *Ceratium*, with conspicuous processes on the shells. *G. cinctum* is well known to observers of pond-life; its smooth case should distinguish it from a similar *Peridinium* with faceted carapax. These genera are represented by several species in American waters. Mr. H. J. Carter has described a most remarkable instance of the coloring red of the waters around the shores of the Island of Bombay by *P. sanguinea*. During its active stages this species is green and translucent; gradually, as the time approaches for it to assume its quiescent or encysted state, refractive oil globules appear within the interior, and the green gives place to red, and thus the water containing them acquires a deep vermilion hue. It is probable that other instances of red waters are due to similar causes. The characteristic *Ceratium* (Fig. 35) appears to be a cosmopolitan infusorian. It has been known a long time as *C. longicorne*, but R. S. Berg has recently indentified it with *Bursaria hirundinella* of O. F. Müller, which, if correct, will change the specific name to the one having priority. *C. hirundinella* occurs often in large numbers in the water-supply of all the cities along the Great Lakes. It is most abun-

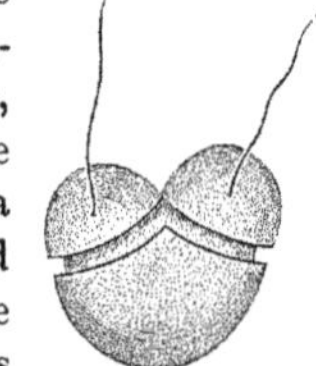

FIG. 34. — *Gymnodinium lachmannii* dividing, greatly enlarged.

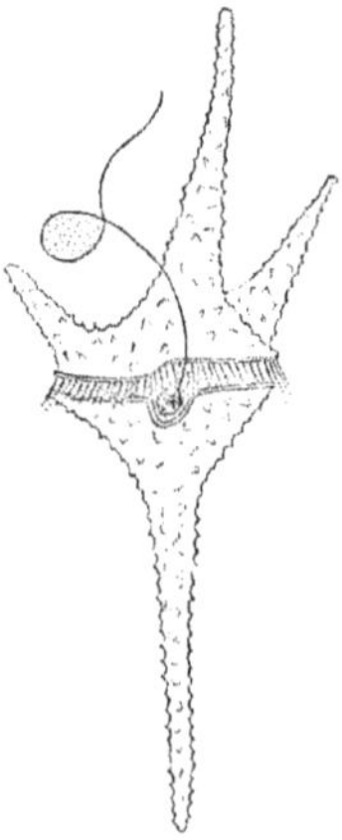

FIG. 35. — *Ceratium hirundinella*, enlarged 300 times.

dant in the fall. It may be said always to occur in these localities, together with a rotifer, *Anuræa longispina*, which has singularly long anterior and posterior spines corresponding in number with those of the infusorian. The resemblance is striking. Mr. J. Levic has recently found the same forms together in Olton Reservoir, near Birmingham, England.

The species of the remaining families have one or more flagella (usually one), with the body more or less clothed with cilia; in some the whole surface bears them, in others only a crown of cilia occurs at the anterior end, the flagellum standing in the midst. *Asthmatos ciliaris* (Fig. 36) exemplifies this structural peculiarity. This species occurs in the mucus from the nasal passages of persons suffering from "hay fever," and is held by Dr. J. H. Salisbury to be the cause of this distressing complaint.

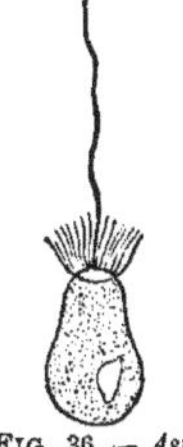
FIG. 36. — *Asthmatos ciliaris*, magnified 500 times.

ORDER II. — CILIATA.

The animalcules of this great order, as the name implies, possess cilia as locomotory organs. They are much more highly organized than the Flagellata, and many of the forms included are generally better known, and are more generally called to mind by the name Infusoria. Stein's division of the order into suborders is as follows: Holotricha, with cilia over the whole surface; Heterotricha, with cilia distributed over the entire surface, having those near or surrounding the mouth longer; Peritricha, cilia mostly in a wreath about the mouth; and Hypotricha, with cilia on the ventral surface only.

SUB-ORDER I. — HOLOTRICHA.

A common type of the first sub-order and of the family PARAMECIDÆ is *Paramecium aurelia* (Fig. 37). It occurs in hosts in vegetable infusions, stagnant pond-water, etc. These active, elongate, animalcules are alike the delight of the amateur microscopist and the joy of the veteran investigator; it is to him what the frog is to the general anatomist and physiologist. It was made for investigation; the comparatively large size and transparent body fit it admirably for study, and it has not been neglected. The anterior third of the body is somewhat flattened and twisted, so that the flattened face resembles a living figure of 8; near the middle of the ventral face — at the posterior extremity of the 8 — the mouth is situated. The rejectamenta issue at a point about half-way from the oral aperture to the posterior extremity of the body. There are two contractile vacuoles near the extremities. When expanded they are round, but when contraction takes place there appear fine radiating streaks, which, as the main portion decreases, gradually broaden, until, when the former is nearly invisible, they are extended over half the length of the body. It has been suggested

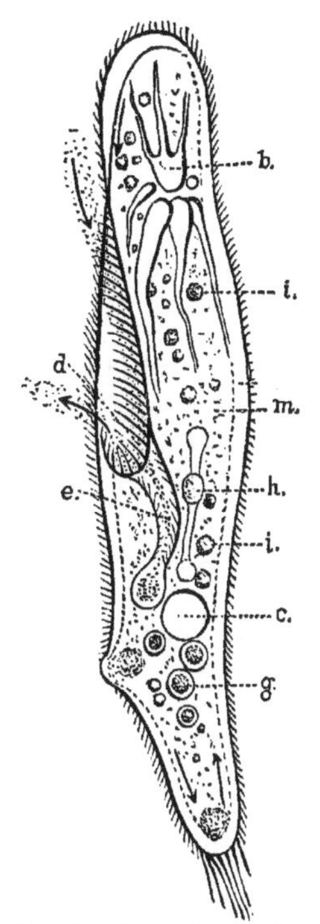

FIG. 37. — *Paramecium aurelia*, greatly enlarged. *b*, *c*. Contractile Vacuoles. *d*. Mouth. *e*. Œsophagus. *g*, *i*. Food vacuoles. *h*. Nucleus. *m*. Endosarc.

that these phenomena are really due to abnormal pressure of the cover glass. *Paramecia* increase by transverse fission. The cortical layer contains numerous vertically disposed rod-like bodies called trichocysts. When a *Paramecium* is treated with very dilute acetic acid these protrude from all parts of the surface, giving the animal the appearance of being clothed with very long cilia. A solution of tannin in glycerine produces a similar effect, although it is claimed by a writer in the Journal of the Royal Microscopical Society that it is due to a hardening of the cilia. These trichocysts have various forms and dispositions in different species. Some regard them as homologous with the thread-cells of the Cœlenterata, and as having a similar function; others regard them as tactile organs. Bütschli has described a species, *Polykrikos schwartzii*, which has trichocysts entirely similar to the thread-cells of the sea-anemone. Since this infusorian inhabits salt-water, and the trichocysts are irregularly disposed, Kent suggests that they may be thread-cells which have been swallowed. *Paramecium bursaria* (Fig. 38) is shorter and broader than *P. aurelia*, and is less flattened; the buccal fossa is funnel-shaped, extending obliquely from left to right. The nucleus is oval and the nucleolus is attached to the side of it. *P. bursaria* is usually colored green by chlorophyll granules, — now held by some to be parasitic algæ, as is also the green color of the fresh-water sponges, and the common green *Hydra*. Owing to the presence of the green corpuscles the circulation of the endoplasm is seen to better advantage than, perhaps, in any other infusorian, although there are forms like *Vorticellæ* which exhibit this phenomenon in a marked degree. This rotation is uniform, ascending on the left side, and descending on the right, when seen from above (indicated in the figure by the arrows). Balbiani has shown that the so-called longitudinal fission is not really a fission, but a phase of the act of conjugation. Two animalcules may remain attached by their anterior extremities for several days; after separation, the nucleus and nucleolus changes, the latter becoming more or less striated, while the former breaks up into a variable number of spheroidal bodies, which finally separate from the parent, and possibly are to be considered as ovules.

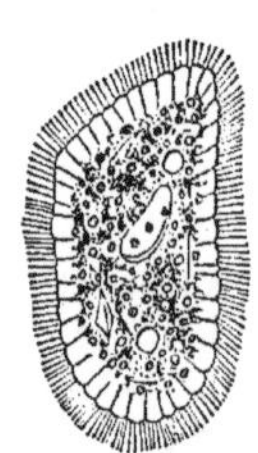

FIG. 38. — *Paramecium bursaria*, magnified 250 times.

Among the most curious of ciliate Infusoria those of the family TRACHELOCERIDÆ are entitled to the front rank. Their flask-shaped bodies are drawn out anteriorly into a long flexible neck, with the oral aperture at its terminus. *Trachelocera olor* (Fig. 39) is the type of the group; it appears to be cosmopolitan, occurring among algæ in ponds and streams. Under examination in the living state it appears to be incessantly exploring for food, thrusting its wonderfully extensile neck right and left into every cranny. As it swims gracefully through the water, with a spiral motion, its form and attitude very naturally suggest the swan. In *Lachrymaria* the neck is only slightly extensible. *Maryna socialis*, as its name implies, affords an instance somewhat rare among the Holotricha, that is, the formation of a *zoocytium.* This structure is branched like a tree, the cup-shaped zooids projecting from the termination of the branches. *Amphileptus gigas* is an elongate compressed animal, which may easily be mistaken for a *Trachelocera* on account of its long neck, which assumes as many shapes as in that genus; it is readily distinguished, however, since the mouth in *Amphileptus* is at the base rather than at the apex of the proboscis. It is said to feed

FIG. 39. — *Trachelocera olor*, enlarged 375 times. c. Contractile vesicle. *m.* Mouth. *n.* Nucleus.

on animalcules, which it takes by means of its trunk, transferring them to its mouth, after the elephant's manner of feeding. It has a number of contractile vacuoles, from ten to fifty, arranged in two longitudinal rows. as mentioned on a previous page. It is one of the largest known Infusoria.

The next family (TRICHONYMPHIDÆ) is characterized by the possession of a membraniform expansion as well as cilia. The type may be illustrated by the interesting *Trychonympha agilis* (Figs. 40 and 41), described by Leidy as parasitic in the digestive canal of the white ant, *Termes flavipes.* He observed that the canal was distended by brown matter, which on examination proved to consist largely of infusorial

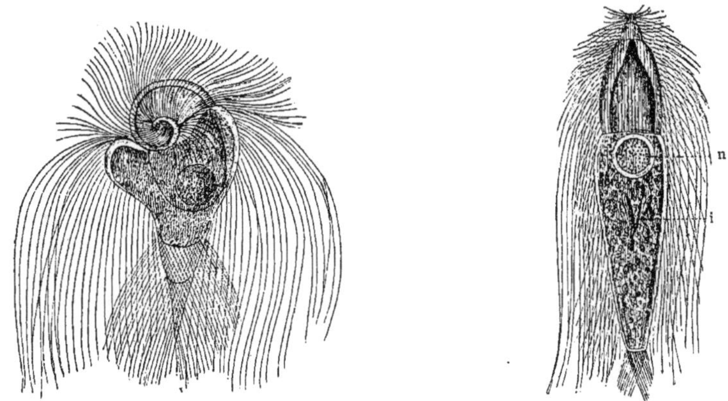

FIGS. 40 and 41.—*Trichonympha agilis*, enlarged 450 times. *n.* Nucleus. *i.* Ingested food particles.

parasites and particles of wood. Three species, belonging to as many genera, were discovered. Fig. 40 represents *T. agilis* in its extended position. As it progresses in its medium it takes on many protean forms. The cilia are arranged apparently in series, some longer than others. The oral aperture is terminal. Until its life-history shall have been made out, its place in the system and its relation to its companions are uncertain.

The mouthless Holotricha (OPALINIDÆ) are all parasitic, degraded forms. They have been taken for the larvæ of *Distomæ.* These now unquestioned Infusoria should be looked for in the intestines of frogs, mollusks, and worms. *Opalina* and *Anoplophrya* are examples.

SUB-ORDER II.—HETEROTRICHA.

As was mentioned on preceding page, the Heterotricha are characterized by the possession of cilia on the whole surface, those surrounding the mouth being longer than those on other portions of the body. In all except *Bursaria* and its allies the above definition holds good; there the oral cilia do not encircle the mouth. With the mention of these exceptions, we may now pass to a consideration of a few of the typical forms.

In *Spirostomum* are met Infusoria which at once arrest the attention, both by their elongate, snake-like form and their remarkable anatomy. *Spirostomum ambiguum* (Fig. 22) may serve to illustrate their striking features. The figure represents the animal somewhat contracted. It is capable of great distention, so as to become fifteen or twenty times longer than broad; it then attains a length of one-twelfth of an inch, or even more. Its cylindrical body, sometimes flattened, is rounded at the extremities, often truncate posteriorly. The single line of peristomal hairs extends down the left side of the anterior ventral face to the oral aperture, situated near the middle of the body. The remarkable contractile vesicle and nucleus have been referred to already. The generic name was given on account of the apparently spiral peristome, as seen when the animal twists itself about its long axis. The writer recently found this species so abundant on a scarcely submerged moss that the water taken up by a dipping-bottle was rendered turbid by them. The solution of tannin in glycerine, previously referred to, appears to be a valuable reagent for studying this animalcule.

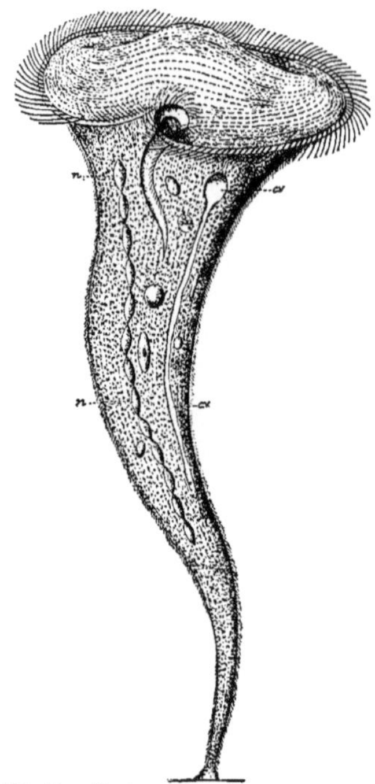

FIG. 42. — *Stentor polymorphus*. *n*. Nucleus. *cv*. Contractile vacuole. Magnified 90 times.

The *Stentors*, or trumpet animalcules, are among the most entertaining heterotrichous Infusoria. They are large, active, and often highly colored, so that a colony of them incessantly extending and retracting their bodies, at the same time driving, by means of their oral cilia, strong currents of water against their peristomial surfaces, presents a scene, when well defined by the microscope, not soon to be forgotten. The excellent cut (Fig. 42) of *Stentor polymorphus*, a widely-distributed form, displays the characteristics of the genus far better than words can do. The Stentors often secrete gelatinous sheaths, which sometimes embrace several individuals.

This group has, too, its species which secrete a lorica, and as in other cases they are very attractive objects. An example may be cited in *Folliculina:* its flask-shaped sheath is attached by the side, the neck being bent upwards. The animal closely resembles a *Stentor*, except that the peristome is two-lobed, instead of nearly circular. The species are defined according to the shape of these lobes. *Folliculina ampulla* has been found in America by Dr. Leidy. In *Chætospira* the two lobes give place to a slender ribbon-like extension of the anterior region, which, when extended, is twisted into a spire; a hyaline expansion also extends laterally along the broad part of the peristome, giving the extended zooid a unique appearance. It is not attached to its surrounding sheath. The preceding loricate forms are sedentary; on the other hand, *Tintinnus* includes free-swimming species. The beautiful tests of these are common in the water-supplies of most American cities.

In the genus *Codonella* there are an outer circle of twenty tentacle-like cilia, and an inner one of lappet-like appendages; the case is similar to that of the preceding.

SUB-ORDER III.—PERITRICHA.

This group resembles the preceding, and indeed there are a few forms whose exact position is doubtful. The typical Peritricha, however, have the cilia confined to a circle around the mouth, while it is only in the aberrant forms that supplementary cilia are found. The members of the sub-order are divided into two not very natural divisions, according as they are free or attached, at least during a portion of the existence of the individual. The attached forms frequently develop elegant cases, or in other instances beautiful branching colonies.

As an example of the first (free) group we may consider the unusually active *Halteriæ*, which are globose forms seen in water from ponds, especially after it has been standing for some days. They have a spiral adoral wreath of cilia for swimming, and usually, in addition, a girdle of long-springing setæ, by means of which they leap comparatively long distances. One may be quite under the observer's eye, when to his annoyance instantly it darts out of the field of view. To facilitate their study Claparède and Lachmann recommended placing under the cover with them a form like *Acineta*. They soon jump against its sucking tentacles, where they stick fast, and then may be conveniently examined. Fig. 43 illustrates *Halteria volvox*. It resembles the more common *H. grandinella* in form and in leaping organs, but has besides an equatorial zone of long, recurved cilia. A singularly aberrant form appears in *Torquatella typica*, described by E. Ray Lankester. Around the front margin there is a membranous expansible frill, which is plaited, and alternately closes up and expands with a twisting motion. It was obtained in salt-water from decaying eggs of the worm *Terebella*.

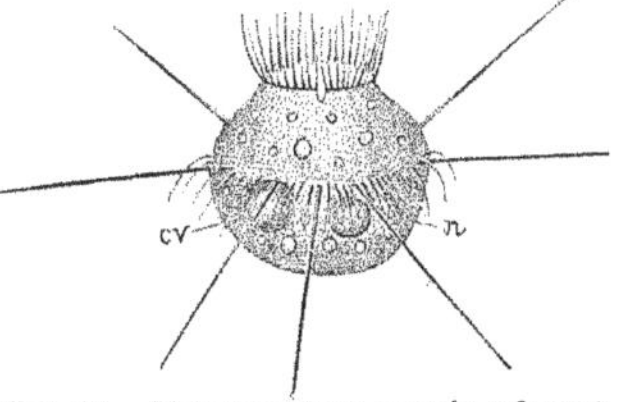

FIG. 43.—*Halteria volvox*, greatly enlarged. *n.* Nucleus. *cv.* Contractile Vacuole.

Any one having examined the common *Hydra*, or the gills of *Necturus lateralis* to any extent with the microscope has doubtless encountered minute bodies gliding over the surface of the hosts, or now and then swimming rapidly away, but soon returning. Seen from above they are discs; from the side, shaped like a dice-box. Prof. H. J. Clark studied carefully their anatomy. He showed that the body surface between the concave extremities is ribbed by the thickenings of the body walls; that the posterior truncated margin produces a thin annular membrane called the "velum," into the base of which, and on its inner side, the posterior fringe is inserted; that the nucleus, examined in the fall, was a moniliform, band-like spiral situated near the truncated base. Besides the vibratile cilia of the border of the posterior disc, there is on its inner border a wreath of stout hairs or uncini, in an outer and an inner series; those of the outer circle are stout and curved, the others, slender, straight, and apparently radiating from the centre of the discal area; they consist of a solid portion and a membrane-like expansion. These forms belong to the genus *Trichodina*.

The family VORTICELLIDÆ includes the attached forms of Peritricha. The three sub-families are Vorticellinæ (naked), Vaginicolinæ (loricate), and Ophrydinæ (in gelatinous covering). The student of protozoic life must ever find the keenest delight in the study of this varied family, examples of which may be found at any time of the year, or at any place where there are natural or artificial bodies of water. In *Vorticella*, the type of the family, each individual is solitary, and consists of an oval body attached by a slender stalk to some foreign support. At the end of the body farthest from the support, a band of cilia surrounds a flattened disc, at one side of which is the opening dignified by the name of mouth. On careful examination it is seen that the band of cilia does not form a complete circle, but rather a spiral, the inner end of which passes into the mouth and down the tube known as the œsophagus. The nucleus is sausage-shaped, and in many forms is coiled in a spiral. The stalk is also an interesting portion of the anatomy, as within the external cuticle one may by careful examination see a core of contractile protoplasm, whose function will appear further on.

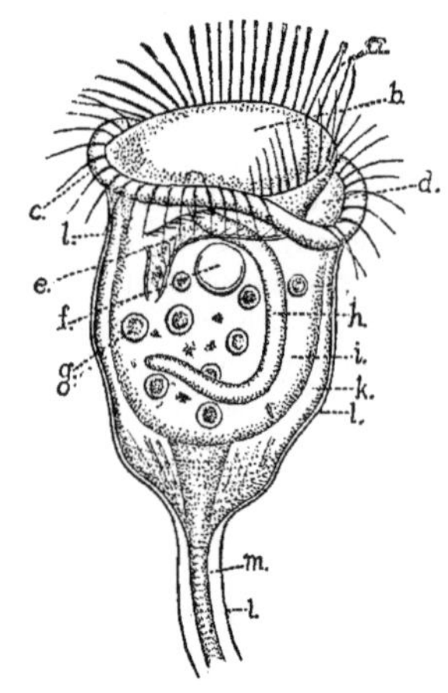

FIG. 44. — *Vorticella nebulifera*, enlarged about 600 times. *a.* Cilia. *b.* Ciliated disc. *c.* Peristome. *d.* Vestibule. *e.* Œsophagus. *f.* Contractile vacuole. *g.* Food Vacuoles. *h.* Nucleus. *i.* Endosarc. *k.* Ectosarc. *m.* Muscle of stem.

Since *Vorticella* is one of the most common of the Infusoria, it may be well to give here a slight account of its method of life, as illustrative of the physiology of the whole class. On placing a little powdered carmine in the water in which the *Vorticellæ* are living, and examining them under the microscope, it will be seen that the motion of the cilia around the mouth creates a current of water, which pursues a constant direction down the œsophagus and then out again. The particles of carmine and other small bodies in the water follow the general course of the current down to the bottom of the œsophagus but are not allowed to go out again. In this way a ball of nutritious matter is formed, which soon forces its way into the protoplasm of the body, where it is known as a food vacuole. These food vacuoles keep up a constant though slow motion through the body, passing down on one side and up on the other, until at last all nutritious substance is digested, when the rejectamenta are forced out into the beginning of the œsophagus and carried away from the body by the outgoing current of water.

It requires considerable time and much patient watching to make out these points in *Vorticella*, for though attached, these animals are far from stationary; every few moments the cilia will be suddenly withdrawn, and the animalcule will itself as suddenly disappear. On moving the slide one readily ascertains the cause of this, for it will be seen that the contractile protoplasm of the stem has exerted its powers, and the long, slender stalk is now coiled in a close spiral. Gradually the stalk straightens out, when the contractile vacuole renews its pulsations, and the cilia begin their vibration as suddenly as they had stopped them a minute before.

The process of binary division in these familiar objects is of high interest; it is longitudinal; first the ciliary disc is withdrawn, and the body assumes a spherical contour; it soon becomes dilated, and a notch appears in the anterior border; a new vestibular cleft and oral system is developed on each side of the median line; a line of division now proceeds from the anterior notch, through the centre of the animal's body, cleaving both the contractile vesicle and nucleus. The result is two animalcules on a single stalk. One zooid remains attached to the original pedicle, the other, with its peristome usually contracted, develops round the posterior region of the body a circle of cilia, by the action of which its attachment to the pedicle is broken, and it swims away, soon to attach itself and acquire a new stalk. Then the temporary girdle of cilia is absorbed, the peristome border is now displayed, and the business of adult life commenced. Another sort of sub-division has been recorded by Stein, and confirmed by others. The body divides into two unequal parts, after which the lesser one is set free, and then enters into genetic union with some other normal *Vorticella*. This union is supposed to produce a rejuvenescence, which means a capacity to continue the process of multiplication by self-division. That the union is followed as in the Flagellata, by encystment and sporular sub-division, has not been demonstrated. Perhaps this word of caution is necessary; the individual *Vorticella* may often be seen with the posterior ciliary wreath. This is not always an indication of recent division, for when these animals become dissatisfied with their surroundings they produce the extra cilia, and remove from the old pedicle, and set up in a more congenial place.

Among the forms allied to *Vorticella* we may notice that in *Spirochona* the attachment is by means of a disc, and the peristome is developed into a hyaline, spirally convolute membranous funnel. *Stylonichia* is similar, except that the body is mounted on a rigid pedicle; in *Rhabdostyla* the body is like that of *Vorticella*, but the pedicle is not contractile, but flexible. In *Carchesium* the zooids are united in social tree-like clusters, but the muscle of the pedicle does not extend through the main trunk; the individuals can withdraw themselves to the point of branching of their stalk, but the colony cannot withdraw itself from its position. In *Zoothamnium*, on the other hand, the muscle is continuous throughout the colony. In this genus there are zooids of more than one form and size in the same colony. In the genus *Epistylis* the branched pedicle is rigid throughout, the base of the body alone being contractile. Members of this genus are, doutless, next to those of *Vorticella*, most frequently met with. Their tree-like colonies are readily seen by a hand lens on aquatic plants. The carapax and gill chambers of the cray-fish, and the shells of aquatic snails, are also rich hunting grounds for these creatures. *Opercularia* differs from *Epistylis* in the fact that the ciliary disc is attached to one side of the oral entrance, and is usually elevated to a considerable distance above the margin of the peristome, like a lid. They are often seen as commensals on aquatic larvæ and Entomostraca.

The loricated Vaginicolinæ is not less rich in surprisingly beautiful forms. *Vaginicola* has the sheath erect, sessile, and open at the top. The animalcule is fastened to its case, protruding its body and spreading its peristome; at the least disturbance in its surroundings it instantly retracts, soon to very cautiously again protrude its body. Those species with a lid to close the case when the animal is withdrawn, have been placed in the genus *Thuricola*. *T. crystalina* is a common species. If the case is pedicellate and open, the form is a *Cothurnia* (Fig. 21); if, in addition, there is a corneous lid, it is a *Pyxicola;* if a fleshy lid, *Pachytrocha*. All these forms may be looked for on Entomostraca and aquatic plants, like *Lemna*, *Anacharis*, and *Myriophyllum*.

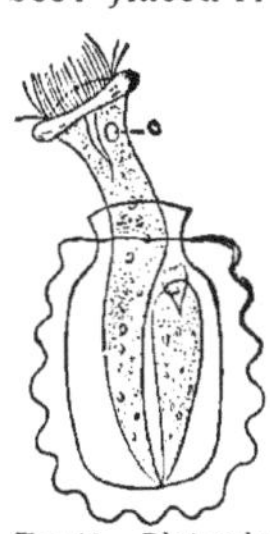

In *Platycola* and *Lagenophrys* the cases rest on one side. Fig. 45 represents the charming *Platycola dilatata*, the brown, laterally attached case occurs on fresh-water plants. The animalcule is quite similar to those of a majority of the loricate forms. There are two genera, whose social individuals inhabit common gelatinous matrices, viz., *Ophionella* and *Ophrydium*. One species of the latter genus, *O. versatile*, may often be seen in shallow fresh and salt water as more or less spherical green masses, sometimes floating or resting on the bottom, and may easily be mistaken for algæ, such as *Tetraspora* or *Nostoc*. These masses are inhabited by

FIG. 45.—*Platycola dilatata*, magnified 300 times.

myriads of green Infusoria whose structure does not materially differ from that of *Epistylis*, except the greater elongation of the canal-like extension of the contracticle vesicle, which ascends to and surrounds the peristome. A notable structure is found in the thread-like pedicles which unite the individuals of the whole colony. These appear to be homologous with the branched stalk of *Epistylis*. In *Ophrydium sessile* the pedicle is wanting, the bodies radiating from one point in the mucilaginous envelope. *O. versatile* and *O. eichornii* are known to inhabit American waters.

Sub-Order IV. — Hypotricha.

This sub-order includes numerous families and genera, nearly all of which are free-swimming; their bodies are smooth above, with variously disposed cilia below; they are usually flattened and elongate. *Chilodon cucullulus* affords another stock form; it is as cosmopolitan as *Paramecium aurelia*, inhabiting both salt and fresh water. It has received many names; its flat, sub-ovate body has the anterior apex turned one side. The cilia on the ventral surface are arranged in parallel lines. Its pharynx is surrounded by a series of rod-like teeth. Its food appears to be diatoms, for these plants are often found in its endoplasm. *Dysteria armata*, described by Huxley, is remarkable for the indurated, complex pharynx. The oral pit is strengthened by a curved rod which terminates in a bifid tooth. This is followed by the pharyngeal apparatus proper, which may be said to consist of two parts — an anterior rounded mass in opposition with a much elongated, styliform, posterior portion. This part is quite complicated, and cannot be clearly defined in a few words. On account of this complicated structure, and the single ventral stylet, it has been considered a rotifer, but recent research has brought to light facts sufficient to warrant the formation of a family with this species as the type. A curious genus is *Stichotricha*, in which the animalcules secrete a domicile; several of the species live singly, but in one the stock is branched, and a social group or colony is the result. This Kent has put in his new genus

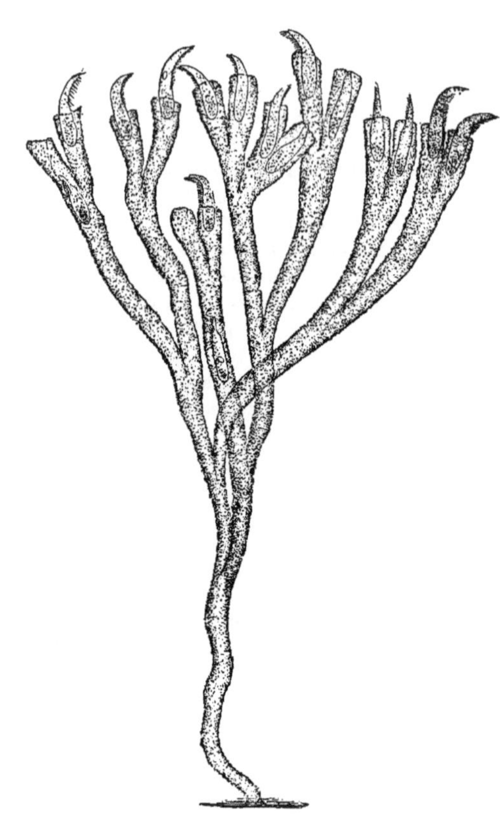

Fig. 46. — *Schizosiphon socialis*, enlarged.

Schizosiphon. *Stylonichia mytilus*, which is abundant in vegetable infusions, illustrates a type of structure somewhat common in this sub-order: the presence of stylets and hooked hairs. It is elongate, elliptical, with a slight left-handed curvature, tapering backwards from the centre; two of the fine anal stylets project beyond the body, the three long caudal setæ are radiating; there are five claw-like ventral stylets and several frontal ones. *Euplotes* also includes well-known forms. They differ from those of the last-mentioned, first in being encuirassed, second in the styles, although frontal, ventral and anal are represented.

Order III. — TENTACULIFERA.

It remains to introduce some typical forms of the order Tentaculifera. Compared with the orders already described, the specific forms are comparatively few, but their remarkable structure renders them as interesting. Until recently the Tentaculifera were not recognized as a distinct order of Infusoria. By Ehrenberg they were arranged with the diatoms and desmids. Stein in his earlier publications regarded them as developmental stages of the Vorticellidæ. To Claparède and Lachmann is due the honor of pointing out their true nature. The term Tentaculifera was proposed by Prof. Huxley, while Suctoria, — the name applied by Claparède and Lachmann, — has, by Kent, been retained for the division in which the tentacles are wholly or partly suctorial. He has also called those whose tentacles are non-suctorial, but merely adhesive, Actinaria. The animalcules in their adult life bear neither flagella nor cilia, their embryos, however, are ciliate. Maupas claims that adults of some *Podophryæ* and all the *Sphærophryæ* are able to resume their cilia and become free. Their food is taken by means of tentacles developed from their cuticle, the tubular sort terminating in a sucking disc; and the protoplasm of the body extending into the tentacles. When an infusorian is caught by an *Acineta* and held at the extremity of one of the tentacles, a rupture is produced in the cuticle of the victim at the point of contact. The axillary substance of the tentacle penetrates this perforation. The tentacle now increases in size, due, doubtless, to a flow of sarcode from the body of the *Acineta*. On penetrating the body of the prey this sarcode, according to Maupas, mingles with the substance of the victim's body, and then returns to its place of departure. A nucleus and one or more contractile vesicles are usually present. The Tentaculifera increase by division and by budding.

Sub-Order I. — Suctoria.

The species of *Acineta* and its allies are numerous; the animalcules have many tentaeles, while in the genera *Rhyncheta* and *Urnula*, there are only one or two; all are stalked, some are loricate, others naked. The *Sphærophryæ* are free forms, frequently parasitic within other Infusoria. *Sphærophrya sol* is found in *Paramecium aurelia*, *S. stentorea* in *Stentor rœselii*, etc. They are spherical, with suctorial tentacles scattered over their surface. The earlier stages of the next genus are free, and may be taken for *Sphærophryæ*, and the latter in turn have been mistaken for the acinetiform embryos of their hosts. The genus *Podophrya* also includes many species. They differ materially from the preceding in that they are pedicillate, while some species differ from others in having the suctorial tentacles in fascicles. To the latter belongs *P. quadripartita*, which has been often seen by the writer on the stalks of *Epistylis plica-*

tilis, whose acinetiform embryo it was once regarded. Its stalk is long, the body ovate, with the upper border divided into four tentacle-bearing lobes in the adult; in the young there is but one lobe; this gives place to two, and finally to the full number. According to Claparède and Lachmann, two sorts of embryos, large and small, are developed; the former enclose a portion only of the nucleus of the parent. According to Bütschli they are liberated through a specially developed orifice; the other forms are produced by the sub-division of the nucleus. In both cases, at the time of liberation, the embryos are ciliated like the peritrichous Infusoria, with an equatorial girdle and anterior tuft. In *Hemiophrya gemmipara* (Fig. 47), we have a remarkable Acinetan. There are two sorts of tentacles, viz., a few, centrally placed, of the usual suctorial type, and a larger number of prehensile ones around the border. When the latter are seen under a high magnification the surface is seen to be not smooth, but nodular, the component particles of externally developed granular protoplasm being usually disposed in a spiral manner around the central axis. The production of embryos by gemmation has been referred to on a previous page. The genus *Acineta* has many representatives inhabiting both salt and fresh water. An interesting species is found in large numbers on the surface of a *Mysis* taken in the Great Lakes.

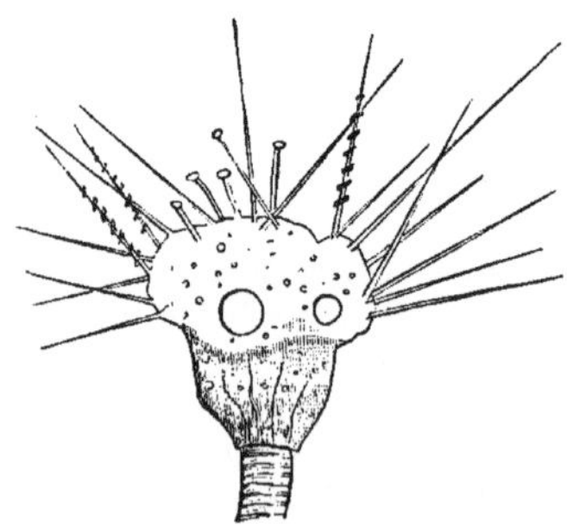
FIG. 47. — *Hemiophrya gemmipara*, magnified 150 times.

SUB-ORDER II. — ACTINARIA.

This sub-order includes a few forms in which the tentacles are filiform and prehensile. They are inhabitants of salt-water, and, like their nearest relatives, are mostly commensal upon aquatic animals.

D. S. KELLICOTT.

BRANCH II. — PORIFERATA.

THE sponges are even now popularly regarded as plants, although for many years naturalists have recognized them as members of the animal kingdom, while the investigations of the past fifteen years have shown them to be animals of by no means the lowest type. In the preceding pages we have seen that the unicellular Protozoa do not reproduce by means of eggs, but by a process of division or segmentation, resulting in a varying number of embryos, germs, or spores. All of the higher animals, including the sponges, are composed of multitudes of cells, each performing its own part in the economy of the individual, and while reproduction by division is frequent in certain groups, all have recourse to specialized cells or eggs for the perpetuation of the species. On account of these differences all multicellular animals have been collectively termed Metazoa, in contradistinction to the single-celled Protozoa. There is here a similar relationship to that which exists between the spore-bearing and the seed-bearing plants. In an egg-bearing animal there is a specialization of some of the cells of the tissues and parts to form the male and female reproductive elements, just as in the flowering plant there is a similar specialization of the tissues and leaves to form the male and female products and the organs of reproduction, and as the latter by the union of the sexual elements form fertile seeds, so in the Metazoa the union of the egg, or female element, with the spermatozoan, or male reproductive product, produces a fertile egg.

In the Poriferata the development of the sexual elements appears in a simple form; parts or cells of the tissues within the body of the same sponge grow larger than the rest, and become eggs while other cells change into spermatozoa. The sponges are, therefore, hermaphrodites, and besides they have no external genital or reproductive apparatus and no special apertures for the extrusion of the young. It has been found, however, that some sponges are female, or at least produce few if any sperm-bearing cells, and these sponges in some cases die soon after giving birth to their broods of young. In most sponges self-fertilization seems to take place; indeed, such would appear to be the inevitable necessity since the male and female elements are enclosed in the same membranes.

Sponges are all aquatic, are found in the waters of every part of the globe, and in suitable locations may be exceedingly abundant. So far as known they are all sedentary animals, constrained with few exceptions to pass all but the earliest stages of their existence fastened to the same submerged object to which they became attached in their early youth. The young possess powers of locomotion and can seek out new places of abode, but the adults must remain in one place and take whatever of food or fortune the passing currents may bring them. Thus they can only live and flourish in places where there are floating clouds of microscopical plants and animals, and their spores. These form their staples of subsistence and must come to them as the rain comes to the plant. They can use for the reception of food only the upper and lateral surfaces of the body, the lower, attached surface, being of course unavailable for such purposes. To this rule there are some exceptions. For instance, *Suberites compacta*, a sand sponge, has no base of attachment and is apparently capable of living with either side uppermost; there are also some wanderers, sponges which have broken away from the base and, still living, are rolled about on the bottom. Some of the commercial sponges are said to be tough enough to stand this.

The sponge is typically, or in its most perfect aspect, a vase contracted at the top. In nature it has none of the usual signs of symmetry observed in other animals, and is in most forms even very irregular. There is absolutely no forward or hinder end, except in the embryo; there is no right or left, except again in the embryo. Being a purely sedentary animal, and having no appendages, it has become and usually is designated as amorphous or formless. The conditions which influence growth have caused not only this degradation in symmetry, but they occasion, also, great differences in form in the same species. Thus, while they may be called formless in respect to symmetry, from another point of view they are really animals with more forms than usual.

Among those which live near the shores and in the varied conditions of the shallow water habitats, there is the strangest diversity. Every change of bottom, every change in the surrounding conditions of the current or the place to which the larva may become attached, has some effect upon their aspect. Thus in the same species we find flattened sheets, irregular lumps and clumps, and branching, bush-like modifications of each of these in every variety, and finally vase-like shapes, either imperfect and open on one side, or perfect and not wholly without grace of outline. If we pass from the varied bottom of the shore-line to one of uniform character, whether the mud bottoms of the deeper waters of the ocean or those nearer shore, or the sandy shallows, where the surroundings and conditions of life are more uniform, we find that the sponges inhabiting these localities are remarkable for greater uniformity of shape within the species.

Sponges exhibit most plainly in their forms the direct action of gravity and the peculiarities of the base of attachment. In a sedentary animal the fluids of nutrition would naturally tend to expend their forces primarily, in the early stages of growth at the lowest points of the periphery, and after building the base, cause the sponge to grow upwards in the direction of least resistance. This is practically what happens, and if the rock is smooth and free from other animals, some species, having no hereditary form, will grow in a broad sheet without branches; but if the base of attachment be small or crowded, the same sponge will take a bushy, plant-like outline. The force of growth which otherwise would have expended itself in increasing the sponge horizontally, is diverted by the strain on the supports or skeleton to the secreting membranes of the threads, and we find they become thicker or denser where the strain is greatest, until in some very old sponges the trunks or bases are almost solid. Above, the branches are arranged so that the form is balanced, and there is the same equal distribution of the weight around a central axis as in plants and in sedentary animals of all kinds. This tendency or response of the animal to the attraction of gravitation by equal growth in horizontal planes, so as to balance one side with another, one lateral organ with another, I have previously termed geomalism. Geomalism appears in its primitive aspect among the sponges since they are comparatively soft and supported by a pliable and primitively fragmentary internal skeleton.

It will be seen from these remarks that the form of the sponge is more largely the result of the character of the base of attachment than any other cause. When this is uniform, as in a mud or sandy bottom, the form is either vase-shaped or branching and comparatively constant; when upon rocks or irregular surfaces, all forms may occur. Another correlation has been frequently noticed by the writer. In rapid tide-ways a species, which is flat or chubby in quiet water, will tend to develop into branching forms.

This plasticity of form in response to environment also correlates with the peculiarities of the digestive system. The sponges have thousands of minute cavities within the body, devoted to performing the functions of digestion. These cavities receive their food from streams of water, circulating through a double system of tubes, and flowing in through the narrow meshes of a network, formed in the outer covering or skin of the body. With this sieve-like structure there is no use for any particular set of external appendages, and no necessity for any fixed symmetry of form. All that the sponge needs is a capability to adapt itself to its surroundings and the sole requisite of success in obtaining food is the presentation of as much surface as possible, thus securing a large supply of water and accompanying food.

Such an organism requires a peculiar skeleton. Since the internal tubes and minute stomachs would be liable to compression by the weight of the soft tissues, after the attainment of a certain size, unless some firmer framework was interposed, we find

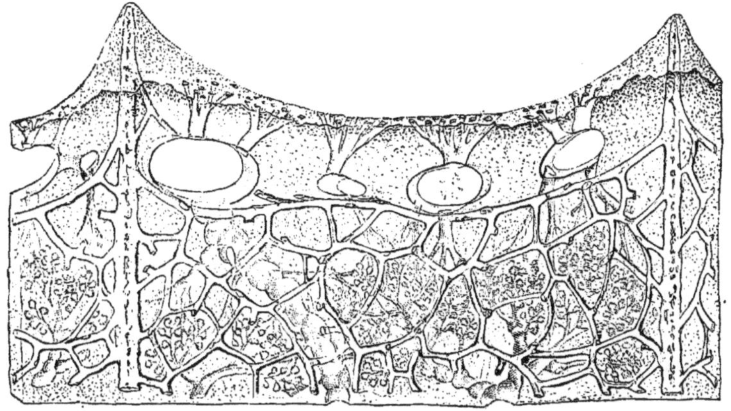

Fig. 48. — Portion of a section of a bath-sponge (*Spongia*), showing the fibrous skeleton, portions of the supply and drainage systems, and the ampullæ.

in most sponges such a supporting skeleton. In some cases this framework is formed by a woven mass of elastic threads, of a horny nature; in others the framework is composed partly of such threads and partly of stiff and unelastic spicules which may be calcareous or silicious, or in still other cases of a network of spicules united by only a small amount of horny or silicious material. The same principle of construction runs throughout the whole of the Poriferata; the skeletons are really networks or scaffolds of spicules, or of threads permeating all parts of the body, in order to support the whole mass and keep open not only the digestive ampullæ, but also the numerous tubes for supply and drainage.

A skeleton is not, however, an absolute essential in all the members of any branch of the animal kingdom; thus there are sponges entirely destitute of spicules or threads, but these are mostly flattened or small vase-like forms, in which the weight is small in proportion to the strength of the tissues.

In the commercial sponges the skeleton is an intricate mass of interwoven elastic horny threads, as may be seen by slicing one through the middle (Fig. 48). This network is permeated by numberless tubes, but these can be reduced into two systems, one leading from the interior outward, and the other leading from the external surface toward the interior. The first or internal system is composed of several large trunk tubes, largest interiorly, but branching and becoming smaller as we approach the exterior. The outer surface of the sponge is ornamented with projecting bunches or ridges of threads. Between these projections there are numerous depressions, the bottoms of which are perforated by openings of medium size, which we can follow as tubes leading into the interior by examination of the cut surface of the section. These are the tubes of the external system. They often terminate abruptly, but here and there are divided into branches, and we can see that they really diminish in size towards the interior. Not infrequently these tubes may be traced directly into the trunks of the internal system, but in this case, their walls are thickly set with the openings of small tubules which lead into systems of tubes diminishing in size internally, and therefore belonging to the external system. The dried skeleton looks as if there was no room for fleshy material between the meshes, but the increase in size upon wetting a sponge shows that when in the natural element and fully expanded there is plenty of room between the threads for all the organs we have to describe.

The surface of the living commercial sponge is of a dark color, and some species, were they smoother, would remind one of a piece of beef liver. On the upper surface we can see large crater-like openings as in the skeleton, but the surface is otherwise quite different. The tufts of fibres and the depressions between them, which are so marked in the skeleton, are more or less covered with a skin which conceals all the cavities and channels. The tufts, however, do show themselves as slight prominences, while the skin over the intervening depressions is smooth and perforated by groups of holes. These small holes may be opened or closed at the will of the animal, and when open they serve to admit water freely to the external or supply system of tubes. These openings may in many sponges entirely disappear, and new apertures be formed when needed. This faculty has, however, been greatly exaggerated.

The superficial cavities are lined with a smooth skin, lighter in color than that of the exterior, while the sides and bottom are perforated by small holes, the openings of the tubules which line the skeletal tubes of the external system and form the fleshy canals of the supply system. These tubes are lined with a light colored skin and branch as they descend into the interior. The tips of the minute branches expand into globular sacs. These little enlargements, the ampullæ, open in turn, into small fleshy tubules which line the internal system of tubes of the skeleton. They constitute what may be called a drainage system, and instead of growing less, they increase in size as they go inward, and by uniting with other similar tubes, they form larger and larger branches until they finally open into one of the central trunks.

These sieve-like openings, the superficial hollows, and the supply system act as feeders, bringing water loaded with nutriment to the ampullæ or digestive sacs. After digestion the refuse is passed out of the ampullæ into the internal system and thence into the large central trunks which finally open on the outside of the sponge in large crater-like orifices. In some sponges these two systems of canals are not distinguishable and there is but one outlet to the ampullæ.

The outermost covering of the body is an extremely delicate membrane composed of a single layer of flat cells, giving a peculiar shade of purple bloom to the living sponge, but being easily abraded by rough handling. This layer is the ectoderm, and is continuous at the edges of the craters with a somewhat similar layer, lining all of the passages of the drainage system, which should be considered as the endoderm. To this latter system the ampullæ belong, but the endoderm which lines them is of a different character. The tubes of the supply system are doubtless of ectodermic origin. The endodermal cells are usually flat and have polygonal outlines, except in the ampullæ, where they give place to oval or even columnar cells, the free ends being crowned by transparent collars, from the centre of which protrudes a long flagellum (Fig. 52). These collared cells have unusually large nuclei. The ectodermal cells vary somewhat in outline, according to position, but are usually hexagonal or quadrangular and rather constant in form. The cells of the endoderm, or the contrary, are subject to extraordinary changes, bulging out into balls on their free side when gorged with food, or extending to hair-like cells of enormous length when stretched across an opening.

Between these two layers lies the middle or fleshy layer of the body, the mesoderm. This is composed of cells, but the intercellular spaces are so abundantly filled with protoplasm that Haeckel and others consider it as a characteristic of the sponges. We are, however, of the opinion that the abundance of intra-cellular substance has been greatly exaggerated, and that the mesodermal cells are numerous and closely aggregated. Such we have found to be the case with the Calcispongiæ and *Chalina*, and Lieberkuhn and Huxley claim the same for *Spongilla*. The cells

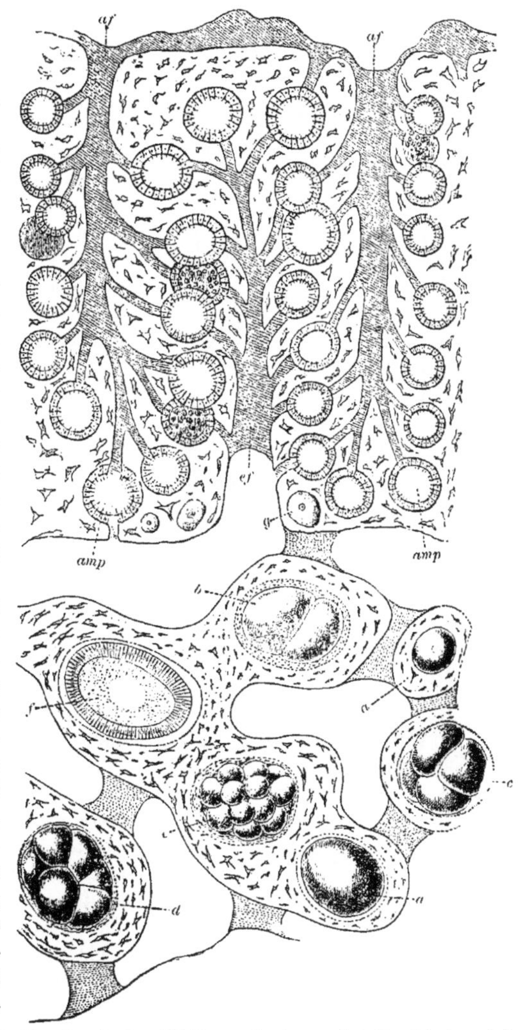

FIG. 49. — Section of *Halisarca*, showing supply (*af*) and drainage (*cf*) systems, the ampullæ (*amp*), and eggs in various stages of development (*a, b, c, d, e, f*).

of the mesoderm vary considerably in character and appearance. They may be transparent, granular or deeply colored, globular or elongated, entire or amœboid in outline, and capable of extensive changes by expansion or contraction. In many sponges there occurs between the undoubted mesoderm and the ectoderm distinct layers, the origin of which is uncertain.

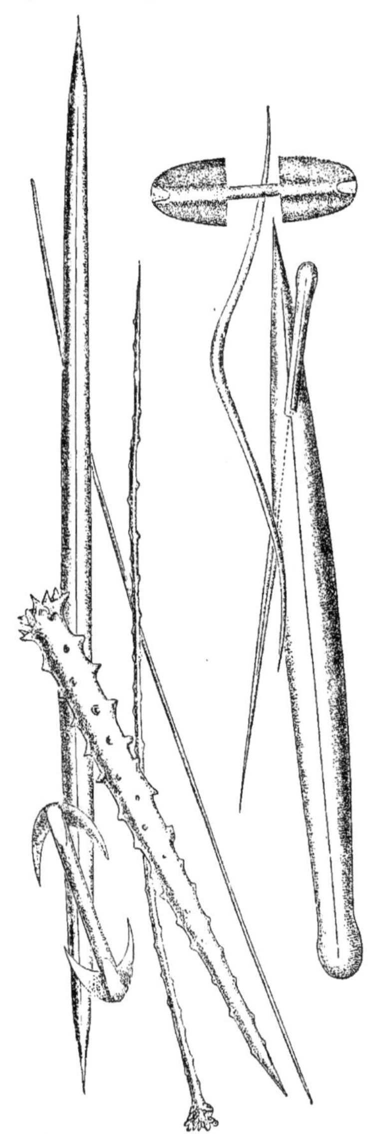

FIG. 50. — Different forms of sponge spicules.

One of the most interesting points to the naturalist lies in the history of the skeleton and its elements. This consists of two parts, the thread of binding substance of horn or keratode and the hard mineralized spicule. All authors apparently agree in considering the spicules as mesodermic, but the origin of the threads has not been so thoroughly worked out. Barrois, however, considers them of ectodermal origin in the silicious sponges, and the author has expressed the same opinion regarding the fibres of the horny sponges. In the Chalininæ the same would also appear to be true. The skeletal threads of *Chalinula* are surrounded by a special membrane, which I have seen in several instances, and which may be called the perifibral membrane. This is composed of flat epithelial cells, either transparent or deeply colored by granules. They somewhat resemble the cells of the ectoderm in outline, but are longer, fusiform in outline, very closely set, and usually spirally arranged around the fibre. These are evidently the cells which secrete the threads, and in one section I followed this sheath and found it continuous with the ectoderm. We can thus readily account for the skeleton of *Chalinula* by the presence of invaginated prolongations of the epiderm which would naturally follow and surround first the vertical threads and then others arising in all directions. The differences in the structure of the inner and the outer portions of the fibres of the Aplysina, and their often hollow condition, can only be accounted for by this explanation as well as the fact that in *Spongia* and its allies the centre of the threads is frequently occupied by foreign matter, carried in from the exterior by the invagination of the ectoderm to form the sheaths and subsequently enveloped by the horny matter secreted.

The form of the spicules varies greatly, and affords good systematic characters. A few of the forms are shown in the adjacent figure. Some are pointed at one end, some have both extremities acute, while others may terminate at one or both ends like anchors. They may be smooth or variously knobbed and ornamented.

We cannot hope to disentangle the intricate relations of the parts in such confused structures as the sponges without studying the history of their development. The young can always be relied upon to present the observer with simpler or more elementary conditions, and generally help us materially in understanding and translating the adult structures.

As we have said, the male and female elements are found within the sponge. After fertilization, the egg undergoes a regular segmentation, and then the two ends of the body become distinguishable, one being composed of smaller cells than the other. The embryo is hollow at this the so-called morula stage, but soon the central hollow, the segmentation cavity of embryologists, becomes filled in the following manner. The cells of one end of the embryo become pushed in, much as one inverts the finger of a glove, and these constitute the inner layer or endoderm of the young sponge. In this, which is called the gastrula stage, there are then two layers. In the calcareous sponges they form a cup with a mouth at one end, but in the carneosponges the gastrula is usually but not invariably solid, the invaginated endoderm completely filling the interior. The mesoderm is developed between these two layers, but from which one is not yet known. The spicules begin to be formed in the mesoderm soon after its appearance, and seem to be due to direct transformation of single cells.

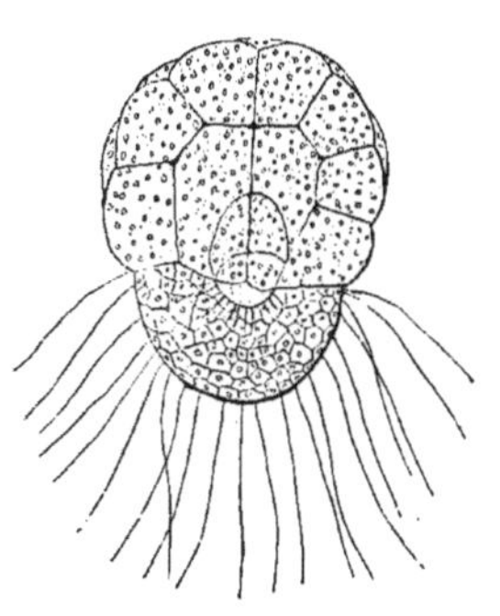

FIG. 51.—Free swimming young of *Sycandra*.

These young larvæ swim rapidly through the water by means of the cilia, or small hairs, which clothe the exterior, and which can be moved like so many oars with force and rapidity at the will of the tiny animal. The smaller end in the larva of the calcareous sponge is foremost as the little creature moves aimlessly about. When it encounters any obstacle it usually exhibits no ability to back off, but manages by keeping its cilia in constant motion to get away by rolling around the obstruction. At last the embryo settles down, with its mouth or blastopore below, upon the space to which it is to become attached. The membranes at this end form a sort of sucker, which spreads itself out and enables the animal to exclude the water between it and the surface to which it is being applied. The pressure of the water holds the sponge in its place, and on some smooth spots this may continue to be its only anchorage, but in rougher situations it naturally acquires additional hold by growing into any cavities or around any projections.

FIG. 52.—Section of attached embryo of *Sycandra*; *a*, primitive stomach; *b*, blastopore; *e*, ectoderm; *h*, endoderm; *s*, segmentation cavity.

On soft, muddy ground fresh-water sponges usually begin to grow upon some small substance, which often is very small, and then the weight of the growing sponge may sink a portion of the stalk into the mud below. This portion then dies, but even when dead it plays its part and forms an anchor for the whole structure. We cannot imagine an ordinary sponge growing upon a muddy surface unless the water was absolutely still or the mud hard; otherwise the tiny creature would be suffocated by the sediment. The deep-water mud sponges of the sea (*Hyalonema*, etc.) have, however, grown so long on soft bottoms that they have developed a system of threads which, protruding below, penetrate deeply into the mud, and may either serve as anchors or bases of support. The most curious case of this kind occurs in *Tethya gravata*, a globular form, in which the threads form a network below, enclosing small stones and gravel. Thus the animal carries ballast, and if turned bottom up in the water it rights itself immediately. When rolled over by the waves upon the muddy bottoms of Buzzard's Bay, where it occurs, it is always sure to end its gyrations right side up, like a bit of leaded pith.

The observations of Schultze on the young of *Sycandra*, one of the calcisponges, show that the ciliated cells, when invaginated, form an ampullaceous sac, confirming the view that we have always held that the typical sponge was a single, isolated ampulla, surrounded by the two layers of the body. A single pore is opened into this sac, and this completes the likeness to one of the Ascones group. The observations of Barrois, Carter, Schultze, and Marshall all seem to show that the ampullæ in the silicious sponges have a different development. After the larva has settled, a hollow space appears in the body of the sponge, lined by a non-ciliated endoderm. The ampullaceous sacs arise as buds from this endoderm, communication with the exterior is formed by tubes, which arise as invaginations of the ectoderm, and grow inward, uniting with the ampullæ.

The evidence at present seems to be in favor of Barrois' opinion, that the water flows in through these lateral pores and accumulates in the interior, assisting to raise the soft tissue into a dome or spire, until, at last, unable to withstand the pressure, the top gives way, and the crater is formed. This accounts for the rise of the spire before the formation of the crater, and gives a reason for its disappearance after the pressure has been relieved by the formation of an adequate outlet. Certain it is that the crater is not in any sense the mouth or blastopore of the sponge, as is usually supposed. Thus the cloacal apertures have no special morphological location, and arise as purely mechanical necessities, as do the excurrent openings of all colonial forms.

The simplest sponges have only a single body cavity, surrounded by ectoderm and mesoderm, and lined by the inner layers. This typical form or vase shape occurs in the young of the calcareous sponges and in the adults of the Ascones. Individuation in these forms is complete and simple; they are each equivalent to a single ampullaceous sac, separated from any other sponge and surrounded by mesoderm and ectoderm. It is evident, therefore, that when a number of these sacs still remain connected with the body cavity, each additional sac must be regarded as a bud or offshoot from the cœlomatic cavity, and the whole can be regarded as a branching gastro-vascular system, through which water and food are circulated and excrement discharged.

The active collared cells of the ampullæ are both structurally and functionally, as was pointed out by H. James Clark, similar to the zoöns of the flagellated Protozoa; they have the same organization, catch their food by means of the same slender lash, swallow it at the same place within the collar, and throw out the refuse matter in precisely the same manner. The Flagellata are individuals, each having the typical structure of the Protozoa, and though in every respect simple cells, with collars and flagella, as in the separate cells of the sponge, they are not shut up in sacs inside of a mass of flesh, but are free or attached animals, getting their food in the open water. This correlation and the aspect and functions of the cells which form the tissues of all structures in the bodies of sponges and higher animals show us that all cellular tissues must be regarded as aggregates or colonies, while the single cells of which the tissues are composed are the exact morphological representatives of the

Protozoa. The sponges are simply less altered than other animals, the cells of the inner layer still retain some traces of their original structure, and we have to rate the Poriferata as intermediate in these characteristics between the Protozoa and the Metazoa.

The word 'individual' leads to many serious misconceptions owing to its popular meaning, and we use the word zoön for any whole animal or part of a colony of animals whose structure can be said to embrace the essential characteristics of the grand division or branch to which it belongs. In this sense the single cell is a zoön, with regard to the whole animal kingdom, or when we wish to contrast the Protozoa with the Metazoa. The young sponge, at the period when it has but a simple cœlomatic cavity and one opening, is also a zoön, but it is only a zoön when we wish to consider the Poriferata by themselves.

We can test this position by comparison with the simplest known forms of sponges, such as the Ascones. The forms of this group have a vase shape, with only one opening above, while the pores for admission of water are formed as wanted. The structure and form of this adult sponge is similar to that of the simplest ampullaceous sac, and is also similar to that of the young when the cœlomatic cavity is first formed, and it shows us that all these three forms contain the essential elements of sponge structure, and can thus be appropriately called spongo-zoöns.

After the three layers are fully formed, the cœlomatic cavity extends itself in every direction by the formation of ampullæ as outgrowths from its sides, but these outgrowths do not carry with them the mesoderm and ectoderm. On the contrary, the outward growth and the formation of a new ampullaceous sac, which is the nearest approach the sponge makes towards the formation of a new zoön, takes place wholly inside of the mesoderm, and the outer layer remains unmodified. This is the case in all the sponges with a thick mesoderm, and even among the higher forms of calcisponges. Among the primitive Ascones, however, a bud from the side carries with it all the membranes of the body, and is a repetition of the original zoön, a complete bud or 'person.'

New craters are formed anywhere as the sponge increases in size, by the conjunction of canals of the drainage system and without the slightest signs of budding, and yet Haeckel and others regard each of these craters as a person or individual. The mass may grow out solidly into a branch with a dozen craters, then, according to these authors, it is one dozen small 'persons,' or as it grows out, the dozen small canals may unite and form one canal and one crater, then it is one 'person.' There are plenty of examples in which such variations occur on the same stock, and we think they prove that the accepted ideas of what constitutes an individual or person among the sponges with a thick mesoderm and branching gastrovascular canals are entirely erroneous and founded on the deceptive resemblances of the branches of a sponge to those of other compound forms, which really arise from true buds and are true zoöns.

Haeckel and others have regarded all the vase-shaped sponges as single individuals or zoöns, but this seems untenable, except in the group of Ascones. It is not uncommon to trace the form of the same species among living sponges from a flattened disc with several craters to a vase shape, the vase being built up by the more rapid growth of the periphery. The inner portion of the ectoderm or top of the animal thus becomes internal, and the opening above, the crater of one large 'person.' Here the so-called zoön is formed by a transformation which can be clearly proved to be the result of the growth of the external parts. It is evident that the mere fact of the existence of a cloacal outlet does not necessarily indicate the presence of an individual. We must regard the whole mass which springs from one base as being an individual, while the buds or branches which may arise from it are not branches, but may be regarded as the prototypes of true buds and branches of colonial animals in other divisions of the animal kingdom. They resemble the branches of other colonies in aspect, and arise from unequal growth of parts in a more or less symmetrical way, and may have any outline essential to the equilibrium of the form, but are no more individuals than are the arms and legs of a human being.

The whole mass is the individual, and the fact that it has a branching gastrovascular system is accounted for by the budding of the cœlomatic cavity just as the gastrovascular system of the Hydrozoa and the water system in echinoderms is formed by the prolongation or budding of the walls of the gastric cavity of the larval forms. In fact the similarity of these parts in the Cœlenterata and Echinodermata indicate to us that the sponges present a much more primitive condition of the gastrovascular system than do any of the higher types. In the echinoderms, the system becomes separated into the gastric cavity and the system of water tubes in the early stages; in the Hydrozoa and Ctenophora, the two remain in connection and a true water system is not developed. In the sponges there are two systems, the supply or water system and the cloacal or gastric system, and these two together make a complete gastrovascular system which, however, is more primitive than either of the other types, combining both the gastric and the water systems in a double set of inter-communicating canals. It is difficult to explain the similarities of the water systems among these animals on any other grounds, and this view enables us to throw some light upon the similarities of the cœlomatic cavity. This cavity is merely the primitive hollow of the body of the embryo, and in many of the lower forms, as the Hydrozoa, it is the digestive cavity, the cells being modified for assimilative purposes. This is only the next stage above the cellular mode of digestion in which each cell performs this function as in the ampullæ of the sponge, and it is an adaptive change both in the structure of the cells and their function.

If our view of the affinities of the sponges is correct, this cavity in the Ascones is directly derived from the communal inlet and outlet of some colonial form of Protozoa, and the water system must have arisen subsequently in response to the budding of the cœlomatic cavity, and the need of special sources of supply for each bud with its ampullæ. The correlations in the structure of the feeding cells of sponges and the aspect and similar functions of the cells which form the tissues of higher animals show that not only the sponges, but all the Metazoa, however highly individualized, must be regarded as aggregates or colonies in which each cell represents a zoön of the Protozoa, and which has derived its structure by inheritance from an ancestral protozoon. That is to say, there is such a phenomenon as the inheritance by the single cells of a metazoon of the peculiarities and even the tendencies of the independent individualized protozoon, and from this results the communal characteristics of the metazoon which appears to be, but in reality is not, a simple individual. The only simple individual in the animal kingdom is the single unicellular protozoon, or a single cell from the tissues of the Metazoa.

This view, though for a time overwhelmed with ridicule, has of late years obtained a quite general acceptance. It dates back to an inspiration of Oken in 1805. The transitions by which it could have taken place have never been satisfactorily stated nor can we here do anything more than add another step towards a final solution. It is now well known that there is no ascertained limit to the action of heredity, and that not only observable characteristics, but even habits and tendencies may be directly transmitted from one generation to another. There is a universal and necessary law of heredity which can be used in bridging the gap between the Protozoa and the Metazoa which may be briefly formulated as follows: All animals exhibit a tendency to inherit the characteristics of their ancestors at earlier stages than those in which these characteristics first appeared. Thus, if an ancestor or radical form acquired a new character or took on a new habit when adult, this would tend to reappear in the descendants at earlier and earlier stages, and in course of time would become carried back to the adolescent, then to the larval stages, and finally either become useless and disappear altogether, or if useful, and therefore retained, become restricted to embryonic stages. This law was first advanced in 1866, by three persons, Haeckel, Cope, and the writer, almost simultaneously.

The Ascones are certainly, so far as known, the simplest or most generalized of the Metazoa, and approximate to the Protozoa in such a way that it is possible with the aid of the law of concentration of development to explain the transformations by which such an organism could have risen from the Protozoa. The egg of all Metazoa is in its first stages a simple cell, and like all other cells is a homologue of an individual or zoön among the Protozoa. This primitive egg cell has but one mode of growth by which it forms tissues. It divides or segments and builds up the primitive tissues of the embryo by a similar process to that by which colonies are formed among the Protozoa. Therefore the egg after segmentation is no longer a single zoön, equivalent to a single zoön among the Protozoa, but a mass of such zoöns, differing from the mass of a free colony of amœboid Protozoa in about the same way that the included cells of an ampulla differ from a colony of Flagellata.

Among colonial Metazoa we frequently find on the same adult stock, individuals devoted to the performance of distinct functions, and having their shape and structure so modified thereby as to differ widely from each other, though in their younger stages they were more nearly alike.

Thus, on the same hydrozoan stock we may find females loaded with eggs, males carrying only sperm cells, sexless polyps devoted wholly to alimentary purposes, others with only defensive functions. We have therefore excellent reasons for assuming that similar transformations took place in the transition from the simple colonies of the Flagellata to the more complex condition of the sponges, and we can make a picture of these changes in strict accord with the laws of morphology.

Throughout the Metazoa as well as in sponges, the external layer or ectoderm is protective and builds the protective armor, scales, etc.; the mesoderm is essentially devoted to the formation of flesh and organs of support, while the endoderm is devoted to the function of digestion, and the elaboration of all the parts concerned in this process; but everywhere these three layers are derived from one cell (the egg). If now we imagine a series of changes, beginning with any flagellate protozoön, and following out the indications of embryology, we should first have a sheet of attached flagellate feeding forms; secondly, these surmounting or arched above a base composed solely of supporting individuals without collars or flagella; thirdly, the outermost losing their flagella and collars, would become simply protective pavement cells, while the central ones retain their digestive functions; the change slowly becoming more complete, and the central ones acquiring a capability of being withdrawn into the interior when alarmed. The last step would be the inheritance of the invaginated condition, and this would give the vase-shaped Ascones. The inheritance of the invaginated stage or of the primitive differentiation of the colony into protective and feeding zoöns in any encysted egg form would be necessarily attended by the formation of a globular shape in which one end would have cells of a different kind from the other, one being composed of endodermal cells inheriting the digestive functions of the original colony, while the other would be formed of ectodermal cells arising from the protective zoöns. This encysted form would be composed of but one layer of cells, and therefore have a hollow interior, and the supporting zoöns or mesoderm would be formed between the other two membranes when it became necessary by the protozoon method of reproduction by fission.

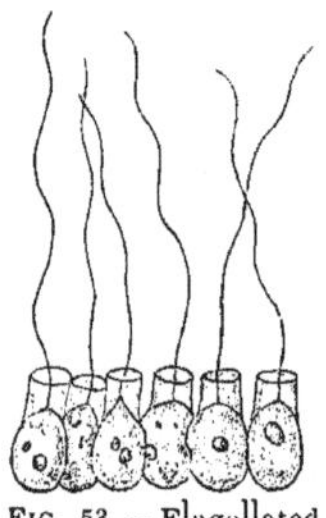

Fig. 53. — Flagellated cells from ampulla of *Sycandra*.

We can also reverse this explanation and imagine a sponge, one of the Ascones, being reduced to a protozoön; losing first the form, then the supporting layer, then the protective cells, and finally becoming converted into a layer of zoöns, each of which would closely resemble those to be seen in Fig. 29. The validity of this comparison may be seen by comparing this figure of *Codosiga* with the flagellated ampullaceous cells of a true sponge shown in Fig. 52; and the comparison will also gain when we recollect that in the young of the flagellated protozoön the stalk is absent.

The normal action of the law of concentration and acceleration of development would alone have caused such changes in the modes of growth of the Metazoa if the latter were really the descendants of the Protozoa, and this series of transformations is included when we say that the Metazoa, in accordance with this law, have inherited the tendency to form colonies or tissues by fissiparity, at an early stage in the existence of the cell or zoön. Thus, the individualized protozoanal stage has become confined to the earliest periods of existence instead of being more or less permanent and characteristic of the later stages of growth as in the Protozoa. When the colony becomes embryonic the process of multiplying by division is, as a necessary consequence, also accelerated and concentrated, and tissues are rapidly formed for different purposes. We can therefore, without calling to our aid any but the well-known effects of habit and the law of concentration of development, account for the segmentation of the egg, and the subsequent tendency of the primitive tissue to give rise to the three layers of the Metazoa. The fact that certain cells become differentiated into eggs, and others from the same or other layers, into spermatozoa, is not more remarkable than that certain zoöns of many colonial Metazoa, like the hydroids, are exclusively egg-bearers, while others are solely sperm-bearers. The fact that the male element seeks out the egg and becomes merged in it, is paralleled by the process of conjugation among the Protozoa. The sperm cells of the Metazoa, like the germs of the Protozoa, arise by division of a single cell, and, although frequently of similar shape to and swimming freely like many protozoon germs, they do not wait until maturity before conjugating with the females. This is plainly only the inheritance of the tendency to conjugate at an earlier stage, and is a natural result of the law already laid down.

It is well known that there is a tendency to reproduce after conjugation, and that conjugation is performed by Protozoa of different sexes, and also that there are sexual colonies among the higher Protozoa. The result of differentiation or progress is evidently towards the formation of sexual differences in the Protozoa as in other branches of the animal kingdom, and if our view is correct, we ought to expect that the Metazoa, springing from the Protozoa, would show similar tendencies toward differentiation of the colonies. If, as in the sponges, the lower forms had male and female cells in the same body, then the progress of differentiation should lead to a more decided separation of these functions so that some would produce only female and others only male cells. In other words, the complete separation of the sexes would take place by a perfectly natural transition, and we should have male metazoöns and female metazoöns.

The sponges are frequently regarded as degraded Metozoa, but to the author this view seems highly improbable. Huxley first recognized the systematic importance of the sponges, but contrasted them as a division with the rest of the Metazoa, while MacAllister, and subsequently the author, gave them their true taxonomic rank as an independent branch of the animal kingdom.

Class I. — CALCISPONGIÆ.

This division is somewhat inappropriately named for the reason that some of the genera have no skeletons, but this objection might, with equal justice, be made with regard to the names applied to the other groups. The animals of this class have fusiform or cylindrical bodies which may be single with one cloacal aperture, or branching with an aperture at the end of each branch, or more or less solid as in the other sponges. When a skeleton is present, the spicules which compose it consist of carbonate of lime, and their longer axes are arranged in lines parallel with the canals, that is at right angles to the inner and outer walls of the sponge.

Order I. — PHYSEMARIA.

This order contains the remarkable genera, *Haliphysema* and *Gastrophysema*, which, according to Haeckel, are nearer in form and structure to his archetypal animal form, the gastrula, than are any other adult animals. They are small and vase-shaped in *Haliphysema*, while *Gastrophysema* may have from two to five chambers. There is but one aperture above, and the water is drawn into this by ciliary action. According to Haeckel, the body wall consists of but two layers, the ectoderm and the endoderm, but it is evident that the ectoderm of the German savant can be nothing else than mesoderm, for it is composed of loose cells and intercellular protoplasm, while the true ectoderm of all sponges is a simple pavement epithelium and never a compound tissue. In Haeckel's figure, the whole interior of *Halisphysema* is paved with ciliated cells, among which are interspersed amœboid cells. Both Haeckel and Bowerbank deny the existence of pores, and it is not likely that even transitory openings would have escaped their observation.

The strangest part of the history of the Physemaria is that both Carter and Saville-Kent claim that *Haliphysema* is a true protozoön, and Kent's figure, which is as specific as Haeckel's, depicts a true foraminifer. These observations render it very uncertain whether the group should be referred to the sponges or to the Protozoa. *Gastrophysema* may be a true sponge, and we therefore describe the order in this connection. Mr. J. A. Ryder describes as an American representative of the group, a curious club-shaped animal with a tough cortex and a cellular interior, under the name *Camaraphysema*.

ORDER II. — OLYNTHOIDEA.

These forms differ from those of the last order, in having a skeleton of calcareous spicules. These may be either straight and needle-like, or one end may bear three or four rays. The spicules, which are of mesodermal origin, are arranged at right angles to the inner and outer walls of the body in the tube or vase-shaped forms, and the rays being interlaced, afford a firm scaffolding for the support of the walls. Here, as among the Physemaria, Haeckel claims that an outer epithelial membrane is absent, but Grane, Schultze, Metschnikoff, and others have repeatedly demonstrated the existence of a true ectoderm, and the writer has seen this membrane several times in the living sponges. The species are usually colorless, generally of small size, and although abundant along our coasts must be looked for carefully under stones and upon sea-weeds. They are found exclusively in shallow water, and, with few exceptions, do not occur on muddy bottoms.

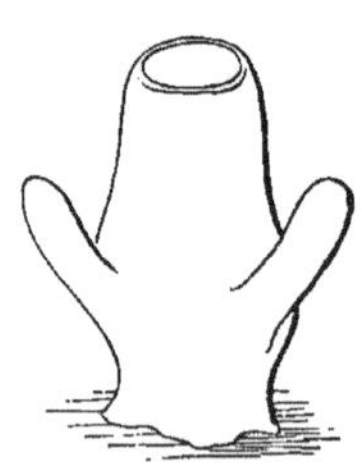

FIG. 54. — *Ascaltis*, one of the Ascones group.

SUB-ORDER I. — ASCONES.

These forms have a vase-like shape, are thin-walled, and have a distinct skeleton formed of a single layer of triradiate spicules, their bases outward, while the pores of the supply system are formed as they are needed, through the sides. The inner or cœlomatic cavity is lined with flagellated and collared cells, while these, of course, are not found in the transient supply-canals, which, according to Haeckel, are but temporary openings in the sides of the sponge.

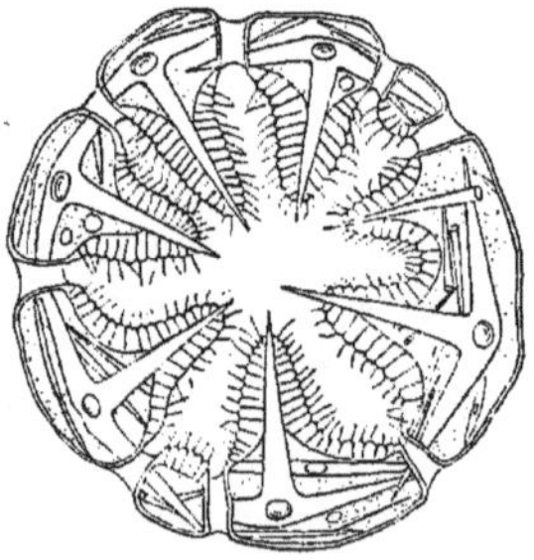

FIG. 55. — Diagrammatic section of *Ascaltis*.

SUB-ORDER II. — SYCONES.

The typical form for the members of this group is that shown in the figure of *Scycandra ciliata*. The individuals are attached by the base or small end, and are very like those of the Ascones, but are stouter and more frequently spindle-shaped, while the walls are thicker and more opaque. They are, however, quite distinct in their structure. The flagellated and collared cells are confined to the cavities of the permanent supply canals, where they occupy special cavities, the ampullaceous sacs. The cells of the cœlomatic cavity are flattened and similar to those of the ectoderm. The mesoderm is very thick, and the canals radiate with great regularity from the cavity of the sponge to the exterior. The outer part of each canal represents the supply, and the inner, the drainage system. The spicules are usually in two rows, their radiated bases being turned, respectively, inwards and outwards. While the living species are somewhat numerous, but one fossil genus is known, and this is of Jurassic age.

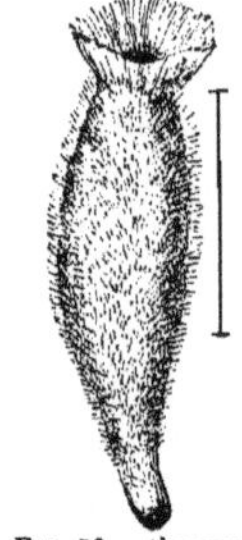

FIG. 56. — *Scycandra ciliata*.

Sub-Order III. — Leucones.

The higher position of this group is shown by its greater complexity. The mesoderm is thicker than in the last sub-order. The canals of the supply system are irregularly branched and frequently anastomose with each other, forming cavities near the outer surface. The collared cells are distributed along the smaller canals in the lower forms, while in the higher they are confined to the ampullaceous sacs. The Leucones represent the massive growths of the Keratosa and Silicea, but usually have a common or united cloacal aperture and are composed of consolidated tubes. Zittel considers this group a lineal descendant of the next, a view which does not seem to be justified by the morphology or the mode of development of the individuals of the group. No fossil forms are known.

Sub-Order IV. — Pharetrones.

This division was established to contain a number of forms which occur as fossils in the rocks between the Devonian and the end of the cretaceous period. The author is inclined to consider the genus, *Trichonella*, which is represented by a species in Australia, as a living member of this group. The spicules are so united as to form irregular threads, and sometimes a very intricate network. The canal system was branching and irregular, while the mesoderm must have been very thick.

Class II. — CARNEOSPONGIÆ.

With increasing knowledge the multitude of forms comprised in this class will doubtless be separated. The common characters are a very thick mesoderm, the ectoderm and endoderm similar to that of the Leucones, and the supply and drainage system as described above in the commercial sponge. The skeleton may be either composed of horny material (keratode) or partly or entirely of silicious spicules. The skeletal elements are radiately or irregularly arranged according to the plan of canal system which it supports. One order has no skeleton, but the form and structure show it to belong to this class.

Order I. — HALISARCOIDEA.

This order, which Haeckel calls Myxospongiæ, embraces but a single genus of fleshy sponge, known as *Halisarca*. One species is common on our shores, and also on those of northern Europe. The animals grow usually in flat masses or little bunches of a dull color, coating rocks or surrounding the stems of marine plants. The general structure can be seen in Fig. 49. With a fleshy nature, of course no fossils of this order can occur.

Order II. — GUMMININÆ.

These are tough and leathery sponges, the external layer forming a cortex which is partly composed of fibres, which also permeate the central mass surrounding the canals, and also penetrate the mesoderm. Their composition is still unknown. The genus *Chondrilla* has star-shaped silicious bodies in the cortex which are not found in *Gumminia* and in the other genera. No fossils are known.

ORDER III. — KERATOIDEA.

These are the true horny sponges, and from an economic point of view, are the only ones which have any practical value. The skeleton consists of fibres of sponge-horn or keratode, forming a network in the mesoderm. They are littoral forms not usually found in water more than seventy-five fathoms in depth. They generally avoid sandy or muddy localities, preferring rocky ground or coral reefs. Passing by the genus *Darwinella*, for which a sub-order has been formed, we come to the sponges of commerce.

SUB-ORDER I. — SPONGINÆ.

The Sponginæ are characterized by having the fibres of the skeleton solid, but in places where the water is filled with floating matter, they usually have a core of foreign material, a fact which we have previously mentioned.

The marketable kinds are all of one genus, *Spongia*, that from which all the sponges derive their common name. There are only six species with, however, numerous varieties, which are offered for sale, and in fact these may be reduced to three species, if one so chooses. Three of the species are from the Mediterranean and the Red Sea, and three from the Bahamas and Florida. Other species of this genus have a very general distribution, but they are all confined to the equatorial and temperate zones within an area on either side of the equator which is limited by the isotherm or average temperature for January of 50° F. The *Spongia graminea* and *Spongia cerebriformis* are occasionally used in Florida and Bermuda, but are not exported.

The marketable sponges owe their excellence to the closeness, fineness, and resiliency of the interwoven fibres of the skeleton. The Mediterranean appears to be particularly favorable to the production of specimens with skeletons possessing these desirable qualities in the greatest perfection. Those from the Red Sea are next in rank, while those of our own shores, though corresponding species to species with these and the Mediterranean forms, are coarser and less durable. Thus *Spongia equina*, the Horse or Bath Sponge of the Mediterranean, is finer than the *Spongia gossypina*, the Wool Sponge of Florida and Nassau, though it otherwise resembles it closely. *Spongia zimocca*, the Zimocca Sponge, represents in the Mediterranean waters the much coarser *Spongia corlosia* and *Spongia dura*, the Yellow Sponge and Hard-head, on the American side. *Spongia adriatica*, the Turkey Cup-Sponge and Levant Toilette-Sponge of the Mediterranean, answers to the finest though not the best of our sponges, *Spongia tubulifera*.

It is probable that the Red Sea and the Mediterranean were both colonized by sponges from the Caribbean Sea, and, strictly speaking, the six marketable species ought to be classed as three species with six principal varieties, differing from each other according to their habitat. This conclusion is borne out by the facts that the Caribbean Sea contains more species of this genus than any other locality, that no marketable sponges are found in the Indian or Pacific oceans, and that the differences in quality cited above are occasioned in these and other sponges with fibrous skeletons by any change from shallower to deeper water, or from water loaded with sediment to clearer waters. In each of these cases a finer sponge is the result, and this correlates directly with the fact that even in the Mediterranean the marketable kinds are found in waters which are probably very rarely reduced, even during the month of January, to 55°, and perhaps for the best qualities not below 60° F.

Marketable sponges are found in the Mediterranean on the coast between Ceuta on the African side, and Trieste on the Adriatic. None are found in the Black Sea, on the coasts of Italy, France, or Spain, or the Islands of Corsica, Sardinia, the Balearic Islands, or even Sicily. The species do not usually appear in water deeper than thirty fathoms. They are gathered by means of hooks on long poles, or directly by the hands of divers, or, as in the ease of some of the coarser kinds, dragged up roughly by dredges. When secured they are exposed to the air for a limited time, either in the boats or on shore, and then thrown in heaps into the water again in pens or tanks built for the purpose. Decay takes place with great rapidity, and when fully decayed they are fished up again, and the animal matter beaten, squeezed, or washed out, leaving the cleaned skeleton ready for the market. In this condition, after being dried and sorted, they are sold to the dealers who have them trimmed, re-sorted, and put up in bales or on strings ready for exportation. There are many modifications of these processes in different places, but in a general way these are the essential steps through which the sponge passes before it is considered suitable for domestic purposes. Bleaching-powders or acids are sometimes used to lighten the color, but these, unless very delicately handled, injure the durability of the fibres.

The first fossils which undoubtedly belong to this group occur in the carboniferous rocks, Carter having described a *Dysidea* from this period, but masses supposed to belong to the Spongínæ have been found in much older rocks.

SUB-ORDER II. — APLYSINÆ.

These sponges have a skeleton composed of fibres which are hollow or filled with a soft, friable core, and there are no foreign materials introduced from the exterior. The fibres are not as elastic as those of the Sponginæ and generally are much larger and coarser. The sponges are dendritic or grow in sheets, often of considerable size. The skeletons are more open in structure than in the last sub-order, and frequently the fibres have a fan-like arrangement. No fossils are known.

ORDER IV. — KERATO-SILICOIDEA.

As the name implies, this division forms a transition between the horny and the silicious sponges. The skeletons are formed of solid keratose fibres and silicious spicules.

SUB-ORDER I. — RHAPHIDONEMATA.

In this group the spicules are of one kind only, usually with pointed ends, and are loosely arranged in the vertical and horizontal fibres of the skeleton, and covered by the keratode, though often but slightly bound together. The keratode is light-colored or transparent. This division is represented on our eastern coast by the well-known Dead-man's-finger Sponge, *Chalinula oculata*. This is a bushy form, common on piles or rocks, especially in tide-ways where there is a considerable current of clear water. It sometimes grows to a height of two feet. The cloacal openings are small and irregularly scattered over the surface of the branches, while the pores are imperceptible to the naked eye. The color is brown, sometimes softened by a warm undertone of pink. The fibres of the skeleton are light brown, and friable when dry. The fleshy parts disintegrate so readily after death that these sponges cannot be kept for any length of time even in the strongest alcohol. *Tuba* is another genus of this group. No fossils are known.

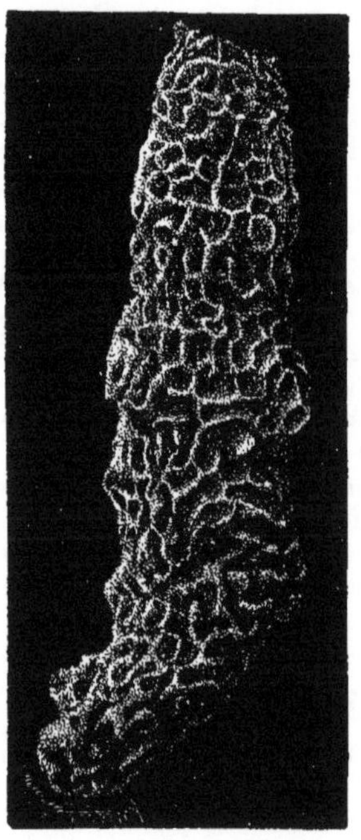

FIG. 57.— *Tuba labyrinthiformis.*

SUB-ORDER II. — ECHINONEMATA.

The spicules of the Echinonemata are of two or more kinds. The simple smooth or double-pointed ones usually lie in the fibre, the rough single-pointed ones, with a more or less expanded base, stand out from the keratode, leaving the point bare. Defensive surface spicules are often present.

This sub-order is represented on our coasts by *Microciona prolifera*, which grows abundantly in the pools and tideways south of Cape Cod. When in still water and on a smooth surface it forms a thin, smooth sheet, but under other conditions it tends to grow upright, and form branching masses a few inches in height. The color is a bright orange red, producing rich effects in pools where much of it is present. No fossil Echinonemata are known.

SUB-ORDER III. — MONACTINELLINÆ.

In this group the fibres are composed of straight silicious spicules, while the amount of keratode is very slight. The most common form is the Crumb-of-bread Sponge, *Halichondria panicea*, which has a world-wide distribution, and occurs plentifully in a dried state upon our beaches. It is almost as light as dried bread, and when well bleached is very white.

Another form, *Suberites compacta*, occurs on the coast south of Cape Cod, and is the only form in that region which is able to live upon the shifting sands. The pores are so small, and the structure so dense, that the sand cannot obtain an entrance, while its lightness keeps it from being buried. Specimens securely anchored have been found, and evidently the usual free condition is an acquired adaptation to a habitat on a sandy bottom. They are washed ashore in considerable numbers, and so fine and homogeneous are their spicules that the skeletons are said to have been formerly used for polishing silver. It grows in flattened masses of a yellow color, but the skeleton when bleached is white. Another species of this genus also frequents the sands north of Cape Cod, but finds more congenial accommodation on the shell of a species of gasteropod, nearly all of which, in certain localities, bear a sponge.

Some of this group have accustomed themselves to lead a life of borers, and though not successful with hard rocks they are very destructive to the shells of various molluscs, and even to limestone and marble. *Cliona sulphurea*, a very common form, is the most remarkable of these borers. It penetrates and excavates chambers in the shell of a mussel for example, and then, after causing the death of the animal, it will entirely enclose and resorb what is left of the shell. Not content with this conquest it often proceeds to grow around stones, or to take in sand until its flesh is full of such indigestible ballast. Such specimens will sometimes be a foot long, and weigh several pounds. Occasionally this form is found attached in the usual manner, and when the locality is free from stones or sand the specimen is clean and free from such eucumbrances. In the Mediterranean this genus plays no small part in the disintegration of the limestone rocks of the shores. It is evident that this sponge is a borer from inclination and not from necessity, and also that the inclusion of sand and stones is not needed, but is probably due to the effort to become attached to any object within reach. It is difficult to explain the boring except as a chemical process, but no one as yet has been able to detect any acidity in the secretions. It may be, however, that it is accomplished by the silicious spicules on the surface. That they are not the sole means of boring is shown by the recent observations of Nassonow, who ascertained that the young began to bore before the formation of any skeletal structures.

The Monactinellidan forms in the palæozoic rocks are uncertain, though Zittel records *Cliona* from the Silurian, and two genera in the Carboniferous. The first undoubted forms occur in the Jurassic.

SUB-ORDER IV. — POTAMOSPONGIÆ.

The Fresh-water Sponges, in our opinion, form a group of sub-ordinal rank. The skeleton is similar to that found in the last group, but a very important difference is found in the reproduction. In these fresh-water forms there are found what are known as winter buds or statoblasts. These are protected by an outer coat of spicules of a peculiar form, wholly unlike anything else found in the sponges. They may be as simple as the spicules of the sponge skeleton, and arranged flatwise in the corneous wall of the statoblast, or they may be shaped like a collar stud (birotulate), and arranged vertically. There seems to be a definite line between these two types, and, in fact the sub-order has been divided into two families upon these characters, and named the Lacustridæ and the Fluviatilidæ respectively. About ten genera have been described by authors from the fresh waters of all parts of the globe. They are usually green in color when exposed to the light, but when found under stones or in shaded localities are of a brownish hue.

They have a decided affection for clean water and hard bottoms, being in large part attached to stones, logs or plants, but will grow sometimes on muddy bottoms. In such cases the young anchor themselves to small sticks or stones, and thus secure themselves from being choked by the mud.

The sponge dies during some cold spell in the autumn, and their quick decay in large quantities is one of the principal causes by which the water supply of even a large city may be vitiated. They seem to be the cause of the peculiar smell known as the "cucumber odor," and render the water extremely disagreeable as a beverage.

The preservation of the species is accomplished by the statoblasts which retain their vitality through the winter, usually enclosed in the skeleton at the base of the colony. They develop in the spring, producing new colonies. Mr. Potts, of Philadelphia, accounts for the large size and rapid growth of the sponges in the spring by the coalescence of numbers of the young which develop within the meshes of the same old base. This author asserts that he has repeatedly observed that the young sponges from the statoblasts build upon the undecayed remnants of the last year's skeleton as if it were a trellis, which, when once constructed, could be used repeatedly.

This is not hard to believe, since two branches of the same sponge will unite if brought in contact, and even two sponges of the same species will not infrequently combine to form a single specimen. A certain proportion of some of the fresh-water sponges does outlast the winter, and these old skeletal frames frequently contain many statoblasts.

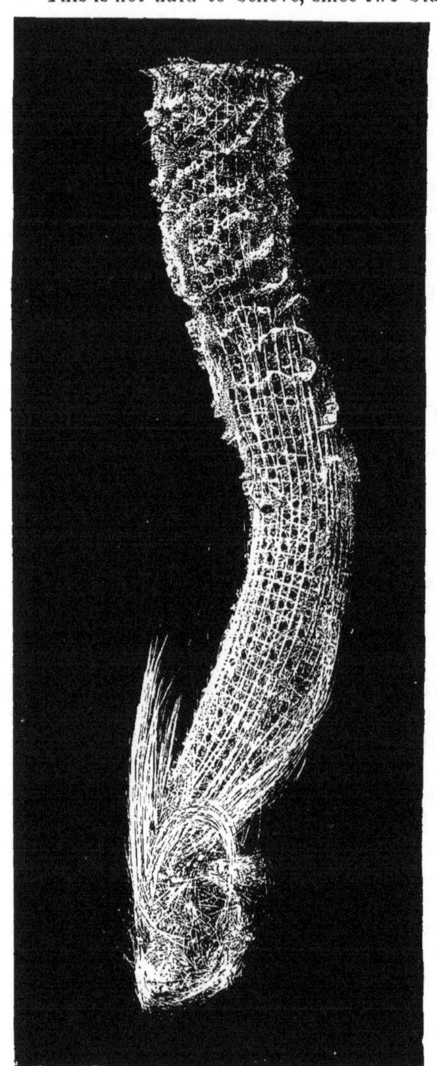

FIG. 58. — *Euplectella aspergillum*, Venus flower-basket.

Certain of the fresh-water sponges in tropical countries have to pass through a dry season, and it is supposed, with a considerable amount of probability, that their statoblasts can undergo dessication without loss of vitality, and even that they may be carried by the winds, thus affording the starting points for colonies in new localities when the rainy season sets in. Some forms are described as having no statoblasts.

ORDER II. — SILICOIDEA.

This, the highest order of the sponges, is characterized by having the skeleton almost entirely composed of silicious spicules.

SUB-ORDER I.—TETRACTINELLINÆ.

This group can be represented by *Tethya*, in which the skeleton is radiatory. The typical spicules have a long, straight axis and three curved arms, reminding one of an anchor, or more accurately, a grapnel. There are also long, straight spicules with both ends alike, and star-shaped silicious bodies. By the latter these sponges are allied to the Gumminінæ. *Geodia* is another remarkable type in this group, with extremely thick and unusually large internal spicules. When dried, these sponges are as hard as if carved out of wood. According to Zittel, the greatest authority on fossil sponges, this sub-order first appeared in the carboniferous, but was represented only by isolated spicules until the genus *Geodia* appeared in the Jurassic.

SUB-ORDER II. — LITHISTINÆ.

This group is composed of fossil forms in which the skeleton is made up of rather irregular star-shaped radiating bodies, firmly united. Thus, a very strong and solid skeleton was constructed, which has consequently been well preserved in the rocks. The normal form is a mass with numerous cloacal apertures of average size on the upper surface, but forms quite as often grow in vase-shapes, with the cloacal apertures on the inside, or like a pear, with the apertures on top. The large opening in the vase-shaped forms is usually described as a cloaca, though, as we have seen, it is not so in reality. The type appears in the genus *Aulacopium* of the Silurian, though Zittel thinks that some of the Cambrian forms may belong here.

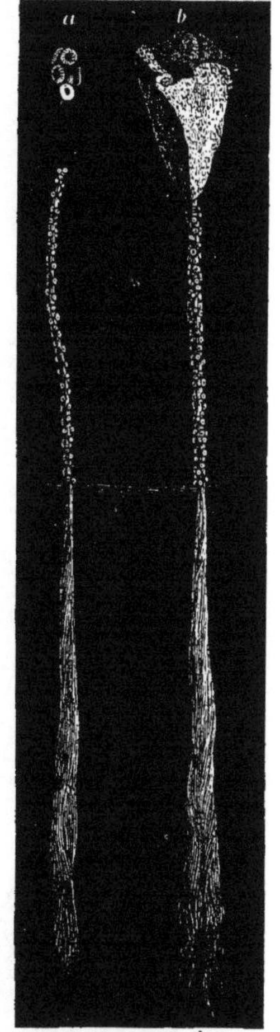

FIG. 59. — *Hyalonema*, glass-rope sponge. The stems are covered with parasitic polyps. *a*, polyps enlarged; *b*, perfect sponge.

SUB-ORDER III. — HEXACTINELLINÆ.

The glass sponges are remarkable for possessing six-armed spicules. Two of the arms may be almost indefinitely lengthened and bound together with others in threads closely resembling spun glass. In others they may be shortened and split into the semblance of flowers with narrow petals. The glass sponges remind the observer of the calcareous sponges, but the resemblance is merely superficial, and not so important as it at first appears. Though the *Euplectella* is hollow and has apertures through the wall as do the Calcispongiæ, they do not lead into radiating canals, but into areolar tissue and communicate with the ampullæ by means of numerous apertures in the walls of the sacs. The outlets of the sacs are large and open internally into the tube. The external and internal walls are supported by the interlacing arms of the crosses or hilts of the spicules, and as these are arranged with great regularity, the surface of the skeleton is divided into squares. The pores of the outer surface are usually situated one in each of the quadrangular intervals, and the cloaca occupy a similar position on the inner wall. The top of the sponge is closed with a network of threads, between which occur, as in *Hyalonema*, the true cloacal outlets. In fact *Euplectella* may be regarded as a hollow *Hyalonema*.

Hyalonema was at first known only by the stem which was highly prized as an ornament. The natives were in the habit of cleaning off the sponge body from the upper part of the stem, and then reversing it in a suitable standard. It was sold to strangers as the skeleton of the parasitic polyps (*Palythoa*) which live habitually on the stem. Scientific men were at first deceived, and the true character was not discovered until 1860, when Max Schultze found the sponge tissues, and showed that the polyps were but commensal parasites, having nothing to do with the formation of the long stem of silicious threads which resembles a plume of spun glass.

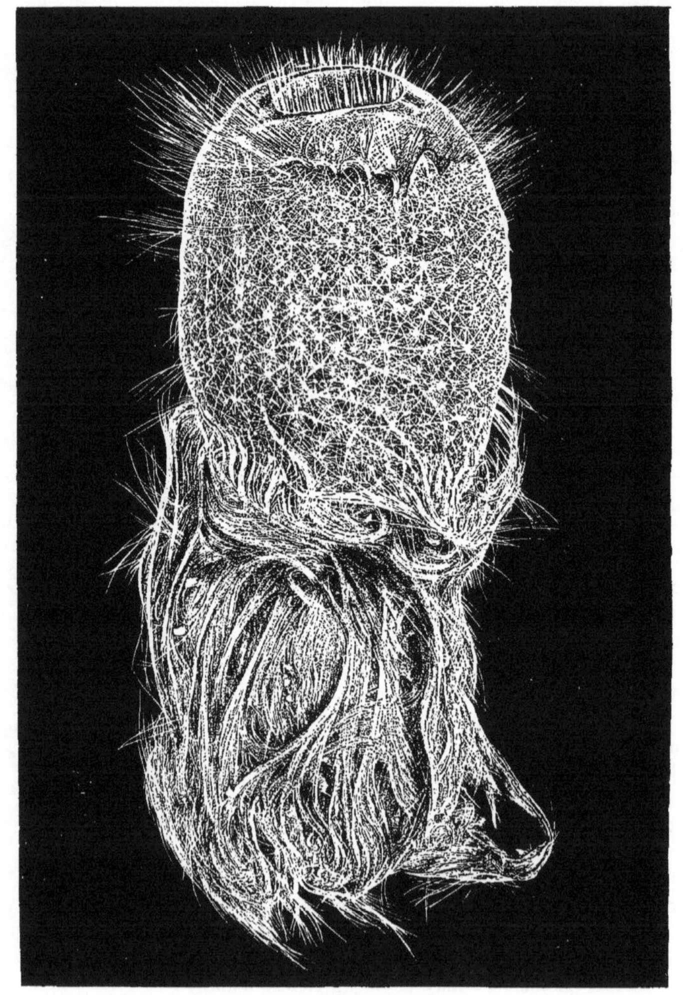

FIG. 60. — *Holtenia carpenteria.*

This genus may be expected in depths varying from forty to one hundred fathoms in northern seas, and in deeper water as we go towards the tropics, apparently requiring an average temperature below 40° F. The sponge itself in the natural state, is not as attractive as *Euplectella*, being of a light-brown color, and friable when dry. The top is usually occupied with a number of cloacal apertures surrounding a central prominence which is in reality the end of the stem. The stem is spun by the tissues as a supporting column of elongated spicules bound together and growing in a spiral as the animal progresses upwards.

The lower end of the stem becomes frayed out, and sinks into the mud as the animal grows, but constant additions to the upper end compensate for this and form a column which sometimes reaches a foot in length. In Fig. 59 we see on the right a perfect specimen. The stem in the living sponge is always enveloped in the fleshy tissues.

In *Holtenia* we have a different type of sponge, similar in shape to the members of the Calcarea, but the resemblance goes no further. The star-like beauty of the external covering of spicules, and the singular profusion of anchoring threads which are formed below, are shown in the adjacent figure. *Dactylocalyx* is another of the open vase forms which occur in this suborder.

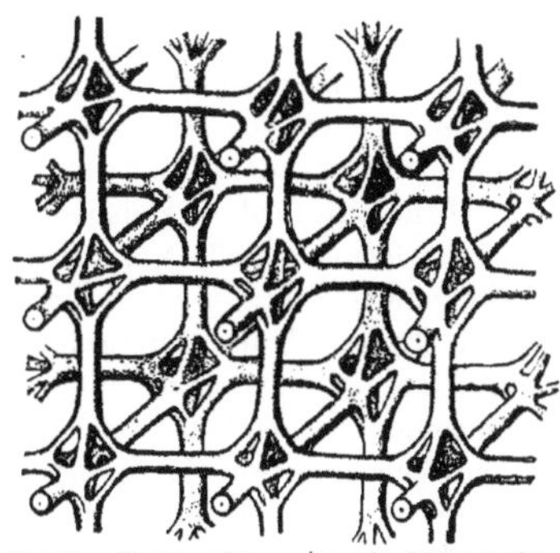

FIG. 61. — Section of the outer wall of *Ventriculites simplex*, showing the structure of the silicious network.

The fossils are very numerous, and it is supposed that several of the Cambrian sponges may belong here, though Zittel cites only certain Silurian genera like *Astylospongia* and *Protospongia* as undoubted Hexactinellids. One of the best known of the fossil types is *Ventriculites*, our figures of which show, not only the general shape, but the structure of the skeleton as well.

ALPHEUS HYATT.

FIG. 62. — Spicule of *Pheronema*.

BRANCH III. — CŒLENTERATA.

THE Cœlenterata embrace the jelly-fishes and corals, or more accurately speaking, the Hydrozoa, Actinozoa, and Ctenophora. In the first and last of these divisions fall most of those animals which are commonly known as the Medusæ, while the Actinozoa include the true corals and their relatives. The endless variety of names which one encounters in this group need not lead to confusion, and if considered in the light of the historical development of the study, indicates those various characteristics which have from time to time attracted the attention of students of these animals.

Of general terms used to designate the group, that of Zoöphytes is one of the oldest. In the infancy of natural science, when superficial observations took the place of more accurate anatomical studies, it is not to be wondered at that the likeness of these animals to plants led to the present name. One of the first comparisons which the novice makes, on seeing these animals for the first time, is that they resemble closely members of the plant world, and in maturer studies we are continually meeting similar resemblances of a deeper-seated nature

The Cœlenterata include two of the large divisions of the Radiata of Cuvier, who first outlined their characteristics in the masterly manner which marks all his works as models of zoölogical research. The name Cœlenterata dates back over a quarter of a century (1847), to the profound investigations of these animals by Frey and Leuckart, by whom it was first used.

The limits of the subordinate group of Hydrozoa are in many particulars obscure, and while many naturalists prefer to include in it a large group of gelatinous animals called the "sea-lungs," comb-bearing medusæ known as Ctenophora, others, from the close likeness of their young to the larvæ of the star-fishes, set these apart as a separate group. The Hydrozoa as here considered include the Hydroidea, the Discophora, and the Siphonophora, and contain by far the larger part of the true Medusæ.

The term Acalephæ, common in many writings on these animals, is almost synonymous with that of Hydrozoa as here used. By many it is also made to embrace the Ctenophora. The term was long ago used by Aristotle, and refers to the stinging powers which many of the Medusæ have. Given by many authors a greater or by others a less extension, it has been wholly abandoned by most of the leading students of these animals.

The Actinozoa or corals are marshalled under two divisions, the Actinoid, or true reef builders and their allies, and the "sea-fans" and "sea-whips," which are called, from more or less fanciful reasons, the Halcyonoids.

The single anatomical feature which is common to the groups mentioned above, to which, in point of fact, they owe the name of Cœlenterata, is the identity of a stomach and the body cavity. In the simplest forms these cannot be distinguished from each other, and in the higher genera there is but a slight differentiation of one from the other.

J. WALTER FEWKES.

Class I.—HYDROZOA.

Order I.—HYDROIDEA.

In the year **1703**, that charming old scientific gentleman, Anthony Van Leeuwhoek, of Delft, sent a very interesting paper to the Royal Society of London. In this article he tells us that "the water of the river Maes is brought by means of a sluice during the Summer flood, directly into our town, and it is as clear as if the river itself ran through the town. With this water comes in also a green stuff of a vegetable nature, of which, in a half hour's fishing, I got thirty pieces, and put them into an earthen pot together with a large quantity of their own water. I took out several of these weeds from the pot, one by one, with a needle very nicely, and put them into a glass tube of a finger's breadth, filled with water, and also into a lesser tube, and caused the roots of the weeds to subside leisurely; then viewing them with my microscope, I observed a great many and different kinds of animalcula. About the middle of the body of one of these animalcula, which I conceived to be the lower part of its belly, there was another of the same kind, but smaller, the tail of which seemed to be fastened to the other."

Our author, in the latter part of his article, assures us that he saw the smaller animalculum separate itself from the larger, and enter upon an independent existence; moreover, that he also determined by his microscope, the formation of a minute bud upon one side of the animalculum, which grew into an animal, perfect in shape, size, and all particulars, and then detaching itself from its parents, floated free in the water. That was the first discovery, so far as all the records give evidence, of the very wonderful animal, which is now called *Hydra*, and which in many respects, both in structure and in mode of life, is a very good type of its order, the Hydroidea, and at the same time of the class of Hydrozoa. The body of *Hydra*, which is entirely soft, having no skeleton without or within, easily changes shape, and when entirely contracted, has the appearance of a small dot or particle of gelatinous matter resting on the surface of the aquatic plant, chip, stone, or whatever may be the object in the water to which this small creature has attached itself. Watching it slowly expand in a dish of fresh water, it is seen to display a long, slender cylindrical body, which, in *Hydra viridis*, is bright green, while in *H. fusca* the color is light-brown. The base, or that end by which *Hydra* fastens itself, is termed the disk or foot, and the external cells of this part of the body secrete a gelatinous substance, which, hardening somewhat in the water, enables it to attach itself at will. Toward the anterior or free end of the body, are a variable number of long, slender processes, the tentacles, which are arranged in a single circle or wreath. Within the ring formed by the bases of the tentacles, the body tapers to a rounded elevation, where the mouth is found, and this tapering portion of the body which extends beyond the retracted tentacles, is known as the proboscis or hypostome.

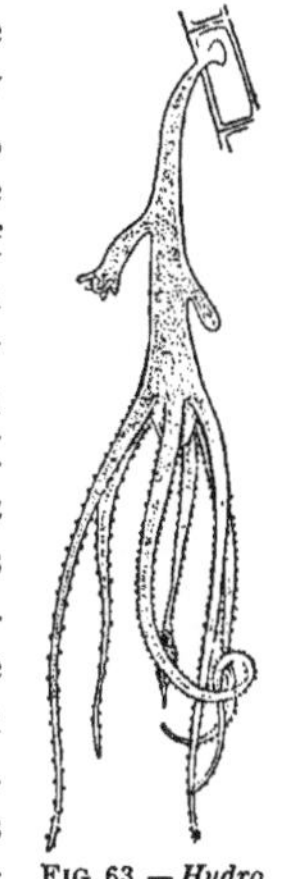

Fig. 63.—*Hydro fusca* with young budding from it.

Within the body there is a cavity extending from one end to the other, from the base to the mouth, and, as these processes are hollow in *Hydra*, to the tips of the tentacles. Not only the body, but also the tentacles are very expansive and contractile, and seldom retain the same shape and position for more than a few minutes.

When they are fully contracted they appear as so many knobs or bosses on the distal end of the body, and when fully expanded, I have seen them three and even four times the length of the fully elongated body. The tentacles are very sensitive, and if touched by some foreign object in the water, they rapidly contract, and the body also sharing in the contraction, the entire creature is withdrawn as much as possible from the area of disturbance and danger. *Hydra* has been observed in two or three rare instances to move from place to place by standing on its head, so to speak, using its tentacles as feet, by which it attaches itself, then it arches the body and attaches the foot-disk, releases the tentacles, straightens the body to arch it again, and so hitches along like a measuring-worm or geometrid larva. Another very peculiar form of locomotion is described by Marshall, of Leipzig, as seen by him in certain *Hydræ* found in brackish water. In this case the *Hydra* lies upon one side, and uses two tubercles as large, lobate, pseudopodial processes which give a creeping motion to the creature. Every one who has watched *Hydra* in aquaria has probably seen it creep or glide slowly over the surface of a leaf or of the glass. It keeps its normal position, attached by the foot-disk, but glides slowly, and with a very uniform motion, over the surface to which it is attached; much as a snail creeps, only with a much slower movement. This power of changing place is due to the cells in the foot-disk. Watching under a microscope, this part of *Hydra*, when it is in motion, it will be found that the external cells throw out pseudopodial processes, which extend in the direction in which the animal is travelling; so that *Hydra* can move by pseudopodia as truly as *Amœba* does. In the position which *Hydra* so often assumes, that of complete expansion with the tentacles extended to their utmost, and forming a very large circle, its chances for getting food in the well-populated, often semi-stagnant waters in which it is so frequently found, are very great. Any luckless crustacean of small size, such as *Cypris* or *Daphnia*, that happens to strike against one of those delicate tentacles is pretty sure to be used as food by the *Hydra*. The tentacle against which the crustacean has touched, curls around him, and after a few struggles his limbs fall powerless, and he acts as though it had been paralyzed.

This peculiar paralyzing or stupefying effect is caused by the action of certain stinging or cnidocells (also called lasso-cells), which are most abundant in the tentacles, but are also found in other parts of the body. Each one consists of a comparatively large body-part, from which stretch away interiorly one or more slender protoplasmic processes to connect with a deeper layer of the body-wall; on the outer end of the cell is usually found a small protoplasmic process which projects into the surrounding water, but is too small to be seen with the unaided eye; this latter process is termed a cnidocil, and probably receives and conveys stimuli from the external objects to the cnidocell; within the body of the cnidocell is the capsule, a more or less ovate structure, consisting of an outer wall which is perfect and complete, and an inner wall which is folded in upon itself at one end to form a tube, which for a very short distance is of some considerable diameter, and then decreases in size and forms a long, thread-like tube, coiled up in the cavity of the capsule; within the larger, shorter part of this tube, attached to its wall, are a number of recurved hook-like processes which vary in

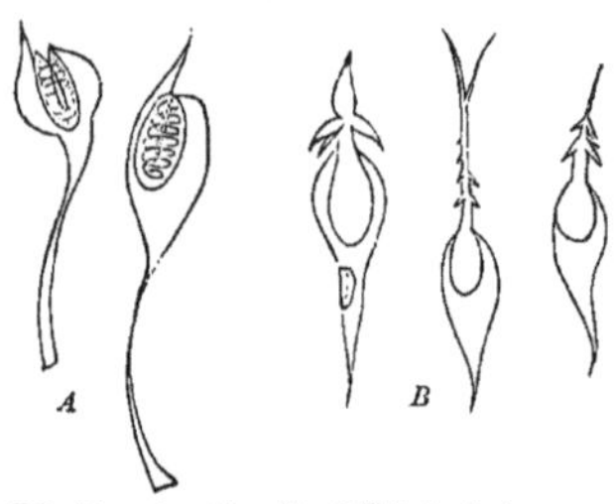

FIG. 64.—*A*, cnidocells of *Tubularia larynx*; *B*, cnidocells of *Hydra viridis*.

number, shape, and position in different species; the remainder of the cavity of the capsule is filled with a liquid very similar to, if not identical with, formic acid. Now, when any stimulus brings a cnidocell into activity, it forcibly ejects the larger part of the tube by a process of evagination or a turning of this part of the tube inside out, as one turns the finger of a glove; this movement is quickly followed by the ejection of the smaller part of the tube in the same manner, by evagination. If the body of some animal has touched the cnidocil, then that body is penetrated by the thread-like tube, and also possibly by a portion of the larger tube with its recurved hooks, and then the formic acid of the capsule pours into the tissues of the prey and produces the general paralysis above mentioned. This paralysis, of course, is not the effect of the formic acid from one capsule, but from many. Once used, the capsule is useless, as the tube cannot be withdrawn into it again.

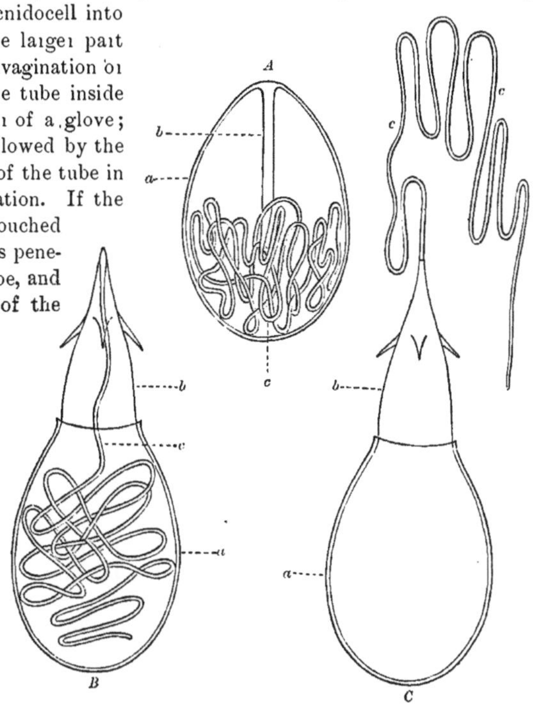

Other tentacles also close around the prey, and by their combined action it is conveyed through the mouth into the general cavity; here it may be seen, with microscopic aid, to break down and go to pieces, the products of the disintegration being a fluid, evidently a nutritive one, which then flows to all parts of the body, and the remnants of the hard chitinous skeleton which are ejected by the mouth or through an opening which may be extemporized anywhere in the wall of the proboscis. This form of *Hydra* in which it is unconnected with any other individual or zooid is termed the solitary condition.

FIG. 65. — Diagrams of cnidocells; *A*, previous to emission of contents; *B*, first stage of emission; *C*, filament completely extended; *a*, wall of capsule; b, barbed sac; *c*, filament.

When the surroundings are favorable for its vegetative life, one usually may find one or more *Hydræ* attached to the body of what appears to be a main stem or parent form. These attached or appended zooids have been produced by a process of budding from the parent individual, and each one of them ultimately separates from its parent by a constriction at its base and becomes a free and independent solitary *Hydra*. A bud starts as a small, rounded swelling on the side of the body; the swelling being hollow, and its cavity being directly continuous with the general body-cavity of the parent; by ordinary growth it attains considerable size, and from its distal end a number of small swellings or prominences appear, which elongating, develop into tentacles; the portion of the bud anterior or distal to the tentacles becomes the proboscis or hypostome, and a mouth is formed in its distal end. Being structurally complete, it catches and digests food, and performs all its functions while still attached to its parent. After a time a constriction separates it from its parent, but the opening at its base never entirely closes (at least in some species), and is known as the porus abdominalis. It does not function as an anus, however, and cannot be so considered. Before the first bud is set free, a second one may appear, and even a third and fourth on the parent body. Moreover, a secondary bud may appear on the body of the first bud, a tertiary on the body of the second, and a fourth on the body of the third before the first bud has become free. This is known as the compound or colonial condition.

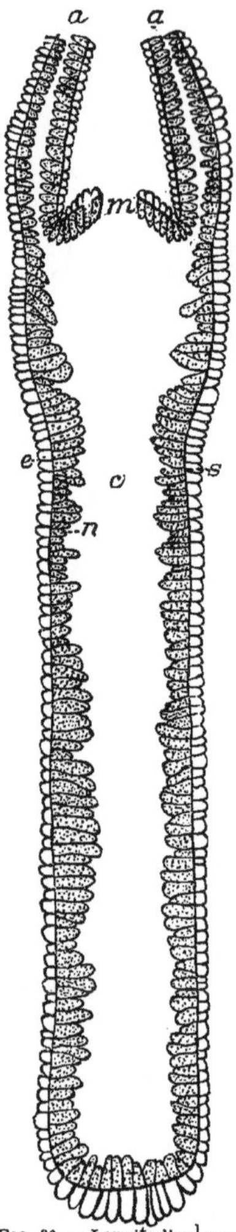

FIG. 66. — Longitudinal section of *Hydra*, greatly enlarged; *a*, tentacles; *c*, body cavity; *e*, ectoderm; *n*, endoderm; *m*, mouth; *s*, supporting lamella.

Another method of increase which rarely occurs in *Hydra* is division or fission, in which the entire animal divides into two parts, each developing all the parts necessary to make it a complete *Hydra*. Trembley observed this method, Rösel also witnessed it, and Marshall has seen three cases of it. In this country the process has been seen by Mr. T. B. Jennings, of Springfield, Ill. The wonderful power which *Hydra* possesses of reproducing lost parts was first discovered and made known by Trembley, of Geneva, in the first half of the eighteenth century. He determined that even a small piece of *Hydra vulgaris* possesses the power, under favorable conditions, of developing into a perfect animal. His experiments were very varied, and many of them have been often repeated with the same results, since his day. Baker repeated nearly all of them. The most remarkable of his experiments in this line, was the turning of the hollow, cylindrical body of a *Hydra* inside out; so that the inner layer which before did the digesting, now performed the functions of the cuticle, and *vice versa*. This experiment, which requires very skilful manipulation, has been, I believe, repeated but by one biologist, Professor Mitsukuri, of the University of Tokio, Japan.

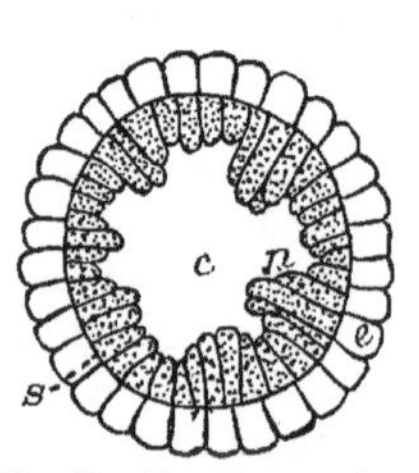

FIG. 67. — Transverse section of *Hydra*, greatly enlarged; letters as in fig. 66.

In a limited region on the body of *Hydra*, just below the tentacles, there appear under certain conditions, small outgrowths of the body-wall which prove to be the spermaries; in them being developed the spermatozoa. Lower down on the body, in another limited zone, larger, rounded swellings are developed, which are the ovaries. Just how fertilization is accomplished is unknown; but the egg having been fertilized passes through a morula stage in which the outer cells become prismatic, forming a definite membrane around the interior; a chitinous coat is developed about it, and then there occurs a retrograde step, as the entire embryo fuses into a simple, non-cellular mass; within this mass a small cavity appears, the first formation of the body cavity. In this condition it remains quiescent for a time, and then the outer shell breaking away, the embryo, still with a delicate shell around it, escapes into the water; a cleft appears in the body-wall, which becomes the mouth; the tentacles are developed, and the embryo bursting its thin shell, appears as a young *Hydra*. The development of *Hydra* is thus seen to be simple and continuous; there are no great or sudden changes such as occur in the life-histories of so many other animals.

There are a number of so-called species of *Hydra* found in the United States, the most common of which are a green one known as *Hydra viridis*, and a light-brown one called *Hydra fusca*. The latter often attains a much larger size than the former, and on account of its being much more translucent, is a better kind for study. They are found in slow or stagnant water, and are sometimes so very abundant as to form a delicate, fringe-like covering over every submerged object, in quite a large pool. *Hydra* has also been found once in a brackish arm of the sea in Germany, by Marshall. Having obtained a general idea of one hydroid, we may now take up the systematic arrangement of the group, considering the various sub-orders and a few of the most prominent families.

SUB-ORDER I. — ELEUTHEROBLASTEA.

This, the lowest sub-order, has for its type the genus *Hydra*, which has already been described at length. No other genus belonging to this group is known. This sub-order is destitute of a hardened body-envelope, and the zooids of the body, or troph-osome, are never firmly attached. Even more simple than *Hydra* is the peculiar genus *Protohydra* found by Greef in the ocean at Ostend, Belgium. It can be best described by saying that it closely resembles *Hydra*, except that it entirely lacks the tentacles so prominent in that form. It reproduces by transverse fission. So little is known of the structure and growth of *Protohydra* that the position which it is made to occupy in our classification must be regarded as provisional.

SUB-ORDER II. — GYMNOBLASTEA.

All the members of this division have a hardened body-envelope called the perisarc, and live in colonies which are always attached to some foreign support. From the next division of the same rank, they are separated by never having the reproductive and nutritive portions enclosed in a chitinous capsule, and the generative zooids do not usually become free, independently developing organisms. The generative zooid, escaped from its parent, may have a medusa form, from which ultimately a large number of ova are dropped, or it may assume the condition called the actinula, an oval body floating passively about or creeping on the bottom. In those hydroids which have an actinula this body develops directly, without intermediate metamorphosis, into a hydroid of the same form as that from which it sprung. In some of the gymnoblastic hydroids there are no free medusæ and no actinulæ, properly so called, but a locomotive zooid, called a sporosac, which performs the same function. The sporosac is a ciliated body, capable of active locomotion, and possessed of two tentacles. It carries in its cavity a single ovum. In many of the young gymnoblastic hydroids, the embryo leaves the mother's care as a planula, which develops directly into a hydroid similar to that from which it originated.

With this sub-order a new feature is introduced. In *Hydra* we found the nutritive and reproductive systems united in the same individual, but here we find certain portions of the colony set apart for the capture and digestion of food, while other portions have for their only function the perpetuation of the species. It must be remembered that the following account is a general one, and that there are many exceptions to it, some of which will be subsequently mentioned.

We can best understand the structure of a colony by following it briefly in its development. From the egg there hatches out an elongated young, known as a planula, which freely swims by means of the cilia with which the surface is covered. This finally attaches itself to some submerged object, loses its cilia and begins to develop the true hydroid condition. Around the upper (free) end appear the rudiments of the tentacles, while the base begins to divide up and send out processes.

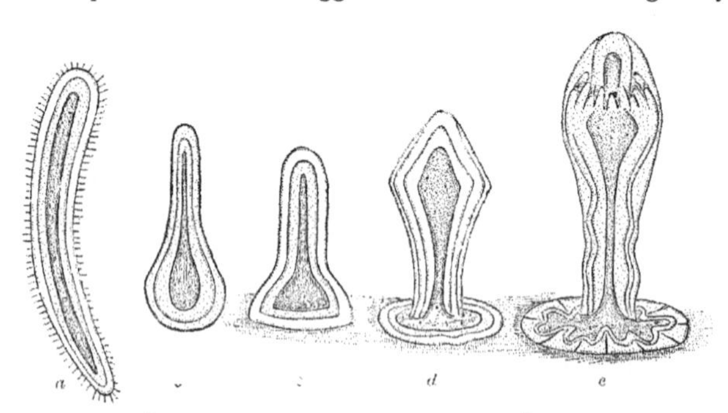

FIG. 68.—Development of *Eudendrium*; *a*, free-swimming planula; *b*, about to be attached; *c*, *d*, attached; *e*, beginning of hydrorhiza and hydranth.

These latter grow and ramify in a manner strikingly like that of the roots of a tree, and produce what is technically known as the hydrorhiza. From this root-like portion other individuals or zooids develop, some of which are like the first, and from their greater or less resemblance to flowers, are called hydranths. These hydranths form the nutritive portions of the colony. They may be either stalked or sessile upon the hydrorhiza. Other zooids are also developed from the hydrorhiza or from the hydranth itself, but these never possess the tentacles and digestive organs of the hydranths, but have only reproductive functions, and are called gonangia. In these latter are developed small zooids which in some cases become free, in others they never separate from the parent. These medusæ or medusa-buds develop the male and female elements (eggs and spermatozoa) which in turn produce other colonies similar to that described.

Here some very interesting questions arise, the most prominent of which is what constitutes an individual? From a single egg there is developed a number of zooids from which there escape quantities of medusæ, which are frequently capable of feeding and of reproduction. Are each of these jelly fishes, reproductive sacs, and feeding portions to be regarded as separate individuals or as parts of one individual? The latter is the true course; an individual embraces all the products of a single egg, and the name zooid is applied to the various more or less independent portions, which, whatever their form may be, arise by budding or fission, but never by a new ovarian reproduction. This distinction is somewhat different from that found in the sponges.

In a number of places in Europe and America, there has been found, besides *Hydra*, another hydroid, living in fresh or brackish waters, known as *Cordylophora lacustris*. It is a compound form, attaining a length of two inches in good specimens, and is usually attached to some water-weed or to the stones in the bottom of a stream. I have seen it flourishing in a stream where the current is very swift. Again it has been found in an old well. These two, ***Hydra*** **and** ***Cordylophora***, are the only hydroids known to live in fresh-water. A third, imperfectly known form, allied to ***Cordylophora***, has been described by Professor Cope, from a lake in Oregon.

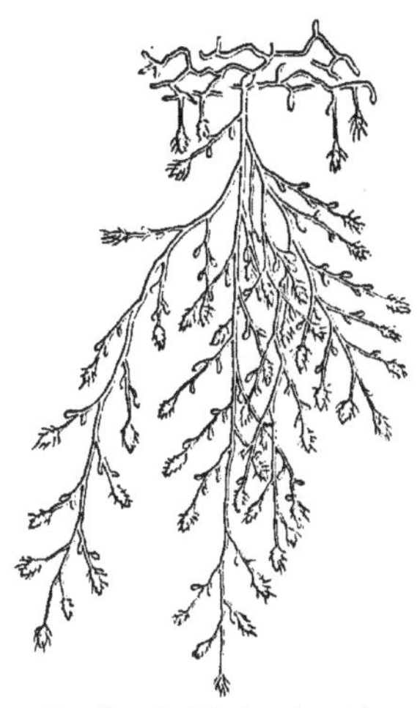

Fig. 69. — *Cordylophora lacustris.*

In the oceans, hydroids are very abundant, and there are at least several hundred species. All of them may be arranged in a few groups, most of which are represented on our shores. Our first example of the marine forms will be *Clava leptostyla*, a beautiful reddish species which occurs on our coast from Long Island Sound northward. Its most common habitat is at or near low water mark, attached to the rockweed (***Fucus***), where it forms colonies consisting of numerous individuals attached to a common rhizome or branching base. It is about a half of an inch in length, and the "head" bears from fifteen to thirty irregularly arranged slender tentacles. Beneath the tentacles, at the breeding season, the small reproductive buds are arranged in groups as shown in the figure. The reproduction is essentially like that of the next species. One of the most common forms found in shallow water (one to twenty fathoms) from Vineyard Sound northward, is known as ***Eudendrium dispar***. It grows in colonies from two to nearly four inches in length, and the parts of the colony which correspond in appearance to the stems and branches of a plant are dark-brown or black. At the tip of each branch and branchlet is a hydra-like animal, or zooid, which is directly connected with every other one in the colony, for the whole colony is strictly comparable with a much-budded ***Hydra*** grown to an equal height, and the general cavity of the body is continuous through all the stems and branches into every zooid. When taken out of the water, however, ***Eudendrium*** retains its shape, which ***Hydra*** cannot do. This stability or rigidity is due to the existence of a nearly complete coat or covering of horny material, chitin, which is secreted by the animal, and which extends over all the colony, with the exception of the zooids; they remain unprotected. During the summer months two kinds of ***Eudendrium*** may be found along the New England coast, which are exactly alike in the characters given, but differ in color, one having white zooids, the other yellow. A little careful examination will show that upon the bodies of the white zooids are a series of structures arranged in a circle just beneath the tentacles; each one of these is in shape like a short string of beads, which are supposed to be male organs, showing that the white colonies are male. The yellow ones are colored by a number of simple bud-like processes which are irregularly scattered on the body of the zooids; they are the female reproductive organs or ovaries. In ***Eudendrium*** then, the sexes are in different colonies. An egg having been fertil-

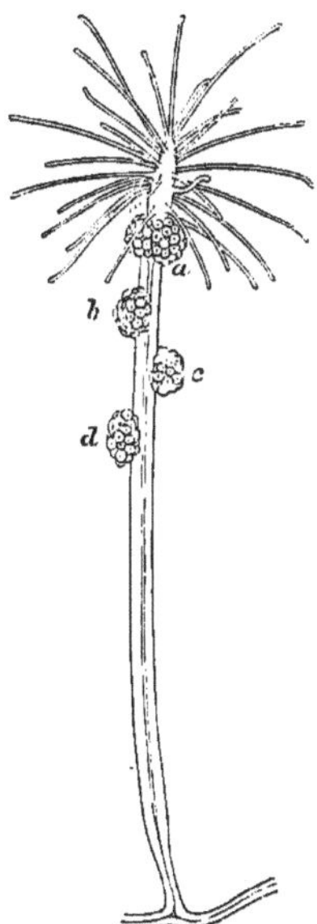

Fig. 70. — *Clava leptostyla*, enlarged; *a*, *b*, *c*, *d*, medusæ buds.

ized, passes through the process of segmentation, a cavity appears within it, then it assumes an elongated form, possesses a double wall about the central cavity, develops cilia upon the outer surface, and breaking through the containing wall, escapes into the water where it leads a free life for a brief time (see Fig. 67). Before long it enlarges at one end, settles down, becomes attached by its larger end, loses its cilia, and proceeds to develop a new colony of *Eudendrium* in the following way: It enlarges at its free or distal end, and around this enlargement appear a number of smaller swellings which develop into a wreath of tentacles; a mouth forms in the extremity of the proboscis and a layer of chitin is secreted around the body. Then by the simple processes of growth, combined with budding, a new colony is formed quite like the one from which the germ came. In this case the medusa buds do not develop into free-swimming jelly-fishes, but discharge their reproductive elements without leaving the parent colony.

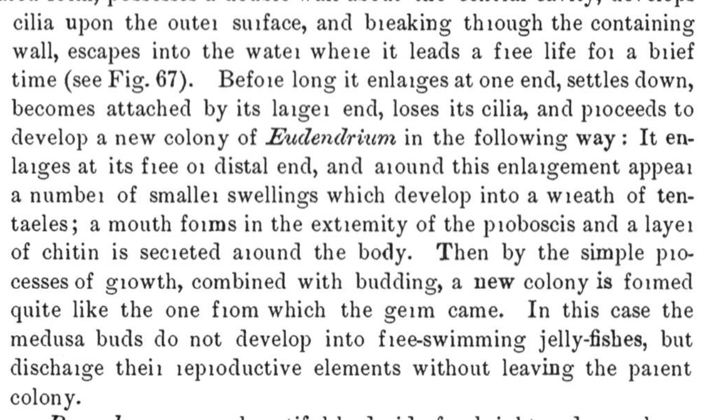

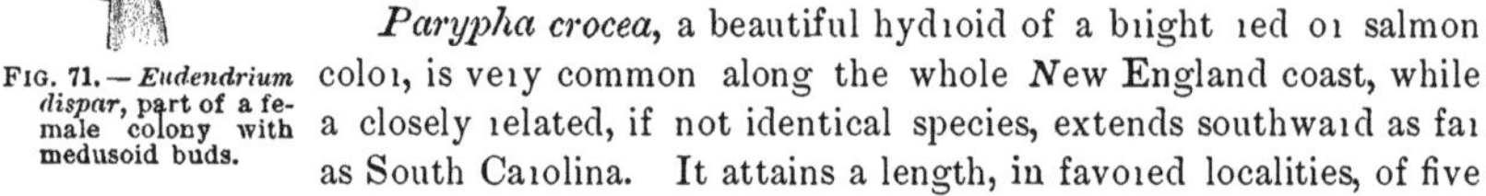

FIG. 71.—*Eudendrium dispar*, part of a female colony with medusoid buds.

Parypha crocea, a beautiful hydroid of a bright red or salmon color, is very common along the whole New England coast, while a closely related, if not identical species, extends southward as far as South Carolina. It attains a length, in favored localities, of five or six inches, and grows in great luxuriance on the piles of wharves or bridges, especially where the water is slightly brackish. The outer or lower circle of tentacles are long, and just within them arise the medusæ buds resembling clusters of small, bright-red grapes. In each colony the sexes are distinct, and in these buds the eggs or spermatozoa are developed. The young escape in the actinula condition, and creep about, finally attaching themselves, and then by budding and branching, large colonies are formed, which in turn produce medusa buds, thus completing the life cycle.

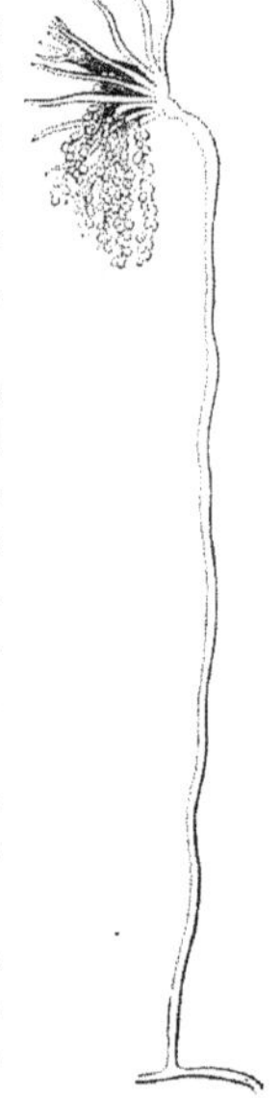

FIG. 72.—*Parypha crocea*, natural size.

Another common form on our Atlantic coast from South Carolina to the Gulf of Maine, is *Pennaria tiarella*. It grows in colonies equal in size or a little larger than those of *Eudendrium*, and is found attached to rocks and eel grass, and often to floating algæ. The zooids are usually a roseate color, and the species is remarkable for its beauty. In general structure *Pennaria* is like *Eudendrium*, but differs in having, in addition to the one row of large tentacles, a number of smaller capitate tentacles, arranged, more or less definitely in two circles near the anterior end of the proboscis; it also differs in its mode of branching, and in its method of reproduction. In the summer months there may be found growing out of the lower part of the proboscis, one or more oval bodies which finally develop a deep bell-shaped body with a considerable opening at the free end, about which are a number of rudimentary tentacles; within the cavity of the bell-shaped zooid is a process corresponding in shape and position with the clapper of a bell, it is in fact the proboscis, and at its free end is the mouth. By means of a sort of gullet or œsophagus passing through the proboscis, the mouth communicates with the central digestive cavity located at the base of the proboscis in the upper part of the umbrella; from this central cavity four ducts at four equidistant points stretch away to the rim of the bell, where they are all connected by a tube passing around the rim. By means of these gastrovascular canals nutritive matter from the stomach is carried all over the body. Stretching partly across the opening into the hell, is a thin, centrally-perforated membrane called the velum or veil. After one of these medusæ has been completely developed on the proboscis of the hydroid of *Pennaria*, it is freed from the proboscis by a constriction which cuts in two the small peduncle by which it had been attached, and the medusa floats away free in the water. It is not left to the mercy of currents, however, but is provided with a rather peculiar locomotor apparatus. The cavity of the bell being filled with water, its muscular walls are powerfully contracted, and the water being ejected from the opening in the velum, the medusa is forced through the water in the opposite direction; then expanding its bell by other muscles, it is ready to contract again and send itself still farther on its way. The tentacles and outer surface of the medusa are well supplied with cnidocells with which they defend themselves and kill their prey in the same manner as *Hydra*, and as the zooids in the hydroid colony. The medusæ are sexual zooids, the sexes being separate, and in the case of *Pennaria*, the male and female elements are developed within the walls of the proboscis. From a fertilized egg a planula is developed, which in turn gives rise to a hydroid colony of the *Pennaria* kind. The life-cycle is thus more complicated than in *Eudendrium* by the introduction of the medusa stage. The length of an average *Pennaria* medusa is about one-sixteenth of an inch.

Objects of more exquisite beauty than some of these hydroid-medusæ do not perhaps exist. Each minute crystal chalice with its beautifully curved outline, elongated, delicate tentacles gently coiling and uncoiling, and its slender proboscis which hangs like a lamp in its centre, lighting it with a soft phosphorescent glow as it swims with most perfect grace at the surface of the ocean, is the very type of delicate beauty, suggesting the wonders of fairy-land.

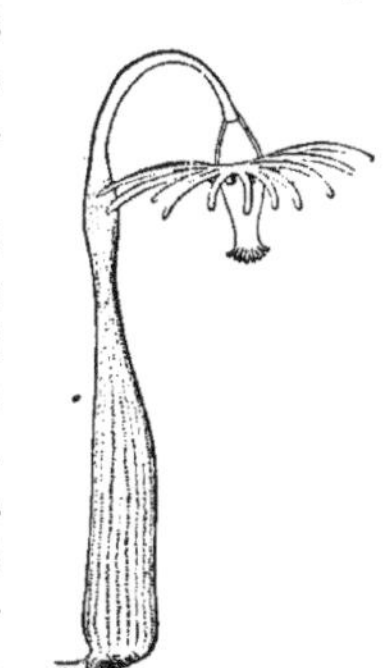

FIG. 73.—*Monocaulis pendula*, natural size.

The dredge frequently brings up delicate pink or flesh-colored hydroids consisting of single stems, each supporting a single hydranth. This hydranth bears two sets of arms, those around the free end of the proboscis being much shorter than those nearer the base. This form was called by Agassiz *Corymorpha pendula.* It lives with the base imbedded in the mud, and grows to a length of four inches. The investing envelope is very soft, and the animal is able to greatly modify the shape of the stalk and proboscis. The medusa buds never become free-swimming jelly-fishes, while the hydroid stem always bears a single head or hydranth, a fact which led Allman to refer it to the genus *Monocaulis.*

The genus *Tubularia* and the closely allied *Thamnocnidia*, are represented on our coasts by several species. The hydranths are borne on slender stems, and form colonies reaching sometimes a height of eight or ten inches. Under a low power of the microscope, the beauties of the animals stand revealed, far exceeding the power of any pen to describe or brush to paint. The hydranth is surrounded with two circles of tentacles, and from between the lower ones the reproductive zooids hang down like bunches of grapes, or they cluster around the proboscis inside the outer circle of tentacles, so that it requires no very vivid imagination to imagine the whole a delicate fruit-dish filled with the most beautiful fruit. From these raceme-like clusters the young come forth in an actinula condition, presenting distant resemblances to a jelly-fish. The body is long and surrounded by a single circle of tentacles. This larva soon becomes attached and then develops into a form like the parent.

Many of the small spiral shells found in the shallow salt-water just below the water's edge, are found to be inhabited by hermit crabs, which travel about very actively by protruding their legs from the aperture of the shell. On the backs of many of these shells is what appears to the eye, a white, delicate, mossy growth, covering most all of the shell, excepting that part which drags on the bottom as the crab travels. Under the microscope, this mossy growth proves to be a colony of very beautiful hydroids named *Hydractinia*. They live in colonies, but instead of forming a colony by branching in the ordinary way, the hydrorhiza, or part which attaches the colony, spreads out farther and farther, and sends up more and more buds, each one of which becomes a zooid, but which does not bud and is not covered by chitin. The hydrorhiza is covered by a layer of chitin, and at irregular intervals the chitin is developed into a large projecting spine. The zooids are very contractile, and when withdrawn to their utmost, the hard chitinous spines project slightly beyond and protect them. Examining carefully the zooids of *Hydractinia* it is found that there are the ordinary feeding zooids, the reproductive zooids, male and female, and a third kind which are destitute of true tentacles, have very slender, much elongated bodies, and are powerfully armed with strong batteries of cnidocells with which they perform their duty of protecting the colony. From a fertilized egg of *Hydractinia* is developed a planula, which in time gives rise to a *Hydractinia* colony. There are a large number of jelly-fishes known, which, from their structure, are classed among the Gymnoblastea, although nothing is known of their attached hydroid condition, or even if they pass through such a stage.

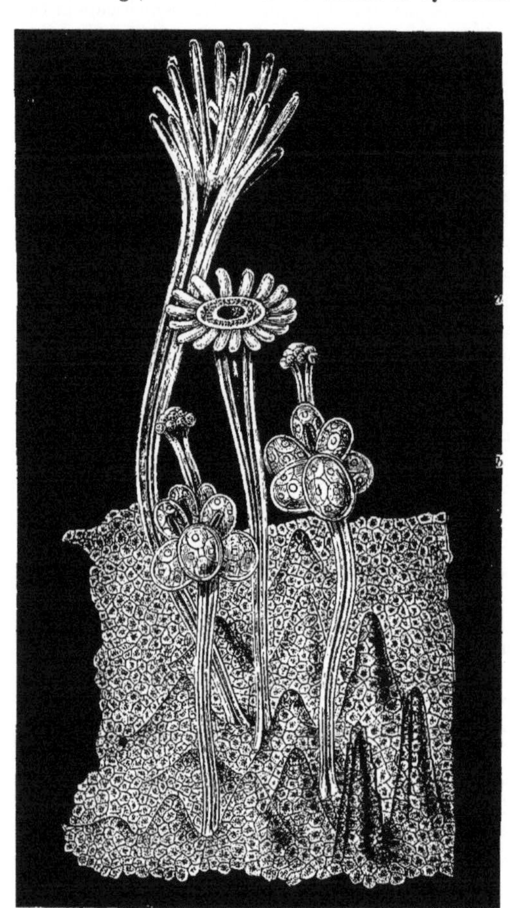

FIG. 74.—*Hydractinia echinata*, enlarged; *a*, nutritive, b, female reproductive zooids.

In the form known as *Lizzia*, it is the jelly-fish itself that produces the medusa buds. In our figure, which represents the young of *L. octopunctata*, may be seen younger jelly-fishes budding from the sides of the proboscis of the parent, and frequently in life, one can see still younger buds in these embryos before they free themselves from the parent. When arrived at a moderate size, these buds begin their contractions and struggles which finally end in their breaking loose from the parent, and the beginning of life on their own account. With age and increasing size, the tentacles grow much longer, those arising opposite the radial canals being in bunches of five, while those at the intermediate points are in threes, so that there are thirty-two in all.

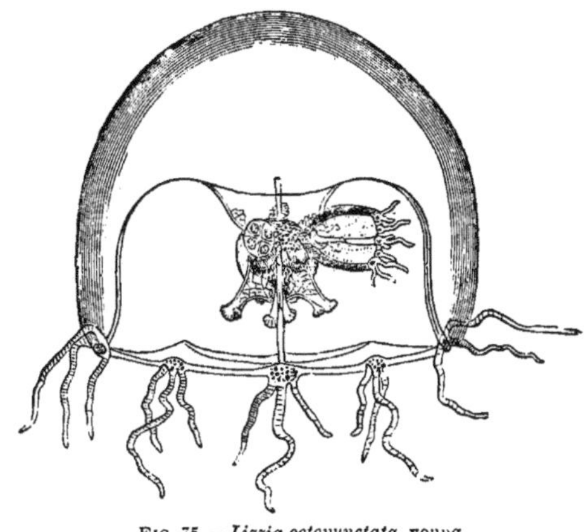

FIG. 75.—*Lizzia octopunctata*, young.

Here also belongs the genus *Stomatoca*, with its two long, marginal tentacles. In confinement our *S. apicata* seems to prefer the bottom of the aquarium, and but rarely comes to the surface.

SUB-ORDER III.—CALYPTOBLASTEA.

Nearly all the many species of Hydroids on the American coast which have bell-shaped hydrothecæ, belong to the large family CAMPANULARIDÆ. One of the finest representatives in American waters of this family of hydroids is *Obelia longissima*. It lives in shallow water, and down to a depth of about twenty fathoms, from Long Island Sound to the Bay of Fundy. The colonies are often quite large, measuring eight to twelve inches in length, and are of great beauty; at the tip of each stem and branch is developed a zooid, and about the zooid is a cup of chitin, called hydrotheca, into which the zooid may nearly or completely retract itself, and out of which it may stretch and unfurl its single wreath of tentacles; the rim of the hydrotheca is cut into a number—twelve to sixteen blunt teeth; the proboscis is very large and very mobile, constantly changing shape. In the axils of the branches are developed other chitinous cups (gonothecæ) larger and always of a different shape from the hydrothecæ, in each of which there is a long, simple zooid, destitute of mouth and tentacles (a blastostyle); on its sides are produced small buds, from eighteen to twenty-four, which develop into medusæ. They are found escaping from the genethecæ from April to June.

These medusæ are sexual, and bear either male or female elements along the radial canals; each fertilized egg develops into a ciliated planula, and this gives rise to a colony of *Obelia longissima*. One finds certain points of difference between this medusa and that of *Pennaria*. The contracting wall which subserves the function of locomotion is not bell-shaped, but is nearly a flat disk, and tentacles exist all round the edge of the disk, there being from twenty to thirty, while the medusa of *Pennaria* has only four rudimentary ones. On the edge of the disk, at equidistant points, are a number of globular bodies containing a cavity in which is a bristly ridge, and which is nearly filled with a clear liquid, in which are a few small calcareous particles that strike against the bristles when any disturbance in the water outside sets the liquid of the sac in motion. These are known as otocysts, and are supposed to be auditory organs. The medusæ of *Obelia longissima* are very minute, measuring only one-sixtieth of an inch across the disk, and one-fortieth across the outstretched tentacles.

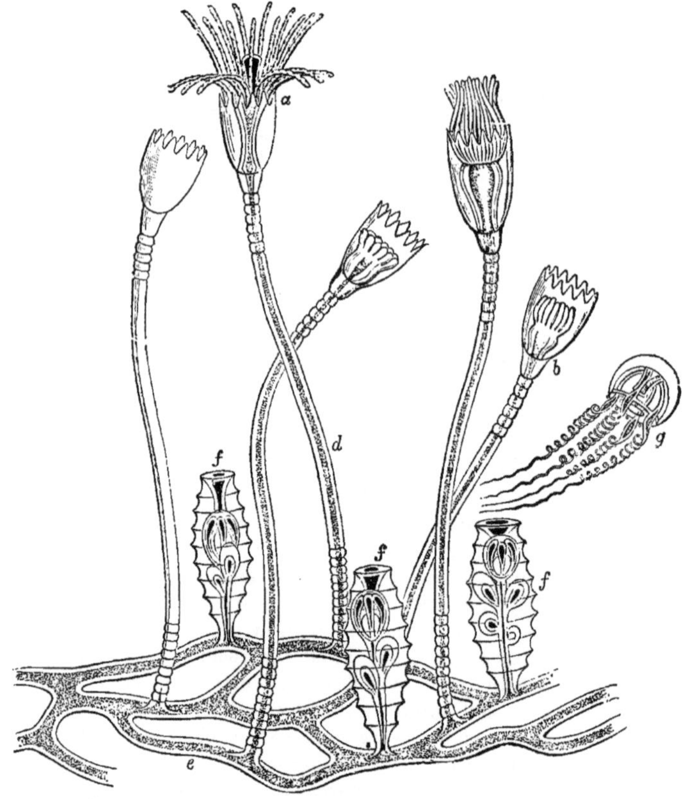

FIG. 76.—Campanularian hydroid; *a*, b, hydranths; *e*, hydrorhiza; *f*, gonangium; *g*, medusa.

A less conspicuous but very beautiful hydroid of special interest, and belonging to the same family as *Obelia*, is represented by several species on the New England coast. They belong to the genus *Gonothyrea*, and, at a hasty glance, look like diminutive or young specimens of *Obelia*. In height they do not exceed an inch and a half or two inches; the hydrothecæ in the most common species, *G. hyalina*, are long, of very thin texture, and the rim is cut into numerous shallow teeth of castellated form. The gonothecæ spring from the axils of the branches, and contain a blastostyle upon which are formed a number of buds that develop in regular sequence from above downward; when the uppermost one is fully grown, it pushes out of the top of the gonotheca, but still remains attached to the blastostyle by a slender peduncle; this zooid is now seen to be sexual, and contains within the walls of its proboscis, the sexual elements; the outline is nearly spherical, being cut off at the farther end where there is an opening into the cavity of the zooid; about this opening is a wreath of tentacles, and pendent in the cavity of the bell is a proboscis destitute of a mouth; the cavity of the blastostyle is directly continuous with a central cavity in this meconidium, as this kind of zooid is termed, and from this central cavity four radial canals pass out to four equidistant points on the edge or rim of the bell, where they all join a circular canal; these meconidia never become free, but after discharging their contents, they die and disintegrate. The fertilized eggs develop into ciliated planulæ which finally form colonies of *Gonothyrea hyalina*.

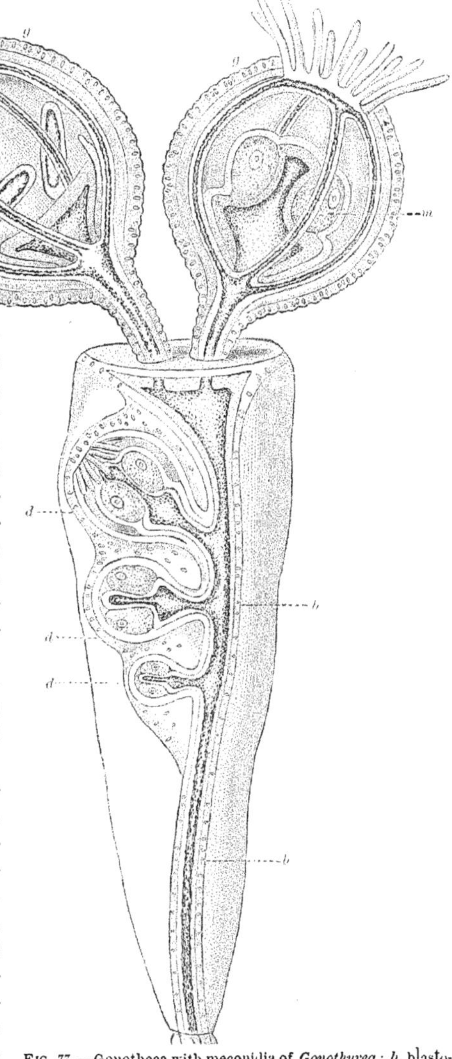

Fig. 77.—Gonotheca with meconidia of *Gonothyrea*; *b*, blastostyle; *d*, gonophores in various stages of development; *g*, meconidia; *m*, ovum; *o*, embryos.

These meconidæ which are evidently medusæ that never become free, are of great interest, being, in all probability, degenerate forms.

Another large family, the Sertularidæ, belonging to this same group, is represented in America by a beautiful species, *Sertularia argentea*, so-called from its light silvery color. The colonies are often a foot in height, and the shoots usually grow in clusters; the branches have a subverticillate arrangement, giving the colony an arborescent appearance. If a small portion of a colony be examined with a magnifier one discovers very peculiar hydrothecæ, which are very differently arranged from any described above; they are nearly tubular, somewhat narrowed at the top, with pointed lips, and are either free or set into the sides of the stems and branches.

The gonothecæ are developed on the branches and are elongated, somewhat urn-shaped, aperture central, terminal with usually two, occasionally one long horn at the anterior end. The eggs are partly developed within the gonotheca, and then pass into a sac which projects from the orifice of the gonophore, where they finally become planulæ; these after living a free life become attached and form a new colony. *Sertularia argentea* is found from the New Jersey coast to the Arctic Ocean from low-water mark to a depth of over one hundred fathoms. It is very widely distributed, being found on both sides of the Atlantic and on the Pacific shore of the United States. Like many other Hydroids it is often collected as sea-moss and is not infrequently seen at the florists for decorative purposes.

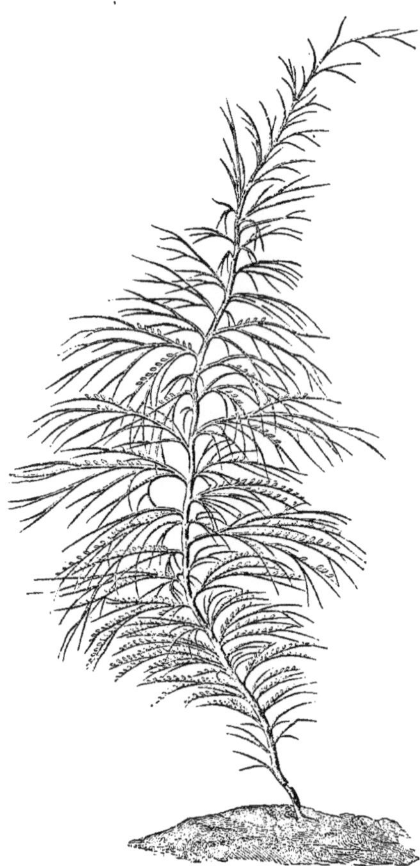

FIG. 78. — *Sertularia argentea*, natural size.

FIG. 79. — A fragment of *Sertularia*, enlarged, showing the cups or hydrotheca.

Another very common species of *Sertularia* is *S. pumila*, a very much smaller hydroid, not over one inch and a half long, and often found in abundance on the common Fucus or dark brown rock-weed. The hydrothecæ are opposite one another on the stem, giving it a compactness of structure and regularity of outline not possessed by *S. argentea*. The colonies are sexually perfect from May to September on the New England coast. The method of reproduction is very similar to that of *S. argentea*.

A third large family comprises the feathery forms known as the PLUMULARIDÆ. They are represented on the New England coast by *Plumularia tenella*, *Plumularia verrillii*, and *Aglaophenia arborea;* the last species was described by Desor in 1848, and has, I believe, never been found since. Other species of *Aglaophenia* and *Plumularia* are found on the Carolina coast, and still others in the Californian waters. Perhaps the most elegant in appearance of all the American hydroids is the ostrich plume of our Pacific coast, *Aglaophenia struthionides*. It varies much in size and color, but always retains the appearance of a diminutive ostrich feather. Microscopic study shows that the hydrothecæ are arranged in a single row on one side of each branch or pinna, and that the branch is divided into very short joints, one to each hydrotheca. Each hydrotheca has its rim ornamented with a number of sharply pointed teeth, and three minute tubular processes are disposed about its mouth, one on each side and one on the outer or anterior surface. These processes are termed nematophores, are filled with processes of the body substance, and in structure and development are believed by Hamann to give evidence of being degenerate zooids. Certain of the branches or pinnæ are at times replaced by cylindrical structures which are covered with rows of nematophores, and are the cups or baskets in which the generative zooids are developed; they are termed corbulæ, and in some genera are metamorphosed branches, while in others they are modified pinnæ. A pinna is smaller than a branch, and differs from it in the character of the zooids formed upon it. The egg develops into a planula, which becoming attached forms a new hydroid colony.

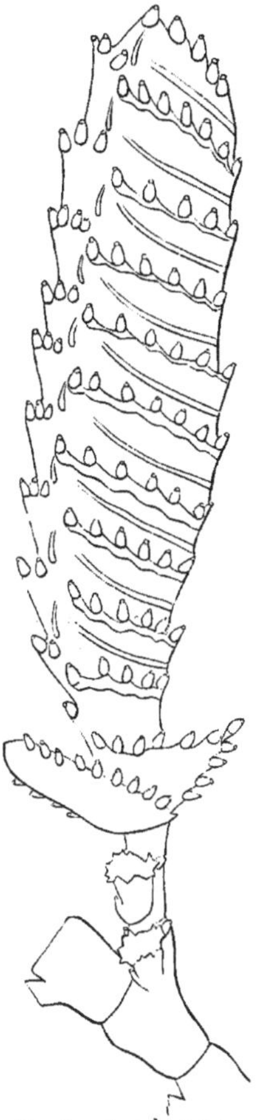

These three great families, represented here by the genera *Sertularia*, *Obelia*, *Gonothyrea*, and *Aglaophenia*, are all members of a sub-order of hydroids distinguished by having the hydranths surrounded by chitinous cups, and the possession of longitudinal ridges in the body cavity. This group has been variously termed Thecata by Hincks, Calyptoblastea by Allman, and Intæniolata by Hamann.

As among the Gymnoblastea, we find here medusæ which agree in structure with those which are undoubtedly calyptoblastic, but of whose early development we know nothing. We can mention but one example. One of our larger jellyfishes is *Zygodactyla grönlandica*, which sometimes acquires a diameter of even eleven inches. In color it is a light violet, with numerous brownish reproductive organs. The numerous tentacles which fringe the margin of the umbrella hang down a yard or more when fully extended. Concerning the habits of these animals Mrs. Agassiz has written: — "The motion of these jelly-fishes is very slow and sluggish. Like all of their kind, they move by the alternate dilation and contraction of the disk, but in the *Zygodactyla* these undulations have a certain graceful indolence, very unlike the more rapid movements of many of the medusæ. It often remains quite motionless for a long time and then, if you try to excite it by disturbing the water in the tank, or by touching it, it heaves a slow, lazy sigh, with the whole body rising slowly as it does so, and then relapses into its former inactivity. Indeed, one cannot help being reminded, when watching the variety in the motions of the different kinds of jelly-fishes, of the difference in temperament in human beings.

FIG. 80. — Corbula of *Aglaophenia struthionides*, enlarged.

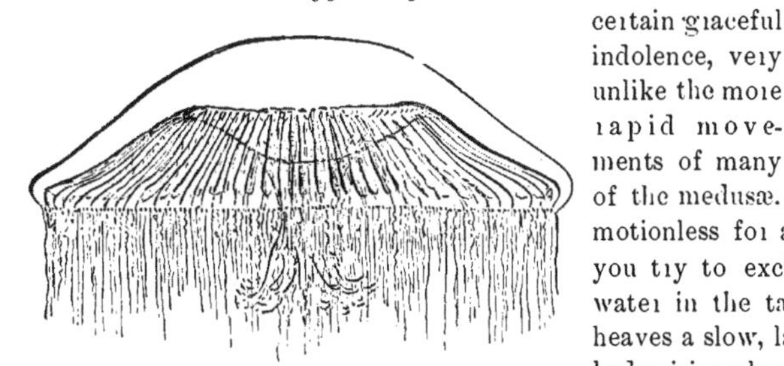

FIG. 81. — *Zygodactyla grönlandica.*

There are the alert and active ones, ever on the watch, ready to seize the opportunity as it comes, but missing it sometimes from too great impatience; and the slow, steady people, with very regular movements, not so quick perhaps, but as successful in the long run; and the dreamy, indolent characters, of which the *Zygodactyla* is one, always floating languidly about, and rarely surprised into any sudden or abrupt expression."

Nothing is known of the development of this form, as all attempts to raise the eggs have proved futile, and it is unknown whether it has a hydroid stage or not.

SUB-ORDER IV. — TRACHYMEDUSÆ.

The Trachymedusæ are usually considered a distinct group of Hydroidea, especially characterized by having a direct development; that is, they are jelly-fish, which, in general structure, are like the medusæ developed from hydroid colonies, but their eggs develop directly into new medusæ, and not into hydroids. They have then no hydroid state. They are represented in our waters by a number of species, among them *Trachynema digitale*. The full-grown medusæ of this species are from an inch to an inch and a half in length, the walls are very thin, and not much used for locomotion; the latter function being performed principally by the muscular velum which pushes itself outward with considerable force. The proboscis is very large and has four lip-like expansions about the mouth; the tentacles are numerous, and four garnet-colored otocysts are present at equidistant points on the margin of the bell. The ovaries develop from the upper parts of the radial canals, are cylindrical and much elongated, hanging pendent in the bell, and reaching nearly to the velum.

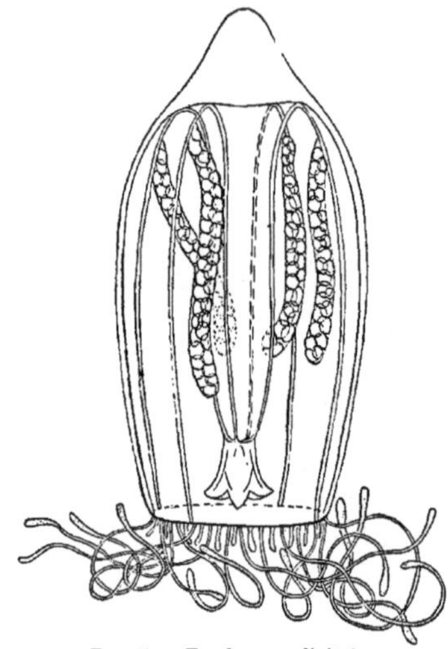

FIG. 82. — *Trachynema digitale.*

Cunina, another genus of this group, though not so common as *Trachynema*, is not rare on our coasts, where it is represented by two species, *C. octonaria* and *C. discoides*. An interesting fact in connection with these forms is that they live on *Turritopsis*, a jelly-fish allied to the *Stomatoca* mentioned on a preceding page.

SUB-ORDER V. — HYDROCORALLINÆ.

Among the many creatures that contribute to the building of a coral reef, are to be counted certain hydroids. For many years there was no suspicion but that the Millepore corals were built by true coral polyps. Professor Agassiz discovered their true nature twenty-five years ago. Later, some other coral-making hydroids have been thoroughly studied by Professor Moseley, of Oxford, late of the "Challenger" expedition. They are very beautiful forms, and there are three kinds of zooids; the ordinary feeding forms, the reproductive kind, and the dactylozooids. The latter have no mouth and gastric cavity, and possess only a tentacular function.

These forms, represented on our southern coasts by *Millepora alcicornis*, form a coral which, however, is composed of calcareous fibres and is traversed in all directions by canals. The little cups occupied by the polyps are shallow, but as one polyp dies another succeeds it, forming a partition separating the new cup from the old, so that in time the pits of the coral become deep but are divided up by a series of transverse partitions. A similar structure exists in some of the true corals.

Among the fossil forms referred to, the hydroids, the Graptolites of the Silurian period are most prominent. In these forms we have an approximation in general appearance to the Sertularian. These, so far as we are able to discover their anatomy, consisted of a narrow tube bearing, on one or both sides, a series of hollow teeth, through which the tube communicates with the exterior. It is supposed that each of these teeth was occupied by a zooid similar to that found in *Sertularia* or *Plumularia*. Haeckel has described, from the lithographic stone of Solenhofen (Jurassic age), two fossil medusæ, which he refers to the Trachymedusæ, and from later beds portions of the incrusting hydrorhiza of *Hydractinia* have been found, and true Sertularians occur in the pleistocene of England. Other forms referred with more or less doubt to this group occur in the Cambrian, carboniferous, etc.

Fig. 83.—Portion of a Graptolite, enlarged.

Samuel F. Clarke.

Order II.—DISCOPHORA.

Among the Medusæ which attain the greatest size and probably are the most commonly observed are those called the Discophora, "sea-nettles," "sea-blubs," or "jelly-fishes." The members of this group are very characteristic, and are named from the disk-shaped outline of their bodies. Although the group of Discophora is not a large one, there being barely a half dozen genera found in our waters, it presents some of the most interesting features of all the known Medusæ.

The bodies of all the jelly-fishes are very gelatinous, composed for the most part of water, and when taken from their native element speedily melt away, leaving a thin film behind. Although these animals are not the only ones which have gelatinous bodies, they excel all in the amount of fluid in their tissues, and are consequently among the most transparent of animals.

The habits of the Discophora are very curious. Many swim at or near the surface, sometimes protruding their bodies a little out of the water. Some are confined to the deep seas and are drifted only by accident into shallow waters. A rough sea sends many below the reach of the agitated waters, and in a calm they rise to the surface, to come into the range of the sun's rays. Whatever their purpose is in this latter habit, we may trace to it the name of "sun-fish," which they bear in some localities. Most of these animals in their adult condition are free swimming, while many are attached to the ground in their younger stages of growth in a way which will be treated of in our account of the development. *Cassiopea* is attached in the adult state in the following peculiar manner. This genus is very common along the Florida reefs where, instead of swimming about in the water, it lies at the bottom on the coral sand, lazily flapping its disk in a monotonous manner. It is not firmly anchored to the bottom, yet rarely changes its position any considerable distance.

Probably all are marine and many genera gregarious, either seeking each other's company, or huddled together by ocean currents or tide eddies.

At spawning-time they are said to don brighter colors, or at least their ovaries and spermaries at that time become more highly colored. Like the hydroids, their organs for defence and offence are the stinging-cells by which their bodies are covered. Many sting with great violence, while others can be handled with impunity. None, however, are destitute of stinging-cells.

All are phosphorescent, especially when irritated, while the color and intensity of the emitted light varies with the genus.

One of the most common Discophores in New England waters is called *Cyanea*, and is a representative of a family of moderate size known as the CYANEDÆ. The most striking peculiarity of *Cyanea* is its disk-shaped body, which varies in size from that of a penny to several feet in diameter. Its color is reddish brown, but when dead and washed about for some time it becomes light blue. The body-disk is divided into two well-marked regions, called the aboral and the oral, or the upper and lower surfaces as the animal naturally swims in the water. The upper surface is smooth and without appendages, save little filaments which are remnants of bodies of considerable size in the young of the animal. The under or oral surface is so called from the fact that from it hangs not only the stomach with its mouth, but also many other important structures. The thickness of the disk in its centre is much greater than at the periphery, where it becomes very thin and flexible, and capable of considerable motion. Around the margin of the bell are found at regular intervals eight sense-bodies, which lie in deep incisions in the rim. Each sense-body is a small sac or cyst mounted on a short peduncle, and in the interior there are a number of rhombohedral otoliths of calcareous composition. Each sense-body is covered by a thin, gelatinous wall stretched above it, which is known as the "hood." From the existence of this hood in the Discophora, and its absence in the Hydroidea, Siphonophora, and a few other jelly-fishes, these animals are called the hooded-eyed medusæ, while the latter are sometimes, especially in older writings, designated the naked-eyed jelly-fishes.

The most prominent of the several appendages which hang from the oral surface of the disk is a thin, curtain-like body of great breadth, which is thrown into a great number of folds and frills. This curtain is open below, and its inner walls make the walls of the stomach. It hangs far down below the oral surface of the bell, extending far beyond it as the medusa, by strokes of the margin of the disk, is driven along through the water.

The tentacles of *Cyanea* are found in bundles, in each of which there is a great number of these organs. Each tentacle is long, thread-like, and very contractile, possessing stinging cells which, however, are rather feeble in their action in the genus *Cyanea*. The tentacles in larger specimens of the genus reach an extraordinary length, and, as in other Discophora, have for their function the capture of the food.

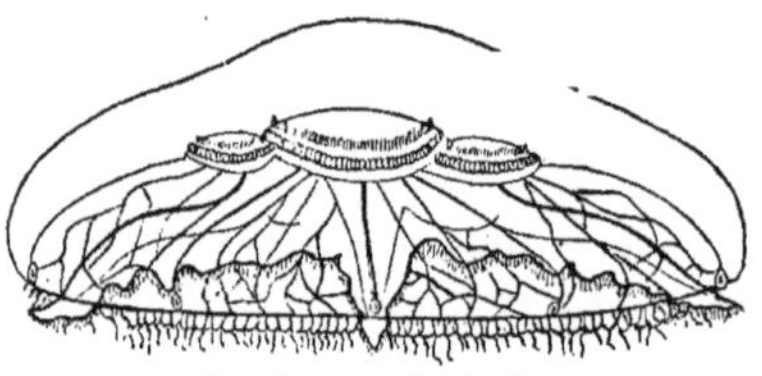

FIG. 84. — *Aurelia flavidula.*

The genus *Aurelia*, the type of the family AURELIIDÆ, is, next to *Cyanea*, one of the most common representatives of the Discophora in New England waters. Although it never reaches the great size attained by the former, it may well be ranked as one of the largest of our Acalephs.

The body of *Aurelia*, like that of *Cyanea*, is disk-shaped; but has a creamy-white color. There are in this genus as in the last, eight marginal sense-bodies, each covered by a hood to which reference has already been made. A great difference between the two genera is in the development of the appendages to the oral side of the body, and instead of there being eight clusters or bundles of tentacles as in *Cyanea*, there is in *Aurelia* a simple continuous circle of short filaments set around the disk-margin. The tentacles are relatively much smaller than those of *Cyanea*.

One of the most important characteristics of *Aurelia* is to be found in the structure of the mouth and oral appendages. Instead of a curtain hanging down from the middle of the disk, the mouth of *Aurelia* is formed in the following manner. From the central region of the oral surface of the disk the oral appendages are suspended by four gelatinous pillars. Between each pair of these pillars there is a circular opening which communicates with a central cavity in which the sexual organs lie. Below the sexual openings the pillars fuse, forming a gelatinous ring surrounding the mouth and serving as a basis of attachment for certain organs, developments of the lips, called the oral tentacles. These oral tentacles are four in number, and are commonly carried extended radially from a central mouth-opening. Each oral tentacle has no resemblance to a marginal tentacle such as is found on the edge of the disk in *Cyanea* or *Aurelia*, but is short and thick, smooth above, and bearing on its under side a deep groove which extends the whole length of the oral arm from its distal tip to the central mouth. On the ridges which enclose this groove are found at intervals peculiar, small, suctatorial mouths. The entrance to the stomach or the large mouth in *Aurelia* is centrally placed on the oral side of the disk, and communicates directly with a disk-shaped cavity, the stomach, which lies directly above it. The lower floor of the stomach is formed by the oral surface of the bell, a muscular layer, from which the four cylindrical bodies which support the oral gelatinous ring are suspended. The roof of the stomach, or the gelatinous wall of the bell, is continued just above the mouth into a pyramidal jelly-like projection, which, however, does not protrude outside the mouth-opening.

The marginal sense-bodies of *Aurelia* are accompanied on either side by a gelatinous extension or lappet which extends outward and hangs slightly downward. On the aboral surface of the bell, in the neighborhood of the hood which covers the sense-body, there is a raised circular area of doubtful function which is not found in the vicinity of the sense-organs in *Cyanea*. This disk is called the *sinnespolster*, and is, as its name signifies, probably an organ of sensation.

Of the many extraordinary genera of Discophorous medusæ, one of the most peculiar is the genus *Cassiopea*, especially a species called *C. frondosa* found about the Florida Keys. This we may consider as the type of the family CASSIOPEIDÆ. Apart from its curious habitat, being attached to the coral mud as has been mentioned above, it is remarkable in the peculiar arrangement of the complicated oral appendages which, although differing greatly from similar organs in the two genera already mentioned, are typical of several of genera belonging to the same great group.

Cassiopea frondosa is found lying upon its aboral surface on the mud near coral islands in Florida and elsewhere in tropical seas. As one floats in a boat over these curious jelly-fishes, they look very similar to an algons growth on the sea-bottom, and are easily confounded with some of the forms of corallines which abound on the neighboring sheltered submarine banks. If, however, the medusa be closely scanned, it will be found to move portions of its body voluntarily, and a throbbing or vibration, especially of the edges of its disk, can be plainly seen. Although fastened to the ground, it still keeps up a flapping motion of its bell probably for purposes of breathing, just as is the case with free-swimming animals of closely allied genera.

One of the functions of the marginal tentacles of the Discophora is the capture of the food. They wind themselves about their prey, sting it to death, and then, by contraction, draw it to the mouth. In a medusa which is fastened to the ground, tentacles would seem to be necessary if the food was large and capable of movement. The construction of the mouth of *Cassiopea* shows that its food is of very small size. The medusa feeds upon the animal and plant life which drifts past it, or which is caused to move over it by the slow flapping of the bell margin. It is therefore evident that tentaeles would be of little service to an animal with this mode of life, and accordingly we find its bell margin is wholly destitute of those filaments called tentacles, which form such a prominent feature in the adults of *Cyanea*, *Aurelia*, and several other genera.

Throughout the animal world there are several examples which might be cited of animals which upon becoming attached to the ground, after a free larval existence, having no use for well-developed sense organs, lose the same or suffer a degeneration in their complication. This can well be illustrated in the development of some well-known genera of Ascidians, where the free larva has higher affinities throughout than the adult, and where a highly-developed organ of sense is formed in a larva to be lost in the fully-grown animal. The organs of sensation on the margin of the bell in *Cassiopea* are, however, as highly developed as in any of its relatives which swim freely in the water. Abnormal as its mode of life is, the otocysts, or organs of sensation, found on the rim of the bell, have not disappeared, neither has their number diminished. In *Cassiopea* there are sixteen of these bodies in normal specimens, and we also often find monstrosities by which this number is increased to eighteen. Professor Agassiz found twelve of these structures in *Polyclonia*, a closely related or identical genus.

The structure of the mouth of *Cassiopea* is somewhat as follows: In the centre of the oral surface of the bell there is a gelatinous cylinder in which there is a central cavity, but no external opening, in a position which corresponds to the mouth of other Discophora. On the side of this cylinder, however, there are openings, four in number, leading into as many cavities partitioned by a thin membrane from the main cavity in which the sexual products are formed, and perhaps through which they pass when mature. From the oral cylinder there arise eight long arms which are commonly extended at right angles to the cylinder parallel with the lower floor or aboral side of the bell. Their tips extend a little beyond the bell margin, while the side adjoining the bell is smooth. Each appendage is branched, and from its aboral surface there is formed a great number of curious appendages of various functions. Two kinds of appendages can be recognized. The former are simply little feeding mouths surrounded by a circle of tentacles and resembling little *Hydræ*. Of these there are a large number on the oral appendages, and each and all open into a system of vessels which pass through the appendages, and ultimately pour their contents into the central cavity of the oral cylinder. All of these *Hydræ* together make up the mouth of the medusa, for they are the orifices through which food is taken into the stomach. The second prominent appendages to the oral arms are small, flask-shaped, and ovoid bodies, with a central cavity which opens into the vessels passing through the arms. They are, however, without an opening into the external water, and their true function is not yet definitely known.

A most interesting family, the PELAGIDÆ, is represented in our waters by two genera called *Pelagia* and *Dactylometra*. In *Pelagia* we have a spherical-shaped medusa of pinkish color and eight marginal sense bodies, alternating with as many tentacles on the bell margin. From the under side of the bell the oral appendages hang far outside of the bell cavity, resembling in many particulars the oral tentacles of the genus *Aurelia*. *Pelagia* is not a large medusa, and is very remarkable in its development, as will be explained more at length later in our account of this part of the subject. *Dactylometra* is closely allied to *Pelagia*, but has a larger number of tentacles around the bell rim. The sense body of both these genera differs in important particulars from those of the families already described.

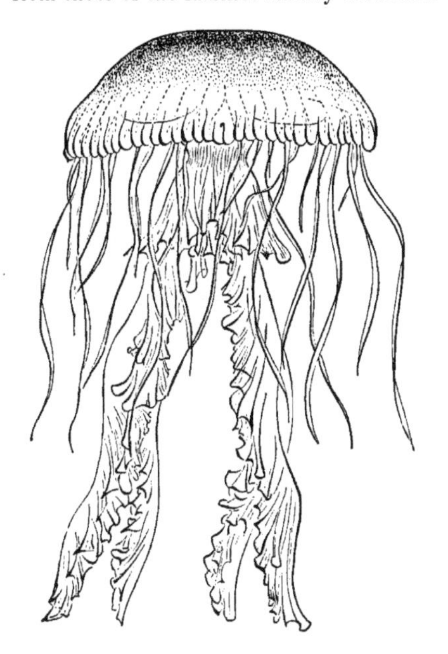

FIG. 85. — *Dactylometra quinquecirra.*

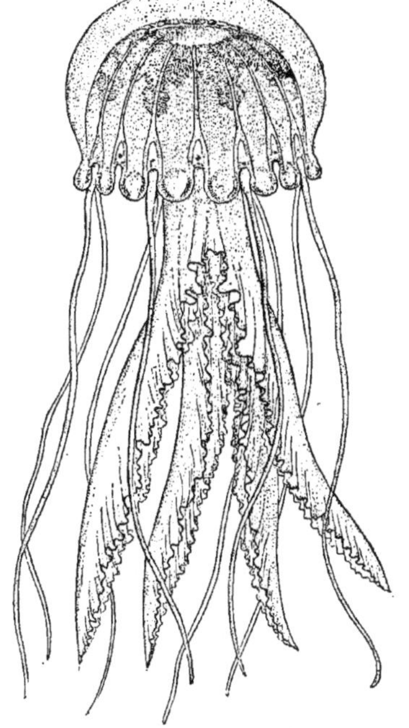

FIG. 86. — *Pelagia cyanella.*

The aberrant families of the Discophora are among the most wonderful of this group. A mention of a few of these may not be without interest. One of the most abundant medusæ at times in the neighborhood of the Florida Keys is a Discophore, called by naturalists *Linerges*, and known to fishermen there as the "thimble-fish," "mutton-fish thimble," and by similar designations. Under proper conditions the number of individuals of this strange genus is very great, and they may be often seen extending in the water in long lines, where they are thrown by the tide-eddies and ocean currents. The popular name of thimble-fish designates exactly the form which these medusæ assume. The bell is not unlike in size and shape a common thimble, differing considerably in this respect from that of the other jelly-fishes of the Discophorous type. The bell has a brownish color on its lower floor, and its walls have a bluish tinge. Around the bell margin there are sixteen marginal lappets or rounded lobes, between which, alternating with each other, there are eight rudimentary tentacles, and the same number of marginal sense bodies. Each sense body is covered by a gelatinous extension of the bell walls of such a form that when looked at above, it seems more like a cyst surrounding it than a hood serving as its cover. From the inner walls of the bell, hanging into the bell cavity, there are placed sixteen dark-brown pigmented bags which lie in a circle with a radius about one-third of that of the bell. Although the function of these bodies is unknown, it may be predicted that they will be found to serve as receptacles for the elaborated food eaten by the medusa. The stomach of *Linerges* is very simple in its structure and never hangs outside of the cavity enclosed by the bell walls. While the jelly-fish is in the act of swimming, the marginal bell lappets are commonly folded inward, forming a notched veil which distantly resembles the so-called velum of the hydroid medusa. At one time in the history of the nomenclature of the jelly-fishes, the presence or absence of a veil was used in designating the two great groups into which the medusæ were divided. The term Craspedota refers to those in which a well-marked velum is found, the Acraspeda where the same is absent. The Hydroidea and Siphonophora are craspedote, the Discophora are supposed to be destitute of a veil, and are therefore acraspedote.

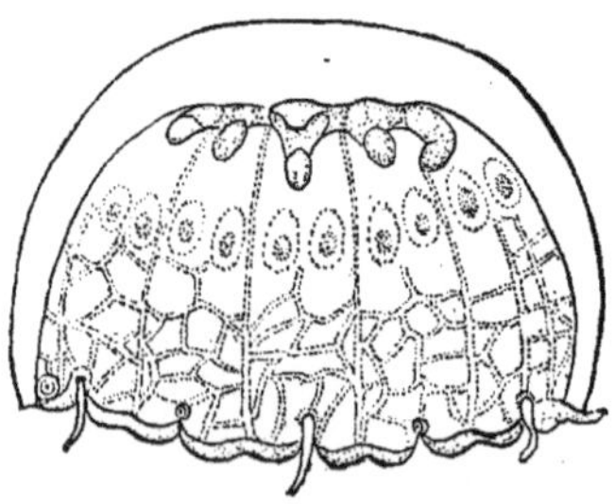

FIG. 87. — *Linerges mercurius*, thimble fish.

Of the many aberrant families of the Discophora, none differ more widely from the genera which we have already considered, than that of the LUCERNARIDÆ, or CALYCOZOA as they are sometimes called. In *Lucernaria*, the best known genus of this family, we have a trumpet-shaped animal of comparatively small size, which is attached by the smaller end, but has the enlarged extremity free. The free end has a disk-shaped form, and in the centre there is an opening into the body cavity which is the stomach. Around the edge of the disk there are arranged at intervals eight bundles of short tentacles or tentacular bodies of doubtful function. The body walls of one of our common species has a greenish color.

Several theories of the relationship between the Lucernaridæ and the other Discophora have been suggested, and their relations to this group are not recognized by all naturalists. Of these theories there are two which seem to the writer the nearest approximations to truth in regard to the affinities of the family. Several naturalists, considering the attached mode of life of *Lucernaria*, but more especially its anatomy and what little is known of its development, have supposed that *Lucernaria* is in reality an adult in an arrested form or stage of development, and that its nearest ally must be looked for in the young of other Discophores. The young of many genera pass through a condition in the progress of its development when it is attached to the ground, and the allies of *Lucernaria* are by many naturalists recognized in these forms.

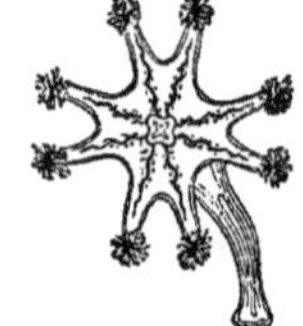

FIG. 88. — *Lucernaria auricula.*

A second interpretation, suggested by E. Haeckel, has even more plausibility than that already mentioned. It has this in its favor, that it refers the *Lucernaria* to the adult and not to the young of another genus. A beautiful medusa was found by A. Agassiz and by the Fish Commission in the Gulf Stream, and has been referred to a genus long ago described under the name *Periphylla*. *Periphylla* is in fact a type of a family called the PERIPHYLLIDÆ, and is in many respects one of the most aberrant of the many genera which make up the Discophora.

The resemblances between *Lucernaria* and *Periphylla* are for the most part anatomical in character, and so little is known of the development of both that there is little possibility of a comparison in this particular. The comparison, step by step, of the many likenesses between the two genera would take us too far into special studies of the peculiar anatomy of them both, but these points of likeness are of a most important character, and show that, notwithstanding one form is attached and the other free, they may be closely allied to each other.

The character of the development, and the different larval conditions which the Discophora pass through in that growth, present some of the most interesting facts in regard to these animals. In the progress of research into the anatomy and classification of the lower forms of animals three curious zoöphytes, placed in three genera, had been described by different naturalists. These genera were called Scyphistoma or Scyphostoma, Strobila, and Ephyra. It was suspected that they were not adults, but in the early days of the history of marine zoology no one had any idea that these animals had close relationships with one another. The first and most important step in a true understanding of the nature of the larvæ of the medusæ was made by Michael Sars, by whom it was found that these three genera were one and the same, and Steenstrup, shortly after, recognized that there exists in the medusæ a true alternation of development such as the poet Chamisso had pointed out is found in the forms of the Ascidian genus *Salpa*, known as the "chain form," and the solitary or asexual individual.

In late summer and autumn specimens of *Cyanea* of large size are often taken in which the membranous fold which hangs downward from the oral region of the disk is loaded with white packets or bundles. These bundles are composed of ova, and if they are examined with a microscope of even low magnifying power will be found to have already entered upon the first steps in their development. In other words the genus *Cyanea* carries its young about and protects them in the folds of the mouth, from the very youngest to some of the higher larval conditions. The highest condition which it has in its career in the mouth-folds of the parent is what is known as a planula. The planula is an elongated, spheroidal body whose walls are formed of two or perhaps three layers, within which is a small cavity, and whose outer surface is covered with vibratile cilia. The function of the vibratile cilia is that of progression through the water, and, as a consequence, immediately on attaining this condition it swims away from the fostering care of the parent, and shifts for itself in the water. In this free-swimming or planula stage it remains until, freighted by the weight of increased age, it can no longer swim through the water by the ciliary movements. When that age comes in the progressive growth of the *Cyanea*, the embryo, which was formerly spheroidal in shape with symmetrical poles, becomes pear-shaped, presenting an obtuse and a pointed pole which can easily be distinguished. The larva next attaches itself by one of these poles to some fixed object, and the two following stages in its growth are passed through in that condition.

Immediately after attachment there forms at the free end of the body a circle of little protuberances which, as the growth goes on, become more and more elongated, while in the centre of the circle, in the periphery of which they lie, an opening is found leading into a cavity in the interior of the body. The resemblance of the young animal, in this first of the attached forms, to the common fresh-water *Hydra*, which has been described, is very striking. The larva was one of those three supposed genera mentioned above which were formerly thought to be widely different from any of the

Discophora. It was then called Scyphistoma or Scyphostoma, and, notwithstanding we now recognize that it is part of the life history of the young of another genus, it is convenient to retain the name as characteristic of the first of the attached larvæ of these animals.

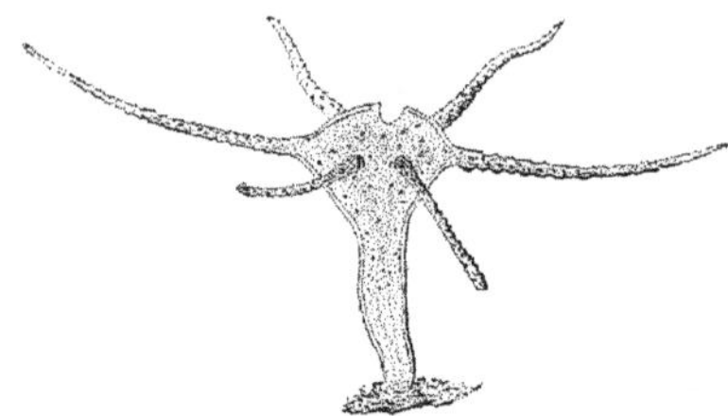

FIG. 89. — Scyphistoma of *Aurelia flavidula.*

The Scyphistoma larva of *Aurelia*, for the following larva has not been observed in our *Cyanea*, although there is no doubt that its development is identical with that of *Aurelia*, is followed by one called the Strobila, which like the former is still attached to some fixed object. In the growth of the Scyphistoma, that part of the free end of the larva situated inside the circle of tentacles, and in which the mouth lies, gradually rises higher and higher, forming an elongated cylinder of great relative size as compared with that of the original body of the Scyphistoma, which lies at its base, and upon which it is borne. There next forms on the outer wall of this cylinder a number of parallel constrictions which encircle the body of the cylinder in waving lines. These constrictions become deeper as the larva gets older, imparting to it a remote likeness, as Professor Agassiz has pointed out, to a "pile of saucers" resting below on the remnant of the Scyphistoma body and increasing gradually in size from the lowest member to the saucer which caps the pile. The next change in the progress of the development of *Aurelia*, after the Strobila just mentioned, is one in which the attached condition is abandoned and a free locomotor larva again adopted. This condition, for a reason identical with that mentioned with regard to the Scyphistoma, may be called the Ephyra, and more closely approaches that assumed by the adult than any of the others. The whole fixed Strobila, however, does not break from its attachment and swim away as an Ephyra, but fragments of the same, or individual saucers which compose the pile, in consecutive order one by one drop from their attachment and swim away as perfect little Ephyræ. The cycle is now complete, and although the Ephyra differs greatly in form from the adult, yet still there are few important additions, and no departure from a direct growth in passing from one into the other.

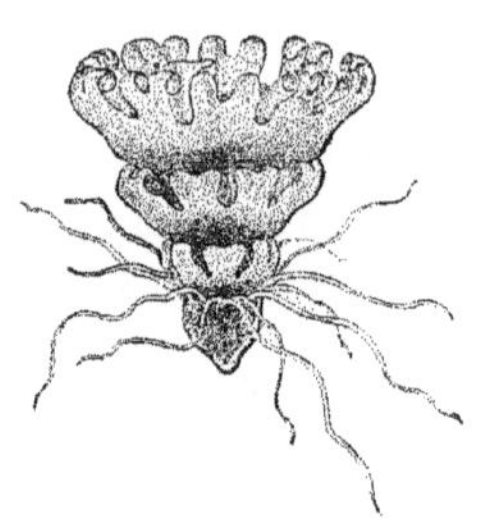

FIG. 90. — Strobilia of *Aurelia flavidula.*

By reviewing the history which has just been considered it will be seen that in two of the intermediate larval conditions, known technically as the Scyphistoma and Strobila, between the egg and the parent, we have a wide departure from the adult in mode of life as well as external shape. We have seen also that the Scyphistoma does not pass directly as a whole into the Ephyra, but that it divides into fragments, each of which becomes a perfect adult. From one Strobila a number of Ephyræ are produced without any conjugation of sexes in the attached animal. From this latter fact the mode of reproduction is said to be asexual and the Strobila an asexual individual. Gathering together the whole history of the development into one chain we find it presents this remarkable circumstance. Between the egg and the female Discophore from which it came there is an asexual, sessile larva which multiplies in an asexual way by simple division, thus producing from one egg a numerous progeny, each of which has no known differences from the parents which produced the egg or spermatozoon. The principle is a wide-spread one in the animal kingdom, and is known as the alternation of generation. It is evident that the Scyphistoma and Strobila, more especially the latter, have a wide difference in shape from the form of the adult *Cyanea*. They develop directly from the egg and are asexual, while the adults which are developed from them are sexual. Sexual animals produce ova which develop into Strobilæ as before. Here then is an alternation of sexual with asexual forms of the same animal, and the technical name of the anomalous development is "Alternation of Generations," nowhere better illustrated than in the Hydroidea and Discophora.

FIG. 91. — Ephyra of *Aurelia flavidula*.

The development of the ovum of *Cyanea* into the adult by a process of alternate generation, in which intermediate larvæ are fixed to some foreign body and reproduce the adult by self-division, is not found in all the Discophora. As this method of growth may be said to be indirect in character, another, called the direct from the absence of these intermediate asexual conditions, also exists. In a direct development among the discophorous medusæ we have simply a continuous growth from the egg to the adult. One egg produces only one adult. Such a development takes place in *Pelagia* and one or two related genera.

CLASS III. — SIPHONOPHORA.

Among the most beautiful of all the medusæ is the group called the Siphonophora, the tube-like jelly-fishes. These animals are all marine and free swimming, and although they often have a hydroid-like shape, which resemblance becomes more marked when we study their anatomy, they are never attached to the ground as are the members of the Hydroidea. They are found in all oceans, although the tropics seem to be richest in the variety of these animals, and those from the Mediterranean have up to the present time been the most carefully studied and described.

As their name signifies, the Siphonophora are characterized by a tube-like body, which is generally so much elongated that it takes the form of a small axis or stem. Although there are several genera in the group where the body does not assume a tubular form (of which one of the most common is *Physalia*), a tubular body seems as a rule characteristic of the group.

The relationships of the Siphonophora to other medusæ have been variously interpreted by different authors. By the majority they are regarded as comparable to the Hydroidea, and are often called the free-swimming hydroids, in distinction from those already considered which are fixed. Others still compare them with the gonophores of the hydroids, some of which as the genus *Lizzia* bud off from the side of their manubrium new individuals, which later develop into medusæ like their parent. The Siphonophora would be regarded by them as similar to the parent with many attached young. While many facts can be mentioned in support of either of these theories, it may be said that the differences which exist between a free medusa and an attached hydroid are not very great, and although at first sight it might seem as if the two theories involve very different comparisons, they are in reality identical.

ORDER I. — PHYSOPHORÆ.

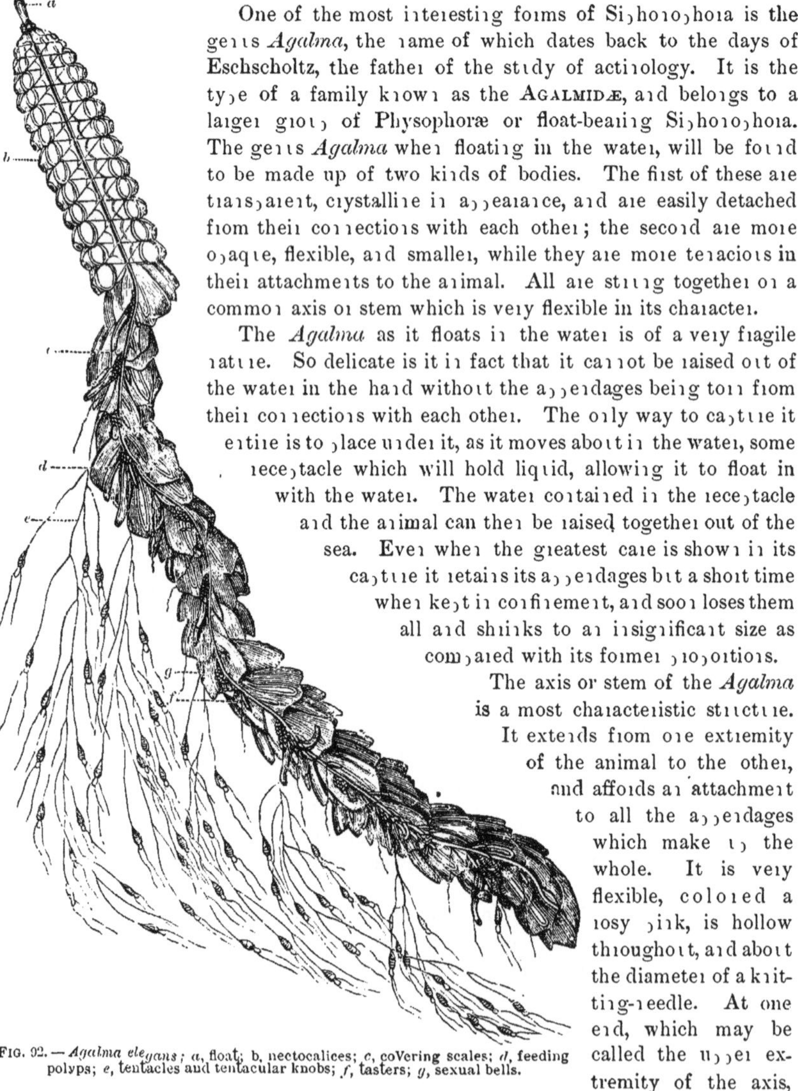

FIG. 92. — *Agalma elegans*; *a*, float; b, nectocalices; *c*, coVering scales; *d*, feeding polyps; *e*, tentacles and tentacular knobs; *f*, tasters; *g*, sexual bells.

One of the most interesting forms of Siphonophora is the genus *Agalma*, the name of which dates back to the days of Eschscholtz, the father of the study of actinology. It is the type of a family known as the AGALMIDÆ, and belongs to a larger group of Physophoræ or float-bearing Siphonophora. The genus *Agalma* when floating in the water, will be found to be made up of two kinds of bodies. The first of these are transparent, crystalline in appearance, and are easily detached from their connections with each other; the second are more opaque, flexible, and smaller, while they are more tenacious in their attachments to the animal. All are strung together on a common axis or stem which is very flexible in its character.

The *Agalma* as it floats in the water is of a very fragile nature. So delicate is it in fact that it cannot be raised out of the water in the hand without the appendages being torn from their connections with each other. The only way to capture it entire is to place under it, as it moves about in the water, some receptacle which will hold liquid, allowing it to float in with the water. The water contained in the receptacle and the animal can then be raised together out of the sea. Even when the greatest care is shown in its capture it retains its appendages but a short time when kept in confinement, and soon loses them all and shrinks to an insignificant size as compared with its former proportions.

The axis or stem of the *Agalma* is a most characteristic structure. It extends from one extremity of the animal to the other, and affords an attachment to all the appendages which make up the whole. It is very flexible, colored a rosy pink, is hollow throughout, and about the diameter of a knitting-needle. At one end, which may be called the upper extremity of the axis, the stem is enlarged into a small globular body which is called the air-bladder or float. This float contains a little sac filled with gas, and in some related genera has for a function the support of the axis in the water. In *Agalma*, however, it is so small that its functional importance in this respect is very slight. The axis of the *Agalma* is divided into two regions, one of which lies adjacent to the float, and is called the nectostem, and the other, more distant from the same, the polyp-stem. In larger specimens the length of the nectostem is about one-third that of the polypstem. The former bears a number of appendages of interesting character called the nectocalices. These bodies are situated in two rows or series, and are glassy clear in their transparency. Their union with the stem is of a very fragile nature, and easily ruptured when the animal is raised out of the water. If we examine a single nectocalyx we shall find that it resembles closely a medusa bell (hydroid gonophore) in which the walls have a more or less polygonal shape. This form is the result of a flattening of two opposite sides of the nectocalyx in order that it may fit closely in the series of which it is a member. Each nectocalyx has a cavity within, which opens into the surrounding water through a circular orifice, partly closed by a thin, washer-shaped body called the veil. The apex of the nectocalyx is situated opposite the external opening, and marks the point of union of the bell and the necto-stem. On either side of the apex, embracing the nectostem, the bell walls are continued into gelatinous horns which closely interlock with similar projections from nectocalices situated in the opposite series.

The arrangement of the nectocalices on the nectostem is as follows: There are two rows or series of these bodies placed diametrically opposite each other on the axis. Each series is composed of a number of nectocalices placed one above the other, fitting closely together by the flat faces on the outside of these bodies. The gelatinous horns already mentioned interlock with corresponding bodies from the opposite series. By the close approximation of adjacent bells on their flat faces, and the interlocking of bells from opposite series, a certain rigidity is given to this portion of the animal, notwithstanding the delicate attachment to the stem.

The disposition of the nectocalices causes all the bell openings in each series to point in the same direction, or almost at right angles to the length of the axis. The action of the nectocalices is as follows: They are, as their name implies, structures for a propulsion of the *Agalma* from place to place through the water. When water is taken into their bell cavities, by a violent contraction of the bell walls it is violently forced out through the opening into that cavity against the surrounding water in which the medusa is floating. The necessary result of this action is that the animal is forced through the water in an opposite direction from that in which the resistance takes place. By a nice adjustment of the different bells, acting in concert or independently, almost any motion in any direction can be imparted to the *Agalma*. Just below the float on the nectostem there is a small cluster of minute buds in which can be found nectocalices of all sizes and in all conditions of growth.

The attachment of the nectocalyx or swimming bell to the nectostem, not only serves to move the animal from place to place, but also renders it possible for the swimming bell to receive its nourishment. Although the nectocalyx resembles very closely a medusa, it is a medusa bell without a mouth or stomach. It is not capable of capturing nourishment for itself, but is dependant upon others for that purpose. The nectocalyx has a system of tubes on its inner bell walls which communicate with the cavity of the nectostem by means of a small vessel which lies in the peduncle by which it is attached. Through this system of tubes the nutritive fluid is supplied to the nectocalyx from a common receptacle, the cavity of the stem. In the largest specimens of *Agalma* which I have studied, there were seventeen pairs of well-developed nectocalices.

The appendages to the polypstem are somewhat different in character from those of the nectostem, and are of several kinds, differing in character, size, and shape. The first and most prominent of these are known as the covering scales. They are transparent, gelatinous bodies, and are found throughout the whole length of the polypstem. Their shape is quadrangular or almost triangular, and they are united to the axis by one angle. The upper and lower faces are flat, and the whole appendage has a thin, leaf-like appearance. Through its walls from the point of attachment to the distal angle there runs a straight unbranched tube which communicates freely with the cavity of the stem. The covering scales are easily detached, and are incapable of voluntary motion. Their function seems to be to shield the structures which lie beneath them.

Below the covering scales three kinds of bodies hang from the polypstem. They are known as the polypites or feeding stomachs, tasters, and sexual bells. The polypites are the most conspicuous of these bodies. They have a flask-like shape, and are united to the polypstem by one extremity, while the free end has a terminal opening which is a mouth. The walls of the cavity of the polypite are crossed longitudinally by rows of cells which have been compared to a liver. In the cavities of the polypites the half-digested food can be seen through the walls. The nutritive fluids formed in these bodies are poured into the cavity of the axis, there to be distributed throughout the different appendages of the animal. When indigestible substances, as the hard parts of Crustacea, are taken into the stomach they are thrown off again through the mouth-opening. I have never seen the polypites more than seventeen in number, and they hang at regular intervals along the whole length of the polypstem.

One of the most prominent bodies next to the nectocalices and covering-scales in the *Agalma* are the so-called tentacles, which hang from the base of the polypites, and which when extended are very long. The tentacles of *Agalma* are long, highly flexible, tubular filaments whose function is the capture of food. At times widely extended their length is little less than that of the *Agalma* axis itself. At other times they are drawn up under the covering-scales at the base of the polypite, and have a very diminutive size. Along their whole length they are dotted with crimson pendants of minute size which are called the tentacular knobs. These will be found, on close study, to be of a very complicated structure. Their true function is somewhat problematical, but they are supposed to assist in the capture of the food. In addition to the well-developed tentacular knobs which dot the whole length of the tentacles, there are many half-grown bodies of the same character clinging to the base of the polypite.

Alternating with the polypites at intervals along the polypstem are found very curious bodies called tasters, which have a close likeness to the flask-shaped feeding zoöids. These bodies are without a mouth-opening at their free extremity, while from their base hangs a long, highly-contractile filament which is destitute of tentacular knobs. The tasters have an internal cavity which is in free communication with that of the axis of the animal. Various functions have been assigned to the tasters, but none without objections seems yet to have been hit upon. Their usual position is in clusters midway between the adjacent polypites. The term taster is somewhat misleading, for these bodies do not have gustatory functions.

The sexual bells are of two kinds, male and female, and both are found in grape-like clusters, the male near the base of the tasters and the female near the polypites. If we isolate one of the members of a cluster, we find that it has a bell-like shape, and that the ova or spermatozoa are found on a proboscis within. Each bell hangs from the cluster by a tender peduncle which arises at its apex, and each female bell contains a single ovum.

The growth of the young *Agalma* from the egg to the adult is of a rather complicated nature. When cast in the water the egg is a tiny, transparent sphere barely visible to the naked eye. After fecundation, and obscure changes similar to a segmentation of the yolk, a slight protuberance arises at one pole. This prominence is formed of two layers between which, in a short time, a third layer is also formed. The outer layer is the ectoderm, the middle the mesoderm, and the internal the endoderm. Between the endoderm and the remainder of the egg there is a cavity called the primitive cavity. As the embryo grows older the elevation at one pole increases in size, and the proportion in thickness of the middle layer, as compared with the ectoderm and endoderm, becomes very large, while the ectoderm becomes very thin. The prominence has now assumed a helmet-like shape, and fits like a cap over the remnant of the yolk. The whole larva in this stage of growth is called the primitive larva or *Lizzia*-stage, and the cap-shaped covering, the primitive scale. The primitive scale is an embryonic organ which is lost in subsequent development of the larva.

Immediately after the primitive larva stage there is found to develop under the primitive scale an air-bladder or float, which first appears as a little bud near the opening into the primitive cavity, which has now taken the form of a tube in the primitive scale lined with endoderm. At about the same time also there appears a covering-scale of very different form from either the cap-like primitive scale or the covering-scale of the adult *Agalma*, which have already been described. The float is the permanent float of the adult, while the second formed covering-scale, like that of the first, is also embryonic and larval in character. The larva has now the following parts: 1, The remnant of the yolk; 2, a cap-shaped covering-scale; 3, a second embryonic covering-scale, and 4, a float. As the larva grows older more covering-scales like the second appear, and the beginnings of a tentacle and tentacular knobs are seen at the adjacent end of the growing larva. At the same time the yolk becomes elongated, and in its walls appear reticulated masses of red or crimson pigment.

The tentacle first formed as well as its pendants, the embryonic tentacular knobs, are transient in character. They differ essentially from the adult knobs, and are confined to this stage in the development of the larva. Meanwhile the primitive scale is lost, and a circle of covering-scales of the second kind appears at the base of the float. This larva is called the Athorybia larva from its remote resemblance to a related adult genus called *Athorybia*. The appendages of this larva are: 1, a float; 2, a crown of embryonic covering-scales; 3, the remainder of the yolk-sac with an attached tentacle and temporary pendants.

The next following larval condition of *Agalma* is one in which the embryonic covering-scales have disappeared and new scales like those of the adult have formed. Four well-developed nectocalices appear on a nectostem, and an adult polypite bearing the characteristic pendants of the adult has grown on the extremity of the short polyp-stem. A remnant of the yolk-sac, however, still persists, and from it depends an embryonic tentacle and its characteristic side branches. The Physophora larva resembles the adult in all particulars, except size, and the presence of the last of temporary organs later to disappear in the growth of the *Agalma*, viz., the embryonic tentacle.

FIG. 93. — Tentacular knob of *Agalmopsis*.

Several other genera of Physophores are so closely allied to *Agalma* that they are placed in the same family. One of the most interesting of these is the genus *Agalmopsis*, which differs from *Agalma* in its slighter form and the intimate structure of its tentacular knobs. *Halistemma* has also a peculiar tentacular pendant which differs from those of *Agalma* or *Agalmopsis*. In the adult knob of *Agalma* the following structures are found: 1, an involucrum; 2, a sacculus; 3, terminal filaments and vesicle. The involucrum is a membranous sac which covers the knob when the other parts are retracted. The great mass of the knob is made up by the sacculus, which is corkscrew-shaped and dark crimson in color. At one extremity it is fastened to the inner walls of the involucrum, the free end bearing two terminal filaments and a vesicle. The various genera of Agalmidæ differ in the character of this knob. *Agalmopsis* has a sacculus and involucrum, but no vesicle, and only one terminal filament. *Halistemma*, probably the type of another family, has no involucrum, while it possesses a spirally-coiled sacculus and a single terminal filament. The genus *Crystallodes* has tentacular knobs like those of *Agalma*, and is by some authors made a species of this genus. It differs, however, from *Agalma* in the rigid nature of the axis, in the shape of the covering-scales, and in minor points in the anatomy of the nectocalices.

FIG. 94. — *Agalmopsis picta.*

The genus *Stephanomia*, a name which has been applied to genera of Siphonophora of widely different form, was given by the elder Milne-Edwards to a Physophore in which there are many series of nectocalices appended to the nectostem. *S. contorta* is one of the most beautiful and graceful of all the Siphonophores as by the combined movement of its many swimming-bells it gaily swims along in the water. It is peculiar in this respect, that the polypites are mounted on a long peduncle, which also bears the covering-scales and the tentacles. The tentacular knobs resemble more closely those of *Halistemma* than of *Agalma*. It may be regarded as the type of the family FORSKALIADÆ.

One of the most beautiful genera of PHYSOPHORIDÆ is the interesting animal known as *Physophora*, called in the dialect of the Messina fishermen, "Boguetti." *Physophora* is remarkable in possessing no polypstem, but in place of this structure the axis is enlarged into a bag-like inflation. There is, however, a well-developed nectostem and two series of nectocalices as in *Agalma*. Around the circumference of the inflation, which takes the place of the polypstem, the tasters with their short, tentacular filaments are arranged side by side. The polypites and sexual

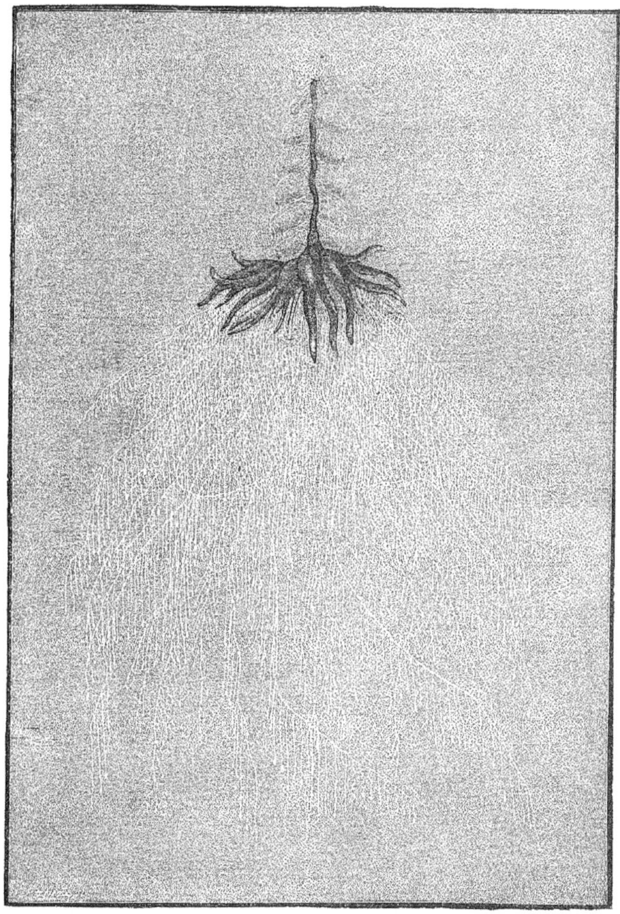

FIG. 95.— *Physophora hydrostatica.*

bodies are found below these structures. No true covering-scales exist in the genus *Physophora*.

One of the largest of all the Physophoræ is the genus *Apolemia*, the type of the APOLEMIADÆ, which is called by the Italian fishermen by the suggestive name of "lana di mare." In this beautiful Siphonophore we have, as in other float-bearing medusæ thus far considered, a double row or series of nectocalices, but unlike the last mentioned genera, there arises from the nectostem, small bodies resembling minute tasters which wave about in the water with great freedom. The polyp-stem also, instead of being covered throughout its whole length with covering-

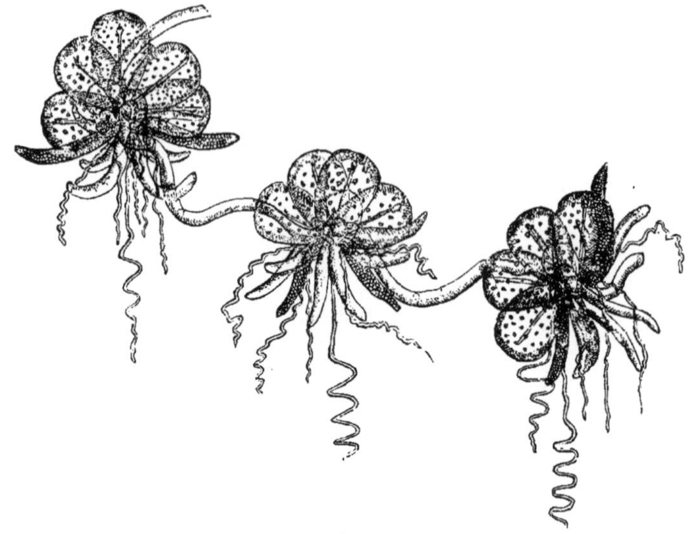

FIG. 96. — Portion of *Apolemia*.

scales, has these structures arranged at intervals and in clusters, each with tasters, polypites, and sexual bodies.

ORDER II. — PNEUMATOPHORÆ.

There are two genera of Siphonophora closely related to each other and to the group of Physophoræ already studied, which are now looked upon as forming a group by themselves. These genera include the well-known Portuguese Man-of-War or *Physalia*, often erroneously called by sailors the Nautilus, and a less common genus, *Rhizophysa*, one of the most bizarre forms of these animals.

Physalia is one of the most common of all the Siphonophora in tropical oceans. The most conspicuous part of this animal, as it floats along on the surface of the water, is an enlarged air-bladder, six or eight inches in length. On the upper side of this float there is a raised crest colored by brilliant blues, yellows, and pinks. On the under side of the same there hang a great variety of appendages of several kinds. There are feeding-mouths or polypites, flask-shaped bodies resembling tasters with long tentacles, which, as the animal floats in the water, extend far behind it in the water as magnificent streamers, and grape-like clusters of sexual bodies. The *Physalia* is wholly destitute of a tube-like axis, and as it floats on the surface of the water, resembles more a bladder with richly variegated walls, than the tube-like forms which we have already considered.

The closest relative to *Physalia*, as far as anatomy goes, to which affinity also what is known of their development adds additional evidence, is the strange genus *Rhizophysa*. *Rhizophysa* is a simple skeleton of a siphonophore. It is the axis of an

Agalma stripped of all its appendages, except feeding polyps and sexual bodies. There is in it no distinction between nectostem and polypstem, and no means of voluntary motion. The float is particularly large and has an apical opening through which its contents communicate with the surrounding water. The feeding polyps hang from the stem at regular intervals when extended, and midway between them appear on the same axis botryoidal clusters which are called the sexual organs. Tentacles hang from the bases of the polypites as in other related siphonophores. The tentacular knobs have, however, a highly characteristic form which varies with different species in number and general anatomy.

FIG. 97.—*Physalia arethusa*, Portuguese man-of-war, one-fifth natural size.

ORDER III.—DIPHYÆ.

In all the genera thus far studied, there is always a float at one extremity of an axis when such was present. In no case is a float missing, although oftentimes it is functionally unimportant. In none of the remaining Siphonophora, on the other hand, is a float present. These last medusæ may conveniently be divided into the Diphyæ, in which there are one or two nectocalices, and the Hippopodiæ, floatless forms in which there are several or more than two swimming-bells.

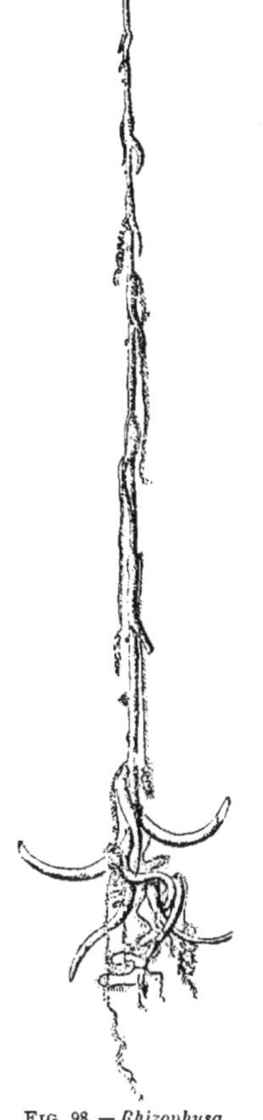

FIG. 98.—*Rhizophysa*.

One of the best marked families of the Diphyæ is the DIPHYIDÆ, of which *Diphyes* is a typical genus. This genus and most of its relatives is smaller than the majority of those already studied, and are easily distinguished from the former by the absence of a float, and the presence of but two nectocalices. The two swimming-bells which are possessed by *Diphyes* are of somewhat different form. The anterior is conical in shape in order to facilitate rapid progression through the water, while the posterior which lies behind it, seems to perform the greater part of the work in the progression of the medusa. As in *Agalma*, onward motion is caused by the resistance of the water as it leaves the bells on the surrounding medium in which the animal swims. The motions of the nectocalices are spasmodic and not long continued as in the *Agalma* and other Physophoræ. The axis of *Diphyes* hanging from the interval between the two bells, is a long, filamentous, flexible structure not unlike that of *Agalma*. It is highly contractile, and has a cavity throughout its entire length. The polypites arise at intervals along the length of the stem, and are in no respect peculiar. Each polypite bears a tentacle and tentacular knobs, or pendant side-branches. At the point of attachment of the polypite to the axis, we also find a transparent bell-shaped covering-scale and a cluster of sexual bells with eggs and spermatozoa. Each cluster of bodies near a polypite ultimately separates from its attachment to the *Diphyes* axis, lives independently, and is called a diphyizoöid.

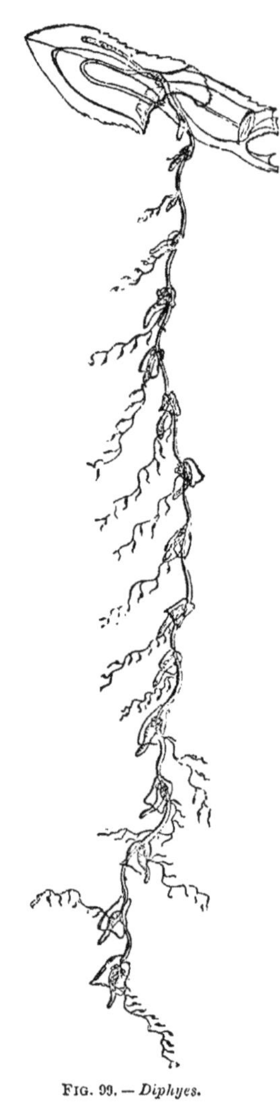

Fig. 99. — *Diphyes.*

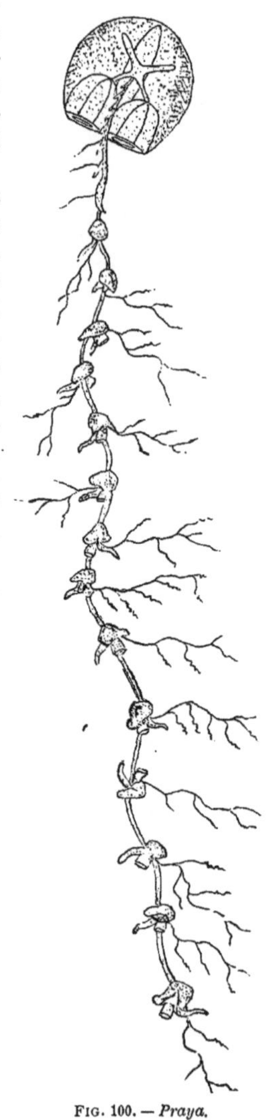

Fig. 100. — *Praya.*

There are several families related to the Diphyidæ which might be mentioned. They differ from it in the character, size, and general anatomy of the two nectocalices. One of the most marked of these is *Praya*, a solitary genus composing a family called the Prayidæ. In *Praya* there are two nectocalices which are of about equal size, and have a rounded or semi-ovate form. The bell walls are not as rigid as those of *Diphyes*, and their motion less spasmodic. The axis is very long and flexible, and the polypites, found at intervals along its length, are protected by a helmet-shaped covering-scale, beneath which are found clusters of sexual bells mounted on short peduncles. The genus is one of the most striking of the many beautiful genera which characterize the Siphonophore fauna of the Mediterranean Sea. I have also observed a fragment of a large *Praya* near Fort Jefferson, Tortugas, Florida.

The fourth of the large groups into which the true Siphonophora may be divided is called the Hippopodiæ from a genus sometimes called *Hippopodius*, which has a highly characteristic and peculiar structure. *Gleba* (*Hippopodius*) is in most respects related to the Diphyæ, but unlike them has more than two nectocalices. There is no float and no extended axis with individuals found at intervals in its length. No polypstem is developed, and the nectostem has little in common with that of *Agalma*. The nectocalices are of characteristic shape and different from those of any other siphonophore. Each bell has the shape of a horse's hoof, and has a very shallow cavity and rigid walls. As far as yet known the Hippopodiæ have no diphyizoöids such as exist in several genera of the Diphyæ.

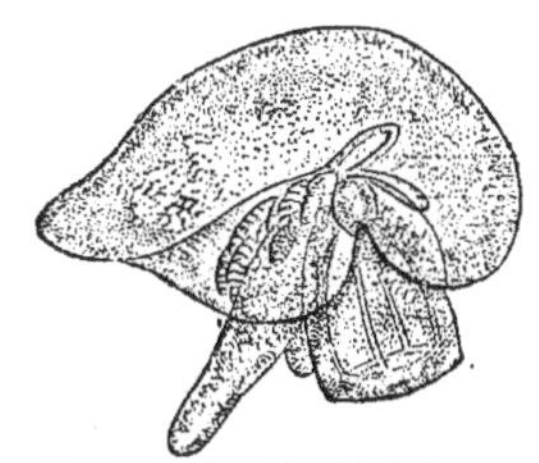

FIG. 101.—Diphyizooid of *Praya*.

ORDER IV.—DISCOIDEÆ.

Among the many interesting forms of Medusæ related to, and by most naturalists included in the Siphonophora, are two beautiful genera called *Velella* and *Porpita*. These, with a genus *Rataria*, which is probably the young of one or the other, make up a group called the Discoideæ.

Velella has borne the name which designates its most striking peculiarity since the middle of the fifteenth century, on account, perhaps, of a somewhat fanciful likeness to a little sail. It is commonly called in Florida, where it is sometimes very abundant, the "float," and is likewise commonly confounded with the *Physalia* or Portuguese man-of-war. The body or disk of *Velella* has an oblong shape, flattened upon its

FIG. 102.—*Velella limbosa.*

upper and lower sides. The float is composed of a number of concentric compartments in free communication with each other, seven of which open externally in a line extending diametrically across the disk. In the whole diameter there are fourteen such openings, seven in each radius.

A triangular sail rises on the upper side of the *Velella* disk and extends diagonally across its surface. It is firmly joined to the upper plate of the float. Over the triangular sail as well as the float, there is stretched a thin, blue-colored membrane, which is continued into a variegated soft rim along its border and around the rim of the float. In our most common American *Velella*, which often reaches a length of four or five inches, the portion of the rim of this membrane around the disk is entire; in some species, however, it is continued into elongated appendages.

The most important appendages are found on the under side of the *Velella* disk which is commonly submerged as the animal floats on the surface of the water. Of these perhaps the most prominent is a centrally placed body which hangs downward below the remaining appendages, and is open at its unattached extremity. This structure is the feeding-mouth or polypite, and is single in both *Velella* and *Porpita*. In the zone just surrounding the polypite we find a large number of small appendages, each of which has a thread-like shape bearing along its sides a number of little transparent buds in all conditions of growth. Each of these little bodies ultimately separates from its attachment, and in the form of a minute jelly-fish, not larger than a pin-head, swims about endowed with independent powers of life for a considerable length of time. The medusa which has thus separated is known as a Chrysomitra. Surrounding the bodies last mentioned on the lower surface of *Velella*, there is a circle of feelers of bluish color which are commonly in constant motion. One of the most prominent superficial differences between *Porpita* and *Velella* is the total absence of a triangular sail in the former genus. They are commonly found associated together, and often accompanied by a third jelly-fish allied to both, called *Rataria*.

Class IV. — CTENOPHORA.

The highest of the jelly-fishes, both on embryological and anatomical grounds, are known as the Ctenophora. In these animals that characteristic of higher forms of life known as bilateral symmetry appears for the first time. An obscure symmetry which has been called by some naturalists bilateral, appears among the Siphonophora, and even in the Hydroidea. In the Ctenophora, however, it is more plainly indicated than in either of these groups.

One of the earliest mentions which we have of the ctenophorous medusæ we owe to Martens. Freiderich Martens was a ship's physician or a ship's barber, as he styles himself, who accompanied Captain Scoresby in a voyage of exploration into the Polar Seas. He first found one of these beautiful medusæ in the neighborhood of the island of Spitzbergen. Eschscholtz was the first to recognize the common likenesses between the different members of the group, and gave to them the name of Ctenophora or comb-bearing medusæ.

The Ctenophora take their name, as he first pointed out, from the existence on the external body walls of eight rows of vibratile plates called combs. These combs are arranged in such a way that in flapping they strike upon the water, and by their motion the jelly-fish is driven along through the water. We find here for the first time since our studies of the Cœlenterata began, a large group of animals where movement in the water is produced both by special locomotive organs and contractions of the body.

The varieties in form in the bodies of different Ctenophora is very great. In some genera they appear as long ribbon or belt-like creatures which move through the water with serpentine movements, in others as transparent caps or globular gelatinous masses over which the rows of combs shine with most lovely iridescent colors. No greater variety of more beautiful genera is to be found anywhere among the medusæ.

Cestus, called also the Venus Girdle, is perhaps the most striking genus of the Ctenophora. Its shape departs the most widely of all the Ctenophora from that of the medusoid types. In *Cestus* the body of the jelly-fish has a girdle or belt-like form, and is moved more by the contractions of the body than by the rows of combs which fringe its edges. The animal is very transparent and extremely tender, so that it is with the greatest difficulty taken from the water without breaking. Its motion through the water is a graceful undulation to which body contractions and vibrations of the combs contribute.

The mouth of *Cestus* is situated midway in its length between the two extremities of the belt-like body. On either side of it there hangs a single short tentacle which protrudes from a tentacular sac. Opposite the mouth there is a sense-body, or otocyst as it is commonly called, in which is situated a compound otolith. The rows of combs upon the external surface of the body of *Cestus* are not as conspicuous as in some other genera, but the course of their lines can be easily traced. The enormous expansion of the two lateral lobes of the body which give a girdle-like form, impart to the rows of combs which lie in these regions an extraordinary development as compared with

Fig. 103. — *Cestus veneris*, Venus girdle.

those which lie in the intermediate regions or on the flattened sides of the body. The adult *Cestus* reaches a length of from two to three feet, and is one of the largest as well as most beautiful ctenophores of the Mediterranean. There is scarcely any color in its body walls.

Of the many genera of medusæ closely or remotely allied to *Cestus*, one of the most interesting and least known is a genus called *Ocyroë* from the Gulf of Mexico and the Caribbean Sea. This genus, like many other ctenophores, and especially like *Cestus*, is very transparent, and has on the external body-surface, eight rows of vibratile combs, the lines of which converge at a point near that pole of the animal in which a sense-organ is situated. One of the most extraordinary things about *Ocyroë* is the great development of two opposite sides of the body into wings. In *Cestus* the opposite sides of the body are so developed that a band or belt-like form is given to the animal, but in *Ocyroë* these lateral developments take on the form of wings or similar bodies. *Cestus* moves through the water by a slight undulation of the body, while *Ocyroë* does the same by a flapping movement of its greatly developed lateral lobes, and by their beating upon the surrounding water. When *Ocyroë* is at rest, the lateral wings are widely extended, giving to the animal a remote likeness to *Cestus.* When, however, motion is attempted, the lobes are raised above the horizontally extended position which they occupy at rest, and then violently swung downward, passing through almost 180°. This flapping of the two wings in concert is continued several times, and in that way the animal is propelled through the water. The function of the rows of combs in the movements of the medusa is secondary to the flapping movements of the lateral wings.

The body of *Ocyroë*, from two sides of which the wings arise, is of oblong, oval shape, with a mouth at one pole, and a cluster of otoliths in an otocyst at the opposite. There is no vestige of tentacular appendages near the mouth, and the lips are undi-

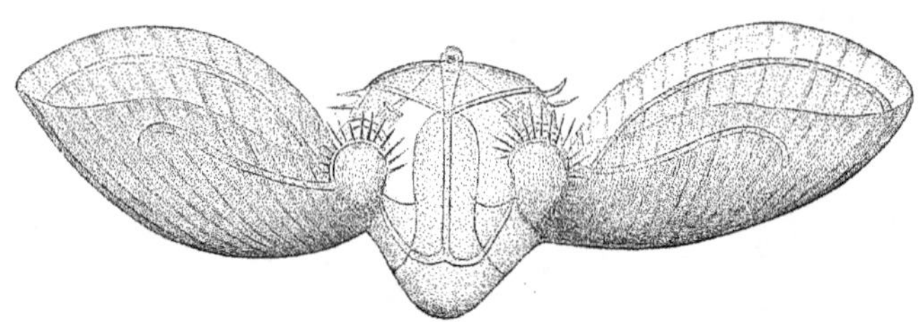

Fig. 104. — *Ocyroë crystallina.*

vided, smooth, and highly flexible. *Ocyroë* thus far has been taken only from the waters of tropical America.

In colder latitudes as on our New England coast, we have a beautiful ctenophore called *Bolina*, and another very closely allied genus known as *Mnemiopsis.* These medusæ are in many respects most closely allied to *Ocyroë*, or rather *Ocyroë* seems an aberrant form of these more northern jelly-fishes. Although the same lateral lobes exist in both genera, their importance in *Bolina*, as far as movement is concerned, is much less than in *Ocyroë.* They are also seldom or never carried extended horizontally at right angles to the body, as in the curious genus *Ocyroë. Bolina* is one of the most transparent of the comb-bearing medusæ. The body is very gelatinous and highly phosphorescent. The sides of the body are developed into two larger lappets or lobes which are carried or hang vertically instead of horizontally. On account of the contractile powers of the body walls, *Bolina* can vary its outlines very considerably; as a rule, however, when the body is seen from the side, it has an oval or elongated form. Eight rows of vibratile combs contribute to the propulsion of the medusa through the water. These lie upon the external surface of the body, and arising from the pole opposite the mouth-opening, extend to the more distal edge of the lateral expansions to the vicinity of the mouth. From the great development of the lateral lobes, the four lines of vibratile combs which cross these bodies are much longer than those which lie on the body regions between them. Two tentacles are found hanging from the sides of the body of *Bolina.* These tentacles are of diminutive size in the adult, and are remnants of structures which in early conditions of growth were very much more developed.

Four other curious appendages called auricles are found on the sides of the body in the grooves which lie on opposite ends of a horizontal diameter of the body, and which are enclosed by the edges of the lateral lobes. These auricles are simply extensions of the body walls on the sides of the mouth, and their edges are skirted by vibratile plates somewhat like those found on the exterior of the body in the lines mentioned above.

The stomach and chyme tubes, vessels analogous to veins and arteries, in *Bolina* do not differ greatly from the same structures in other Ctenophora, but are highly characteristic as compared with similar bodies in other medusæ. The mouth of *Bolina* is a narrow, elongated slit, opening with a correspondingly flattened receptacle called a stomach. From the end of this stomach opposite the mouth, there arises a number of vessels which pass to various regions of the body. One of the most important of these is continued directly to the aboral pole of the medusa, and is known as the funnel. Of the others, four, after a bifurcation, pass to the vicinity of the rows of combs, and following a meridional course, eventually join each other in pairs near the opposite pole of the jelly-fish from the sense-body or in the lateral lobes. There are two tubes which originate from the base of the funnel and extend along the sides of the body to the tentacles which from their characteristic course have always attracted attention. Their function seems to be to convey the nourishing fluid to the tentacles into which they eventually open.

There are several genera of ctenophorous medusæ allied to *Bolina*. One of the most pompous of these, as well as the largest of the comb-bearing jelly-fishes, with enlarged lateral lappets is a genus called *Chiaja*. This genus is remarkable in many particulars, but more especially in the great length of the auricles which appear as long, filament-like tentacles, and the very complicated course of their chymiferous tubes.

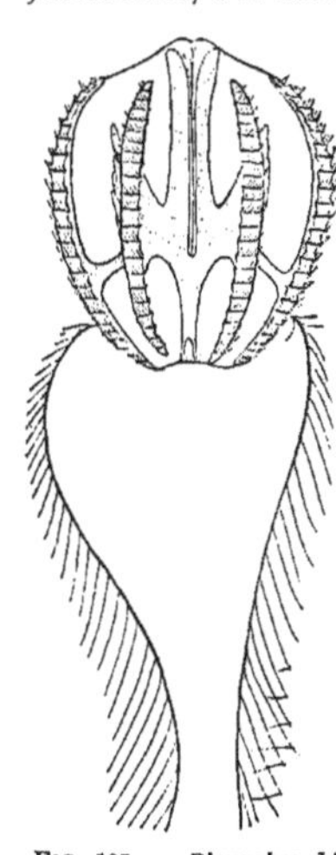

FIG. 105. — *Pleurobrachia rhododactyla*; only a portion of the tentacles shown.

Many classifications have been made of the Ctenophora, but as yet all are open to objections of some kind. One of the last suggestions with this subject in view emphasizes the presence or absence of the tentacles. By this classification we would have the Ctenophora divided into the Tentaculata and the Nuda, accordingly as tentacles exist or are wanting. In the genera which we have already considered, with the exception of *Ocyroë*, well-developed tentacles are found either in the larval or adult condition.

Probably the best example, however, which might be mentioned of a tentaculated ctenophore is the genus *Cydippe* or *Pleurobrachia*. There are often in our New England bays and harbors, after a southeasterly wind, a number — myriads at times — of little transparent gyrating spheres, not larger than a common marble. These little gelatinous spheres move through the water with a great variety of motion, and seldom change in any important particulars the regular spherical form of their bodies in the manner characteristic of the genera of Ctenophora to which we have referred. When the cause of this variety in motion is sought out, it will be found that on the surface of the body there are a number of iridescent lines extending from one pole to another, and that each of these lines is formed of a number of minute comb-like bodies such as exist in other Ctenophora.

It is by the strokes of these bodies on the surrounding water that the jelly-fish is moved about from place to place.

The tentacles of *Pleurobrachia*, oftentimes almost wholly inconspicuous they are so securely packed in little lateral pockets on either hemisphere of the medusa, are most important structures in the economies of this beautiful medusa. It often happens, when the jelly-fish is at rest, that the tentacles are extended from their pockets, stretching far outside of this receptacle. Their length when extended in this way is so great that it seems impossible that they can be retracted into the tentacular pockets. The tentacles are two in number, each tentacle bearing a large number of side branches of brownish color, and in the movements of the medusa are often thrown into the most fantastic shapes.

The Ctenophora Nuda, or those ctenophores which are destitute of tentacles, are represented by a very beautiful medusa called *Beroë*. This is perhaps one of the most remarkable jelly-fish which we have yet considered, although it is without doubt one of the lowest in its organization. The form of the body of this animal is that of a cap or rounded sac of very simple structure.

If we closely consider the structure of this animal it will be noticed that, extending longitudinally across the outside surface of the body, there are eight rows of combs as in the other Ctenophora. At the pole of the body there is a sense-body in a similar position to that of the sense-body of other genera. The whole interior of the body serves as a stomach, and into it there opens a mouth of very great size. The stomach will often be found gorged with food, and the animal swollen to double its natural size by the mass of food collected in this organ. This animal is indeed one of the most ravenous genera of medusæ. There are no tentacles in the genus *Beroë* and no sign of tentacular sacs. The chymiferous tubes have a very simple course in the walls of the body, extending from common origin to the vicinity of the mouth in an almost direct course with, however, many side branches. The color of our common *Beroë* is a delicate pink.

Class II.—ACTINOZOA.

Much discussion, happily now for the most part of the past, hangs about our knowledge of the nature of the Actinozoa or corals. Their relationship to animals was first recognized by Peysonelle, who records his observations in the quaint language of a former century. The word Actinozoa, of comparatively modern date, has an almost exact equivalent in the older term Zoöphyte, and refers to a great variety of animals fixed to the ground like plants, and possessing in common with them certain superficial resemblances. They all have remote likenesses to a plant or flower, with which for a long time in the history of science they were confounded. It includes the great families of reef-building corals, and has an added interest on account of the many questions which suggest themselves in relation to the method of formation of coral reefs and coral islands. In the general structure of their bodies the Actinozoa differ somewhat from the Hydrozoa, but the difference is not of such great importance as to call for a wide separation in a scheme of classification of the two. The differences would appear at first sight very great. Nothing for instance could seem more different than the soft, gelatinous body of the medusa and the stony mass of a coral, and yet this difference has no homological meaning, and as far as general structure is concerned the two are identical. In the medusæ the body is so filmy and transparent that it is often wholly invisible as the animal swims in the water, and the sense of touch is necessary to supplement that of sight in order to know where the animal is. In stony corals on the other hand we find secreted in the animal tissues a very hard, calcareous substance which forms a skeleton, and, when the secretions of a number of individuals are united, an axis or head upon which the corals build or secrete their skeletons. When we view a white branch of coral we see nothing of the animal save its dried and bleached skeleton, from which most of the organic matter has disappeared. If we could go to the tropics and study the medusæ and corals alive many points of likeness might be traced in their general structure.

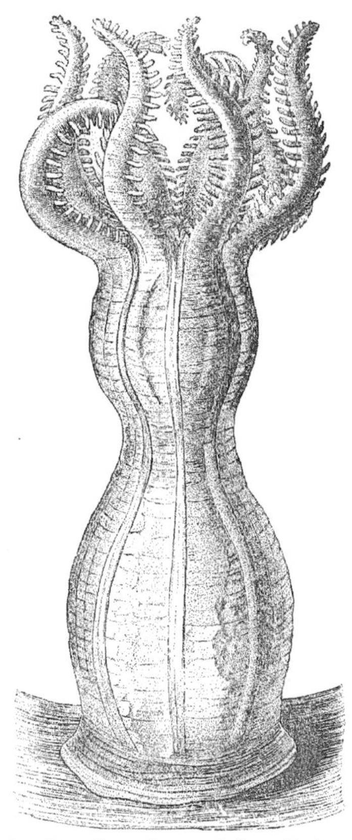

FIG. 106. — *Monoxenia*, young halcyonoid polyp.

We are accustomed to associate with all animals a certain amount of motion from place to place in the medium in which they live. In the Actinozoa, however, we find very few of the adults endowed with locomotive powers, and all, as a general thing, are fixed to some foreign body. The solid carbonate of lime which the majority of the corals secrete in their tissues is, in respect to its method of formation, identical with all secretions in animal bodies. It is not the work of the coral any more than the shell of the clam or the covering of the lobster is the work of the animals which they enclose and protect. It is an internal or external formation by secretion, and as a consequence coral animals are not mechanical builders any more than any other animal can be called the mechanical constructor of its shell or skeleton. Many of the Actinozoa have no power of secreting hard matter in the form of a skeleton, and the bodies of such have a soft, gelatinous character, while in their tissues, as in those of the medusæ, a large proportion of water is found. The solid secretions of coral animals, commonly known as coral, is sometimes erroneously supposed to harden on exposure to the air. This opinion is probably well founded as far as coral rock, a chemical product of the cementation of coral sand in a way to be explained, is concerned, but is wrong as regards the coral in the form secreted by the animal. Although it appears to harden by the loss of animal matter, the skeleton itself changes but very slightly, if at all, in its hardness after the death of the animal from which it came.

The genus *Actinia*, a soft-bodied Actinozoan which has not the power of secreting carbonate of lime, is for various reasons the one commonly taken to illustrate the anatomy and characteristics of the group. *Actinia*, found in almost all seas in the temperate as well as torrid zones, is represented by a large number of species, the majority of which are of comparatively large size. The Sea-Anemone, *Metridium marginatum*, our New England Actinia, has been chosen to illustrate certain of the important general features in the anatomy of the group.

Metridium is common almost everywhere on the New England coast in sheltered pools left by the tides, on spiles of bridges, and on rocks near low-water mark. Under the name of sea-anemone it is known to collectors of marine curiosities as the common zoöphyte from Eastport to New York. In no place have I seen the species larger than on the spiles of Beverly Bridge and at Nahant, but equally giant specimens have probably been taken from other localities. The diameter of the largest *Metridium* found in the former locality measured, when expanded with water in its tissues, a little over ten inches in diameter. Forms of Actinaria allied to *Metridium* from the Florida Keys and Bermuda attain a gigantic size, often fifteen to eighteen inches in diameter. When the Actinia is seen from one side it will be found to have a cylindrical body firmly fixed at one end to some foreign object, and bearing at its free end a circle of tentacles surrounding a central mouth. The tentacles are thickly set together, are very movable, and when the animal is alarmed are quickly drawn to the body. The whole body and the tentacles are very much inflated with water, which at the will of the animal can be expelled from the body through the mouth, the body walls, and the tips of the tentacles, where there are small orifices. When inflated with water the body and its appendages are all expanded, but when this water is expelled the animal shrinks to a shapeless lump, the tentacles are drawn back, and there is little resemblance to its former condition.

The internal structure of *Metridium* is of a character typical for most of the Actinozoa. If we make a horizontal section through the body about one-half the distance between its attached and free extremities the cross-section thus made will present the following characters. In the centre lies a cavity, the stomach, whose wall is held in position by radiating partitions passing from it to the outer walls of the body. Intermediate between these partitions there are radiating walls or septa which arise from the outer body walls but do not extend to those of the stomach. The body walls from which these partitions all arise are, as a general thing, thicker than those forming the partitions, and in their sides there are openings through which the water at times leaves the body cavity.

The method by which the *Metridium* feeds is very simple. The food captured by the tentacles when the anemone is expanded is passed from one member to another through the mouth into the stomach. Here digestion takes place, and after the soft portions have been digested the harder parts, skeletons, shells, and the like, are thrown off again through the mouth by which they entered the stomach. The fluid passes from the stomach through an opening opposite the mouth into the body cavity, bathing the interior of all the organs which lie in that place.

Special organs of respiration are not unknown among genera of Actinaria allied to *Metridium*, but in this genus probably the whole external and internal surfaces of the body contribute their part in the performance of this function. In *Metridium* special organs of sensation are of a very low grade of organization and of the simplest kind, as would naturally be expected from the attached life of the animal.

Reproduction among the Actinozoa presents some of the most interesting features connected with these animals, and in the case of the coral colonies in which the reef-builders live, is the most important factor in the determination of the ultimate form.

Three kinds of reproduction, which are known as generation by fission, by gemmation, and by the laying of eggs, or ovarian, occur in the Actinozoa. The first two methods are asexual in their character, the last sexual. In reproduction by fission we find a simple voluntary self-division of a single individual into two or more secondary animals. In many reef-building corals this method of increase is most natural in order to increase the size and style of growth of a colony of these animals. Among the solitary forms like *Metridium* it is seldom found. A second mode of increase, a reproduction by gemmation or budding, is much more common than that of fission and is found in the solitary as well as the communal forms of Actinozoa. In the formation of a bud we have a very simple method of reproduction. In such a case there simply appears on one side of the base of the body, or, as in some genera, on the disk surrounding the mouth, a small protuberance, which is a simple elevation of the body walls. From this simple beginning of a bud we pass to a more developed condition in which the protuberance has become a small coral animal attached to the

FIG. 107. — *Crambactis arabica*, sea anemone.

parent at one extremity which is its base, and with a free extremity furnished with a mouth surrounded by a circle of tentacles in most respects identical with those of the mother. The bud from the parent has every resemblance to the parent, and can live independently although still attached, and drawing nourishment in part from her through the base of attachment.

All the Actinozoa reproduce by means of eggs. The ova pass through the condition of a ciliated planula which is free-swimming and sometimes parasitic in its youth. Phenomena similar to those of alternation of generations have been found in some genera, but as a rule the development is direct from the egg to the adult. Special features in the development of individual genera will be touched upon as we continue our account of these animals.

The Actinozoa are commonly divided into two great groups, easily distinguished from each other, and known as the Actinoida or Zoantharia, and Halcyonoida or Alcyonaria, which includes, roughly speaking, the reef-builders in the first instance and the "sea-fans," "sea-pens," and their allies in the second. Anatomically, the two groups are distinguished, in part, as follows: In the Actinoid corals we find a large number of internal radial septa and numerous external tentacula about the mouth. When the number of these organs is small they are generally in multiples of six, and in most instances there are no lateral-appendages to the tentacles. When hard matter is secreted in the tissues it is commonly in the form of carbonate of lime. The second great group of Actinozoa, called the Halcyonoida, differs from the former in the possession of eight, or a multiple of eight, tentacles and body septa, while the former almost universally bear side branches and appendages. In those genera where a skeleton exists it is tough and elastic, oftentimes of very great hardness, as in the genus *Isis* and the well-known ornamental coral of commerce.

Order I. — ZOANTHARIA.

The so-called Actinaria, which are referred to the Actinoid corals, include a large number of interesting genera. As a general thing, these genera are solitary in their mode of life, and often reach a great size. One of the best known genera of the Actinaria is the genus *Metridium*, of which we have already spoken.

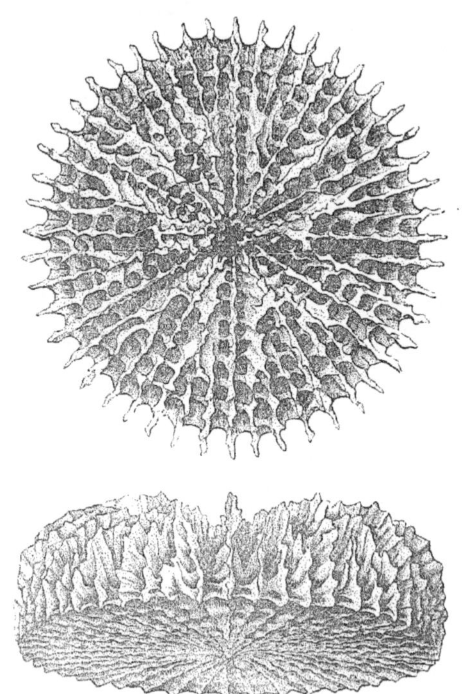

FIG. 108. — *Fungia symmetrica*, three times the natural size.

Many of the Actinaria are either free-swimming in their adult form, are parasitic, or live with their bodies partially hidden in the sand; still others are attached to the ground. They do not, as a rule, secrete a calcareous or horny skeleton, and their bodies are usually very soft, without even the needle-like spicules which occur in the soft forms of the Halcyonoida.

One of the most interesting of the Actinaria is the genus *Edwardsia*, which is not attached to the ground, but lives in the sand in its adult form, while in younger conditions it is free-swimming, even after the tentacles have reached a considerable size. The young *Edwardsia* was at first mistaken for the adult condition of an Actinoid coral, and was described under the name of *Arachnactis*. Later, however, it was shown to be simply the free-moving young of the genus *Edwardsia*. *Peachia*, another genus of Actinaria, is parasitic in the mouth-folds of the discophorous jelly-fish, *Cyanea*.

The Zoantharia, which are reef-building corals, are closely related to the Actinaria, and although generally found in colonies, are sometimes solitary in their habits of life. One of the most nearly related to the Actinaria of all the Actinoids which secrete lime is a beautiful genus, which bears the highly suggestive name of *Fungia*. It is the largest of the solitary lime-secreting corals, and often reaches a diameter of from six to eight inches. It is disk-shaped, with a large number of radiating partitions, or septa, which extend from the centre of the disk to a periphery not bounded by a vertical wall. The tentacles are not placed in a circle around the periphery of the disk, but are found irregularly disposed over its whole upper surface. *Fungia*, in its adult condition, is not attached to the ground, but lies in the coral lagoons in rather sheltered places. In the younger stages, however, of its development it is a fixed form, and passes, according to Semper, through a larval condition comparable to the strobila of the Discophora, exhibiting a true alternation of generation. Reproduction also takes place in *Fungia* by fission and gemmation. The larger species of the genus are found in the Pacific and Indian oceans. A small species, *Fungia symmetrica*, was dredged by Pourtalès, and afterwards by Mr. Agassiz in the "Blake," in the depths of the Straits of Florida and the Caribbean Sea.

There are several very fragile compound corals which in their young conditions resemble *Fungia*, and which on that account naturally come in the neighborhood of this genus. One of these, called *Mycedium*, is one of the most common corals in the sheltered lagoons of the Bermudas and along the Florida reefs. *Mycedium fragile*, called in the Bermudas the Shield Coral, has a thin, fragile disk, easily broken, which hangs to the submarine cliffs a little below low-water mark. This disk has a chocolate-brown color, which bleaches, as do most coral skeletons, into a beautiful white. Upon the upper side of the disk we find two kinds of coral animals. One of these is centrally placed, and is the mother of the colony, or the parent from which the others have formed by a budding. The remaining individuals are smaller than the central, around which they are arranged in concentric circles, which increase in number with the diameter of the disk upon which they are formed. It will, therefore, be seen that we have in *Mycedium* two kinds of individuals, a single, large central animal, which is the oldest in the colony, and many smaller, which it may at once be said are formed by budding from the original. Nothing is yet known of a reproduction in *Mycedium* by self-division or by the deposition of ova, although there is every reason to suspect that both of these methods of formation of new colonies really exist.

One of the most common genera of Actinoid corals is called *Madrepora*. Although rarely, if ever, found as far north as the Bermuda group of islands, it is one of the most common of the reef-builders, and especially near the western termination of the Florida Keys it forms great banks miles in extent, whose upper surface rises to within a few inches of the low-water level.

One of the most abundant species of Madrepore is *M. cervicornis*, a branching species, which sometimes attains a large size. The difficulty of comprehending the structure of *Madrepora* comes from the fact that in each branch of one of the dendritic species, as *cervicornis*, we have a large number of different individuals arising from a common axis. If we take a terminal fragment of such a branch, we can see that its very tip is formed by a small rounded body of calcareous nature, in the interior of which there are radiating partitions passing from centre to circumference, as in the genus *Fungia*. The peripheral ends of all these septa are bounded by a calcareous wall connecting them all, which is absent in the last-named genus. If we study carefully the slight excrescences upon the sides of the branch, we find that although smaller there they have in all important respects a similar structure to the terminal individual.

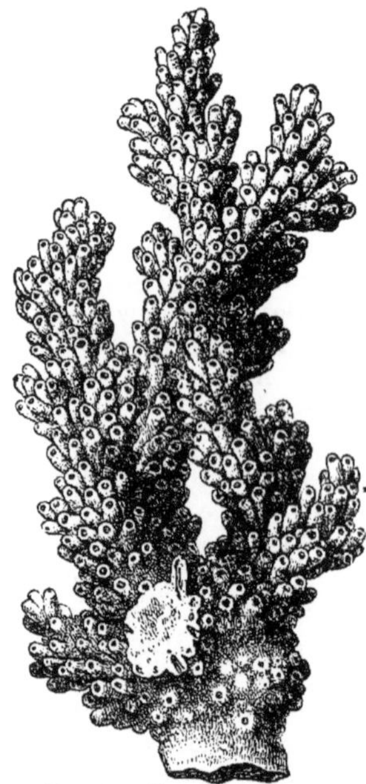

Fig. 109. — *Madrepora verrucosa.*

The branch of Madrepore, when alive, presents an altogether different appearance from that which the bleached skeleton has. By carefully studying the soft portions of a growing coral of this genus, it will be noticed that the terminal and lateral individuals have the general appearance, as far as their soft tissues are concerned, of a minute sea-anemone or a typical Actinozoan. Each individual has a brownish, cylindrical body, composed for the most part of water, with a central stomach, into which opens a mouth at the free end of the animal.

In the species which we are now considering, the less conspicuous bodies, which arise from all sides of the calcareous branch upon which they are formed, originate as buds from the base of the terminal individual. In order to understand the relationships between the large terminal and the smaller lateral individuals, both of which form a colony in the branch of a Madrepore, we must regard the former as a parent form from which all the lateral have budded, and of which they are the young. Suppose that we go back in the development of the branch to a time when there was but one individual where now we find a branch with the colony clinging to its sides. At that time there would be formed a small, single individual, bearing every likeness to that which is now terminally situated on the branch. As growth goes on from that early condition, buds arise one after another from its base and sides in a manner identical with that which we find in the genus *Metridium*, which we have already described. As the coral with its attached bud at the base grows, there is deposited near its attachment a larger amount of lime than is necessary, so that in fact the processes of life are impeded by the surplus of solid matter in the tissues. The result is that the lower part of the animal becomes dead matter, while at the same time the soft portion of the same is growing upward, and is rising above the solid matter, now too much solidified to allow vital processes to go on in that portion of the Madrepore. The original animal, however, still lives, and its place is not wholly taken by the buds which earlier in its history formed upon it. A surplus of solid deposition in the walls of the base of the bud cements it firmly to the parent, while from the live part of the original polyp, now slightly elevated above the plane in which it formerly was, there forms a second bud, a third, a fourth, and so on. A repetition of the formation of too much

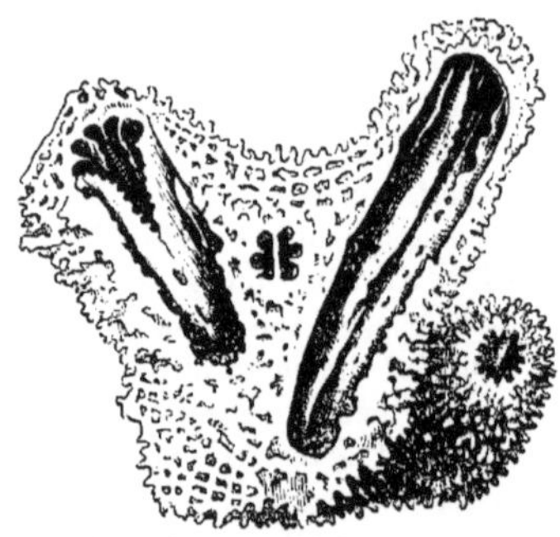

Fig. 110. — Section of *Madreposa verrucosa*, enlarged.

lime for vital functions takes place continually, and with equal pace the live portion of the original individual mounts higher and higher until a branch, similar to that with which we started, is eventually formed. It can thus be seen that there is in the branch of coral a larger individual, which is the original polyp with which the colony starts its growth, and smaller individuals, or lateral buds, which have from time to time, as the first polyp grows, budded out on its base and sides. When, as often happens, a lateral branch is protruded from the sides of another branch, we simply have a bud which partakes of the character of the larger individuals rather than its neighboring lateral buds. It grows by the same laws as the branch from which it originates, and as in the first, so in this, small lateral buds which resemble those on the parent branch are freely formed upon it.

A second species of *Madrepora*, although departing widely in general form from the first, is closely allied to it in generic characters. This species is known as *palmata*, and instead of a branching habit, occurs in flat, massive slabs, on the upper

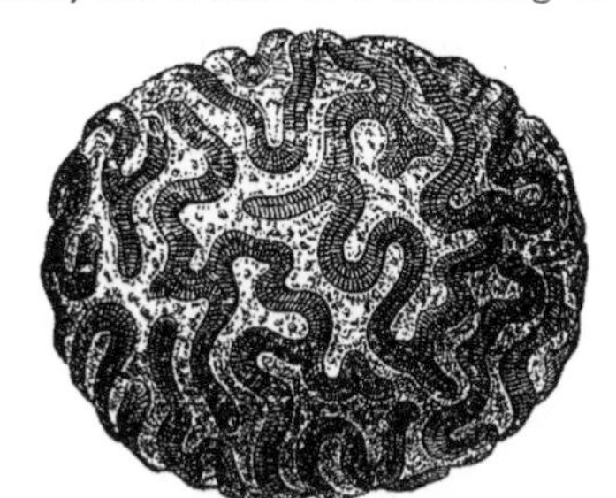

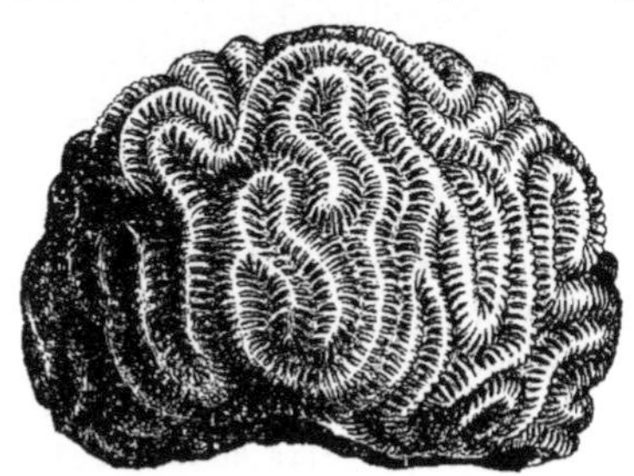

FIGS. 111 and 112. — *Heliastræa heliopora*, brain-coral, with and without fleshy parts.

surface of which the different individuals are irregularly distributed. Blocks of such fragments generally grow at greater depths than *M. cervicornis*, and when waterworn form the so-called coral boulders which make the foundations upon which the reef rests. They play no small part, by their solidity, in the successful resistance of these islands to the encroachments of the sea.

Some of the most massive genera of Actinoid corals are those known as the "brain-corals," or "brain-stones." There are several genera which are commonly confounded in this nomenclature, and all such have many similar anatomical peculiarities. The genus *Diploria* assumes a hemispherical shape, varying in size from a few inches to several feet in diameter. Its external surface is covered with serpentine furrows, which recall vividly the convolutions of the brain, and probably suggested the name of brain-stones for these corals. Over the curved surface of a live brain-stone is stretched the soft organic parts of the coral, while in the superficial furrows lie the stomachs, which have a similar serpentine form to the convolutions in which they lie. Upon the surface of the brain-stone, arranged in lines, are found rows of mouths, each opening into that stomach or part of the stomach which lies just beneath them.

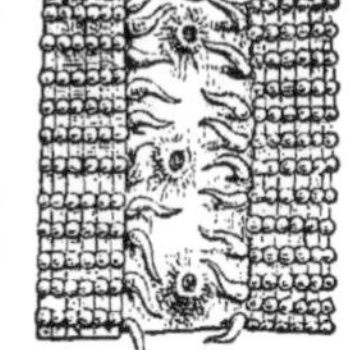

FIG. 113. — Three mouths of *Heliastræa*, enlarged.

Another genus allied to *Meandrina* and *Diploria* is known as *Manicina*. This genus is not commonly as large as either of the former, and does not have the brain-like shape. Its upper surface, however, has the same furrows as the genera above mentioned, in which lies the stomach. *Manicina* is commonly found on the floor of a coral lagoon, but is not attached to the bottom. Here also belongs the genus *Heliastræa.*

Several other genera of Actinoida assume, in the manner of their growth, the shape of hemispherical heads, although they do not have a convoluted surface like the true brain corals. One of the best examples of these is *Astræa*, a coral in which the colony has a globular form, but without superficial canals. The different coral individuals in *Astræa* are placed side by side over the whole surface without being arranged in lines, while every individual is externally almost wholly distinct from every other in the colony. The mode by which animals of this genus reproduce their kind, and increase the size of the community is by simple self-division or fission. When the single individuals, by their growth, exceed a certain size, they spontaneously divide into two similar well-formed smaller individuals.

FIG. 114. — *Astræa pallida.*

The only Actinoid coral which secretes a stony base found in New England waters is a beautiful little genus known as *Astrangia*, which occurs in small patches in the crannies and clefts in the cliffs along the southern shore of New England. At Newport, R. I., on the most southern point of the island, there are several localities where beautiful colonies of these animals are found. This is, however, but one of many localities which might be mentioned. The calcareous base which the *Astrangia* secretes is inconsiderable in size, and forms only a slight crust on the surface of the rocks. It builds no considerable reefs or coral deposits of any size. The animal itself is one of the most delicate and beautiful of all the reef-building corals.

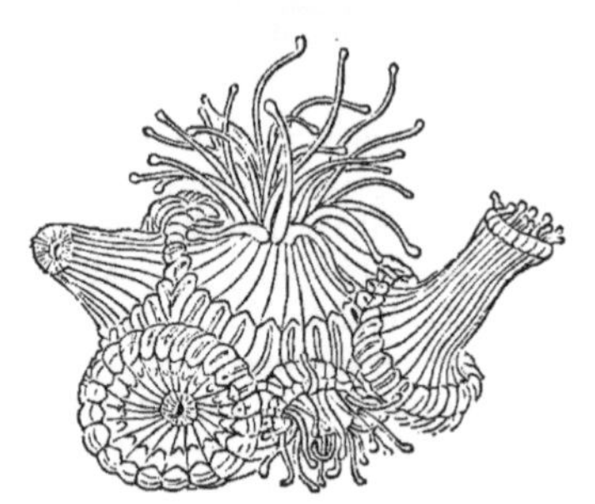
FIG. 115. — *Astrangia danæ.*

No living genus of corals better illustrates the formation of new individuals by self-division than that known as *Mussa*. Here we generally have a limited number of individuals, never branching, but attached to each other at their bases. Almost every fragment of one of these corals shows individuals with evidence of a self-division, either just beginning, partially formed, or wholly completed. In the parent individual before any sign of division begins, the upper extremity or distal end is of circular form and the coral itself has an irregular trumpet-like form, fastened by the smaller end to some foreign object. The first sign of change in the original *Mussa* preparatory to a fission, is the elongation of the disk-like shape or a lengthening of its axis, by which the two opposite sides closely approach each other. This growth ultimately leads to a condition in which the two opposite sides approach, and the disk of the coral is divided into two individuals, both of which have a common stomach and a common basal attachment. The complete fission or division of the original individual into two is accomplished in the last stage, which is one in which we have the same common base of attachment, bearing at its top two perfect individuals, which have resulted from the self-division of the single animal with which we started.

ORDER II. — HALCYONOIDA.

The second large group of Actinozoa is called the Alcyonoida or Halcyonoida from the genus *Halcyonium*, supposed to be the nest of the "Halcyon" or king-fisher of the Greek fables. To this division belong the "sea-fans" and "sea-whips," so-called, and many others, some of which are widely aberrant in general characters. So different in structure from the typical "sea-fans," are many of the genera now thrown among the Halcyonoids, that they are probably destined later, when a more perfect classification is made known, to be the nuclei of new groups of equal rank.

The colonies of Halcyonoids have commonly a branching form, are flattened in fan-like shapes or extended into long, flexible whips. In *Tubipora*, or the organ-pipe coral, we have the hard parts tubular in shape. The amount of deposition of solid matter in their tissues varies considerably. In certain genera all hard inorganic depositions are wanting, and the body is soft and without skeleton. In still others, secretions in the form of spicules are well developed. Of those which have a hard skeleton, we find all degrees of hardness, from a simple horn-like axis of the "sea-whips," to the extreme hardness of the ornamental coral of commerce.

FIG. 116. — Section of red-coral.

In the classification of the Halcyonoids many systems have been attempted, but the subject is still in an unsatisfactory condition, and at present there is little uniformity among naturalists in regard to the limits of families and genera. A well-marked genus called *Antipathes* is separated from the Halcyonoids by almost universal consent, as the family ANTIPATHARIA, and seems to stand between the Actinoids and the group which we are now considering. The likeness of *Antipathes* to the sea-fans is best seen in the branching character of its axis, while the number of tentacles and the absence of side branches to these organs, ally it more closely to the majority of the Actinoids. One of the most interesting species of *Antipathes* is the well-known *A. columnaris*. In this species we have a very interesting case of a worm building a tube from the smaller lateral branches of the coral or by its presence causing a modification in the mode of growth of the coral. Upon one side of the axis of the *Antipathes*, the worm, a true annelid, places itself, and the small lateral branches of the axis in the immediate neighborhood are modified into a columnar network forming the walls of a tube in which the worm lives surrounded by the case, bearing the relationship of a true messmate to the coral upon which it is found.

Of the true Halcyonoida the genus *Halcyonium* of the HALCYONIDÆ is one of the most interesting, which is sometimes designated by the suggestive name of dead-men's-fingers, looking not very unlike a human hand with the fingers remaining as mere stumps. Although in general appearance *Halcyonium* resembles the soft corals, well-marked spicules of beautiful shapes are found regularly arranged in its walls.

The common sea-fan, *Rhipidogorgia flabellum*, which we select as illustrating the GORGONIDÆ, is one of the most common Halcyonoids of our tropical and semi-tropical seas. The fan-like shape which the colony has is the result of the fission of many lateral branches, large and small, forming a network often of great fineness. The sea-fan has a hard central axis, and a still softer rind which can be easily broken off, and at the death of the animal is almost wholly deciduous. In this softer rind are found the spicules, and from it the animals directly arise. There is a great variety in the forms of the different genera of sea-fans, and the colors are sometimes very striking; in many, bright reds and yellows predominate with purples and browns.

FIG. 117. — *Corallium rubrum*, red coral.

The sea-whips, of which there exist a great variety of forms, assume either the shape of low, branching, shrub-like zoöphytes, or long, unbranched, straight or spiral rods. Their colors are sometimes black, often light brown, and occasionally, as in a *Chrysogorgia* from deep water, golden.

One of the best known of all the Halcyonoids is the genus *Corallium* so much prized as the ornamental coral of commerce. The greater quantity of this coral which is used is gathered in the Mediterranean where the most extensive fisheries are situated on the western coasts of Italy, the shores of Sicily and Sardinia. The city of Naples, where the cutting of the coral into ornaments is a great industry, is a great centre of coral commerce, and many pounds of the more precious varieties are yearly sold there. The commercial value of different coral fragments depends upon the size, but more especially on the tinge of color which they have. The pink rose-color is esteemed the most valuable, while "ruby" varieties are ranked among inferior wares. Much, of course, also depends in individual specimens upon freedom from blemish, and purity of color. The coral has from the earliest times been cut into cameos by lapidaries, and its use in ornamentation reaches far back into classical times. The word has been derived from the Greek, Κορή, daughter, a highly fanciful comparison of these most beautiful gems to the daughters of the sea goddesses.

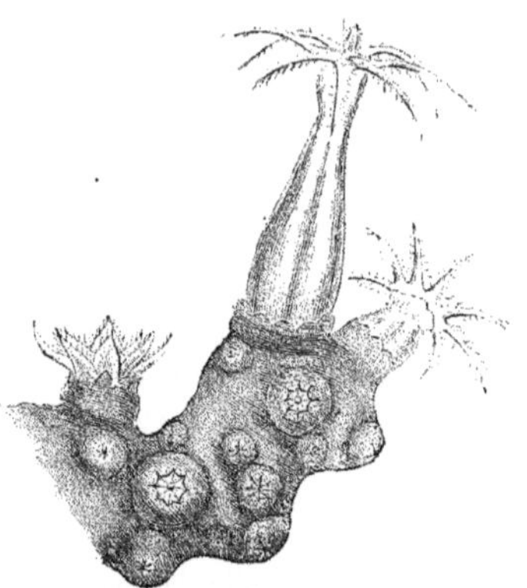
FIG. 118. — Red coral polyps, enlarged.

A genus called *Isis* is closely allied in many respects to *Corallium*, and approaches it very intimately in the great hardness of the axis. While portions of this axis are as hard as that of *Corallium*, the stem is not continuous, but is formed of hard and soft articulations, alternating with each other. The hard joints are securely bound together by softer and more flexible articulations, permitting a slight bending of the axis. The ISIDÆ, including *Isis*, *Mopsea*, and one or two allied genera, are often used for ornamentation, but no considerable traffic with them has taken place. *Isis flexibilis*, which extends from the latitudes of the Caribbean Islands to the coast of Norway, is one of the most graceful of this family. It is sometimes found in deep water, often in the profound depths of the ocean.

There are many genera allied to *Isis*, some of which are found in deep seas, which present most interesting connecting features between the Isidæ and the true sea-fans and sea whips. One of the most interesting of these is a beautiful ochre-colored coral called *Melitœa*. The branches of *Melitœa*, like those of *Isis*, are composed of alternate stony and soft joints, of which the size of the latter are relatively much larger than the former, which appear as bead-like expansions. *M. ochracea* is reddish yellow in color, and has the branching habit of the majority of the true sea-fans. It is found in the Pacific and Indian oceans, many specimens bearing Singapore as the locality from which they were taken. It is of considerable size, and seems to connect structurally the family of Isidæ with the true sea-fans.

Of the many aberrant genera of Halcyonoida, the genus *Tubipora* or the organ-pipe coral, the type of the family TUBIPORIDÆ, departs the widest in general appearance from that of the majority of Halcyonoids. In this genus we find no radiating partitions of hard secretion as in most corals, but instead, a number of tubes arranged side by side separated from each other by a slight space and bound together by horizontal floors. In these tubes live the tubipore coral, and to them it owes its suggestive name. The color is a deep red, and the coral community often reaches a considerable size. It is, however, very fragile, and easily crumbles under the action of the waves, presenting a great contrast to the harder species of *Corallium*, *Isis*, and *Mopsea*.

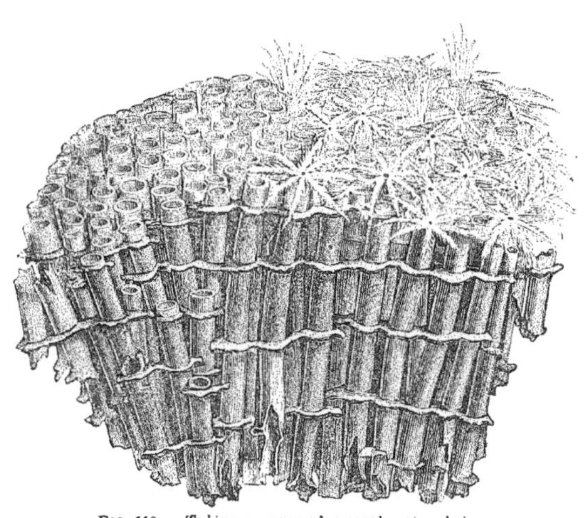

FIG. 119. — *Tubipora*, organ-pipe coral, natural size.

The sea-pens, PENNATULIDÆ, embrace a few most interesting forms of Halcyonoida. In *Pennatula rubrum*, the likeness to a quill-pen is very close. In this coral a central axis extends from one end to the other of the body, inside a sheath from one-third of the length of which there hangs, on either side, opposite each other, rows of leaf-like disks supported by calcareous spicules. Upon the faces and edges, more especially in the latter position in many allied genera, polyps are borne as in all corals. It commonly happens that many of these polyps are aborted in character, those upon the stem especially assuming this form, so that there is present a polymorphism of most simple character.

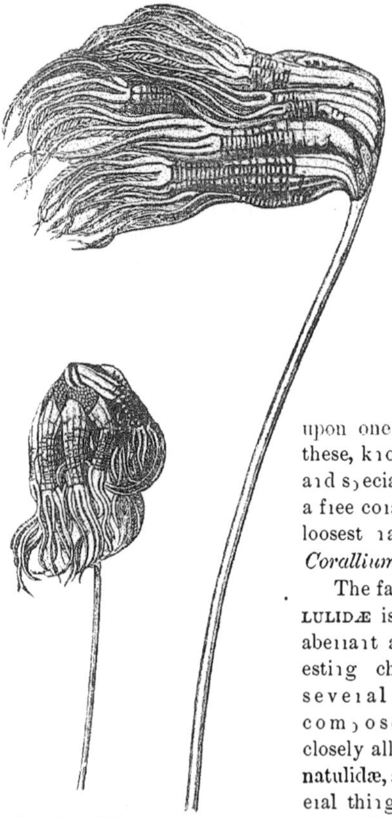

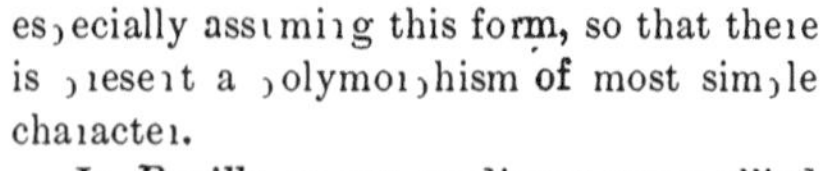

In *Renilla*, an extraordinary genus allied to *Pennatula*, we have a still greater difference between the two kinds of zoöids found on its body, and a still better illustration of the principle of polymorphism so well seen in many of the jelly-fishes. The form of the genus *Renilla* departs widely from that of almost all other Actinozoa. The body has a thin, flat, kidney shape, from which hangs a short, highly flexible hollow stem. There is no hard axis such as is found in some other Halcyonoids, and the body walls are flexible throughout. The polyps are borne upon one surface of the disk-like body, and one of these, known as a *Hauptzoöid*, has a prominent size and specialized character. *Renilla*, like *Pennatula*, is a free coral, and its attachment to the sand is of the loosest nature as compared with the stony base of *Corallium* and *Isis*.

FIG. 120. — *Umbellularia grœnlandica*, natural size.

The family of UMBELLULIDÆ is also of widely aberrant and most interesting character. The several genera which compose it are most closely allied to the Pennatulidæ, and are as a general thing found in deep seas. The genus *Umbellularia* was described long ago, and the accurate figure given of it remained, for many years, the best and only account of the animal. Although discovered in comparatively shallow water, the great exploring expeditions of the past twenty years have again brought the animal to light, this time from profound depths.

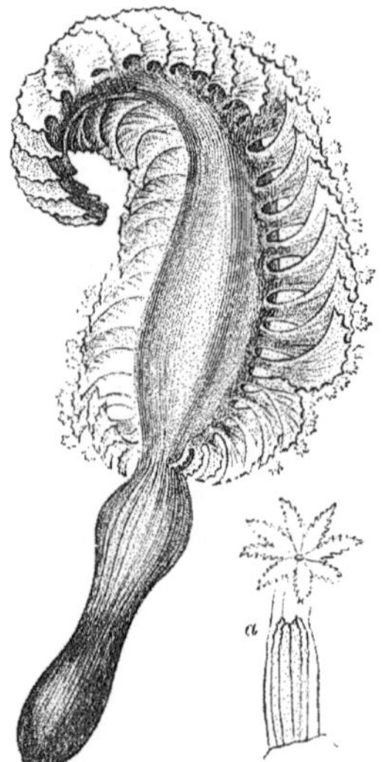

FIG. 121. — *Pteroides*, sea-pen, one-fourth natural size; *a*, a single individual, a little enlarged.

In *Umbellularia* we have a long axis, more or less firmly attached by one extremity, while from the other there arises a cluster of polyps in the form of a terminal tuft. These polyps are of two kinds. Some approach closely the regular form of the Halcyonoids, while others resemble the abortive zoöids of *Pennatula* and *Sarcodictyon*. We have here another expression of a law of polymorphism already pointed out in *Renilla*, and already developed at length in our account of the Siphonophora or tubular medusæ.

CORAL ISLANDS.

The study of the formation of coral islands is one of the most fascinating connected with the Actinozoa. So long as naturalists limited their attention to the coral life of the Mediterranean and other European seas, where no great coral reefs exist, this part of the subject attracted but little attention, but with the awakening of interest in all branches of natural science at the beginning of the present century, more especially at the time of the great exploring expeditions sent out to the tropics by the different governments, the human mind naturally turned to these islands, and the mode of their formation became a subject of scientific interest. The subject has a geological as well as a zoölogical side, and touches on several questions of physical geography. It is, moreover, almost incomprehensible without a clear idea of the chemical nature of the solid lime secretions of the coral body, which plays the most important part in the origin and growth of these characteristic islands.

While the present subject is considered in relation to the coral animals, it must not be supposed that they alone make the coral reefs. Although most conspicuous in this work, there are others of very great importance in their formation from the time the foundations are laid on the sea floor up to that when the reef emerges from the waves in the form of an island. Of these might be mentioned a few genera of stony hydroids as *Millepora*, molluscan shells of all kinds, Bryozoa, Radiolaria, pelagic Protozoa, and genera of marine Algæ. Many instances of coral islands formed in part by any of the last mentioned, will no doubt suggest themselves. A few examples will suffice for illustration. Cooper's Island, one of the large coral islands of the Bermuda group, has a sandy beach which is almost wholly made up of oceanic Foraminifera. The beach adjacent to the landing at Fort Jefferson, on the Tortugas, Florida, is wholly formed of fragments of a large Nullipore, an Alga which secretes lime in its tissues, and is found alive in large clumps in the moat about the fort. Shelly Bay at the Bermudas is composed almost entirely of fragments of lamellibranchiate shells ground into a fine sand. Strata of rock formation in the Bermuda of four or five inches in thickness are made up entirely of the shells of land Helices, some of great size, firmly cemented together. While these and many other animals contribute to the formation of the coral islands, corals seem essential to their growth, for we find characteristic coral reefs confined to the zone which by the thermal conditions of the water limits the home of these animals.

The distribution of reef-building corals follows a number of most interesting laws. In latitude their home is limited north and south of the equator by the water isotherm of 68° Fahr. These lines projected on our globes follow no parallels of latitude, often being widely separated from the equator and then approaching to its immediate vicinity. While the reef-building corals are limited to this zone it must not be supposed that all genera of Actinozoa are hemmed into these narrow limits. Many corals, some of which secrete a calcareous skeleton, are found in all latitudes as far as naturalists have explored the marine life.

The distribution at present of coralline life on the globe is very different from what it has been in the past. While reef-building corals now never venture into latitudes higher than 35°, the evidence drawn from fossil coral banks shows that in older times they were found in very high latitudes. In the North Atlantic Ocean at the present time the northern limit of extensive banks of coral is the lonely Bermuda group in the latitude of 32° N.

The distribution of the coral islands in longitude presents some very interesting facts. As a general thing it may be said that the eastern shores of the continents are richer in coral banks and islands than the western. The eastern coast of North America, for instance, has the Florida reefs, the great Bahama bank, one of the largest in the world, the reefs of Yucatan, Honduras, and Cuba. On the western shore of North and Central America there are no extensive coral reefs. The eastern border of Australia has a wealth of coral life, while the western is almost a desert as far as plantations of these zoöphytes are concerned. The Atlantic coast of Africa has no extensive reefs, while that washed by the Indian Ocean is fringed by very extensive banks. If in studying this distribution in longitude we consider also the limitations in latitude we find the following law to hold good. While in the longitude of the eastern shores of the continents the reefs extend far from the equator, in that of the western border they are, when found, limited to the immediate vicinity of the equator. An explanation of this curious fact in coral distribution is found in the direction of the great equatorial currents of the Atlantic and Pacific Oceans. These currents cross the ocean from east to west, and as they approach the eastern continental borders they divide, one branch flowing to the north the other to the south along the coast lines. Two streams of warm water are thus continually carrying the oceanic isotherm of 68°, which limits the home of the reef-builders, into higher latitudes, and broadening the zone in which these sensitive animals can live. On the western coasts, however, we find an opposite condition of things. The equatorial current is there fed by branches flowing in opposite directions, in which the water is colder since they are setting from higher latitudes towards the equator. The result upon the coral organisms is that they are hemmed into narrow limits on this side of the ocean by the diminution in the breadth of the zone which they can inhabit.

The distribution of corals and the limitation of coral islands by such local phenomena as rainfall, volcanic activity, and the like, present many very curious facts. The almost constant changes in the level of the sea floor produced by volcanoes would necessarily destroy plantations of growing corals in the immediate neighborhood. It commonly happens that coral and volcanic islands are found in intimate association, while it is probably true that all oceanic coral islands rest on volcanic bases. Where the volcanic forces are active, the lava poured into the ocean, or the ashes raining from the air at times of great eruptions, generally destroy the coral banks. A good illustration of this fact may be seen in the Sandwich Islands, where the most southern members of a chain of islands which make up the group are volcanic and destitute of extensive coral banks, while the northern are almost wholly formed of coral. The almost uninterrupted volcanic activity in the southern members of this group has prevented the formation near them of coral banks, while at the north, where active volcanic forces are now unknown, the islands are almost wholly coralline. The eastern members of the Lesser Antilles, as Barbadoes, are coralline, while the western are volcanic. In the island of Guadeloupe we find the same law of local distribution to hold in a single island, the eastern extremity being coralline and the western volcanic. This distribution of corals in the Antilles probably depends upon the direction of the ocean currents in the neighborhood, or perhaps upon the constant direction of the winds by which the ashes from the volcanoes are blown to the leeward, thus killing the growing corals on the western side. It may perhaps be that the profound depths to which the Caribbean Sea sinks to the westward of the lesser Antilles prevents the corals obtaining a foothold on that side, while the shallows on the eastern shores are better suited for the luxuriant growth of these animals. The situation of mountains on or near the coast of the continents or elevated islands, and the direction in which the rivers of such flow to the ocean, also has its influences on the distribution of coral islands in the immediate vicinity. One of the very best illustrations of this limitation of coralline life is seen in the island of New Caledonia. Most of the rivers in this large island, according to Dana, empty themselves into the ocean along the western coasts, while the eastern side is almost wholly destitute of streams of fresh water. Along the eastern border we find an almost continuous coral reef, and no signs of coral growth on the western. Fresh water is destructive to coral life, and rivers bring from the land large quantities of silt, which is also detrimental to its growth. To these causes may perhaps be traced the almost total absence of coral formations on the northern border of South America, near those parts of the coast line where the Amazon and Orinoco pour their great volumes of fresh water into the Atlantic.

There are several very curious facts, many of which are not yet explained, in regard to the geographical distribution of genera of corals. The coral fauna, for instance, of the Bahamas and Florida regions is widely different from that known to exist on the western coasts of Central America. The corals of the latter region have a closer likeness to those of the Indian Ocean, almost its antipodes as far as geographical position goes, than they do to those of the Caribbean Sea, separated from them by only the narrow isthmus of Darien. *Madrepora cervicornis*, a most abundant coral along the Florida Keys, is very rare in the Bermudas.

Corals are found at almost all depths in the ocean. The greatest profusion of life lies in the zone between the surface and the depth of one hundred to one hundred and fifty feet. Deep sea corals, of an interesting character from their relationships with extinct genera, are found at great depths, even in the abysses of the ocean.

The general distribution of coralline life in the different oceans is as follows: —

The Atlantic Ocean has several very considerable coral reefs and coral islands, all of which are confined to its western border. In the South Atlantic a large and peculiar reef is found extending many miles along the southeast coast of Brazil. In the chain of islands called the Lesser Antilles we find that the frequency of coral reefs increases as we go towards the northern members. Along the northern shores of San Domingo, Hayti, and Cuba, we find extensive reefs. There are evidences also of elevated coral banks in these islands. The peninsulas of Yucatan and Honduras are fringed with coral shoals, and large banks exist on their northeastern and northern borders. The line of Florida coral reefs, extending from Loggerhead Key, the extreme western island of the Tortugas, to the southern extremity of the peninsula of Florida, is one of the most instructive collections of coral islands in the Atlantic region. The Bahama Islands are composed wholly of coral, and are the largest continuous bank of growing coral in the North Atlantic Ocean. Its extension is as great in length as the whole eastern coast line of the United States from Maine to Florida. The only example of an oceanic coral island in the North Atlantic is the Bermuda group, which lies in the latitude of Charleston, S. C., about seven hundred miles from Cape Hatteras, the nearest land. This reef is one of the most interesting, from the fact that it is in the highest latitude in which corals are known to flourish with any vigor, an exception which is probably the result of the sheltering action of the well-known Gulf Stream. The most extensive coral plantations are found in the Pacific and Indian oceans. On the eastern border of the former they are, however, very insignificant. Worn fragments of coral are found on the Galapagos, and Col. Grayson mentions a beach of coral sand on Socorro, one of the Rivillagigedo group. The Sandwich islands are in part coralline. The Feejee, Paumotu, Society, and Friendly islands are composed wholly or in part of extensive coral banks. The coral sea along the northeastern coast of Australia is the most extensive coral reef in the world. In the Indian Ocean, again, we find a large number of coral islands and reefs, among which may be mentioned the Laccadives, Maldives, and reefs of the island of Madagascar. Near the entrance into the Red Sea there occur coral banks of considerable size.

The part which the coral plays in the formation of the coral island has been variously estimated, and many opinions have been advanced in regard to this point. The difficulty comes oftentimes from a lack of an accurate knowledge of the relationship of the solid matter of the coral to the animal which secretes it. The carbonate of lime which makes up the great mass of the growing coral bank is the skeleton of the animal, and is simply a secretion of the membranes of the body. The skeleton cannot be said to be the work of the coral, except so far as it is an animal secretion.

A growing coral plantation, with its multitudinous life, oftentimes arises from great depths of the ocean, and the sea-bed upon which it rests is probably a submarine bank or mountain, upon which have lodged and slowly aggregated the hard skeletons of pelagic forms of life. When, through various sources of increase, this submarine bank approaches to the depth of from one hundred to one hundred and fifty feet from the surface of the water, there begins on its top a most wonderful vital activity. It is then within the bathymetric zone of the reef-building corals. Of the many groups of marine life which then take possession of the bank, corals are not the only animals, but they are the most important, as far as its subsequent history goes. As the bank slowly rises by their growth, it at last approaches the surface of the water, and at low tide the tips of the growing branches of coral are exposed to the air. This, however, only takes place in sheltered localities, for long before it has reached this elevation it has begun to be more or less charged and broken by the force of the waves. As the submarine bank approaches the tide-level, the delicate branching forms have to meet a terrific wave action. Fragments of the branching corals are broken off from the bank by the force of the waves, and falling down into the midst of the growing coral below fill up the interstices, and thus render the whole mass more compact. At the same time larger fragments are broken and rolled about by the waves, and are eventually washed up into banks upon the coral plantation, so that the island now appears slightly elevated above the tides. This may be called a first stage in the development of a coral island. It is, however, little more than a low ridge of worn fragments of coral washed by the high tides and swept by the larger waves — a low, narrow island resting on a large submarine bank.

The second stage in the growth of a coral island results from the formation on such a ridge as we have just described of a quantity of fine coral sand. In the grinding of the coral fragments which lie upon the fixed portion of the reef, a large quantity of the finest sand is formed. This sand is sometimes held in a mechanical suspension in the water, and in that way is transported from place to place. It is generally swept along from the locality where it originates, and is ultimately, if not lost in ocean depths, thrown up on the ridge of coral fragments which has been already mentioned. The wind assists the waves, and, taking the sand which they cast from the waters, blows it as it becomes dry higher and higher on the ridge. Thus we have formed the second stage in the development of a coral island, which is simply derived from the former by capping the coral fragments which form the foundation with a layer of sand.

As the amount of sand increases it collects in such quantities as to be a detriment to the growth of the coral animals themselves. It covers not only the foundation of the coral island but extends far and wide over the coral plantation upon which the island rests, and tends to kill the great colonies of life by which this platform is peopled. The part of the reef where the coral life still lives now retreats into the ocean to the greatest possible distance from the sand.

The winds, tides, and rains continue the work which they have taken up. Ocean currents, especially, perform a most important part in the many changes which take place. The problem now becomes a geological one, and wholly independent of those of animal life.

The continual wear and tear resulting from the erosion of the water on the coral island in its second stage of growth is ever increasing the amount of broken fragments near the low-water line, while the winds snatch the sand thrown up by the waves and heap it into high banks, dipping invariably to the sea. Exposure to the air and other causes now exert their influences upon it, and the coral sand is hardened into a soft rock, the well-known coral rock of these islands. The island is now formed? Not yet. To the stability of the precarious foundation thus made few persons could with impunity trust themselves through a tropical hurricane. There are other changes before the coral island becomes firm, and its career has just begun.

In the progress of time the processes of erosion go on, and the waves and rains eat their way into the soft rock so that the island is honeycombed by the erosion. Once more the rock is reduced to fine sand, and scattered far and wide over the submarine flats. The softer, least consolidated layers of the rock, which, from not being exposed to the air, are below the harder, wear away faster than the upper strata, which are thus broken off in large fragments. The products of the erosion are strewn far and wide over the coral platform, and heaped up into a new island, of a different form from that which existed before. One of two things now takes place with the *debris*. Either the sand is again thrown up on the forming island, to again harden into coral rock, or it is swept over neighboring growing coral reefs, raising these one more stage to the level of the surface of the ocean. The movement by the wind of this sand upon a coral island of considerable size often assumes formidable proportions. The constant winds on some coral islands, as the Bermudas, heap the coral sand into dunes of considerable elevation, which slowly move *en masse*, engulfing everything which lies in its path. One of the most interesting of these moving masses of sand is to be seen in Paget Parish, in Bermuda. The sand on the south shore of this parish has in its motion suggested the name of a "sand glacier," and for several years it was slowly making its way inland from the coast, covering to a considerable depth farms, and even a farmhouse, in its course. Artificial means of staying its progress had to be resorted to, and by planting trees in the line of its onward motion the progress of the sand was stopped.

In sheltered coral lagoons, whose floor is formed of coral sand approaching the surface of the water, or with but a moderate depth, the mangrove trees oftentimes furnish a nucleus around which characteristic coral islands, known in Florida as mangrove keys, are formed. A small mangrove, sending its root into the submerged sand, forms an obstruction upon which catch floating seaweeds and similar organisms. As the mangrove increases in size, and its spreading branches send down rootlets which fasten themselves more firmly in the sand below, the obstruction is increased in size, and the island grows with every increment to its size, until eventually we see a coral island of peculiar character formed on the coral shallows and flats. This kind of coral reef is very common in sheltered localities in the Florida waters.

A coral island, however formed, is continually changing its contour, on account of the method of its formation and the liability to erosion of the soft rock of which it is composed. A study of the causes of the different shapes which coral islands assume is highly interesting and instructive. In some cases their outline is due to the changes in level in the foundations upon which they rest, elevation, and subsidence of the sea-floor; in others to the direction of oceanic currents of the waters out of which they rise.

Coral islands have every variety of form, although elongated, circular, ring-shaped, and crescentic forms predominate. A coral formation skirting the shore, called a fringing reef, follows the contour of the coast except when the continuity is disturbed by local causes. The same may be said of reefs separated from the coast line by a lagoon, and known as barrier reefs.

Circular or crescentic reefs are the most striking in shape, and their mode of formation has been a cause of considerable speculation. The circular reefs are known as

FIG. 122. — Ring-shaped coral island or atoll.

atolls, and are abundant in the Pacific and Indian oceans. Several atolls also occur in the Atlantic and the Gulf of Mexico, of which the Marquesas Islands in the Florida reefs is a well-known example. The Bermudas are, I think, erroneously ranked as atolls by many authors. From circular reefs, or atolls, to the simple elongated or crescentic coral island we find every form of ring-shaped islands. There is nothing to show that all circular coral islands are formed in the same way, or owe their peculiar outlines to identical causes. In some cases they result from a sinking of the sea floor, and in others from the direction of ocean currents in their immediate vicinity.

Many coral atolls have been shown by Darwin and Dana to have been caused by a slow sinking and corresponding growth of corals on a submarine base. Let us suppose an island favorably situated for coral growth to have a narrow, fringing reef around its coast. Suppose also that in the geologic changes of the sea floor this mountain is slowly sinking below the level of the sea. As the mountain settles the animals on the fringing reef cause it to rise, *pari passu*, with the depression. The intensity of coral growth is always at the periphery of the fringing reef on the side turned away from the island, and exposed to the pure sea water. There the coral formation, by the growth of the animals, is kept to the sea level, notwithstanding the sinking of the island. The first effect of the sinking island appears in the formation of a lagoon between the periphery of the reef and the coast line. This lagoon grows in width as the island sinks, while the constant growth of the coral on the outer rim keeps it at the sea level. The submergence does not stop until the top of the mountain sinks below the waves, but even then the outer rim, the peripheral edge of the original coral reef, still holds its place at the water's surface on account of the corresponding growth of the corals at that point. The island has now a true atoll shape, a ring-shaped reef of slight elevation above the sea level, with a diameter equal to that of the base of the mountain when the submergence began, — the diameter, of course, measured between points which lie in the coralline zone of the island. In this way, yet more graphically, Darwin has explained the circular form not only of true atolls but also of many curved or crescentic islands, such as are very common in the Pacific and Indian Oceans.

Many objections have of late been urged against this theory of Darwin, and probably other causes must be sought for to explain the circular outlines of many other reefs which have the form of true atolls. Semper has suggested the direction of ocean currents as an explanation of the circular reefs of the Pelew group. Let us see if a

Fig. 123. — Island with fringing and barrier reefs.

similar cause cannot be found to account for the ring-shaped islands of the Atlantic Ocean and Gulf of Mexico.

Every one who studies a good map of the coral islands of southern Florida will have his attention attracted to the general trend of a long series of islands extending westward from Cape Florida into the Gulf of Mexico. He will perhaps notice that these islands are very narrow in a north and south direction, and elongated from east to west, strung along one after another in an almost direct, yet slightly curved line. About midway in the chain, however, there are a few marked exceptions to this general law, and these include some of the largest members of the group known as the Pine Islands. Upon one of these islands, called Key West, is situated a large city of the same name. Unlike the other islands of the chain, the longer axes of the Pine Islands, as pointed out by L. Agassiz, extend north and south, or at right angles to the others. As a general rule, all the Florida Keys are low, formed of coral rock, without caves of any size, and have no red soil, a characteristic of worn and eroded coral islands. Their highest altitude, nowhere more than a few feet, is on their southern border, and parallel with them throughout their whole course runs a coral reef, separated from them by the Hawk's Channel, which is a half-dozen miles broad in its widest part. South of the reef, which is a succession of dangerous coral banks and small islands, the water becomes deep, forming a trough for a great oceanic current, the Gulf Stream, whose floor at this point is called the Pourtalès Plateau. South of this plateau is the deepest water of the stream which flows hard by the neighborhood of Cuba.

The cause of the general trend of the Florida Keys must be looked for in the direction of the Gulf Stream, or of a current flowing below it in an opposite direction, of which they were once the northern bank. This "oceanic river" flows tangentially to the Florida reefs throughout their whole length, and to it may be traced the general trend of the Keys from the Tortugas to the southern extremity of the peninsula of Florida.

The exceptional position of the longer axes of the Pine Islands is directly due to the Gulf Stream and tide currents about them. The submarine elevation or plateau, upon the southern border of which the Florida Keys lie, has not the great depth of the Gulf of Mexico or the Gulf Stream bed between Florida and Cuba. On this comparatively shallow platform the water rises in tides twice every twenty-four hours, and at times, especially on its western extremity, the waters of the Gulf Stream are forced over it. The water, thus raised above its natural level, must return to deeper channels; and one course which it may take is into the trough of the Gulf Stream at right angles to the direction of its flow. The Pine Islands have their longer axes tangential to several of these subordinate branches or currents. A similar phenomenon is also seen in the channels which separate many of the Bahamas. The last but one of the groups of coral islands which compose the Florida chain, which is called the Marquesas, has a circular or atoll-like shape. In this group we have a resultant of two currents, or a combination of those minor currents which placed the axes of the Pine Islands north and south and the Gulf Stream which gave the general trend to the chain. From the position of the group near the extreme western end of the series looking out into the depths of the Gulf of Mexico, these forces are difficult to separate, and at intervals reinforce each other. The circular form of Marquesas is due to their combined action.

We have now arrived at a point in our discussion of coral islands where it may be possible to appreciate the influence of ocean currents in the formation of atolls. Let us consider the cluster of islands which occupies an irregular triangular space midway between the southern point of Florida, the Bahamas and the island of Cuba. This comparatively small coral bank is known as the Salt Key Bank, a coral plateau which lies in the eddy of three great ocean currents. At most points the bank has a moderate depth below the surface of the ocean, but in places along its outer rim there are several coral islands of moderate size, some in process of formation, and others which show the marks of great erosion. The Salt Key Bank, as far as it has been explored, is a circular coral bank, fringed by islands which are not continuous above water along its border, but which, nevertheless, are parts of a true atoll. On its sides Salt Key Bank is washed by ocean currents, on the north by the Gulf Stream, to which the Florida Keys owe their formation, on the east and south to the Bahama current in the old Bahama channel. Coral islands, as pointed out by Semper, form tangentially to the direction of ocean currents, and the outlines of the Salt Key Bank result from the direction of the three currents which surround it.

There are few evidences of submergence in the Salt Key plateau, and if we cross the Bahama channel to Cuba we find terraced coral banks elevated above the sea, showing a great elevation of the coasts. The Salt Key bank, as well as the Florida reefs and the Bahamas, are explicable without any theoretical supposition of submergence or elevation of the foundation upon which they rest.

The origin of an atoll in mid-ocean where currents are not confined as in the triangle between Cuba, Florida, and the Bahamas, presents a similar although somewhat modified problem. Let us suppose a submarine mountain situated in mid-ocean upon which for ages has rained successive depositions of sediment from the waters above. This sediment is composed in part of shells of pelagic animals, and plants, with which the waters of the tropics or currents from the same are filled. Even at the greatest depths life exists upon such a bank, and the hard portions of the animals which live and die there are being continually added to a growing submarine bank. By increments made for many years the bank slowly rises to the surface of the water. As it rises higher and higher the activity of the life on its crest increases, and when it rises into the bathymetric zone of the reef-builders, between one hundred and one hundred and fifty feet below the surface, a more rapid growth awaits it. The coral plantation as it develops is washed by an ocean current, probably on one side. Such a current in fact is divided by the bank, so that each of the bifurcations flow tangentially along its sides. It is evident in the first place that it is around the border of the bank washed by the current that the most active coral life is to be looked for, and in this region also that the predominant upward growth must be sought. If the bank with which we started from the sea depths is circular the resulting atoll will have the same form, and the part which first raises itself above the water will be ring-shaped, containing a central lagoon.

The luxuriance of the growth of the animals, which form a coral reef, is directly dependent upon the amount of food which the oceanic currents bring to the growing colonies. The waters of the ocean are filled with a wealth of microscopic and other life floating in it upon which the living coral animals feed. It is evident that the possibilities of prolific coral life are greatest where their food is most abundant, and there too we must look for the greatest amount of coral growth. It is clearly in the line of ocean currents that this food, being constantly renewed by the flow of the stream, is most abundant, and along its borders the possibilities of the coral animals of obtaining their food are the greatest. This cause alone is not capable of determining the outlines of coral islands, but it is a most important factor in regulating their growth.

Coral islands in two different conditions ought to be mentioned. We have coral islands in process of formation, and those in conditions of destruction by erosion. The Florida reefs are examples of the former, the Bermudas of the latter. Coral islands, in progress of formation, seldom rise to any great height above the ocean, are composed of coral fragments, sand, or half-formed coral rock. They are but very slightly eroded, have no extensive caves with stalactitic formation, and no red soil. Fully formed coral islands in which the erosive power of the water has had its full effect are generally elevated, honey-combed throughout by caves, and possess a soil of red earth. This last characteristic of coral islands in the progress of erosion is of a problematical nature and origin. By some authors it is regarded as the heavier residuum resulting from the wearing away of the coral rocks, while others have even gone so far as to look upon it as the excrement of birds, mingled with coral sand. The former theory seems most rational, but there is, if this theory be adopted, great difficulty in accounting for its reddish color. Red soil is very abundant in the Bahamas, the Bermudas, and several of the West Indian islands. In the Bermudas it is contained in pockets in the rocks, and in it are grown the well-known early vegetables, potatoes and onions, for which the islands are so justly famous. In some places in the latter islands it is found solidified into a compact red rock in which are found embedded Helices, and other shells belonging to species which are at present alive in Bermuda.

The caves of coral islands are sometimes of most beautiful character and of considerable size. In the Bermudas most of these caves have a submerged floor in which in many cases is sea water which is sensitive to the tides in the neighboring ocean. Caves in coral islands present many very beautiful examples of stalactitic and stalagmitic formation. Many of the stalactites appear to extend from the cave-roof below the level of tide water, indicating either a wide spread or local sinking of the cave-floor. Of the many beautiful effects of the erosion of the sea water on cliffs of coral rock, one of the most interesting is the formation of natural arches and the like out of spurs of the hills projecting into the sea. In the neighborhood of a small Bermudian village called Tucker's Town there is a very fine natural arch which bears a remote resemblance to the famous Arco Naturale of the Island of Capri in the Bay of Naples. A similar arch called the "glass window" is found in the Bahamas.

J. WALTER FEWKES.

Branch IV. — ECHINODERMATA.

The animals embraced in this group were included by Cuvier in his great division of Radiata, along with the cœlenterates. More complete knowledge of the anatomy and especially the embryology has shown that the two groups have nothing in common, except those features which are common to all Metazoa and the absence of a segmentation of the body. The radiate arrangement, really a feature of minor importance, is here very strongly marked in most forms, for crinoids, star-fishes, and serpent-stars have a central disc which branches into five or more arms, which radiate from it like the spokes of a wheel from the hub. In the sea-urchins and holothurians this radial symmetry is less marked, but it may readily be traced, although the radiating arms are apparently lacking.

Though at first sight widely different, the similar structure of a star-fish and a sea-urchin can readily be traced. Let us first examine that of the first-named form. We have a central disc with five radiating arms; in the centre of the lower surface of the disc we find a mouth, hence this is called the oral surface. On the upper or aboral surface we find no opening (or only a very minute one), but a little one side of the middle, between the bases of two of the arms, is a round plate, which, from its peculiarly ornamented appearance, has received the name of madreporic body, in allusion to its resemblance to some of the corals. On the under surface of each arm will be found a series of plates, between which project little tubular suckers. As these sucking tubes are used in walking, they have received the name ambulacra, while each of the series of plates between which they project is known as an ambulacral area.

Turning now to the sea-urchin, we find a nearly spherical body. On the under side we have a mouth, from which radiate a series of plates, between which project ambulacra very similar to but longer than those of the star-fish. These rows of plates continue around the sides of the sphere, and terminate near the centre of the upper surface, where we

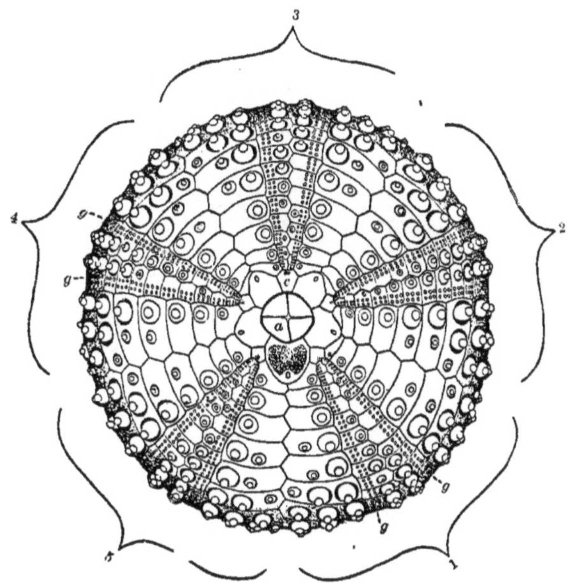

Fig. 124. — Aboral surface of sea-urchin (*Arbacia*); *a*, anal plates; *c*, ocular plates; *g*, ambulacral area; *o*, madreporic body; 1, 2, 3, 4, 5, the five rays.

find a madreporic body like that of the star-fish. It is thus evident that these rows of plates correspond to the ambulacral areas of the star-fish, and we can readily see that were we to bend the arms of the latter form upwards, so as to form a ball (most

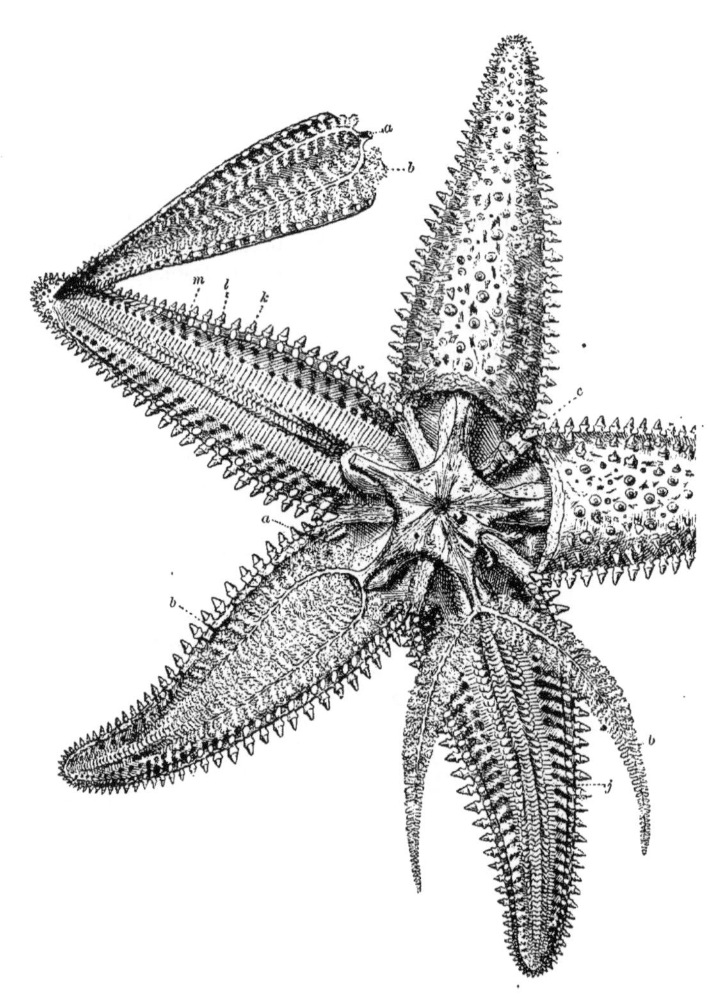

FIG. 125. — Anatomy of star-fish; *a*, duct from liver to stomach; *b*, liver; *c*, madreporic body; *j*, ampullæ; *l*, ambulacral plates: *m*, inter-ambulacral plates.

of the upper surface disappearing during the operation), the star-fish would be converted into a sea-urchin.

The typical number of similar parts (ambulacral and inter-ambulacral areas) which go to make up Echinoderm is five, though frequently this number is exceeded. This radial arrangement is also seen in some of the internal organs, but it is not visible in the alimentary tract. Bilateral symmetry is not so evident in most forms, though it exists in all, and in the young is especially well marked. The line dividing the body into two similar halves passes through the centres of the madreporic body and of the central disc.

The internal organs are far more complex than those of the cœlenterates. The most striking feature is that the digestive canal is entirely distinct from the body cavity. In the higher forms this canal is tubular in form, and as it is much longer than the body cavity, it is coiled in a spiral, which in all, except the serpent-stars, is coiled from left to right. An anus is usually present. Usually there are connected with the functional stomach a number of glandular pouches, which secrete a bitter fluid, and appear to represent the liver of higher animals. Organs which act as teeth are frequently present around the mouth.

The so-called water vascular system is complex and peculiar, presenting several interesting features. From the under internal surface of the madreporic body a canal goes down to a circular tube surrounding the mouth. From the fact that this canal contains lime in its composition, it is known as the stone canal. From the circular circumoral canal (ring canal) a branch follows the centre of each ambulacral area to its extremity. Connected with these radial canals are the ambulacra. These ambulacra are arranged in pairs, and consist of two portions; an outer part terminating in a sucking disc, and an inner sac known as an ampulla. Between the two a tube goes to the radial canal. Another feature is sometimes present. Arising from the ring canal are from one to ten sacs known from their discoverer, Poli, as the Polian vescicles.

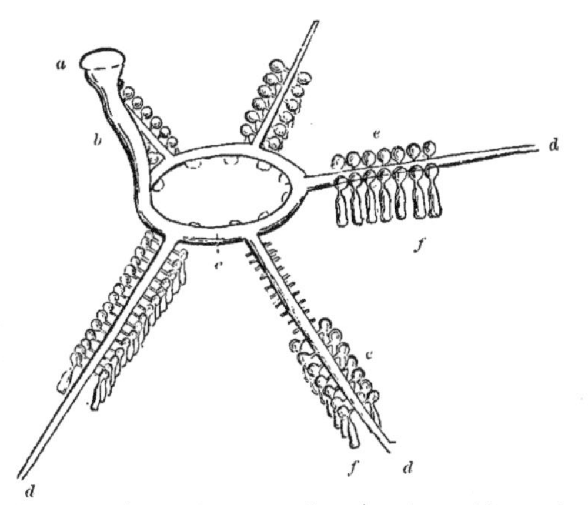

FIG. 126. — Diagram of water vascular system of a star-fish; *a*, madreporic body; *b*, stone canal; *c*, ring canal; *d*, radial canals; *e*, *f*, ambulacra and ampullæ (a few only shown).

The physiology of this water vascular system is not clearly understood. All that can certainly be said is that by the action of the muscular walls of the ampullæ and the Polian vescicles water is forced into and withdrawn from the tubular ambulacra, thus extending and retracting these organs. In the extremity of each ambulacrum is a calcareous plate, to which are attached minute muscles, which by drawing in the external integument form a vacuum similar to that which a boy forms with his wet leather disc and a string. The strength with which these minute feet will cling is remarkable, and the long stalk of the foot may frequently be torn in twain without detaching the foot from the object to which it has become fastened.

Connected with this water vascular system, which is locomotor and possibly also respiratory, is another which is regarded as the true vascular or circulatory system. It consists of two rings, one surrounding the mouth, just below the ring canal, while the upper surrounds the anus at the opposite pole of the body. These two rings give off branches, and are connected by a tube which follows the course of the stone canal. Regarding the functions of this system opinions differ greatly. Some consider it as a true circulatory system, the tube connecting the two rings being regarded as a heart, and described as pulsating in life. Perrier, the French authority on these forms, on the other hand, regards the so-called heart of the star-fishes, brittle-stars, and sea-urchins, and the corresponding dorsal organ of the crinoids, as glandular and the connecting branches as cœca.

The nervous system consists of a ring around the mouth from which radial branches follow the course of each ambulacral area. Subdivisions of these principal nerves supply the ampullæ, ambulacra and other parts of the body. The only sense which appears to be present in all of the group is that of touch, for which various external parts are well adapted. In some of the star fishes and sea urchins rudimentary eyes are found at the extremities of the ambulacral areas, whether at the tip of the arms or at the corresponding position on the aboral surface. (Fig. 124, c.)

The external covering varies greatly in the various forms, but plates of carbonate of lime are usually present. These plates may be firmly united so as to form a solid shell, or they may be separate and imbedded in the integument. In some they may take the form of spicules, wheels, and anchors, those of *Chirodota* and *Synapta* being familiar objects to all workers with the microscope. In the star fishes these plates are very numerous, two series roofing over the ambulacral groove in which the sucking feet are situated. In the serpent stars the ambulacral plates occupy the interior of the arm, which is entirely surrounded by a series of plates. In most of the sea urchins the plates of the ambulacral and inter-ambulacral areas unite to form the solid test in which the body is usually enveloped, the surface of which is covered with little rounded prominences for the attachment of spines. In the crinoids the calyx and the jointed stalk and arms are largely composed of calcareous matter.

Spines, often varying greatly in size in different portions of the same individual, are widely distributed among the Echinodermata, most of which also possess certain fork-like or pincer-like organs, which are modified spines, and which from being stalked in some forms are known as pedicellariæ. These are capable of closing and seizing any object which may come between their jaws, and are supposed at least in some forms to take the excrement from the anus on the upper surface of the body, and pass it along, off from the shell to the ground.

The sexes of most echinoderms are separate, and the genital products are discharged either by a breaking away of the integument or by true genital openings. The young of most of the branch undergo a wonderful metamorphosis in the course of their development, and the embryos are free swimming animals. The 'pluteus' of the sea urchin, the 'bipinnaria' or the 'brachiolaria' of the starfish, and the 'auricularia' of the holothurian, bear no resemblance to the adults, and in fact the name now applied to these larvæ were originally given them under the impression that they were adults. In certain groups the embryo develops into the adult without any metamorphosis. As there is such variation among the different groups we will defer the details of development until treating of the respective forms.

The Echinodermata are all marine, and are found in all the seas of the globe. If ancestry confers respectability these forms should be classed among the nobility, for remains of these animals are found in some of the oldest fossiliferous rocks. At the present time they play an unimportant part in the economy of the world, and are of but slight importance to mankind. A few forms are used as food, while others are injurious to human interests, as they destroy beds of oysters and other shell-fish.

Class I. — CRINOIDEA.

The lowest division of the echinoderms and at the same time the one which has the fewest recent representatives is the Crinoidea. The fossil forms of this class are very numerous in the older rocks, and are commonly known as encrinites and stone-lilies. The recent forms are so little known and so seldom seen by any except those who are familiar with the results of deep-sea explorations that they have received no common name. The greater portion of the species are attached to sub-marine objects by a stem which, frequently, is very long, and made up of a series of joints perforated by a central canal. These joints are among the most numerous fossils in the older rocks, and have received the name of St. Cuthbert's beads. This stem supports a calyx (corresponding to the central disc of the star-fish) which has received its name from its similarity to the calyx of a flower. From this calyx radiate the arms. Some forms have the stalk persistent throughout life, while others possess it only in the early stages, and in a few fossil forms it is said to be lacking in all stages.

The class is usually divided into three orders, the Blastoidea, the Cystidea, and the Brachiata, or true crinoids. The first of these is extinct, the second contains one recent form, and the third, until recently, was thought to be represented, with one exception by free swimming forms. The recent deep-sea explorations have, however, brought to light several species, some of considerable size and others much smaller than some of the fossil forms.

Order I. — BLASTOIDEA.

The members of this extinct group of Crinoids were armless, were supported on a short stalk, and had five double series of pinnules, one along each side of five radiating grooves. The entire animal, in its fossil state, with the oral plates closed, looks like a flower-bud. The most ancient form (*Pentremites*) is found in rocks of upper silurian age, and the group is most abundant in the carboniferous.

Pentremites has the ambulacral and ant-ambulacral regions nearly equal. The calyx is composed of three basal plates, two of which are double. Above these lie five plates deeply cleft above, and in the clefts lie the apices of the ambulacra, the oral portions of which are included between the five interradials which surround the central aperture. This is probably the mouth, and around it are four double pores, and a fifth divided into three. Of these three the middle one is believed to be the anal, while the other two, and the remaining pairs are genital. Each ambulacrum consists of two rows of small plates which are united in the middle line, and bear pinnules at their outer ends.

Order II. — CYSTIDEA.

The Cystidea come near the crinoids, are usually furnished with arms, having jointed pinnules, and have a short stalk. *Caryocystites* has no stalk and no arms, the body being an angulo-spherical ball of solid plates. Several genera (*Edrioaster*, *Agelacrinites*, *Hemicystites*), are also armless and stalkless, but in form resemble such a star-fish as *Pteraster*, except that they are more nearly circular. These forms have five ambulacra, looking like the arms of an ophiurid placed in the midst of the disc, and, like the more normal stalked cystid, they possess, in one of the interambulacral spaces, a cone of small plates called the pyramid, with a central aperture. In ordinary cystids, pinnules are present, which, when the arms are absent, are sessile on the radial plates. An aperture placed in the centre of the calyx at the point of convergence of the ambulacra, is usually regarded as the mouth; a second small one on one side of this is supposed to be the anus; while the opening in the centre of the pyramid is considered to be the reproductive aperture. Thus the Cystidea differ from other echinoderms, the holothurians excepted, in the presence of only one genital aperture.

Hyponome sarsii, from Torres Strait, looks like a small star-fish or Euryale. It has a disc, convex on the oral surface, and flattened on the other, which shows no trace of a calyx, but is covered with a soft and smooth brownish skin. The rays are five, broad and short, yet each divided into two branches, ending in four very short, rounded lobes. The oral surface is covered with rather small, thick-set, irregular scales, with rosettes of six or seven larger ones here and there. These scales extend on to the dorsal surface between the rays, forming triangular spaces pointing to the centre of the disc, and thus reducing the spaces covered by skin to a regular star. The rays have narrow channels, protected by marginal scales, but have no pinnules. Upon the disc the marginal scales form a vault over the channels, and the mouth itself is hidden by the skin. There is a conical anal funnel in one of the inter-radial spaces on the oral surface, and an area in the middle of the dorsal surface is studded by minute pores. Funnel-like passages leading to a concealed mouth are found in palæozoic Brachiata and cystideans. The absence of a calyx excludes *Hyponome* from the former; and it in some respects recalls the genus *Agelacrinites* among fossil cystids.

In opposition to the usual idea about these fossils, Mr. Billings maintains that the large lateral aperture of a cystidean is the mouth, and the small apical orifice an ambulacral aperture. When there is no separate anal aperture, the large lateral aperture is both mouth and anus.

The Cystidea first occur in the cambrian, attained their greatest development in the silurian, and were mostly extinct in the carboniferous period.

Order III. — BRACHIATA.

This order was represented by a multiplicity of forms in the past geological ages, and has latterly been shown to be far more abundant in species in the present age than was supposed. As the name implies, all the species are provided with arms. These arms are composed of numerous calcareous joints, and contain only a small proportion of living matter. Existing crinoids belong to two distinct series. The first of these contains forms which are permanently attached by a stalk to the sea-bottom, while the second is thus attached while young, but finally becomes detached, the cup or calyx with its arms and a pear-shaped centro-dorsal tubercle, formed of the coalesced upper stem-joints, floating off freely. In this stage the animal greatly resembles a star-fish in appearance, but its anatomy is very different. The stalked crinoids, which are principally natives of deep water, may be considered as the representatives of the extinct stone-lilies, while those which become free belong to a more recent and more highly developed type.

The characters which distinguish the crinoids from other echinoderms may be briefly summed up as follows: The animal is attached, during the early portion or the whole of its life, by a stem composed of more or less numerous joints. To the top of the stem are attached a variable number of plates, normally including one or more basals, and three or more sets of radials, which, with interradials, etc., make up a cup or calyx in the hollow of which the internal organs of the animal are accommodated. From the edge of the calyx spring a number of jointed arms, usually five, these again divide once, twice, or more times, and each arm is furnished with pinnules as a feather is set with barbs. The ambulacral feet are situated in the furrows of the calyx and along the arms. No other echinoderms are fixed at any period of their life-history; in no others do the arms subdivide into pinnules, and in no others is the madreporic plate absent, though in most holothurians it is internal. The mouth is situated in the centre of the upper side of the calyx, instead of in the centre of the lower side as is the case in starfishes and sea-urchins, while the anus is placed on a conical projection between the bases of two of the arms. It is thus probable that the upper surface of a crinoid is homologous with the lower surface of a star-fish or sea-urchin. The œsophagus is short, and the intestinal canal is more or less coiled in its passage to the anal extremity. The mouth does not contain any masticatory apparatus, comparable with the so-called 'jaws' and 'teeth' of sea-urchins and serpent-stars. The upper surface of the

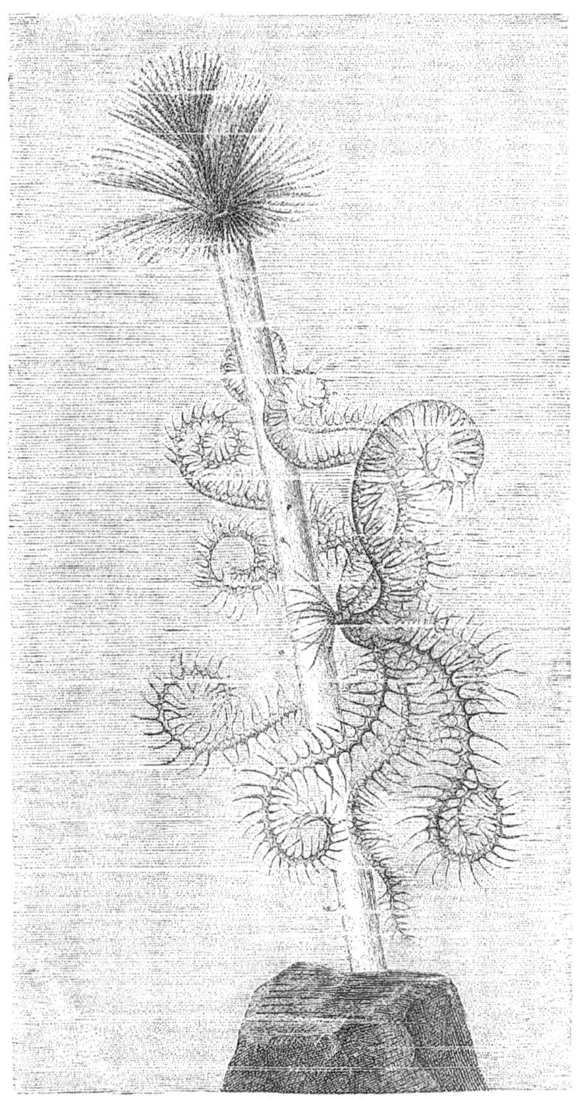

FIG. 127. — *Antedon* on the tube of a worm.

calyx is more or less covered over by the oral plates, usually five in number. These are separated by narrow spaces that are continuous with grooves that run along the upper surface of the arms and pinnules. The water system is upon the usual echinoderm plan, that is, a ring around the mouth, and radial vessels running along the arms. The ambulacral feet upon the oral surface of the calyx are connected with the ring canal. The body cavity extends into the arms, which also contain the greater part of the ovaries, as in the star-fishes.

In *Antedon* and other free crinoids there is a cavity known as the chambered organ, which has its walls and floor formed almost entirely by the centro-dorsal tubercle, and therefore, by what was once a stem segment. This chambered organ is rudimentary in the cystidean or unarmed phase of growth, but becomes developed as the comatula acquires arms and cirri, and is connected with these parts by fibro-cellular cords in the axis of the calcareous part of the arms and cirri. These cords are believed to be nerves, and the chambered organ must therefore be regarded as the centre of the nervous system. The 'ovoid gland,' which some consider to be a heart, is implanted on one of the horizontal floors of the chambered organ.

In *Pentacrinus* the chambered organ is a part of the central space enclosed within the pentagon formed by the radials and basals, while in the Apiocrinidæ (*Rhizocrinus* and its relatives) an intermediate condition exists. The ovaries (in *Antedon*) discharge their ova from openings on the arm pinnules. The eggs are fertilized while attached to the exterior of the opening, undergo total segmentation, and after awhile develop into oval embryos with a surface covered with cilia. When the embryo leaves the egg, it is girded with four zones of cilia, bears a tuft of cilia at one end, has a mouth (surrounded by large cilia), and an anal opening, and is free-swimming. In a few hours or days, traces of the calcareous plates, destined to form the cup of the future crinoid, begin to appear; then the plates of the stalk develop, and lastly the basal plate. As is the case with all echinoderms, there is little in common between the larva and the young of the perfect form. The young crinoid is formed within the larva, and the mouth and digestive cavity of the latter are not converted into those of the former. Two or three days after the appearance of the plates, the larva begins to change its form, the cystid-like young crinoid is seen embedded in the body of the larva, and the latter sinks to the bottom and adheres to some object. The stem becomes more elongate, while the part which will be the calyx still remains short and thick. The broad end of this part becomes five-lobed, each lobe answering to an oral plate, and these plates open like the petals of a flower, showing the oral aperture. Tentacles then appear between the oral plates, eventually arranging themselves in groups of three. Alternating with the basal and oral plates, the five radial plates now appear, and the arms grow out rapidly. The calyx also widens, so that the oral plates become widely separated from the basal, which encircle the stem. The alimentary canal of the young crinoid, which has before been a mere sac, now develops an intestine which opens out on an interradius where an anal plate has now appeared. If the animal is a stalked crinoid, the principal further external alterations are the acquisition upon the stem of whorls of cirri at intervals, the bifurcation of the arms, and the development of pinnules upon them.

In *Antedon* or *Comatula*, *Actinometra*, and *Atelecrinus*, forming the family Comatulidæ, the young are stalked and attached, but the calyx, together with the uppermost joints of the stem, breaks off at a more advanced period of life, and the crinoid swims off freely. Articulated to the lower or aboral face of the centro-dorsal tubercle formed by the upper stem-joints are the numerous cirri with which the animal grasps bodies to which it may desire to become temporarily attached. There are five series of radials, each containing three ossicles. The first or lowest three are closely adherent to each other and to the centro-dorsal tubercle, which conceals them on their outer side. That the central ossicle with the cirri is not the true basal plate is proved by the presence on its upper surface of a basal enclosed between the apices of the first three radials. This basal plate, called the rosette, is formed by the coalescence of the five basals of the larva.

The alimentary canal makes about a turn and a half round the axis of the body, and ends in the projecting inter-radial rectal cone. Included within the coil of the alimentary canal is a sort of cone of connective tissue, which has been called the columella. The five oral valves contain no calcareous plates in *Antedon*. The position of the genital glands in this and other crinoids, namely, in the pinnules, seems very exceptional, yet they are lodged in tissue, which is a continuation of the cellular tissue of the arms, comparable to that in which the ovaries are lodged in star-fishes. The species of Comatulæ are numerous. The "Blake" expeditions have latterly added forty to the twenty previously known from the Caribbean Sea, and about seventy were dredged by the "Challenger" between Cape York and the Philippines, and thence southward to the Admiralty Islands.

The chief distinctions between the two principal genera are as follows:—In *Antedon* the mouth is central, or sub-central, and the ambulacra are equal, the arms are equal in length, grooved, and furnished with tentacles. Red spots, the nature of which is unknown, and which are absent in *Actinometra*, are always present at the sides of the ambulacra. The cirri are numerous, and more or less cover the centro-dorsal tubercle, and the outer faces of the radial plates are relatively high, and inclined to the vertical axis of the calyx. In *Actinometra* the mouth is not in the centre of the disc; the ambulacra are variable in number and unequal in size, two of them always forming a horse-shoe round the anal area; the pinnules of the mouth have combs at their tips; some of the hinder arms may be shorter than the others, ungrooved, and without tentacles; brown spots, thought to be sense-organs, may be present on the dorsal side of the pinnule segments; the cirri are few; and the outer faces of the radials are wide and parallel, or nearly so, to the axis of the calyx. In *Antedon* the ambulacra of the pinnules may be protected by side-plates and covering-plates, but these are absent in *Actinometra*.

In the Caribbean Sea *Actinometra* is represented by more species and more individuals than *Antedon*, and two-thirds of the species of the latter and three-fourths of the former have ten simple arms. In the remaining species the rays rarely divide more than twice; only two species divide four times. *Antedon* and *Actinometra* are about equally represented in the eastern seas, but while half the species of the former genus are ten-armed, only three *Actinometræ* are thus simple. Nearly all the ten-armed *Actinometræ* of the eastern seas have the second and third radials united by a double joint, without muscles, called a 'syzygy,' and each of the first two brachials is a double or syzygial joint. But all the ten-armed *Actinometræ* of the West Indies have the second and third radials joined by ligament, while the first syzygy is on the third brachial.

The two Comatulæ, which, from their abundance, seem to specially characterize the Caribbean Islands, are *Antedon spinifera* and *Actinometra pulchella*. The first has usually thirty arms, but occasionally forks four times. The disc bears a tolerably complete plating, and there is a double row of plates along each edge of the pinnule ambulacra, which are thus covered over. The spread of the arms is about eight inches. *Actinometra pulchella* varies greatly. The arms may be ten, twelve, or even twenty. The mouth is far out of the centre, and the disc is bare, or more or less covered with calcareous plates. It is about ten inches across the arms.

Atelecrinus, of which two species, *cubensis* and *balanoides*, are known, has the first radials visible, and separated from the centro-dorsal tubercle by a complete circlet of basals. No other Comatula, with one doubtful exception, retains its embryonic basals on the exterior of the calyx after the latter part of its existence as a Pentacrinoid, and no other Comatula, recent or extinct, has a complete basal circlet of five pieces.

In *A. balanoides* the acorn-shaped centro-dorsal is marked all over by the horseshoe-shaped sockets of the cirri. The first ten or twelve of the arm-joints are without pinnules. In the characters of the calyx *Atelecrinus* is a permanent larva.

The Comatulæ dredged by the "Blake," were nearly all taken in depths less than 200 fathoms; and as the "Challenger" only found Comatulæ at greater depths than this on twenty occasions, P. H. Carpenter concludes that the group is essentially a shallow water one.

We now come to the stalked crinoids, which, though they so long eluded search, are not really abyssal forms, since they have on only fourteen occasions been dredged from depths exceeding 650 fathoms.

Pentacrinus has a long stem, bearing whorls of unbranched cirri at intervals, and lives attached to rocks in moderate depths. The joints of the stem are pentagonal. There is no distinct basal piece to the calyx, and the arms bifurcate twice, thus giving twenty principal free arms. The first forking occurs at the third radial. The ambulacral grooves of the arms have a series of ossicles along their floor and lamellæ along their margin.

In this genus the third radials, or radial auxiliaries, have their two sides bevelled off like the eaves of a gable, to allow two joints, the first joints of the ten arms, to be seated upon them. The first joints of the arms are split in two by a peculiar joint called a 'syzygy.' The ordinary arm joints are provided with muscles for the various motions, but the syzygies are not so provided, and consequently, when one of the arms is entangled, or caught, it breaks off at one of the syzygies, a beautiful provision for the safety of an animal with so complicated a crown of appendages.

The stem of *Pentacrinus* consists of many flattened calcareous joints, the upper and lower surfaces of each of which show five radiating leaf-like spaces, each surrounded by a border of tiny ridges and grooves. These ridges and grooves fit into each other, so that, in spite of the number of joints, the motion of the stem is very limited. The leaflet-like spaces thus come over each other, and five oval bands of strong fibres pass right through the loosely-arranged calcareous network of these spaces from one end of the stem to the other. There are no muscles between the joints, so that the animal does not appear to be able to move its stalk at will. Each joint has a hole in its centre, and this chain of holes is continuous with others which run through to the plates of the cup, and continue through the axis of each arm, and along every pinnule. In *P. asteria* about every seventeenth joint of the stem bears a whorl of fine, long tendrils, or cirri, which start from shallow grooves in the projecting angles of the pentagonal stem. These cirri usually have thirty-six or thirty-seven short joints, the last of which is sharp and claw-like. Though they have no true muscles, they seem to have some power of contracting around resisting objects which they touch. The cirri become smaller and closer together near the head, for each cirrus-bearing joint develops immediately below the calyx, and the joints which separate the cirrus-bearing joints from each other develop afterwards.

In ***P. asteria*** the arms divide a second time about six arm-joints above the first bifurcation, and again bifurcate seven or eight joints farther, the bifurcations repeating themselves somewhat irregularly until each of the primary arms has divided into twenty or thirty branches, making more than a hundred arms in all.

P. wyville-thomsoni, found upon the coast of Portugal, has the cirri separated by thirty to thirty-five joints. The number of arms is not constant, because in some cases the third radials bear one or two simple arms, while in others there is a third bifurcation.

Some examples of *P. mulleri*, and all of *P. wyville-thomsoni*, that were dredged by Mr. Jeffreys, showed by the smooth and rounded terminal joint of the stem, by the shortness of the lower cirri, and by the small number of joints in the lower internodes between the cirri, that they must have for a long time been free, and Sir Wyville Thomson states his belief that the latter species lives loosely rooted in the soft mud, and can change its place at pleasure by swimming with its pinnated arms.

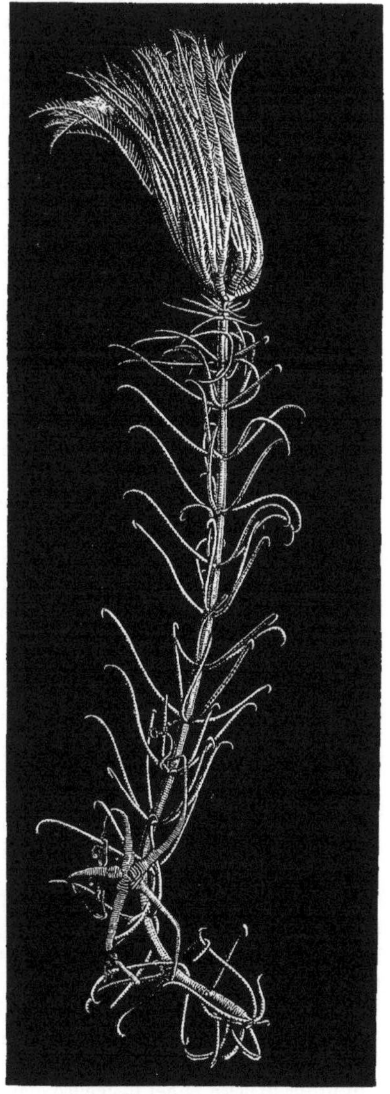

FIG. 128. — *Pentacrinus asteria.*

P. maclearianus, dredged near the coast of Brazil, has a peculiar style of arm-division. Each of the ten primary arms standing upon the radial axillaries, gives off two secondary arms close to its base, so that there would be in all thirty arms, were the arrangement absolutely regular. The arms are very robust, and the joints have a tendency to widen in the middle of the arm. Each arm has about seventy joints. The cirrus-bearing joints of the stem are very short, and much inflated with round, bead-like knobs, and there are only two very thin plates between the nodes. Sir W. Thomson believes that this form floats about unattached.

The most common pentacrinoid of the Caribbean Seas is apparently *P. decorus*, in which the whole of the cirri are separated by ten to twelve internodal joints; the nodal joints are large and projecting, and the two outer radials, and the first two joints beyond them are united by ligament, instead of by muscles or by a syzygy.

P. mulleri, though confounded with this species by Thomson, differs widely. The internodal joints of the stem, are only six or eight; the cirri are stout, and have about forty joints, the outer radials and succeeding joints are united by syzygy, as in *P. asteria*, and the arms fork much as in the latter. Eight species of *Pentacrinus* are now known.

The most widely distributed, and at the same time one of the simplest of living crinoids is *Rhizocrinus lofotensis*, a species which does not exceed three inches in length, and lives at depths of from one hundred to one thousand fathoms in the North Atlantic, and upon the coasts of Florida. The stem is relatively long and many-jointed, some of the basal articulations bear branched, root-like filaments, or cirri, and at its summit is a calyx consisting of a central piece or basal and five first radials, all closely united together and perched upon the enlarged solid, pear-shaped upper joints of the stem. To these five first radials, follow two other series of radials, all included within the calyx, but each following the line of an arm. To the third radials are attached the first of the ossicles of the unbranched but pinnule-bearing arms, which vary in number from four to seven, and have from twenty-eight to thirty-four joints. The pinnules alternate with each other along the arms, and have also a jointed skeleton. The mouth is circular, but is surrounded by the five (or four) oral valves, which close over it when shut. Between the circular lip and the oral valves there is a series of soft, flexible, tentacles, two pairs to each valve. The outer one of each pair is very contractile. Tentacles of a similar character are continued along the deep grooves which traverse the oral surface of the arms and pinnules.

R. rawsoni is readily known from the last species by its more robust appearance and elongated calyx, which is nearly always constricted at the suture with the radials. The greater part of the cup in this genus is formed by the elongated basals, which, in the Norwegian variety of *R. lofotensis* are so completely fused that no sutures are visible, a peculiarity which led to the supposition that this part was formed of enlarged upper stem-joints as in a *Comatula*. *R. rawsoni* is a larger form than *R. lofotensis*.

Hyocrinus bethellianus has much the structure of the palæozoic *Platycrinus*. It has a rigid stem made up of cylindrical joints applied to each other by a close syzygial suture. The cup consists of a basal ring which seems to be formed of two or three pieces, and of a tier of five, thin, broad, spade-shaped radials. The arms are five in number, and are built up of long, cylindrical joints. The first three joints consist of two parts separated by a syzygy, the other joints have two syzygies. From the third and all subsequent joints springs a pinnule, the pinnules alternating on either side of the arms. The lowest pinnules are very long, the succeeding ones becoming shorter.

The outer part of the disc is paved with irregular closely-set plates, bounding the five large oral valves. The œsophagus is short, and is succeeded by a stomach surrounded by brown glandular ridges, the intestine is very short, and contracts rapidly. Round the gullet a rather ill-defined oral ring gives off, opposite each of the oral plates, a group of four tubular tentacles. This species was dredged near the Crozet Islands, and also with *Bathycrinus*, in eighteen hundred and fifty fathoms, off Brazil.

Holopus is a short, stout form with no true stalk, with a broad, encrusting base instead of the branching cirri of *Rhizocrinus*, and ten arms which can be rolled together spirally in such a way as to enclose a closed chamber between them. The pinnules are formed of broad, flat joints, and can be spirally rolled towards the arms.

In *Holopus* there is a marked division into bivium and trivium, as in some holothurians, the three facets of the trivium are larger than the other two facets (the central one largest), and the three arms attached to those facets of the cup are larger than those joined to the opposite side. Another attached genus is *Bathycrinus*, one species of which was dredged with *Hyocrinus* in eighteen hundred and fifty fathoms, off Brazil, while another was taken at two thousand four hundred and thirty-five fathoms in the Bay of Biscay.

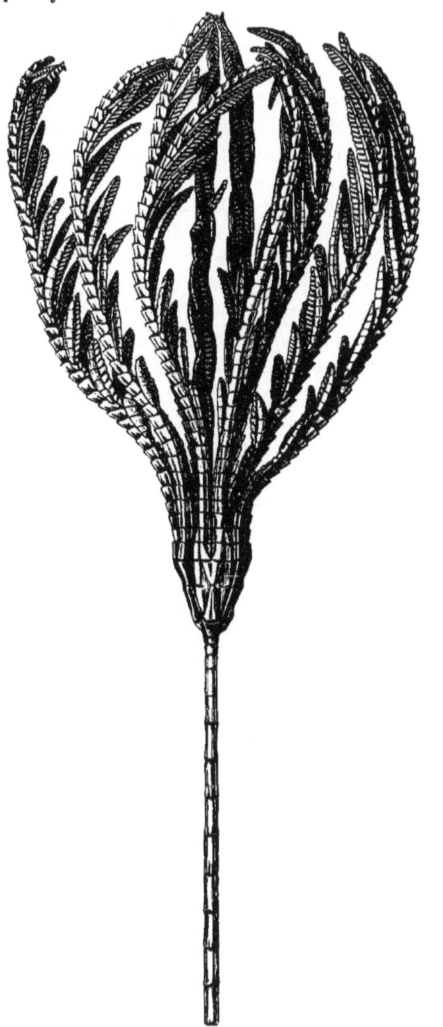

FIG. 120. — *Bathycrinus aldrichianus.*

CLASS II.—STELLERIDA.

The Stellerida are echinoderms with star-shaped or pentagonal bodies, a well-developed water system, and an internal skeleton, consisting of ambulacral plates which are different from those of the Echinoidea, since the nerve cords and radial ambulacral vessels lie outside and below them. There are two orders, Asteroidea and Ophiuroidea.

ORDER I.—OPHIUROIDEA.

The Ophiuroidea are a group of star-fishes, characterized by a more or less sharply-defined central disk, containing a digestive cavity, which may be simple or much plaited, but which does not pass into the arms. There is no anal opening. The arms have an axis composed of calcareous ossicles, usually called arm-bones, which greatly resemble vertebræ, and each of which is made up of two sections. These two sections represent the ambulacral plates of ordinary star-fishes. The axis is cased either with plates, or with a thick skin having rudimentary plates beneath, and the plates upon the sides of the arms usually bear spines. Within the hollow of the arm, covered by the under arm-plates, yet in the same relative position to each other and to the ambulacral plates or arm-bones, that obtains in the true star-fishes, run the nerve, the neuval canal, and the ambulacral vessel of the water system. There are no ampullæ in connection with the water-feet, which are simple tentacles without suckers at their tips. They make their exit between the lateral plates of the arm-covering. Each of the five angles of the mouth is formed of five pieces. The two halves of one or two arm-bones (Lyman says two, because there are two sockets for tentacles) are modified into strong mouth-frames, movably articulated and swung apart from each other. The extremities of these mouth-frames are soldered to a jaw or inter-ambulacral piece, and to the inner edges of each pair of jaws is articulated a long, narrow jaw-plate, which supports a variable number of processes, which are called teeth, and doubtless serve the purpose of such.

On either side of the base of each arm are the radial shield above, and the genital plate below. These are joined at the margin of the disc, and connected by an adductor muscle. On the under side, in the space between the arms, are one or two genital openings, parallel with and close to each arm. These, in the great majority of species, enter a peculiar sac, the genital bursa, with which the tubes from the ovaries or spermaries communicate. Apostolides considers that these bursæ should be regarded as respiratory sacs, as they may be seen to alternately contract and dilate. Each inner angle of the mouth is usually covered by a plate, called the mouth-shield, and one of these usually serves as the madreporic body.

The nervous and circulatory systems, and arrangement of the water system, are upon the usual star-fish plan. The body cavity consists of an enlarged portion surrounding the digestive tube, and a flattened portion in the dorsal region. The nerves have been found to contain cells with large nuclei, somewhat resembling the pigment cells of vertebrates, and also delicate fibrils, with pale, bi-polar cells not collected into ganglia.

The Ophiuroidea fall into two families. In the first and larger family, OPHIURIDÆ, the axis of each arm is encased in a greater or less number of plates, the principal of which, from their position, are known as the dorsal, ventral, and lateral plates. The lateral plates bear a more or less numerous series of spines, and are usually considered homologous with the ad-ambulacral plates of the arm of an ordinary star-fish; the ventral and dorsal plates are primarily unpaired, and the former, at least, are peculiar to the group. Mouth-shields are always present, and there are often two other superficial plates, the side mouth-shields, one on each of the outer sides of each mouth-shield.

The Ophiuridæ rarely have more than five arms, and these are in all cases unbranched, but in the other family, the Astrophytidæ, the five arms usually divide and sub-divide into a very large number of branches. The latter family is destitute of the regular covering of plates that protect the arms of the Ophiuridæ, but in its place has a thick skin, under which are plates, usually of an irregular and elementary character. The arm ossicles consist of a vertical and a horizontal hour-glass-like projection, fitted one on the other. There are no spines on the sides of the arms, and mouth-shields are often absent in the branched species, in which the madreporic plates, sometimes one and sometimes five in number, may be found in various parts of the lower inter-brachial spaces.

The arm ossicles of the Ophiuridæ, according to Ludwig, are originally double, the first rudiment consists of two calcareous pieces symmetrically placed on either side the middle line of the arm; each triangular piece is formed of three rays, two directed orally, the third ab orally; the latter increases considerably in length, and the two others gradually become fused together. Till a late stage of growth, there is in the middle of the ossicle a space with concave sides.

In typical Ophiurans the mouth, just above the teeth, opens by a round, contractile aperture into a large, flattened stomach, which spreads over the basis of the arms and into the inter-brachial spaces. Though sometimes a little wrinkled, it is usually destitute of pouches, convolutions, or cæcal appendages. Between the stomach and the disc wall lie the reproductive organs, consisting of elongated bags communicating with closed tubes, which bear the ova or spermatozoa. In the Astrophytidæ the upper part of the stomach is surrounded by numerous radiating folds or bags, which are attached to the roof of the disc, to the genital organs, and at ten points encircling the mouth. The body cavity would thus be divided into ten parts were it not for the open space or canal which runs around the mouth, and corresponds to the ring-canal of a true Ophiuran, but differs in being a continuation of the body cavity instead of a closed, annular tube. There is no closed bag for the genital products, but the body cavity is the genital cavity, and an ovarial lobe opens into each compartment.

Lyman enumerates about five hundred species of Ophiuroidea, forty-nine of which are Astrophytidæ. Although but few of the sub-order can be regarded as littoral, more than half, or two hundred and seventy-eight species, are found above the depth of 30 fathoms, and two hundred and twenty-six of these do not occur in deeper water. Thirty-eight of the remaining coast species do not descend beyond 150 fathoms, twelve others reach to 500, and only two go lower than 500, but do not reach 1,000 fathoms. Between 30 and 150 fathoms one hundred and fifty-one species occur, sixty-nine of which are not found either above or below this range. Between 150 and 500 fathoms one hundred and thirty-seven species occur, seventy of which are confined within these limits, while thirteen descend to below 1,000 fathoms, and twenty to below 500. Only sixty-nine species occur below 1,000 fathoms, and of these fifty do not occur above that limit. This number may of course be increased by subsequent dredgings, but even now we know of fifty exclusively deep-water species, living in water cold almost to freezing, and in entire absence of sunlight.

In the genus *Ophiura* the disc is covered with small granulations, which more or less covers the small, oblong, separated radial shields; the jaws are set with teeth; the spines, which are on the outer edges of the side arm-plates, and parallel to them, are smooth, flat, and shorter than the arm-joints; side mouth-shields are present, and there are four genital openings. A fine species is *O. teres*, from Lower California and the west coast of Central America.

Pectinura is another large genus, distinguished from *Ophiura* by the absence of the adhesion of the edges of the genital openings that, in the latter genus, doubles their number. In *Ophiozona* the larger scales of the disc are intermingled with lines of smaller ones, while *Ophioceramis* has none of these small scales, but is known by large mouth-frames, developed into wing-like projections, and by a very long and large first arm-bone of unusual form. *Ophioglypha* is a genus with fifty-eight known species, all of which have numerous tentacle scales, while the pair of tentacle pores nearest the disc are slits of comparatively large size, surrounded by numerous tentacle scales, and opening diagonally into the mouth slits.

In *Ophiocten* the side arm-plates are large, meeting below the arm, but not above; while in *Ophiomusium* both upper and under arm-plates are so small that the side arm-plates meet both above and below, while the radial shields and plates of the upper surface of the disc are intimately soldered together, forming a surface like porcelain. *O. flabellum* is a curious little form, with very short, rapidly tapering arms, and the first pair of side arm-plates of each arm so large that they meet in the inter-brachial spaces, and thus seem to form part of the disc. It was found off Port Jackson in 30 to 35 fathoms.

In all the foregoing genera, as well as in several others, the arm-spines are situated on the outer edges of the side arm-plates, and are parallel to the arm, but in the remaining genera the spines are set on the sides of the side arm-plates, and stand out at a large angle with the arm. The spines are thus much more conspicuous than in the first group, and in many cases they are not only long, but adorned with rows of small, pointed teeth. In *Ophiopholis* the upper arm-plates are surrounded by a row of supplementary pieces, and the lowest spine of the outer arm-joints is a hook. A well-known species is *O. aculeata*, which is found at various depths, up to 400 fathoms, on the coast of northeastern America and northern Europe, as well as in the Arctic seas. It is often called *O. bellis*, but was first described under the name before given. *Ophiactis*, with twenty-four species, resembles the last genus, but the upper arm-plates are without the ring of supplementary pieces. In both genera the arm-spines are stout and smooth. *O. savignyi* (*O. virescens*) is found along the west coast of North America, from Panama to Cape St. Lucas.

Fig. 130. — *Ophiopholis aculeata*.

Amphiura is the largest genus of the order, since it contains eighty-nine known species, all of which have a small and delicate disc, covered with over-lapping scales and showing the radial shields, and long, slender, more or less flattened arms with short spines of uniform size. *A. maxima*, obtained by the Challenger expedition in 9° 59′ S. lat. and 139° 42′ E. long., at a depth of twenty-eight fathoms, measures nearly a foot across from end to end of arms, though the disc only slightly exceeds half an inch in diameter. *Ophiocnida* differs from *Amphiura* principally in the presence of small spines upon the scales of the disc. There are, in fact, several genera that are distinguished from *Amphiura* by but slight characters, though the differences between the species contained in those genera are curiously well marked and constant. *Hemipholis*

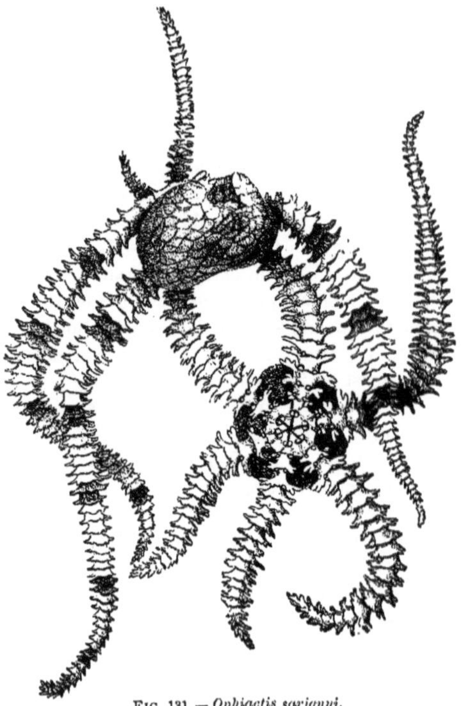

Fig. 131. — *Ophiactis savignyi*.

cordifera, a member of a genus nearly related to *Amphiura*, is plentiful in the harbor of Charleston, S. C., and is apparently viviparous, since it may be found with minute young clinging to the arms and disc. It occurs also in the West Indies.

Fig. 132. — *Ophiacantha chelys.*

Ophiocymbium cavernosum, taken by the 'Challenger' east of Kerguelen Island, is remarkable for the manner in which the disc, which is scarcely attached to the arms, and is entirely covered with small scales, overlies the arms "like a Basque cap." *Ophiocoma æthiops* and *O. alexandri*, are large species from the west coast of North America. Among the genera with numerous long, usually rough or thorny, arm spines the principal are *Ophiacantha* with thirty-nine described species, and *Ophiothrix*, with fifty-six species. The species of *Ophiacantha* can be readily distinguished from each other by points of internal and external structure. The arm spines vary from four to eleven in number at each arm joint. *O. vivipara*, which is widely spread in the southern ocean, occurring at depths varying from 20 to 600 fathoms, carries its young, till they are quite large, in the ovarial bursa or pouch, whence they often thrust an arm through the genital opening. The bursæ are plented bags having lime scales in their substance and adhering to the thickened wall of the digestive cavity. Some of the basal arm spines of *Ophiomitra dipsacos* dredged by the 'Challenger' off Culebra, West Indies, in 390 fathoms, are from five to seven times the length of the arm-joint. The species of *Ophiothrix* are exceedingly difficult to distinguish from each other, although the genus is well marked by the very large three-sided radial shields, disc set with thorny grains; five to ten glassy spines (beset with thorns, and often three times as long as the joints), on each arm joint; single, small, spine-like tentacle scales on each arm-joint; swollen interbrachial spaces; and want of perfect union between the halves of each mouth-frame. The vertebræ of the arms have a peculiar projection which interlocks into a slot in the preceding bone, thus giving a fulcrum (says Lyman) for the

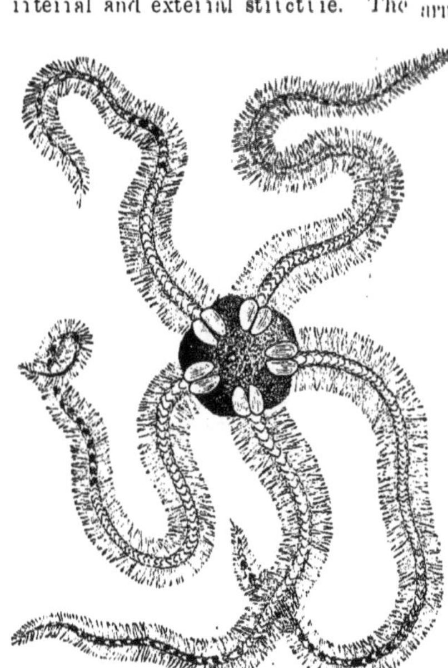

Fig. 133. — *Ophiothrix fragilis.*

powerful muscular action called for in the rapid whip-like motion of the arm. *O. lineata* occurs in Florida, and three or four species are found in southern and Lower California. *Ophiociasma attenuatum* is a curious form found in the South Atlantic. The examples taken had six arms, the disc is covered with thick soft skin, and the arms, which are very long and slender, have the lower and side plates imperfectly calcified, and no upper plates. The spines are short, three on each joint. The two known species of *Ophiohelus* are remarkable for the curious minute spines or pedicellariæ, having the form of a long-handled parasol, that take the place of the true arm-spines on the outer arm-joints. One of the most singular of ophiurans is *Ophiothelia supplicans*, in which the disc is shaped like a high sugar-loaf, and the arms can be raised upwards. The arms bear upon all the joints beyond the ninth a cluster of three or four minute parasol-like pedicellariæ, set a little inside the true arm-spines, which continue to the end of the arm. The under surface of the disc is curiously ornamented. A frill of long, flat, curved papillæ is set upon each mouth-angle, which, ending inwardly in a sharp tooth, resemble so many birds' heads with a pointed bill. Externally to the mouth papillæ are three parallel rows of regular club-shaped flat papillæ. This form is only known from south-west of Juan Fernandez, and was taken at a depth of one thousand eight hundred and twenty-five fathoms. *Ophyomyces* has a very peculiar arrangement of mouth papillæ, and two of its species occur between five hundred and one thousand fathoms.

Two or three genera of Ophiuridæ contain species with cylindrical arms, while the entire animal is clothed with a thick skin, and the arm-plates are imperfectly developed. They thus approach the Astrophytidæ. In *Ophiobyrsa rudis*, which was taken off the entrance to Port Philip, in thirty-eight fathoms, each arm is about twelve inches long, though the disc measures only about an inch.

Ophiomyxa vivipara is, as the name implies, a viviparous species.

The best known genus of **Astrophytidæ** is *Astrophyton* itself, of which seven species are known. Though there are no arm-spines, the outer branches of the arms have spine-like tentacle scales. The long bar-like radial shields of the disc are covered by the thick skin, but show as elevated radiating ribs reaching to its centre. Under the skin of the arms there are side arm-plates, which cover the lower side of the arms, but there is only a basal under arm-plate, and there are no upper arm-plates at all. *Gorgonocephalus* has also branching arms, but the plates under the skin of the disc are differently arranged, and the arms, wide at their base in *Astrophyton*, are here narrow, while the forkings are less numerous than in that genus. The young has at first a flat disc covered with plates like that of an ordinary ophiuran, this first becomes covered a close granulation; then both the granulation and the disc-plates atrophy, except those of the margin, which continue to grow and multiply; and finally the radial shields acquire their great length and height. In *Euryale* no proper arm-spines are present, and in *Astroclon* and *Trichaster* the forks are but few. The total length of an arm of *T. propugnatoris* is about sixteen inches. Among the Astrophytidæ with unbranched arms the principal genus is *Astrochema*, which has well-formed under arm-plates like an ophiuran. *Ophiocreas abyssicola* was taken in twenty-three hundred fathoms.

Order II. — ASTEROIDEA.

The differences which distinguish the star-fishes from the ophiuroids are scarcely less important than those which separate either from the sea-urchins, yet the external form is in both cases that of a star of five or more rays. In the star-fishes there is no such well-marked distinction between the disc and the arms as there is in an ophiuroid, for the stomach, and the ovaries or spermaries, run into the arms. Along the underside of each arm runs a deep ambulacral furrow, from the depths of which project the ambulacral feet, which are provided at their ends with suckers, by means of which the animal moves. At the base of each sucker-foot within the arm is a vesicle or ampulla, connected with the radial or ambulacral canal of the water system. This arrangement seems very different from the solidly plated arm of a serpent-star, with its enclosed row of vertebral ossicles, and its tentacle-like feet unprovided with sucking discs or with vesicles at their base, yet it is demonstrable that the halves of the vertebræ of an ophiuran are to be considered identical in their nature with the calcareous ambulacral pieces which form the deeper parts of the sides of the ambulacral furrows of a star-fish. The development of the interambulacral areas varies greatly, so that while some star-fishes, as *Brisinga*, closely approach the serpent-stars in the distinctness of the arms, others have the angles between the arms more or less filled up by the development of the interambulacra, which in many forms are so extensive that the creature is nearly or quite a pentagon, and the existence of arms can only be traced by the lines of the ambulacral furrows on the oral aspect.

Star-fishes have a very complex skeleton, varying greatly in the different groups, but always consisting of plates or thick rods composed of a dense calcareous net-work. The sides of the ambulacral furrows are bounded by two rows of regularly-placed and similar ambulacral ossicles, which abut against each other like the two sides of a gable, while at their outer ends they abut against a row of short, thick, adambulacral or interambulacral ossicles, which form the borders of the furrow. These are the constant parts of the skeleton, but the sides of the arms are in many cases enclosed by plates of considerable size, known as marginal plates. The net-work of rods which forms the support of the upper part of the arms and disc is very variously developed in the different families and genera. The pores through which the sucker-feet pass are each formed by two ambulacral plates, one half of the pore consisting of a notch upon the outer side of one plate, while the other half is on the inner side of the next plate. Around the mouth the ambulacral and interambulacral plates are modified so as to form a pentagon or polygon. The pieces of this calcareous ring that correspond to the interambulacral plates of the arms are furnished with strong papillæ or spines which form a sort of imperfect dentary apparatus. The spines which project from the surface, both above and below, and which usually form regular rows on each side of the ambulacral furrows, are more or less movably united with the skeleton, but are without the regular joints found in the sea-urchins. Pedicellariæ are present. Above, that is to say, internal to the dentary papillæ, the gullet opens into a wide stomach, the oral or cardiac part of which is produced into sacs that sometimes extend into the cavity of the arm to which they correspond. On the aboral side of these sacs, the alimentary canal again narrows somewhat, and widens again into a pyloric sac, the angles of which are again prolonged into tubes which run along the aboral side of the rays. The upper or pyloric part of the stomach is attached by a mesenteric membrane to the aboral wall of the body, while the cardiac sacs are similarly connected with the ossicles or calcareous plates around the mouth. Beyond the pyloric sac is a short tubular intestine, which ends in a more or less minute anal pore, apparently

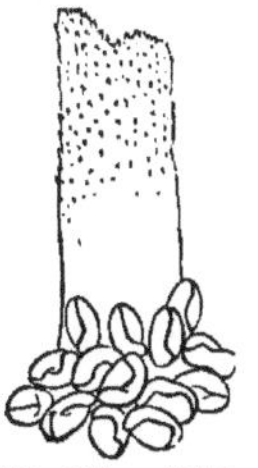

Fig. 134. — Spine of star-fish, with pedicellaria at the base.

rudimentary. Into the intestine opens a duct which divides in two main branches. These branches subdivide into numerous follicles, the whole forming what is supposed to be the liver, since it contains a bitter fluid like bile. The nervous system consists of a nerve which runs, as a longitudinal ridge, along each ambulacral groove, and of a ring around the mouth. At the end of each arm a rudimentary eye, continuous with the ambulacral nerve, is placed at the end of a modified tentacle. The water system consists of a canal running the entire length of each arm. This canal is placed in the angle formed by the ambulacral plates, and is thus external to them. It is separated by a strong partition from a second canal which intervenes between it and the nervous band which forms the bottom of the ambulacral groove in the living animal. This lower canal,

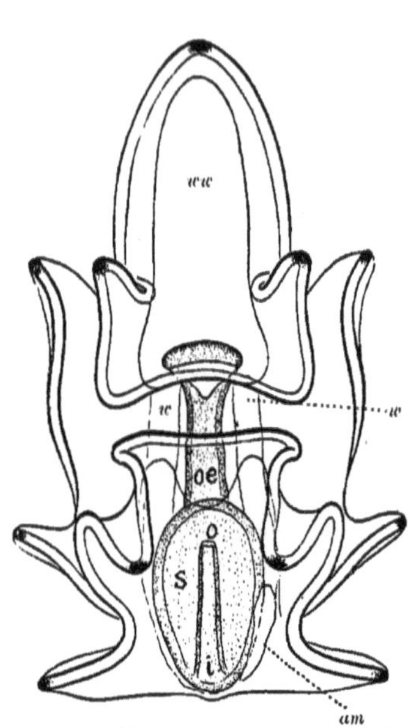

FIG. 135.—Bipinnaria of star-fish; *am*, rudimentary ambulacra; *i*, intestine; *o*, anus; *æ*, œsophagus; *s*, stomach; *w*, water tubes.

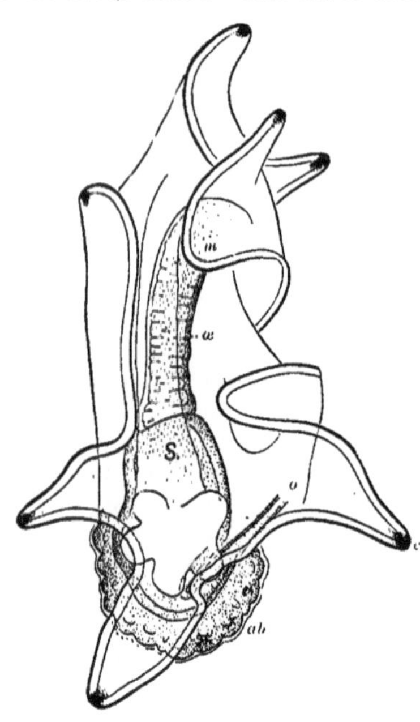

FIG. 136.—Later larva, with star-fish forming; *m*, mouth; *ab*, aboral portion of star-fish.

which is itself divided into two halves, is called the ambulacral neural canal, and communicates with a circular canal around the mouth. From opposite sides of the ambulacral canal of the water system, short branches pass up between the ambulacral ossicles to the ampullæ, which are sacs with muscular walls lying above the ambulacral plates, and within the cavity of the arm. Each ampulla communicates with a pedicel or sucker by the pores in the ambulacral plates, as before described. The cavity of the body and rays is filled with a watery fluid containing corpuscles, evidently representing the blood of higher animals. Pores, by which water can enter the body cavity, are often present upon the aboral aspect of a star-fish.

The sexes are distinct, but can only be distinguished by a microscopic examination of the glands, which are situated on each side of the interior of the arms, or at the junction of the body with the rays. As the plates which enclose the base of the arms in the long-armed star-fishes are interradial, and are homologous with those which fill up the space between the arms in the pentagonal species, the ovaries are actually interradial or interambulacral in position, as in ophiuroids. The eggs pass out by a pore on each side of the base of the arms, situated between two plates, and difficult to detect.

The embryo of a star-fish is usually a free-swimming animal with arms and ciliated bands, and has been called a Brachiolaria, or, in other forms, Bipinnaria. The Brachiolaria, when it has attained its full development, has thirteen arms — a medial anal pair, a dorsal anal pair, a ventral anal pair, a dorsal oral pair, a ventral oral pair, an odd anterior arm, bearing an odd brachiolar arm, and a pair of smaller brachiolar arms connected with the oral ventral pair of arms. The brachiolar arms have wart-like appendages at the tip, whereas all the other arms are surrounded by chords of vibratile cilia. The median anal arms appear first and are largest (in *Asteracanthion pallidus*), and the odd brachiolar arm precedes the pair of similar arms. The adult larvæ move about rapidly by means of the cilia, with the oral extremity in advance. In the Bipinnaria the arms are fewer in number and are more slender, and all are ciliated.

The first commencement of the growth of the young star-fish is by the appearance, on the anal side of the left water-tube of the larva, of five slight folds; while on the other side of the anal extremity appear five limestone rods. The folds are the first rudiments of the rows of suckers, the rods of the aboral skeleton. The rays are indicated as lobes upon the growing dorsal part of the disc while the suckers are still widely separated from them. The young star-fish, thus growing on the opposite surfaces of the two water-tubes, soon loads down the anal end of the embryo, the larva then becomes sluggish, the body changes from transparent to cloudy and opaque, the arms contract and become constricted into cells, and soon nothing remains but the brachiolar arms, brought close to the young star-fish by the shrinkage of the body. These finally follow the rest. Not a single part is dropped off, the whole of the larva passes into the star-fish. The first steps subsequent to this resorption are the approach of the oral and aboral sides by the contraction of the water-tubes, and the approach to symmetry of the at first

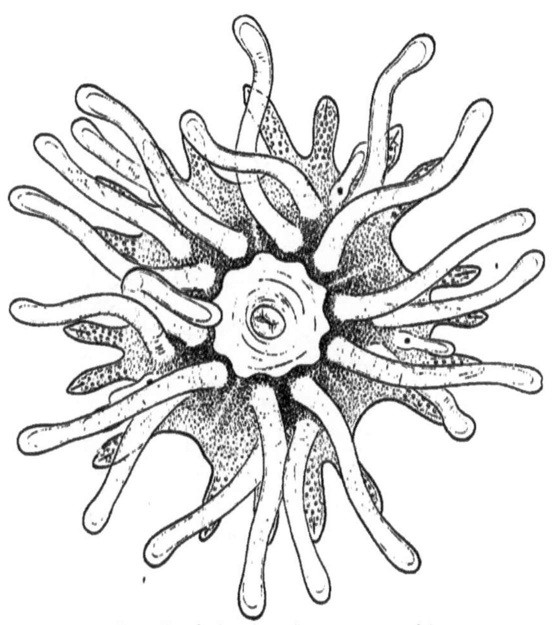

FIG. 137. — Oral surface of very young star-fish.

unsymmetrical arms. From this stage the arms gradually lengthen, and the spines and suckers develop and increase in number until the adult form is reached.

In some star-fishes the embryo develops directly without passing through a free-swimming stage, and this is especially the case with species of the southern seas.

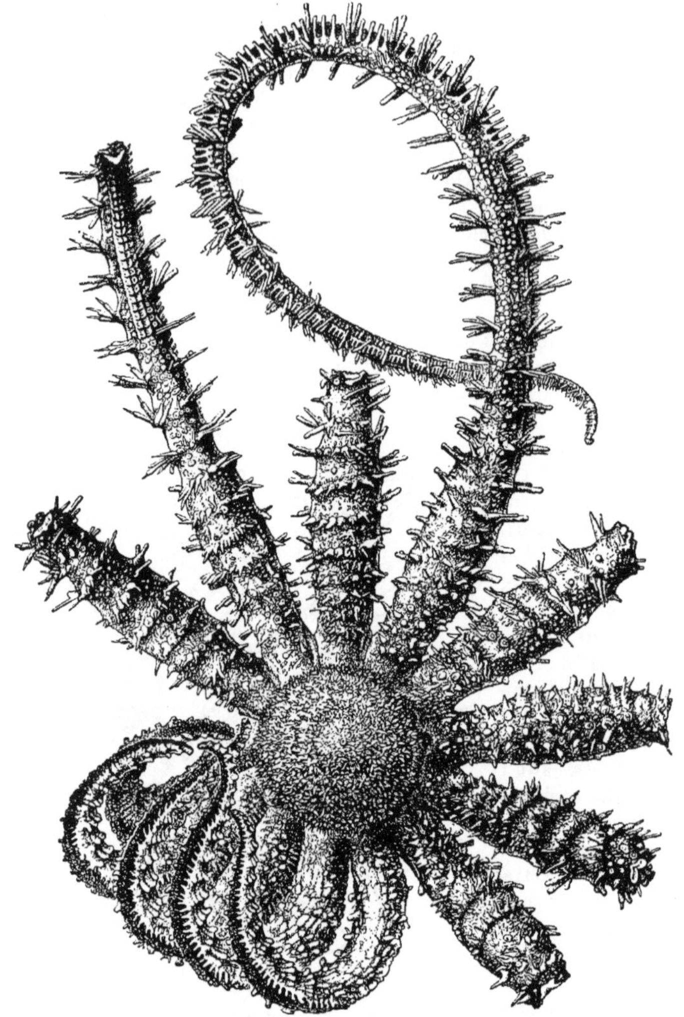

FIG. 138. — *Brisinga coronata.*

Various adaptations of the spines and external membrane form marsupia for the protection of the young while attached to the parent. A large species of *Asterias*, dredged in Stanley Harbor, formed a shelter for its young by drawing its arms inwards and forwards over the mouth. *Echinaster sarsii* does the same.

The family BRISINGIDÆ includes only two known species, both inhabitants of deep water, where they have an exceedingly wide distribution. The genus approaches *Pycnopodia* and *Crossaster*, but in appearance is intermediate between the ophiurids and the star-fishes, since it has a distinctly circumscribed disc like the former, though the long arms are soft as in the latter. *Brisinga endecacnemos* has eleven arms, which are nearly smooth, while the nine to thirteen arms of *B. coronata* have transverse crests of spines. The arms are sometimes a foot long, they are narrow at their insertion into the disc, enlarge considerably towards the middle, where the ovaries are developed, and taper thence to the tip. *B. coronata* has a single generative organ, with a single opening, on each side of each arm, while *B. endecacnemos* has a great number of separate small organs, each with its own opening, in a similar position. The water-feet are furnished with sucking-discs. Rows of long spines covered with soft skin border the ambulacral grooves; this skin is full of small pedicellariæ, and groups of pedicellariæ are scattered over arms and disc. The creature is extremely fragile, and always breaks to pieces before it can be handled.

The PTERASTERIDÆ are furnished with groups of diverging spines, on the tips of which a membrane is carried. *Hymenaster pellucidus* was first dredged in five hundred fathoms off the north of Scotland, and it has since been found that, with perhaps the exception of *Archaster*, *Hymenaster* is the most widely distributed genus of deep-water asteriids, varying from four hundred to two thousand five hundred fathoms.

Hymenaster nobilis is rather a large star-fish, ten inches across. The five arms, each two inches wide, are united to each other by a broad web attached to the outer row of the spines which fringe the ambulacral groove, thus transforming the animal into a large pentagon. The entire upper surface, the web excepted, is set with bunches of four to six diverging spines about an eighth of an inch long. These spines support, clear of the surface of the disc, a tolerably strong membrane, like the canvas of a tent. Something similar to this occurs in *Pteraster*. In the centre of the back this arrangement is modified to form a brood pouch for the young. From five calcareous supports, arising from the ambulacral plates below, spring a double series of spines, the outer series of three or four diverging in the ordinary way beneath the tent cover, while the inner series of six or eight bend inwards, and have a special membrane stretched between them in such a way that each series forms a fan-like valve, the spines closing when the valve is closed and separating when it is opened. In this central pouch the young are carried. This species was taken in eighteen hundred fathoms, about eleven hundred miles southwest of Cape Otway, Australia.

Korethraster hispidus was dredged near the Shetland Isles. It is a small star-fish with the whole of its upper surface covered with long spines (paxillæ) like sable brushes, masking its true outline. Rows of delicate, spoon-shaped spines border the ambulacral grooves. Another species of this genus, *K. palmatus*, has been dredged in the Caribbean Sea.

Pteraster is the principal genus of this essentially deep-water family. In *P. multipes*, of the Norwegian seas, there are four rows of water-feet, as in *Asterias*.

The ASTROPECTINIDÆ are forms with long rays ending in a point, thus resembling in general shape the Asteriidæ, from which they differ in having two rows of ambulacral tentacles, which are usually without sucking discs at their extremity, and also in the character of the skeleton, which is formed, at least in the dorsal region, of contiguous ossicles, often shaped like an hour-glass, and bearing tubercles which are crowned with small tufted spines called paxilli. They are most numerous in the hot seas.

The genus *Astropecten* is a large one, and many species occur in deep water. *A. articulatus* ranges from New Jersey to the West Indies. Several species are known to occur on the west coast of North America.

Nearly allied to *Astropecten*, but differing from it in the very conspicuous feature of the absence of any large marginal plates on the sides of the rays, is *Luidia*, the brittle star-fish, celebrated for its power of breaking into fragments when it is brought from the bottom. *L. clathrata* ranges from New Jersey to the West Indies, and is common in the Carolinas. *L. tessellata*, a fine species, attaining a diameter of more than a foot, ranges from the Gulf of California to Peru. *Ctenodiscus crispatus* is another North Atlantic species, occurring in the North Sea, Greenland, and New England. It is almost pentagonal from the shortness of the rays.

Leptychaster kerguelenensis has its dorsal surface covered with a tessellated pavement of paxilli or spines, with large heads. These paxilli form with their approximated heads a sort of hexagonal mosaic on the surface, but between their slender shafts arcade-like spaces are left, and into these spaces the eggs pass from the genital openings. Eggs and young in all the early stages of development occupy these spaces at once, but when at least six suckers have formed on each arm, they push their way out between the paxilli, usually in the angles between the arms. The young escape mouth uppermost, and disengage their arms one by one. After this they remain attached to the parent for some time. In the young, the madreporic plate can be seen near the margin of the disc, but in the adult it is hidden by the paxilli.

Archaster vexillifer, taken in three hundred and forty-four fathoms, west of the Shetland Isles, is a fine species about ten inches across, remarkable for the arrangement of the ambulacral spines, which form combs that increase in size toward the base of the arms. Each plate has a double row of spines, and each spine has a second short spine at its end. The ambulacral grooves are much wider, and the ambulacral tubes larger in proportion to the animal than is usual.

Archaster bifrons, taken in deep water north of the Hebrides, measures five inches across, and is of a rich cream or light rose tint.

Porcellanaster cæruleus was dredged by the 'Challenger' in one thousand two hundred and forty fathoms, in the Gulf Stream. The arms are rather shorter than the diameter of the disc; the ambulacral plates are large, and furnished with two flattened spines; and the plates forming the angles of the mouth are unusually flattened and expanded. There are two rows of large marginal plates, and the two on the end of each arm are fused together, and bear three spines, one on each side below, and a central one above. Vertical rows of small flattened scales cover the two central pairs of marginal plates in the re-entering angles between the arms, and look like a little brush between each pair of arms. The upper surface is covered with narrow calcareous plates, and is of a delicate cobalt blue color. The excretory opening is very distinct in the centre of the disc. It has been found in the North Pacific and near Tristan d'Acunha.

The Asteriniḋæ are pentagonal, more or less elevated in the centre of the disk, and always sharp at the edge. Spines are present, at least on the ventral surface, and there are two rows of pedicels in the ambulacral grooves.

Asteropsis imbricata, a species which ranges from Vancouver's Island to near San Francisco, is a good example of this family. The skeleton of the aboral side is a reticulation of flat, irregular, star-shaped plates, from which diverge longer flat pieces connecting the others together; the plates and their connecting links are all imbricated, that is, overlap each other.

Asterina folium occurs in the West Indies and in Florida. *Asterina* is a large genus, almost world-wide in its distribution. The skeleton is formed of imbricated or over-lapping and notched ossicula, and this structure is usual in the family, an exception being *Disasterina abnormalis*, in which singular species, a native of the coast of New Caledonia, the plates of the back are disjoined, leaving between them membranous spaces, most of which are pierced by a tentacle. In the genus *Patiria*, the plates of the back are rounded, and simply touch each other. *P. crassa* is from Western Australia.

In the GONIASTERIDÆ, the skeleton, at least on the lower face, is formed of rounded or polygonal ossicles, forming a kind of pavement, and there are usually two rows of marginal plates of comparatively large size. The body approaches more or less to the pentagonal shape, the arms projecting but slightly, and there are two rows of suckers. Some very fine examples of this family are found upon the west coast of this continent. Prominent among them is *Oreaster occidentalis*, which measures eight or nine inches across, and is also of considerable thickness in the centre of the disc. It ranges northward to Lower California.

A very beautiful species, with long, sharp spines upon a bright red disc, is *Nidorellia armata*, a pentagonal, cake-like star some five or six inches across. *Amphiaster insignis*, another Lower Californian species, has short, flat arms and a flat disc, and the regular arrangement of its spines and tesselated plates renders it exceedingly beautiful. A more northern species is *Mediaster equalis*, which has been found as far north as Puget Sound.

Astrogonium phrygianum is a large, bright-red pentagonal star-fish, which resides at depths of from twenty to fifty fathoms, on rocky bottoms, in the Gulf of Maine and northward. *A. granulare* is from Scandinavia. *Pentaceros reticulatus* is about eight inches across the arms, and is found on both sides of the Atlantic, at the Cape Verde Islands, and in the West Indies. In *Culcita* the shape is pentagonal, and the tips of the ambulacra appear on the dorsal surface. The Goniasteridæ are more numerous in types than any other family of star-fishes. Their principal centre of distribution seems to be the west coast of Australia and the Malaysian and Melanesian archipelagos.

The LINCKIADÆ have a skeleton composed of rounded or elliptical ossicles, either contiguous or united by rods. There are no spines, but the surface of the body is smooth or uniformly granular. The species are most numerous in hot regions.

Linckia unifascialis, a fine species ranging from Lower California to Peru, and *Linckia guildingii*, of the West Indies and Florida, are examples of the typical genus. *Ophidiaster pyramidalis* is a large species with the same range as *L. unifascialis*.

The ECHINASTERIDÆ, as defined by Perrier, have a skeleton composed of a network of lengthened ossicles, and have two rows of ambulacral feet. Spines arise from the ossicles of the dorsal surface.

Echinaster sentus is a five-armed species, abundant in the West Indies and Florida, and extending north to New Jersey. The spines are completely sheathed in membrane, and occur only at the angles of the limestone polygons of the dorsal surface. *Solaster endeca* has eleven or fewer smooth arms, and is widely spread in the

North Atlantic. Its place in the North Pacific is taken by *S. decemradiata.* The reticulation of the skeleton is close, and the arms less flattened than in the next species.

Crossaster papposus is common to both sides of the Atlantic, and is found in Norway, Denmark, Britain, France, and in America as far south as Massachusetts Bay. It has twelve arms, the spaces between which are largely filled up upon the oral side by a membrane, while the upper surface is an open network of limestone rods, carrying, at the points of junction, club-shaped processes which bear tufts of small spines.

Cribrella sex-radiata, from the Antilles, is remarkable from its possession of six arms, an exceptional character in this genus. It has also the faculty of reproduction by division into two halves, so that most examples show three larger and three smaller arms. This power is also possessed by several of the many-armed *Asterias*, by some *Linckias*, and some *Asterinas*. *Cribrella sanguinolenta* is common on the New England coast below low-water mark.

The ASTERIDÆ are star-fishes which usually have the arms well-developed, and have four rows of water-feet, each ending in a sucking-disc, along the ambulacra. One of the largest genera is *Asterias*, or *Asteracanthion*, species of which are found everywhere in the northern hemisphere. *A. rubens* is a common European form. *A. berylinus* extends from Halifax, Nova Scotia, to Florida, while *A. vulgaris* ranges from Long Island Sound to Labrador. The last two species are both common in Massachusetts Bay.

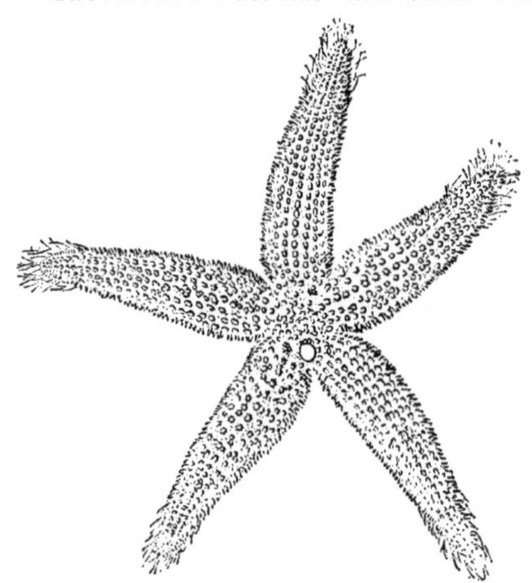

FIG. 139. — *Asterias vulgaris.*

In *A. ochracea*, which ranges from Sitka to San Diego, and is the most common star-fish of the Californian coast, the arms and the ambulacra are wider than in most species of the genus, and the calcareous network which covers in the sides and back of the arms is exceedingly solid:

Several other species of *Asterias* occur upon the Pacific coast of the United States, but the most conspicuous is the large six-armed *A. gigantea*, which attains a diameter of more than two feet. Nearly allied is *Pycnopodia helianthoides*, a gigantic form with more than twenty arms, common on the Pacific coast of North America, from Cape Mendocino to Alaska; the calcareous skeleton of the upper surface is reduced to a few small rods at the base of the spines, and hence a large well-preserved specimen is a rarity. This species attains the diameter of three feet, or thereabouts, and is of a bright red color in life. Professor A. Agassiz considers that this species, as well as *Crossaster papposus*, are in many respects allied to *Brisinga.*

The many-armed Asteridæ are, for the most part, included in the genus *Heliaster*, or sun-star, two species of which, *H. kubinyi* and *H. microbrachia*, occur upon the west coast of North America, from Panama to Cape St. Lucas. The latter form has more than thirty arms, and the free portions of the arms are very short.

Zoroaster fulgens, dredged northwest of the Hebrides, has immensely long arms and a very small disc, not one-twelfth of the total diameter of the animal, which measures ten inches across. It closely resembles *Ophidiaster*, but has four rows of water-feet in the ambulacral grooves. *Zoroaster sigsbeei* and *Z. ackleyi* are two species dredged by the 'Blake' in or near the Caribbean Seas, in 1878–79. The last attains a diameter of about nine inches across the arms. Though the suckers are ranged in four rows at the base of the arms, there are but two rows at the extremity. It greatly resembles an *Ophidiaster* in general appearance.

The small size of the sucking-discs of the water-feet, and the general aspect of the animals, suggest that the genus should be placed with the Astropectinidæ, near to *Luidia.*

Caulaster is a singular star-fish, furnished with a peduncle in the centre of the disc, suggestive of the centro-dorsal tubercle of a comatula. It was taken by the French exploring expedition.

Class III. — ECHINOIDEA.

The Sea-Urchins are echinoderms without arms and without a stalk. They are protected externally by a calcareous test, composed of plates known as the coronal plates. These plates form ten distinct areas, five of which are perforated with pores for the exit of the suckers, and are known as the ambulacral areas, while the five intermediate areas bear no suckers, and are known as interambulacral. The variations in the form and size of these plates, and of the areas they compose, give rise to the varying shapes of the different genera. The mouth is always situated upon the lower or actinal aspect, which is applied in progression to the surface upon which the animal moves; but the position of the anus varies in different families. The surface of the plates, except where the pores are situated, is covered with spines, which occur upon every species of echinoid, but vary greatly in their number, structure, and size, so that they form one of the best characters by which species can be distinguished. There seems at first sight to be little resemblance between the huge bat-like spines of *Heterocentrotus*, the sharp, hollow, brittle spines of the Diadematidæ, the solid-fluted spines of the Echinidæ, and the slender, delicate spines, usually short, but sometimes long and silky, of the spatangoids or irregular sea-urchins, yet A. Agassiz assures us that in their early stages the young spines of Echinids are much alike. They are polygonal, made up of rectangular meshes placed in regular stories one above the other. There is no difference in the typical structure of the spine of the young of *Cidaris*, *Echinus*, *Strongylocentrotus*, *Arbacia*, *Echinocyamus*, or *Schizaster*. Some recent genera, especially the spatangoids, retain a type approaching that of all young echini, while among many older genera, as in *Cidaris* and other regular sea-urchins, complicated types occur.

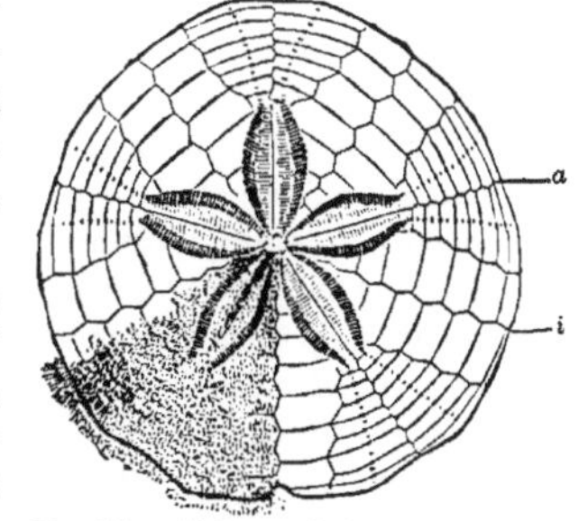

FIG. 140. — *Echinarachnius parma*; *a*, ambulacral; *i*, inter-ambulacral areas.

The coronal plates are more or less pentagonal, and are usually firmly united at their edges. Twenty principal longitudinal series, two in each ambulacral, and two in each interambulacral area, make up the mass of the test, and a series or rosette of ten single plates form a ring round the aboral or apical margin. The apical extremities of the ambulacra abut upon the five smaller of these plates, each of which is perforated, supports the eye-spot, and is called the ocular plate. The apical ends of the inter-ambulacra correspond to the five larger plates, which are perforated with a larger aperture for the escape of the generative products, and are called genital plates. One of these genital plates, lying in what is recognizable as the right anterior inter-ambulacrum, is larger than the others, and has a porous, convex surface. This is the madreporic body, and communicates with the water-system. Within the circle formed by the genital and ocular plates are a number of small plates, of which one, the anal, is larger than the others. The anus lies slightly out of the centre, between the anal plate and the posterior margin of the anal area. The space around the mouth (peristome) is usually strengthened for some distance by irregular oral plates; and ten rounded plates, supporting as many suckers, and perforated by their canals, are placed in pairs close to the lip.

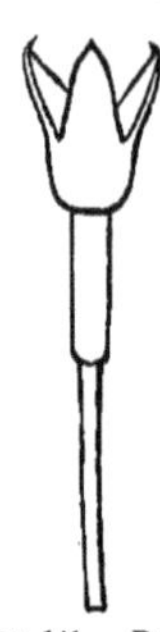
FIG. 141. – Pedicellaria of sea-urchin.

Each of the double series of coronal plates presents a zigzag suture in the middle line. Each ambulacral plate is subdivided by a greater or less series of sutures into a number of smaller plates, which are perforated by the pores of the suckers, and are called pore-plates. These are the primitive ambulacral plates, and in the Cidaridæ do not coalesce into larger ambulacral plates, but simply enlarge.

Scattered over the body, especially near the mouth, are the pronged pedicellariæ; and on most living echini, *Cidaris* excepted, small button-like bodies called sphæridia are found. These are situated upon a short stalk, and are thought to be possibly organs of taste. The spines or other appendages of the test are mounted upon tubercles, the size of which is proportioned to that of the spines, so that the empty and denuded test of a sea-urchin, covered with pores and tubercles, tells much respecting the affinities of its former habitant.

In a large portion of the class, the regular urchins and the cake-urchins, the mouth is furnished with five pyramids or jaws moved by powerful muscles, and in the regular sea-urchins each pyramid is composed of eight pieces, making a total of forty pieces. The digestive canal consists of a narrow gullet, a stomach of considerable length, passing from left to right around the interior of the body, and then turning up and curving back in the opposite direction; and of a terminal intestine. The stomach forms two series of loops, partly enclosing the ovaries, and is held in place by a broad, thin membrane, the mesentery. The sexes are distinct, and there are five ovaries or spermaries, opening outward by the openings in the genital plates. The single madreporic canal extends from the madreporic plate to the circular vessel around the mouth.

The majority of the Echinoidea undergo a metamorphosis, the early stages of which are similar to those of the star-fish. The embryo sea-urchin is called a pluteus, and is furnished with eight long arms supported by slender calcareous rods. These arms and rods are the locomotive apparatus of the young animal, which progresses by opening and closing them like an umbrella. The body is also provided with a curiously-curved band of vibratile cilia. Anything more unlike a sea-urchin than one of these plutei, with its complex sprawling array of arms and bilaterally symmetrical, but not radiated body, cannot well be imagined.

In *Strongylocentrotus dröbachiensis*, the common urchin of the Atlantic coast, the rudiments of the first tentacles appear in twenty-three days, by which time the pluteus has acquired its complete external form, but the shape of the larval digestive cavity is concealed by the growing sea-urchin within. The body of the pluteus is gradually absorbed, and the spines and suckers of the young sea-urchin increase in size and number. When the pluteus has finally disappeared, the young sea-urchin is more like the adult than is the young star-fish, but the plates of the apical region are not only more conspicuous in relation to the size of the test, but differ somewhat in their arrangement from those of the adult. The anus is at first wanting, and the anal plate is relatively large, is in the centre of the apical area, and is united by its edges with the five plates which, though imperforate in the young, become the genital plates in the adult. The other five plates that surround the apical system (the ocular plates) are also imperforate, and form a circle outside that formed by the genital plates, the spaces between them being occupied by interambulacral plates. In this stage the homology

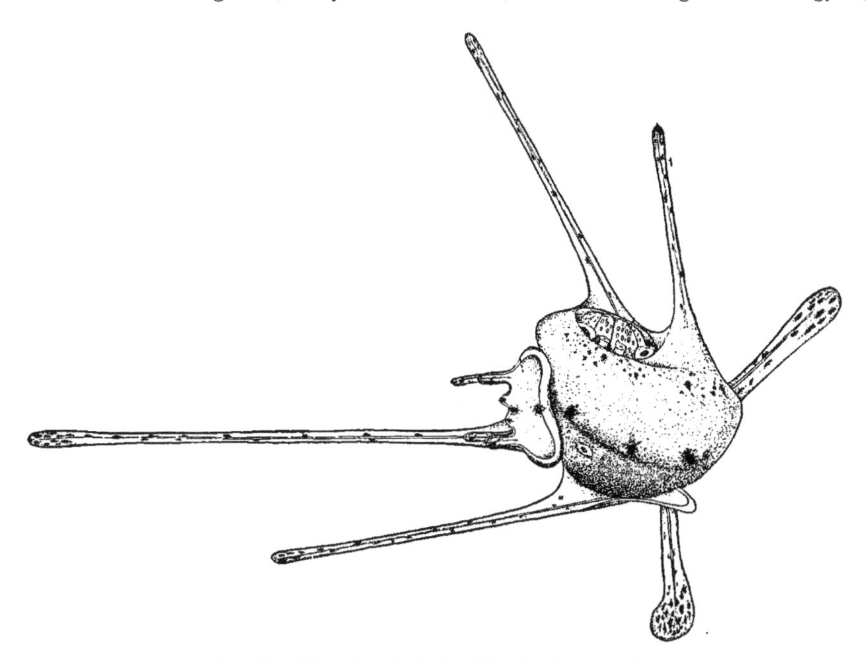

FIG. 142.—Old pluteus of *Arbacia*, with developing sea-urchin.

between the apical region of a sea-urchin, and the calyx of a crinoid is strongly brought out; the anal plate representing the basalia, the genital plates the parabasalia, and the ocular plates the first radials. The ambulacra may therefore be taken to represent the arms of a crinoid, and the interambulacral plates of the echinids are homologous with the interradial plates of the Crinoidea. The calcareous skeleton of the pluteus undergoes resorption, but the remainder of the larva passes into the growing sea-urchin.

The pluteus form is not universal among echinoids, since several forms from the southern hemisphere (*Hemiaster philippii*, *H. cavernosus*, *Anochanus sinensis*, *Cidaris nutrix*, etc.), develop directly into sea-urchins without, or with only traces of, a metamorphosis. Notwithstanding this great difference in the mode of development, the species which develop directly are very nearly related to species which live in the northern seas and pass through the pluteus stage. The 'Challenger' expedition did not find any free-swimming echinoderm larvæ in the southern ocean. In the species which develop directly, some of the plates and spines are modified so as to afford protection to the ova and embryos, which remain attached to the parent during the early stages of growth.

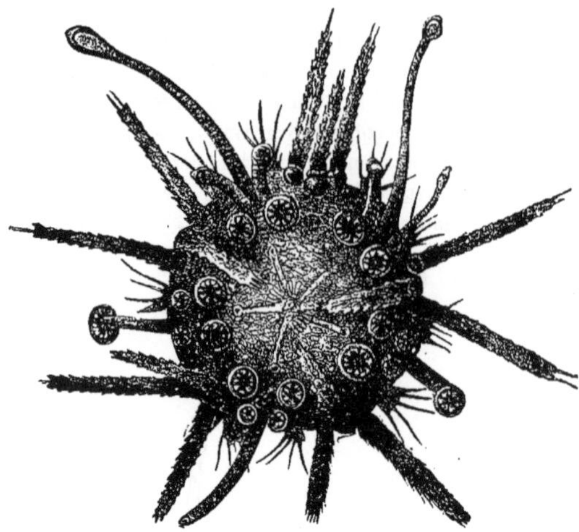

FIG. 143.—Young of *Strongylocentrotus.*

Most of the regular sea-urchins affect rocky coasts, and many of them, by what agency is not known, burrow into limestone rocks and coral reefs until they lie in a cavity which fits their bodies.

The total number of genera of Echinoidea known is not more than two hundred and twenty-five, represented by about two thousand fossil, and less than three hundred recent species.

Twenty-four genera of echini now living, including several spatangoid forms, were already existing at the time of the earliest tertiary formations, and some of these date back to the Jurassic beds, or even to the lias and trias. In tertiary times occur thirty-eight additional genera which have come down to the present time. The tertiary fossil echinids of the European beds are so similar to those now living in the West Indies, that it is really impossible to distinguish the species.

The southern ocean is the home of most of the deep sea or abyssal species, some fifty in all, and only one of these, *Pourtalesia phiale*, extends into Europe in deep water, though a comparatively large number of *Pourtalesiæ* and Echinothuridæ extend into the North Pacific. Twelve of the abyssal species extend beyond two thousand fathoms. Forty-six species may be called continental, occupying an intermediate position between the littoral species and the abyssal forms. Ten of these species extend to great depths.

The orders of the Echinoidea adopted by A. Agassiz in his report upon the results of the 'Challenger' expedition are the Palæoechinoidea (extinct), the *Desmosticha* or regular sea-urchins, the *Clypeastridæ* or cake-urchins, and the *Petalosticha* or irregular sea-urchins.

ORDER I.—DESMOSTICHA.

This order includes those sea-urchins which have a perfectly regular form, the ambulacra commencing at the aperture of the mouth and continuing around the test, which is more or less globose, until they reach the apical system in the centre of the upper aspect of the test. The mouth and anus are thus in this order always to be found upon opposite aspects, the ambulacra divide the circle of the test at five equal angles, and, except in a very few instances (in the Echinometridæ) there is no difference in length between the two equatorial diameters of the body.

All these regular sea-urchins, as they are commonly called, have a highly complex mouth apparatus. Each of the five pyramids consists of a hollow, wedge-shaped alveolus, composed of four pieces, or rather of two lateral halves, each formed of a superior and inferior portion; and of a long, slender tooth, shaped somewhat like the incisor of a rat or other rodent. The five alveoli, with their teeth, form a cone, and the parts are united together by strong transverse muscular fibres and also by long pieces applied radially to their upper edges. These radial rods are the rotulæ. To the inner end of each rotula is articulated a slender arcuated rod, with a free forked extremity. These are the radii. Thus the entire apparatus, usually called 'Aristotle's lantern,' consists of twenty principal parts, five teeth, five alveoli, five radii, and five rotulæ, and as each alveolus consists of four, and each radius of two parts, the total number of pieces is forty. When in position, the alveoli and the teeth face the interambulacra, the radii and rotulæ the ambulacra. For the attachment of this dentary apparatus to the test, the coronal plates of the margin of the mouth are produced into five perpendicular perforated processes, called the auricles. These usually arch over the ambulacra. From the auricles and interambulacral spaces at the margin of the mouth arise a complex system of muscles, protractor, retractor, and oblique, inserted into various parts of the mouth apparatus, at once attaching it firmly and regulating the movements of the parts.

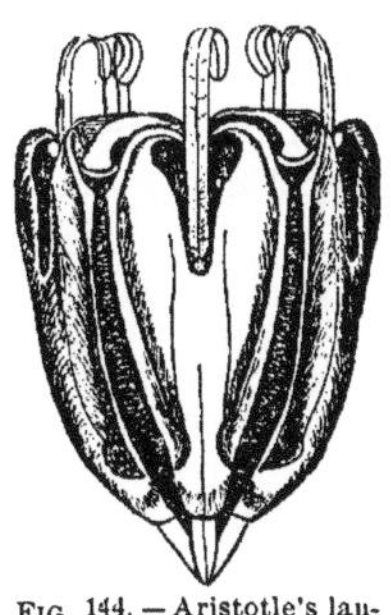

Fig. 144. — Aristotle's lantern of *Arbacia*.

There is great variation in the size, shape, and surface of the spines of the *Desmosticha*, but they are never so delicate and silky as in the other orders of echinoids. Usually certain large spines which form a continuous series from one end of an interambulacrum or ambulacrum to the other may be distinguished as primary spines, while the smaller spines forming less complete series are known as secondary or tertiary. The tubercles to which these spines are movably attached are in some cases marked by a central pit, into which, and into a corresponding pit on the head of the spine, a ligament of attachment is inserted.

The radial ambulacral vessels reach the ambulacra from the circular canal around the mouth by passing beneath the rotulæ and through the arches of the auricles. There are large ambulacral vesicles at the bases of the suckers, which are usually expanded into a sucking disc at their tips, where they are strengthened by a calcareous plate; but in some genera the pedicels of the apical part of the test are flattened, pectinated, and gill-like.

The family Cidaridæ has a large number of small plates in the ambulacral areas, and the pores are arranged in single pairs. The interambulacral regions are very wide, with only a small number of tubercles, each of which is large and perforated, and bears a massive solid spine. There are no secondary spines, but the entire surface of the test between the primary spines is filled in with small papillæ, which extend also over the oral membrane. The areas occupied by the oral and anal systems are larger than in other regular sea-urchins; the jaws are less complicated than in the Echinidæ and Diadematidæ, the teeth are gauge-shaped, and the auricles are inserted in the interambulacral instead of the ambulacral areas. Processes developed from the ambulacral plates form a sort of wall on each side of the ambulacral canal. The ambulacral plates are continued on the peristome to the margin of the mouth, where their edges overlap, producing a structure somewhat like that of the entire test of the

Echinothuridæ. *Cidaris metularia* occurs in the Pacific and East India oceans, *C. thouarsii* on the west coast of North America, and *C. tribuloides* on both coasts of the Atlantic. Fine specimens of the latter measure five inches across the spines. *Dorocidaris papillata* occurs at depths of from one hundred to six hundred fathoms, and has extremely long fluted spines, so that an example with a test about an inch across will measure eight or nine inches from tip to tip of spines. It occurs in the Mediterranean, and the Atlantic, while examples collected at the Philippines cannot be distinguished from it. *D. bracteata*, a Pacific species, has the flutes of the spines set with serrations.

In the species of *Phyllacanthus* the spines are often ornamented with frill-like lamellæ, but vary greatly in shape and decoration. *P. gigantea*, of the Sandwich Islands, has ten spines in a series, and six or eight lamellæ on each spine. *Goniocidaris canaliculata* is a species with elegant spines, tolerably abundant in the southern ocean. It has been dredged in sixteen hundred fathoms. In this species the upper part of the test is quite flat, and the two first series of spines, which are much larger than the spines of *Cidaris* usually are on that part of the test, lean over towards the anal opening, and form an open tent for the protection of the young. These spines are cylindrical and nearly smooth, the outer series longer and shorter than the inner. A somewhat similar arrangement obtains in *Cidaris nutrix*. Sometimes the young creep out, with the aid of their first few pairs of suckers, upon the long spines of the mother, and return to the marsupium.

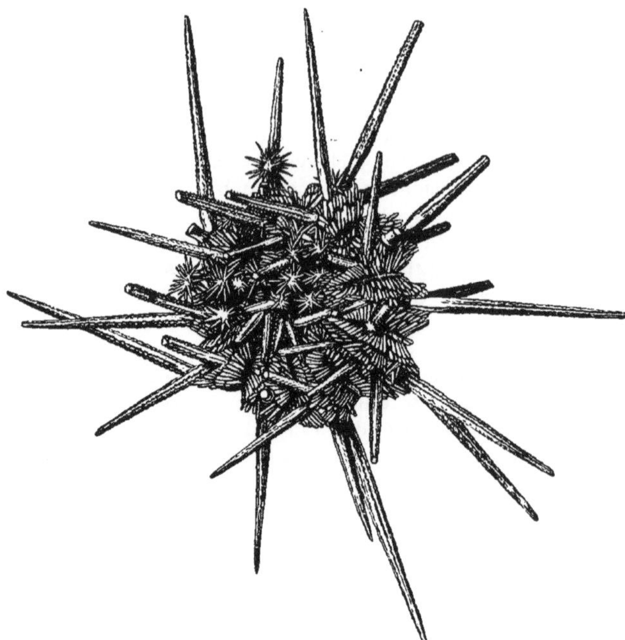

FIG. 145. — *Cidaris nutrix.*

Porocidaris purpurata has several rows of peculiar paddle-shaped spines round the mouth. These spines are flattened, longitudinally grooved, and serrated upon the edges.

Goniocidaris florigera is remarkable for the shape of the primary spines set around the anal area. These spines are dilated at the tip in such a manner as frequently to form a flattened cap, equalling in width one-third the diameter of the test. The oldest species of *Cidaris* occur in the Trias, and are small forms with smooth tubercles.

In the family ARBACIIDÆ the median interambulacral spaces show as so many bare bands, and the structure of the jaws, teeth, auricles, and spines, is intermediate between the Cidaridæ and Echinidæ. The species are few. *A. punctulata* occurs upon the eastern, and *A. nigra* upon the western coast of this country. In *Cœlopleurus* the spines of the primary tubercles are immense, three times as long as the diameter of the test, and taper gradually to a fine point. One species occurs on the coast of Florida.

The SALENIDÆ are a small tribe with spines like those of the Cidaridæ in structure; and with the anal and genital plates soldered together. *Salenia varispina* is quite common in the Caribbean Sea at depths from three hundred and fifty to one thousand six hundred and seventy-five fathoms. This species has a small purple body and long white serrated spines, and in appearance resembles *Dorocidaris*. The character which removes it into another family seems a very small one, yet is one in which it differs from all regular sea-urchins, except its own immediate relatives, which, so far as we know, commenced to live upon this earth in Jurassic times, and have continued through cretaceous and tertiary to the present day. Instead of having five ocular and five genital plates in its rosette, this little urchin has eleven, the additional one large, crescent-shaped, and occupying a central position. This plate thrusts the anus quite out of the centre of the rosette.

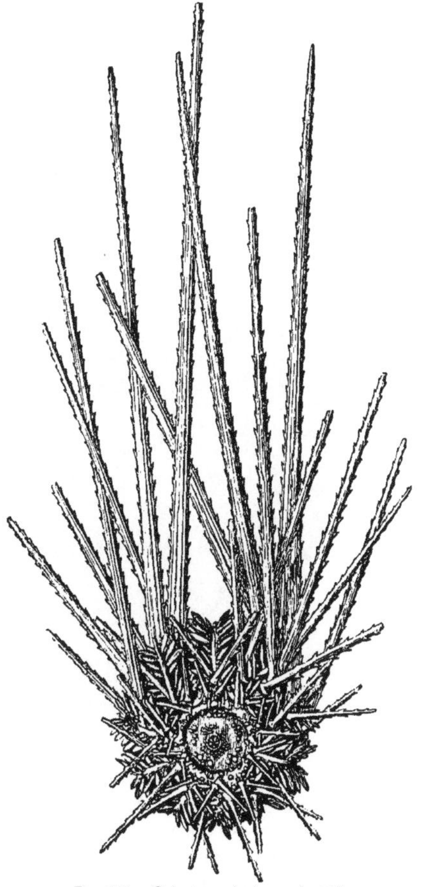

FIG. 146. — *Salenia varispina*, enlarged.

In the DIADEMATIDÆ the spines are hollow, long, and set with rings or verticillations. The test is thin, and the spines delicate, so that it is very difficult to preserve a specimen entire. *D. mexicanus* occurs on the west coast of Mexico, while *D. setosum* is found in both oceans. In *Echinothrix* the test is stouter than in *Diadema*, and there are many vertical rows of very small tubercles instead of the larger tubercles of uniform size which characterize *Diadema* and *Astropyga*. *E. desorii* of the Pacific and Red Sea attains a diameter of about five inches, while the spines do not exceed half the diameter of the test, and are often banded with greenish yellow.

In *Astropyga* the test is so thin as to be more or less flexible, and is greatly depressed, the height usually not exceeding one-third, or even one-fourth of the diameter. In life the colors are very bright, the ambulacral plates have pits or depressions of a brilliant sky-blue, and the spines of *A. pulvinata* are flesh-colored, with brownish, purple bands. The species named occurs on the west coast of Central America and Lower California.

The **Echinothuridæ** are a family distinguished, among other peculiarities, by the flexibility of the test. This flexibility results from the arrangement of the plates; which, instead of meeting and uniting at their edges, overlap, are separated by membrane, and are thus free to move. Prof. A. Agassiz points out that this character is well developed in ***Astropyga***, which may be considered as a connecting link between the Diadematidæ and Echinothuridæ. The latter also present many points of resemblance to the extinct Palæechinidæ. All the species have extremely depressed tests, resembling at first sight those of a cake-urchin or Clypeastroid, a group which they also simulate in the comparative shortness and small size of the spines. In structure they are, however, regular sea-urchins with the oral and anal systems on opposite sides of the test.

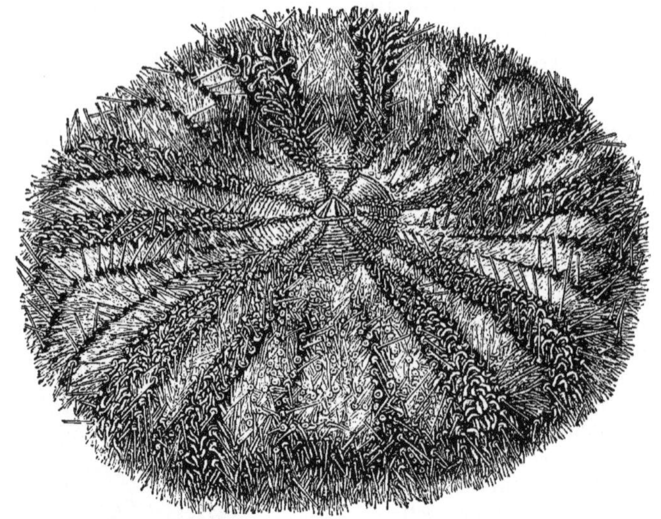

Fig. 147. — *Asthenosoma hystrix.*

The genus ***Asthenosoma*** contains six species, and occurs at various depths from ten to fourteen hundred fathoms. Although the test is so depressed in preserved examples, living specimens, even when brought up from the moderate depth of one hundred fathoms, are nearly globular, as if the test had been blown up like a foot ball. In ***Phormosoma***, of which seven species are known, there is in life a great contrast between the flattened oral side, and the high and globose anal aspect of the test. ***P. luculentum*** is perhaps the most striking species of the group. The test is of a beautiful light violet, forming a brilliant contrast to the white lines, indicating the sutures of the coronal plates, to the comparatively long, smooth, shining, primary spines, and to the silvery white, thick, hoof-like tips that terminate many of the primary spines on the oral surface of the test.

The structure of the plates upon the area round the mouth, which remains flexible in all echini, is in this family so similar to that of the plates of the test itself, as to suggest that they are primarily of similar nature.

Sir W. Thomson gives a graphic account of the surprise occasioned by the movements that passed through the test of ***Asthenosoma hystrix*** as it assumed upon the deck what seemed its usual form and attitude. The test moved and shrank from the touch when handled, and felt like a star-fish. It is quite dangerous to handle the species of this family when alive, as the wounds they make with their numerous sharp stinging spines produce a pain and numbness as unpleasant as that occasioned by the stinging of a Portuguese man-of-war. The tests of some species (as *P. tenue*) attain a diameter of six inches. This species has been taken in two thousand seven hundred and fifty fathoms. Judging from the large size of the eggs and of the genital openings, this group of sea-urchins is probably viviparous.

The largest family of regular sea-urchins is that of the ECHINIDÆ, which may be divided into the two groups of Echinometridæ, in which the pores of the ambulacral areas are arranged in arcs of several pairs of pores, and Echinidæ proper, which have arcs consisting of only three pairs of pores. This difference is greater than it seems, for the mode of growth of the bands of pores is quite unlike in the two groups.

Colobocentrotus atratus is covered by a pavement of closely packed hexagonal spines, completely concealing the surface of the test. Those at the edge of the test are rather longer and cylindrical or club-shaped. This species occurs at the Bonin Islands, adhering to the perpendicular faces of rocks exposed to the ocean swells. In this genus, as also in *Heterocentrotus* and *Echinometra*, the outline of the test, viewed from above, is elliptical. The two species of *Heterocentrotus* have immense club-shaped or angular spines, frequently twice as long as the transverse diameter of the test; certainly the most striking productions in the way of spines to be found in the entire class. These spines are apparently smooth, but actually finely striated. Those immediately round the mouth are flattened, while on the upper surface of the test the secondary spines sometimes form a close pavement.

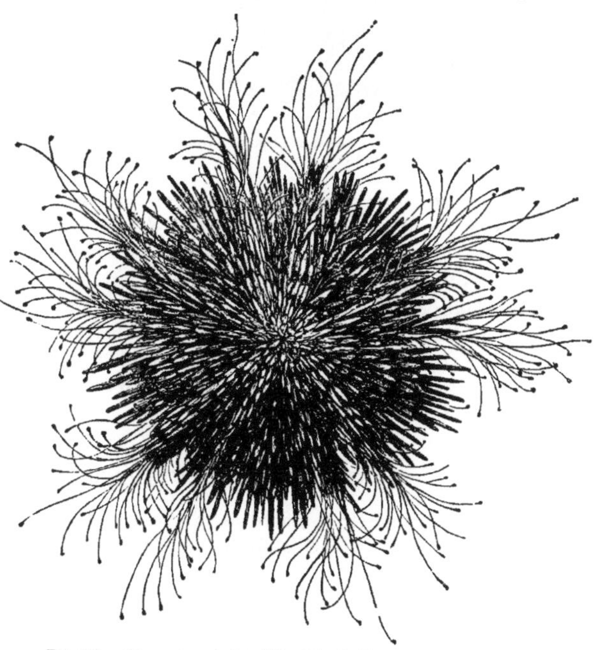

FIG. 148. — *Strongylocentrotus dröbachiensis*, New England sea-urchin.

The auricles are tall and slender, with a large opening. Both species inhabit the Pacific and Indian Oceans, spreading eastward as far as the Sandwich Islands.

H. mammilatus has ten to eleven pairs of pores in each arc, while ***H. trigonarius*** has fifteen to seventeen pairs. The spines of the former are usually stout and bat-shaped, and in color vary from uniform ash-gray or light brown, with white wings at the end, to nearly black. In *H. trigonarius* the spines are usually longer, tapering, and more or less triangular, but Agassiz states that they vary so much that the two species cannot be distinguished by the spines. When a spine of *H. mammilatus* is broken off at the base, it is replaced by a long tapering triangular spine like that of the other species.

The genus to which the cumbrous name of *Strongylocentrotus* has been given contains species with a circular or pentagonal, slightly depressed test, with pores arranged in arcs of at least four or five pairs. There are several species, of which the best known is the *S. dröbachiensis* of the north-eastern coast of North America, and of Alaska. *S. mexicanus* occurs in the Gulf of California, and the test reaches a diameter of nearly three inches; but these dimensions are far behind those of *S. franciscanus*, of the west coast of the United States. In this form the test alone is five or six inches across, and the spines are large, so that fine examples measure a foot. In the Echinidæ proper, a group which contains several genera and species, the test is often nearly globular, — in *Amblypneustes* the height equals the width. *Echinus esculentus* is one of the best known forms, and is found on the coasts of Norway and of England. The test is of a brownish or brick red color. As its name implies, it is occasionally used as food. *Hipponoë depressa* is a large species from the western coast of Mexico and the Gulf of California.

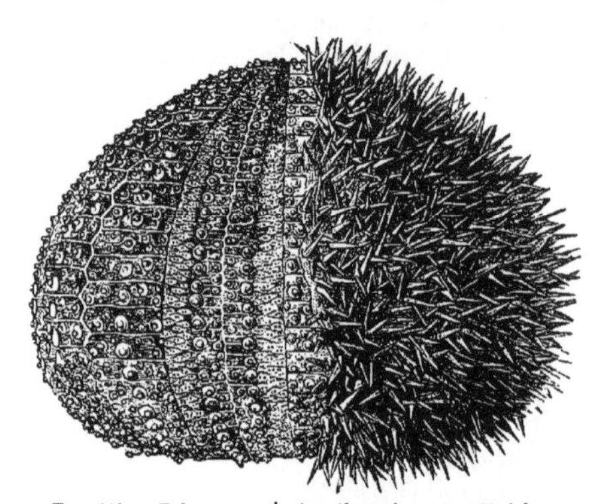

FIG. 149. — *Echinus esculentus*, the spines removed from half of the test.

Prionechinus sagittiger, a species found by the 'Challenger' expedition at depths varying from seven hundred to one thousand and seventy fathoms in the seas around Australia and the East Indian islands, is remarkable from the presence upon the spines of serrations resembling those of *Salenia varispina*, instead of the regular fluting characteristic of most Echinidæ.

ORDER II. — CLYPEASTRIDÆ.

In this order the mouth is placed as in the regular sea-urchins, but the anal opening occupies a position immediately opposite to the odd ambulacrum, and often on the under side. The genital pores retain their position at the summit of the upper surface of the test, which is exceedingly depressed, with its edge or ambitus more or less sharp, so that the upper and under surfaces are entirely cut off from each other, and are of quite different character. The rows of pores for the exit of the suckers do not extend around this sharp edge, but form five pairs of curves, arranged somewhat like the petals of a flower, upon the upper surface only; while on the oral surface the ambulacra are marked by furrows that converge toward the mouth. The pores upon the oral surface are either scattered widely over the ambulacral and sometimes over the inter-ambulacral plates, forming *pore areæ*, or they are arranged in bands which ramify over both the ambulacral and the interambulacral plates. The difference between the anterior and posterior extremity is well marked in this order, the former being known by the odd ambulacrum of the three ambulacra composing the trivium, the latter by the position of the anus between the posterior pair of ambulacra or bivium.

The jaws of the clypeastroids are much simpler than those of the regular echini, and articulate upon the auricles, instead of being held in place by muscles, as in the latter. They are V-shaped, and are placed horizontally. The teeth are secured in a groove corresponding to the line of junction of the arms of the V. The spines are in all cases delicate and short, often almost velvety in their fineness. The water system of the clypeastroids is without Polian vessels, but there are large vesicles at the bases of the suckers. The species mostly live upon sandy or muddy bottoms.

In the EUCLYPEASTRIDÆ the upper and lower floors of the test are connected and strengthened by pillars, needles, or radiating partitions of calcareous matter. *Echinocyamus pusillus* and a few other species are almost as globular as the Echinidæ, but are true Clypeastroids in structure, with simple partitions extending inwards from the circumference. In *Clypeaster* and *Echinanthus* the floors are connected by pillars, slender in the former genus, massive in the latter. The species are large, and, though flattened, have a rounded ambitus, while the oral surface is concave. *C. rotundus* occurs on the west coast of North America, as far north as San Diego, while *Echinanthus rosaceus* is tolerably common about the West India Islands and Florida. *Laganum* and its near relatives have the floors connected by walls that run parallel to the edge of the test.

In the SCUTELLIDÆ the test is extremely flat, and is usually more or less circular. The great quantity of calcareous matter forming the flattened edge is in many species lessened somewhat by the presence of cuts or openings in the ambulacral or interambulacral areas. The furrows of the under surface, which are straight in *Clypeaster*, are in this group more or less branching, and the upper and lower floors are supported by partitions that radiate from single points.

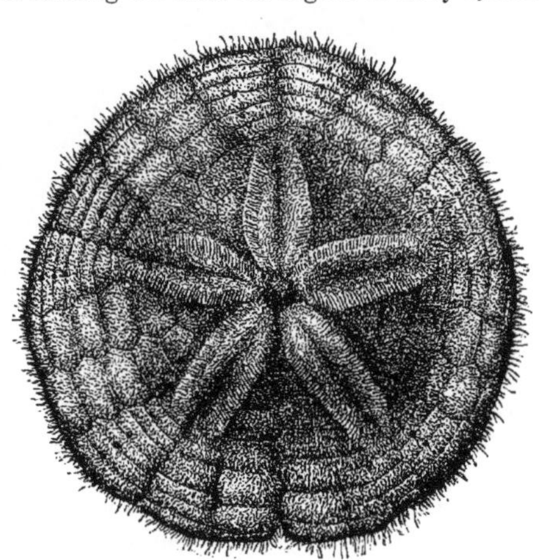

FIG. 150. — *Echinarachnius parma*, sand-dollar.

Some of the best known forms of Scutellidæ belong to the genus *Echinarachnius*, and are without cuts or lunules. *E. parma*, the Sand Dollar, is found on the Atlantic coast of the United States, and also on the Pacific coast as far south as Vancouver Island, and in Asia as far as Japan. *E. excentricus* is the common cake-urchin of the Pacific Coast, from Monterey northward, and occurs also in Kamtschatka. It is very common in San Francisco Bay, where it lives upon the sand at a depth of five to seven fathoms.

In *Mellita* the test becomes large and heavy, and the edges present deep cuts opposite to the ambulacra. In *Encope* the massiveness of the test increases, reaching its fullest development in *E. grandis*, a native of the west coast of Mexico and of the Gulf of California. It is hard to believe that the mass of calcareous material forming the test of this sea-urchin ever contained a living animal. The edges of the test are half as thick as the thickest part, which is at the anus. There is a huge lunule between the posterior ambulacra, beside five cuts opposite the ambulacra.

ORDER III. — PETALOSTICHA.

These sea-urchins, more commonly known as irregular sea-urchins or Spatangoids, have no dental apparatus; the test is variable in form, though usually more or less

FIG. 151. — *Perinopsis lyrifera.*

elliptical; and the anal system is placed between the two posterior ambulacra (bivium). Certain parts of the test and spines are greatly specialized; and the radiate form is accompanied with an evident bilaterality.

Neither the oral nor the anal apertures are in the centre of the test, the former being displaced anteriorly, and situated beneath the odd anterior ambulacrum, while the latter is situated beneath or between the petals of the bivium. The ambulacra in this order vary greatly, but are always petaloid in character upon the upper surface, the series of pores not being continuous around the edge of the test to the under surface. The anterior petal or ambulacrum often becomes more or less abortive, so that there are only four petals visible above; while in other cases it is much enlarged.

The spines of the Petalosticha vary greatly in size, not only in different species, but in the same individual. They are always delicate and silky, though in some species they may be of great length. When the test is cleaned, its surface is in many cases found to be marked by one or more symmetrical bands of close-set tubercles, so small that a powerful lens is needed to distinguish them. During life these tubercles bear slender spines, the heads of which are enlarged, while the shafts are set with cilia, and shaft and head alike are covered with a thick skin. These fascioles lie beneath or around the anus; they surround the outer extremities of the petaloid ambulacra, or, as in *Amphidotus*, they encircle the inner or apical terminations of the ambulacra.

Fascioles, as such, are recognized only among spatangoids, but it is probable, says Prof. Agassiz, that the accumulation of miliary tubercles on the edge of some *Phormosomas* must be regarded as the first trace of them. The earliest spatangoids, like the Dysasteridæ, have no fascioles.

Throughout all these changes of position of mouth and anus, the genital and ocular plates retain their central position; but in some genera the genital orifices are reduced to four.

The circular ambulacral vessel in this order has no Polian vesicles, and there are no vesicular appendages to the bases of the pedicels or suckers, of which there are four kinds. These are single locomotive pedicels without any sucking disc; locomotive pedicels containing a skeleton, and provided with terminal suckers; tactile pedicels, with papillose, expanded lips; and triangular, flattened, more or less comb-like lamellæ. Two or three of these kinds of feet may occur in any ambulacrum, and those which occur upon a fasciole are always different from the others.

In the Cassidulidæ there are no fascioles, and the form of the test is sub-globular, approaching that of the regular echini. It includes the sub-families Echineinæ and Nucleolinæ.

The mouth in this group is placed centrally, or near the centre. *Rhyncopygus pacificus* of the western coast of Mexico belongs here, as does also *Catopygus recens*, which was found at a depth of one hundred and twenty-nine fathoms south of New Guinea, and has an elevated test, the height about equal to the width, heart-shaped when looked at from behind, and pointed in front.

In the Spatangidæ, the principal family of the order, fascioles or bands of crowded miliary spines occur, and a plastron, or space without large spines, surrounds the oral opening, and is bordered by pores. This group is divided into several others, the Pourtalesiæ, Ananchytinæ, Spatangina, Leskiinæ, and Brissina, all distinguished mainly by peculiarities in the petals and fascioles.

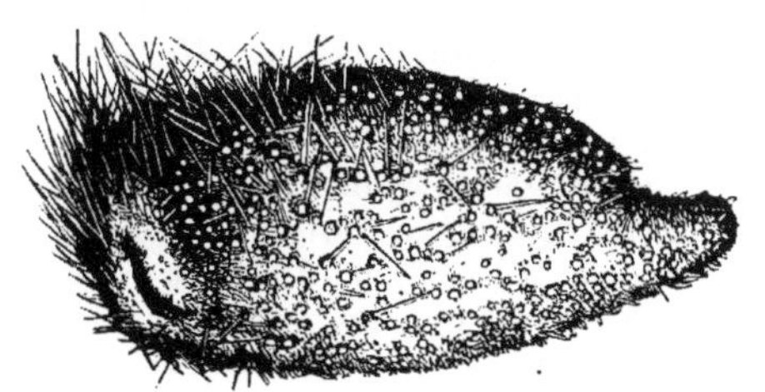
Fig. 152. — *Pourtalesia jeffreysii.*

In *Pourtalesia* and its allies the ambulacral system is simple, and the plates which compose it are large. The mouth, a large opening situated in a groove, is elliptical, and is covered by a membrane strengthened by an outer row of plates. The species of *Pourtalesia* have a curious snout, on the upper side of which the anal opening is situated. This snout gives to the test a most peculiar appearance when viewed from the side, while, viewed from above, some have a more or less bottle-shaped outline, and others are triangular. *P. miranda* has been dredged in the Florida Straits, and near the Shetland Islands.

P. carinata is a large species with a rather stout test and a bottle-shaped form. It is about four inches long, of a light claret color with whitish pink spines, and is a native of deep water, as it was dredged at from sixteen hundred to two thousand two hundred and twenty-five fathoms in the Antarctic Ocean. *P. ceratopyga* is remarkable for the great width of the anterior extremity and the narrowness of the anal snout, giving to the test a triangular shape. Viewed laterally and in the rear, the posterior projection has considerable resemblance to the head of a turtle. It occurs in the same localities with the preceding species, and, judging from fragments of some large specimens, must attain a length of about seven inches. *P. hispida* has a very short anal snout, and the shape, viewed from above, is that of a short, broad bottle. The transverse section is rounded, instead of obtusely triangular, as in the last species. *P. laganculа* and *P. phiale* are both bottle-shaped species, more or less circular in transverse section. The former occurs both north and south of the equator, from three hundred and forty-five to two thousand nine hundred fathoms, while the latter, which is a peculiarly slender and small species, was found in 62° 26′ south latitude, has an extremely thin test, and is of a light yellowish pink color. A peculiar form related to *Pourtalesia* is *Spatangocystis challengeri*, in which the anal snout is a small projection at the posterior end of a sharp keel which runs along the under side from the mouth backwards. In *Echinocrepis cuneata*, the general outline, whether viewed from above, laterally, or endwise, approaches a triangle, and there is no anal snout, the anal system appearing on the lower or actinal surface. The two last forms were both taken in deep water in the southern ocean. Other southern *Pourtalesiæ* without an anal snout are *Urechinus naresianus* and *Cystechinus*, of which two species, *vesica* and *wyvillii*, are known. *C. vesica* is the only spatangoid thus far known, which can evidently expand and contract its test. *Cystechinus* is elliptical in plan and irregularly triangular in profile.

Calymere has simple ambulacral pores; two of the ovaries are in the trivium, and the others are not developed. Its outline is elliptical, and there is a slight keel on the under surface. *C. relicta* has been taken at Tristan d'Acunha, and near Fayal, in two thousand six hundred and fifty fathoms. The central portion of the oral surface, and the apical surface near the posterior pole, has groups of paddle-shaped spines.

To the Ananchytinæ belong, besides many fossil forms, a few recent genera, among which *Homolampas* has very rudimentary petaloid ambulacra, a flattened test, and a well-developed sub-anal fasciole. *H. fulva* is about four inches long, of a light straw color, and heart-like in shape. Large curved spines are scattered at intervals among the short ones that cover the test. It was dredged in two thousand four hundred and twenty-five fathoms, by the 'Challenger' expedition.

To the Spatanginæ belong the typical *Spatangus*, species of which are common in Europe, and the nearly allied *Maretia* and *Lovenia*. *Lovenia cordiformis* is elongate, heart-shaped, flat upon the oral surface, and provided with numerous very long primary spines, half as long as the test. It occurs on the west coast of Mexico and in California as far north as Point Conception. Here also belongs the fine species *Breynia australasiæ*, which attains a length of four inches and a width of more than three, and is found in China, Australia, and Japan, and also *Echinocardium cordatum* of both coasts of the Atlantic.

In the sub-family Brissina, the genus *Hemiaster* is one of the most interesting, on account of the manner in which it carries its young, which develop without a metamorphosis, in receptacles on the apical surface.

H. philippii, a somewhat heart-shaped species found at Kerguelen Island, has certain of the ambulacral plates greatly expanded and depressed, so as to form four deep, thin-walled, oval cups which encroach upon the cavity of the test, and form marsupia or brood-pouches for the protection of the young. The spines are so arranged that a kind of covered way leads from the ovarial opening to the pouch, and in this passage the eggs are kept in place by the spines, two or three of which bend over each egg. In this way they are passed to the marsupium. The embryos stay within the pouch until the plates of the test have developed. Young echini in all stages of development are formed within the pouch, and the small ovaries contain some well-developed eggs ready to escape into it as soon as there is room. In the young the anal opening is nearly central, so that it looks almost like a regular urchin. *H. cavernosus* is regarded by Agassiz as identical with *H. phillippii*, but *H. zonatus* and *H. gibbosus* are distinct.

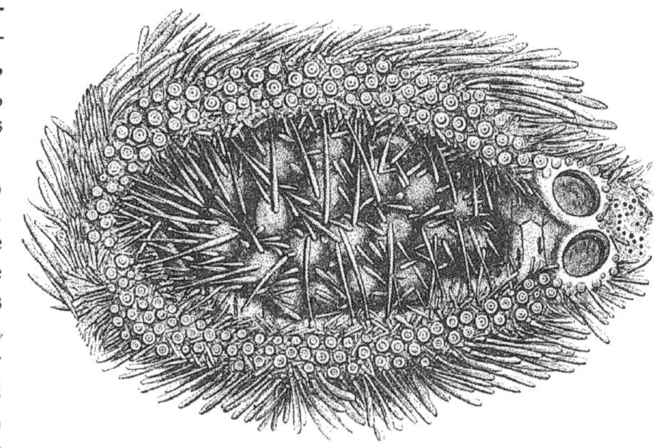

FIG. 153. — Brood pouch of *Hemiaster philippii* containing eggs, enlarged.

Aeropese rostrata is remarkable for its length and narrowness, for its deeply sunken, odd anterior ambulacrum, and for the eight great sucking discs upon the latter. *Aceste bellidifera* is near *Aerope*, yet is in appearance one of the strangest of sea-urchins, for nearly the whole of its upper surface is occupied by the deeply sunken odd anterior ambulacrum, surrounded by a narrow fasciole, from within which spring large, flattened, paddle-shaped spines. These spines curve over the great hollow of the ambulacrum, and underneath them may be seen a number of huge disc-

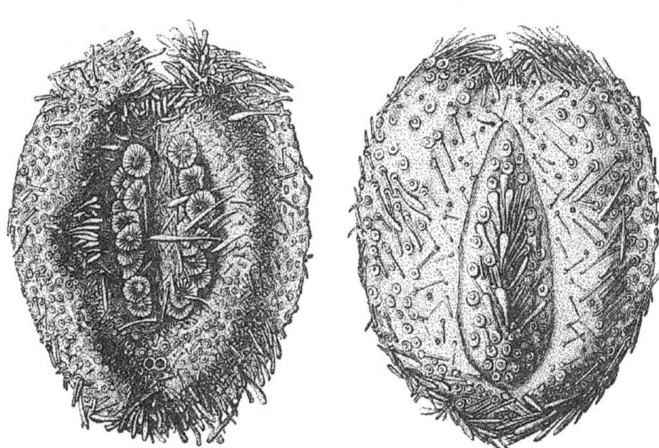

FIGS. 154 and 155. — Upper and under surfaces of *Aceste bellidifera*.

shaped suckers. The excretory opening is at the posterior extremity. Only two ovaries are developed, and the eggs appear to be very large at the time of expulsion, corresponding to the large size of the ovarial openings.

The genus *Schizaster* includes several species of almost circular outline when viewed from above, but with a deeply sunk anterior ambulacrum, and the other ambulacra depressed. In the young the odd anterior ambulacrum, as in *Aceste*, occupies the greater part of the upper surface, and the suckers of this ambulacrum are very large. Thus, *Aceste* may be regarded as a permanent form of the young of *Schizaster*. Other Brissina are the well-known large species *Brissus carinatus*, which is widely spread in the Pacific, and reaches a length of seven inches, and a width of nearly six; the pretty little egg-like and delicate *Agassizia scrobiculata* of the west coast of Mexico, and *Perinopsis lyrifera* of European seas.

CLASS IV.—HOLOTHUROIDEA.

The Holothurians or Sea Cucumbers are the least radiate and least typical of echinoderms, approaching the worms in the length and usually cylindrical form of the body (which is elongated in the direction of the axis), of the oral and aboral systems, and is without arms. The mouth is surrounded by a circle of branched tentacles, and the body-wall is muscular and leathery, instead of presenting a calcareous test or system of calcareous plates or ossicles, as is the case with other echinoderms. But though there is usually no continuous calcareous armature, the integument contains numerous calcareous bodies of varying form. The body-wall consists of an external skin, within which is a layer of connective tissue, and inside this a layer of muscular fibres, some of which are disposed in circles around the body, while others form five longitudinal bands. These bands are attached to a calcareous ring surrounding the mouth. The calcareous plates forming this ring are ten or twelve in number, and the longitudinal muscles are attached to five of these. These five plates are also notched for the passage of the nerves and ambulacral canals.

The water-system varies greatly in degree of development. In one section of the class (Apoda) there are no ambulacral feet or ambulacral canals, and the only trace of the water-system is to be found in the ring-canal round the gullet, with its greater or less number of Polian vesicles internally, and the circlet of comparatively simple tentacles externally, and in the madreporic or stone canal or canals, which run downwards into the body and terminate in a calcareous network, the homologue of the madreporic body of other echinoderms. In all the class, with two exceptions, the madreporic body is internal, and by its openings the water-system communicates, not with the external water, but with the large body-cavity that intervenes between the intestinal canal and the body-wall.

In the higher Holothuria or Pedata the circular vessel of the ambulacral system not only gives origin to the Polian vesicles, stone-canals, and tentacles, but to five ambulacral canals, which pass through holes or notches in the plates to which the longitudinal muscles are attached, and run backward along the median line of each of those muscles, immediately interior to the longitudinal nerve. In most cases each of these ambulacral vessels is furnished with ampullæ, connected with processes of the body-wall which form suckers or ambulacral feet, much as in a sea-urchin. In other genera some of the rows of suckers are suppressed, the other rows forming a surface upon which the animal creeps.

There is no dental apparatus in the holothurids, but a short pharynx leads into a stomach, and this into an intestine usually much longer than the body, and often several times its length. In the higher holothurids the intestine terminates in a distinct chamber or cloaca, often of large size. Into this cloaca open the stems of two (sometimes one) hollow, much-branched organs called the 'respiratory trees.' The sea-water enters into and is expelled out of these organs, which thus fill the body-cavity with water that is taken up by the madreporic body and carried into the water-system. The respiratory trees are also believed to be organs of excretion.

In some holothurians the cloaca or the respiratory tree is also provided with simple or branched appendages the interior of which is occupied by a solid or viscid substance. The use of these is not certainly known, but, as they are readily thrown out when the animal is disturbed, Semper supposes that they are organs of defence. These are known as the Cuvierian organs.

The nervous system is of the usual echinoderm pattern, consisting of a mouth-ring placed above the ring-canal of the water-system, and of five principal ambulacral cords passing through notches in the plates around the œsophagus. The system which is supposed to be analogous to the circulatory system of higher animals is very complex in many of the higher holothurids, extends over the alimentary canal, and enmeshes one of the respiratory trees.

The genital organ is in many cases single, and in the Synaptidæ contains both ova and spermatozoa, so that these forms are

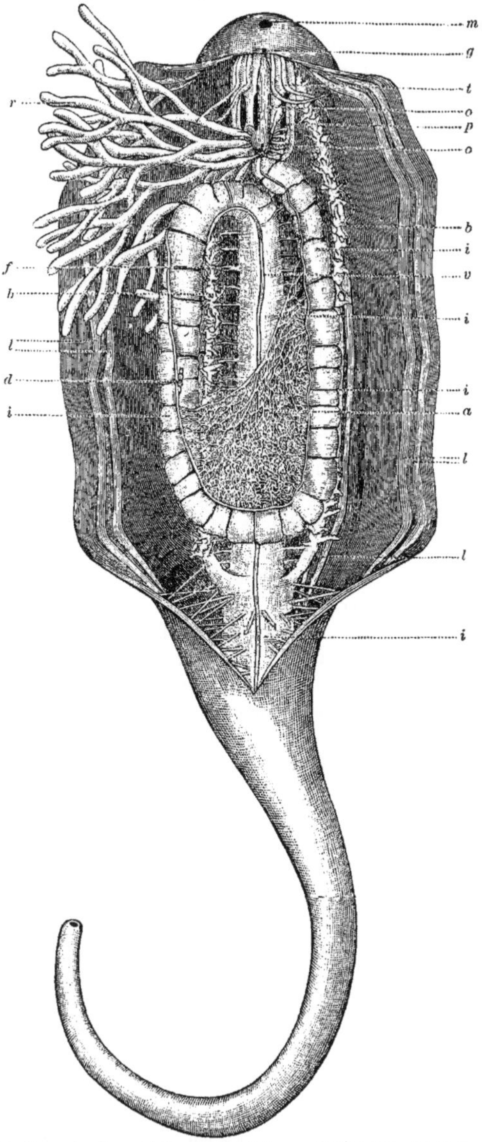

FIG. 156.—Anatomy of *Caudina arenata*; *a*, anastomoses of dorsal blood-vessels; *b*, branchial tree; *d*, dorsal blood-vessel; *f*, mesenterial filaments; *g*, genital opening; *i*, alimentary canal; *l*, longitudinal muscles; *m*, mouth; *o*, genital duct; *p*, pharyngeal ring; *r*, reproductive organs, cut away on right side; *t*, tentacular ampullæ; *v*, ventral blood-vessel.

hermaphrodite. In the majority of the class the sexes are distinct. The oviduct opens near the mouth. The ovum, after segmentation, becomes, by the invagination or turning inwards of a part of the external surface, converted into a hollow gastrula, the opening of which becomes the anus, while a mouth and gullet are produced by the invagination of the outer layer of tissue or ectoderm. The completed alimentary canal of the larva consists of a gullet, a rounded stomach, and an intestine, and the cilia of the external surface become restricted to a number of hoops or bands, from one to five in number, bent upon themselves, yet passing all round the body. These are accompanied by certain ear-like projections, from which the young is known as an Auricularia. Before the auricularia is fully formed the young holothurian buds out near the side of the stomach, and gradually develops its spicules. The ear-like processes disappear, the auricularia becomes cylindrical, the body of the embryo elongates, tentacles are developed around the mouth, and the young holothurian is complete. The entire course of development is largely parallel to that of the other echinoderms, but the holothurian is more directly developed from the larva than is the case in sea-urchins and star-fish. Some holothurians, like some star-fishes and sea-urchins, have the larval stages suppressed or only slightly indicated, the young developing in a sort of marsupium.

The holothurians are of little economic value, and with us are regarded merely as objects of scientific interest. In the East they are more important, and as 'trepang' play a prominent part in the diet of the Chinese and other oriental peoples. The trade of preparing the trepang is almost entirely in the hands of the Malays, and every year large fleets set sail from Macassar and the Philippines to the south seas to catch the 'Bêche-de-Mer.' They are split open, boiled, dried in the sun, and then smoked and packed in bags. The annual catch is estimated at about four hundred tons, and the price varies according to quality from seven to fifty cents a pound. Trepang is very gelatinous, and is used as an ingredient in soups.

Although the holothurians must evidently be classed with the Echinoderms, their simplest forms, as *Eupyrgus*, present nothing of the radiate arrangement except the circle of tentacles; and their affinities to such worms as *Sipunculus* and its allies, which have also a complete circle of tentacles, a ring-canal, and a kind of water-system, seem in many respects close.

The Holothuroidea have been said to feed on living coral, but this seems disproved by recent observations. The manner in which they feed is well illustrated by the following account of the habits of a species of *Cucumaria*, common upon the coast of Cornwall. When in full feed the tentacles were observed to be in constant motion, each separate tree-like plume, after a brief extension, being inverted and thrust bodily nearly to its base in the cavity of the pharynx, bearing with it such fragments of sand and shelly matter as it had succeeded in grasping. No particular order was followed, but the meal continued for hours. One might imagine a child with ten arms, like an ancient Buddha, grasping its food with every hand, and thrusting hand and arm down the gullet with each handful. These animals were kept in a tank with living corals without in any way interfering with them. The nutriment must be furnished by the Infusoria, diatoms, and other microscopic animals and vegetables which always more or less cover the débris at the bottom of the water. Probably the shell or coral débris is triturated by the teeth of the pharynx. It has been calculated that fifteen or sixteen of these creatures will remove about eighteen cubic feet of coral per annum.

ORDER I.—ELASIPODA.

The Elasmapoda, or Elasipoda, are true deep-water forms, none of which are known to exist at a less depth than fifty-eight fathoms, at which level *Elpidia glacialis* has been taken in the Arctic Ocean. The same species has been dredged in warmer seas in two thousand six hundred fathoms. The first example of this group was discovered in the Kara Sea over seven years ago, but from the results of the recent deep-sea exploring expeditions over fifty species are now known.

In the Elasmapoda the adult and larval forms agree more closely than in other holothuroids, and they are therefore placed by some naturalists low in the scale, while others place them high on account of the distinct bi-lateral symmetry of their bodies, the well-marked distinction of the dorsal from the ventral surface, and the frequent specialization of a cephalic or 'head' portion. The ventral ambulacræ alone are fitted for locomotion.

Kolga hyalina is a small example of the Elasmapoda, with the oral disc facing the ventral surface, and the anal orifice facing the dorsal surface, thus producing an arrangement similar to that which prevails in the star-fishes and echini. The anal orifice has a dorsal collar, bearing sucker-like contractile papillæ, communicating with the body-cavity instead of with the water-system. The sand canal (stone canal), instead of hanging free within, opens on the exterior in front of the dorsal collar. This is simply a persistence of the condition of things found in the larvæ. Two other genera, *Trochostoma* and *Irpa*, have the outer end of the sand canal attached to the skin, with a madreporic plate at the point of attachment, but the canal does not communicate with the exterior. In *Elpidia* the arrangement is similar, but the madreporic plate is rudimentary or wanting. *Kolga* is diœcious, but has no respiratory tree. The two dorsal nerve-trunks furnish an offshoot to each of a pair of large vesicles containing otoliths—a rudimentary organ of hearing.

ORDER II.—APODA.

In this order there are no ambulacral feet, and the water-system is therefore restricted to the ring around the gullet, the circlet of tentacles around the mouth, and the canal communicating with the madreporic body. The Apoda are again divided into the Apneumonia, which are destitute of a respiratory tree, have no proper cloaca, and are without Cuvierian organs, and the Pneumophora.

The simplest of the footless holothurians without breathing organs, and, indeed, the simplest of all known holothurians, is *Eupyrgus scaber*, a species less than half an inch in length, provided with a circle of fifteen unbranched tentacles round the mouth, and covered with soft papillæ bearing calcareous plates. The longitudinal muscles are weak and small. It is an Arctic species, and has been taken on the coasts of Labrador, Greenland, and Norway.

FIG. 157.—'Wheel' from skin of *Myriotrochus rinkii.*

Myriotrochus rinkii has also been found in shoal water on the coast of Labrador, and has a transparent skin dotted with minute white spots. These spots, when magnified, are seen to be wheel-like calcareous plates.

Another very worm-like species, common in Labrador, is *Chirodota læve.* It is whitish-gray, with wheel-like plates, like the last species, showing as white spots.

The two last mentioned species belong to the family SYNAPTIDÆ, in which the sexes are united in the same individual, the tentacles are finger-like, or lobulated, and the form elongated and worm-like.

In *Synapta* the integument contains numerous perforated, flat, calcareous plates, to which are attached protruding anchor-like hooks. *Myriotrochus* has a single Polian vesicle, but *Synapta* has several. *Synapta girardii* is a common species upon the Atlantic sea-coast, living in sand at low tide level. Its exceedingly attenuated body breaks up into several pieces when the animal is disturbed. *Synapta similis* is said to live in brackish water. *Synaptula vivipara* is one of those species in which the young develop directly, and are protected in a marsupium.

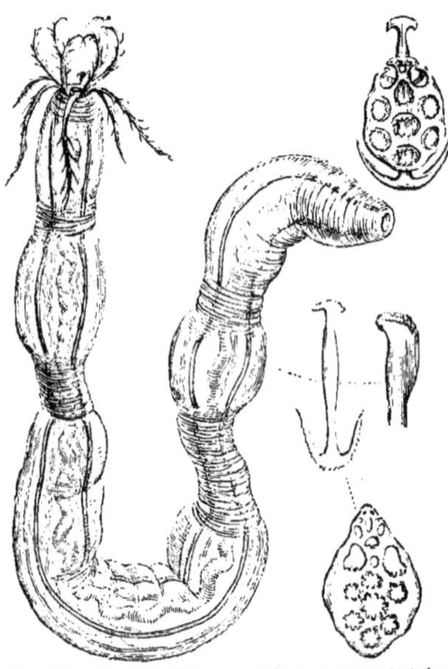

FIG. 158. — *Synapta inhærens*, with anchors and plates.

Oligotrochus vitreus is a species living at depths of from fifty to two hundred fathoms on the coast of Norway, less vermiform than *Synapta*, since its length is only about four times its thickness, and sluggish in its motions on account of the slight contractility and stiffness of its body, which is transparent and colorless. The tentacles are twelve in number, and are less developed than even in *Chirodota*. The calcareous plates, which are wheel-like, are very few, and deeply sunk in the skin. The calcareous ring round the gullet shines snow-white through the body-wall, which also allows the intestines, generative organs, etc., to be clearly seen.

The next sub-order (Pneumophora) has a respiratory tree arising from the cloaca. *Caudina* is so called from the long, tail-like prolongation in which the body ends. The skin is rough externally, from the calcareous pieces imbedded in it. *C. arenata*, the only species, has fifteen four-pronged tentacles, and is a somewhat common object in Massachusetts Bay.

Molpadia turgida, is a deep-water form that has been taken in over a hundred fathoms in the Gulf of Maine, and is known to range southward to Florida. It is a tailed species with fifteen tentacles and an anterior extremity shaped like the neck of a bottle.

Most of the Apoda are natives of the cold seas of the Arctic regions, but some genera, as *Synapta* and *Chirodota*, are almost cosmopolitan.

ORDER III. — PEDATA.

The Pedata, as their name implies, are always furnished with a greater or less number of ambulacral feet; a respiratory tree is always present, and the sexes are distinct.

In the Dendrochirotæ the tentacles are tree-like or branching, and there are no Cuvierian organs. The pharynx has retractor muscles. Many of the species inhabit northern regions. *Pentacta frondosa*, the common sea-cucumber north of Cape Cod, and extending through the Arctic regions to Great Britain, is from six inches to a foot in length, of a tan brown color, and suggests a cucumber by its shape and the corrugations of its skin. The pharynx is muscular; the stomach short, with well-marked transverse folds within; the intestine several times longer than the body, and the respiratory tree with but one main stem. Connected with the ring-canal are two Polian vesicles, nearly two-thirds as long as the body. The single ovary is made up of a mass of long tubes, which are larger than the branches of the respiratory tree, and are tangled up with them. The genus *Cucumaria*, or *Pentacta*, has many suckers in the ambulacral area, and some forms have them also in the interambulacral areas. The body is short and somewhat five-angled, and there are ten tentacles, of which the two belonging to the odd ambulacrum of the bivium are often smaller than the others.

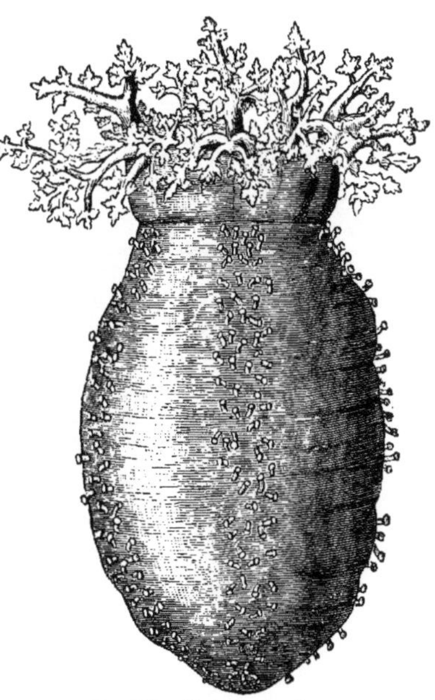

Fig. 159. — *Pentacta frondosa.*

In *Colochirus* the feet are ranged in three rows upon the ventral surface, the remaining ambulacra of the back having only papillæ. The two central tentacles of the lower surface are smaller than the others. In *C. cœruleus*, from the Philippines, the papillæ of the back are very long.

In *Psolus* the ventral surface is clearly separated from the rest, and there are no suckers upon the upper surface. The surface of the body is covered with comparatively large plates, of which *Psolus complanatus* has fourteen to sixteen, in a transverse row. This species and another occurs in the Philippines, *P. squamatus* is found at the Kurile Islands, *P. phantopus* and *P. fabricii* in the North Atlantic, and *P. antarcticus* in the Straits of Magellan.

The female of *Psolus ephippifer* has upon the upper or dorsal surface a group of larger tessellated plates, each carried, like the head of a mushroom, upon a pedicel imbedded in the skin. The spaces left underneath these tiny vaults are utilized for the protection of the young, which develop directly into sea-cucumbers. The males have the plates of the back similarly arranged, though there is no marsupium.

In *Thyone*, *Thyonidium*, and other related genera, the suckers of the entire body are alike, and seldom show traces of arrangement in rows. *Thyonidium* has five pairs of large and five of small tentacles.

Thyone briareus lives just below low tide from Long Island Sound to Florida, and is very common. In a specimen little more than three inches long the alimentary canal is about seven feet in length, though the oval stomach is less than an inch. The tentacles can be deeply retracted. There are three Polian vesicles, the respiratory trees divide at once into two very bushy branches, and the Cuvierian tubes form a brush-like tuft about an inch in length.

FIG. 160.—*Psolus fabricii*, showing under surface with three rows of ambulacra.

Cladodactyla crocea was found adhering to the huge tangle in the southern seas, and is abundant at the Falkland Isles. It is of a bright saffron color. The three anterior ambulacral vessels are near together, and bear numerous well-developed sucking-feet for locomotion, while the two ambulacra of the bivium are also near together on the back. In the females these latter ambulacra have very short, tentacular feet, which, though provided with sucking discs, seem, from the rudimentary condition of the rosette of calcareous plates at their tip, scarcely fitted for locomotion. In the males there is rather less difference between the dorsal and ventral ambulacra. In the females the young were found closely packed in two continuous fringes adhering to the water-feet of the dorsal ambulacra. "Some of the mothers with older families," says Sir Wyville Thomson, "had a most grotesque appearance,—their bodies entirely hidden by the couple of rows, of a dozen or so each, of yellow vesicles like ripe, yellow plums ranged along their backs, each surmounted by its expanded crown of oral tentacles."

In the Aspidochirotæ, or holothurians with disc or shield-shaped tentacles furnished with tentacular ampullæ, the left respiratory tree is bound to the body-walls, there are no retractor muscles to the pharynx, and Cuvierian organs are present. These are the highest type of Holothuroidea, and are mainly tropical in their distribution.

In *Stichopus* the tentacles are eighteen or twenty in number, the body is more or less quadrilateral in section, and the ambulacral feet project from papillæ. Three distinct rows of these can usually be traced upon the flat ventral surface. *S. variegatus*, which has been found at the Samoan Islands, and at the Philippines, attains an enormous size. Examples three feet or more in length, and eight inches thick, have been taken.

Mulleria has five calcareous plates or teeth at the anal extremity.

In the typical genus *Holothuria* the feet are scattered all over the surface of the body, usually without distinction into rows. Some of the species attain large dimensions. *H. marmorata*, from the East Indian islands, Fijis, etc., reaches a foot in length, and *H. tenuissima* attains a length of two feet, and a thickness of six or seven inches.

In some species of *Holothuria*, as in *H. marmorata*, the ambulacral processes of the lower surface only are truly ambulacral feet, the others are papillæ: in another group, including *H. tenuissima*, the suckers or papillæ are all alike, and in still another the ambulacral feet upon the ventral surface are much closer together than those upon the back.

Holothuria floridana is abundant on the Florida reefs just below low-water mark, and reaches a length of fifteen inches. The calcareous pharynx leads to an alimentary canal which is about three times the length of the body, and ends in a large cloaca. The branch of the respiratory tree which is attached to the body-walls extends to the

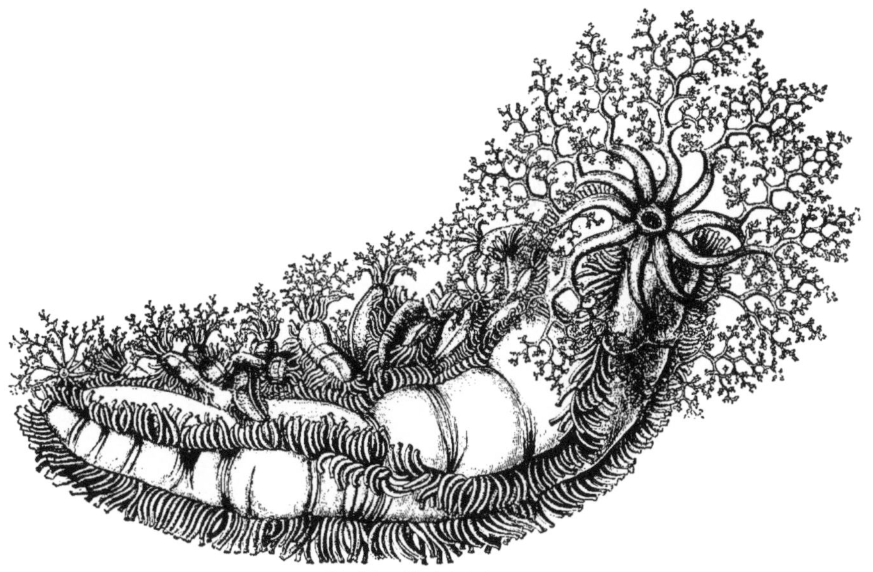

FIG. 161. — *Cladodactyla crocea.*

pharynx. The Polian vesicles are numerous, the largest an inch in length, and the madreporic body has upon it a group of about thirty stalked processes, the largest about a quarter of an inch in length. The tentacular ampullæ are twenty in number, long and slender.

ORDER IV. — DIPLOSTOMIDEA.

This order, or sub-class, established by Semper to contain the singular *Rhopalodina lageniformis*, is characterized by a nearly spherical body, with the mouth and anus close together, and ten ambulacra. Semper regards it as the type of a fifth class of echinoderms.

Rhopalodina lageniformis has a flask-shaped body, and the mouth and anus are at the narrow end of the flask, the former surrounded by ten tentacles, the latter by ten papillæ and by as many calcareous plates. A ring of ten calcareous plates surrounds the gullet, and between the latter and the cloaca the genital disc is situated. The ten ambulacra diverge from the centre of the enlarged or aboral end of the body, and extend, like so many meridians, to near the neck of the flask. Each ambulacrum has its own longitudinal muscular band, five of which are attached to the circle around the anus, and five to that around the mouth. The species is found upon the Congo coast.

W. N. Lockington.

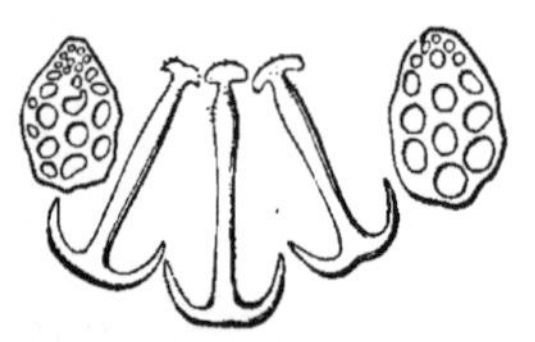

Fig. 162. — Anchors and plates of *Synapta girardii*.

BRANCH V. — VERMES.

THE great and varied assemblage of animals which are put together under the common designation of worms does not present a homogeneous group for study. On the contrary many distinct types have been thrown together to make the branch of worms. Indeed it has been a long-standing current joke among zoologists that this part of the zoological system was the garret, or as the German has it the Rumpelkammer, into which everything was carelessly thrown that did not properly belong elsewhere, and had been therefore rejected from the other portions of the system of classification. The worms have thus come to be a collection of forms whose outside affinities extend to nearly all other animals, while among themselves they fall into classes not closely related with one another. These classes shade off in some cases towards other branches; thus the rotifers approach in their organization the molluscan type, while the Annelida proper show in some respects unmistakable similarity with the insects. Other classes, like the Acanthocephali and Enteropneusti (*Balanoglossus*) attain an anatomical configuration which gives them a certain independence, a place apart, in the zoological system. In brief, as the limits of the branch of worms are vague, and its components multifarious, therefore it is difficult to define the worms with an accuracy corresponding to the requirements of a rigorous science. The following definition is the most satisfactory I am able to give: —

A worm is a bilaterally symmetrical animal, with a distinct head characterized by the presence of the principal nervous centre or so-called brain. It is distinguished from molluscs by the absence of a shell and of that modification of the skin, named the shell gland, which forms the shell and is present at least in a rudimentary condition in all true molluscs. It is distinguished from Crustacea and insects by the want of jointed limbs, and finally from the tunicates and vertebrates by the lack of a structural axis, the so-called notochord or chorda dorsalis, which gives the name of Chordata to the divisions last mentioned. As far as at present known no worm has a true liver, a calcified internal skeleton, an organ homologous with the endostyle of ascidians and thyroid gland of vertebrates, any tracheal tubes like those performing respiration in insects, or finally any unicellular hairs. In fact a worm must be recognized as such rather by the process of exclusion than by the observation of positive characteristics.

To the scientific zoologist the worms are most interesting subjects of study, not only from their manifold variety and strange life histories, but also from their relationship with the higher types, the ancestral forms of which are with good reason supposed to be more nearly represented by certain worms than by any other animals now existent. The mind links in imagination these obscure and humble creatures with the most exalted organisms, and finds in the secrets of their low organization the key to the complex structure of the higher animals. We, however, shall not enter upon these difficult discussions, where debate is still active, and the final decision uncertain. Instead we shall be sufficiently occupied with studying the principal and most interesting forms of vermian life as to their appearances and habits. Although a worm is by popular fancy a loathsome thing, yet only some of them deserve opprobrium, while many others are objects of great beauty, and others again are quaint; a few are of great utility to man, and yet others are among man's most dreaded enemies.

All worms appear to require moisture, and the majority of them are aquatic, inhabiting ponds and rivers and peopling the sea. The adults frequently exhibit a marked preference for a living burial and inhume themselves in sand and mud, some at the bottom of stagnant waters or running streams, others in the floor of the ocean at all depths; but they are found in the greatest variety and number on sandy beaches, which the changing tides alternately cover and expose. Under stones or sunken in the ooze the collector gathers them in astonishing abundance. In the moist earth, especially in vegetable humus, and in manured soil live the common earth worms. Some species, like those of the genus *Sagitta*, are called pelagic, for they swim about upon the ocean surface in company with the embryos and larvæ of worms of many kinds and a marvellous society of other living things. Rotifers and others swim in fresh water as well. Finally is to be mentioned the parasitic life adopted by a large number of the members of the group; in the infested hosts they find all the necessary conditions for their existence. Such parasites are more common in the intestine than in any other organ, but they attack every part of the body. In the following pages the habitats are considered with no little detail, so that we need not occupy ourselves longer with the general subject.

The worms fall naturally into a number of distinct classes, some of which comprise but a single genus, while others are large groups and present a multitude of forms. The *Echinorhynchus* is so isolated among living worms that usually it is placed from anatomical reasons by itself; the jointed worms or annelids, on the contrary, have more representations in the earth's present fauna than the majority of classes among animals. Yet these two classes are regarded by most zoologists as peers, notwithstanding the disparity of numbers between them, because the anatomy of the *Echinorhynchus* entitles it to as distinct a rank as is given to the annelids collectively. The reader therefore must not wonder at the inequality in size of the co-ordinate divisions of worms.

The classification here adopted is that which appears to me to best accord with our present knowledge of the animals concerned. All the forms are bound together by a hypothetical link, the Trochozoon, which is also the starting point of the Mollusca and all bilaterally symmetrical animals. This Trochozoon must have been similar in organization to those little creatures, the wheel animalcules or Rotifera, and in the course of their metamorphoses the young of many worms and annelids pass through what is known as the Trochozoon or Trochophora stage, so that the life history of the individual proves that the adult is derived by modification of the Trochozoon type; hence the induction is probable that the ancestors of these animals were Trochozoa. This necessitates placing the hypothetical animal at the base of the system. It is, however, still questionable whether the low types known as the Plathelminthia, that is the tape-worms and their allies, are not derived from something still simpler than the Trochozoon; the doubt as to the affinities of this class renders it desirable to isolate it somewhat, as is done in the adjoining diagram. The rotifers come very near the ancestral forms. Next we place a very simply organized small group, the Gastrotricha, by way of which we pass off from the main line of progress to the nematodes, to *Gordius* and *Mermis*. The parasitic *Echinorhynchi* are usually associated with the nematodes rather than with any other group by systematists, but their true affinities are by no means definitely settled. Keeping on we come to the *Sagitta*, and the cognate *Chætosoma* and *Desmoscolex*. Returning now to the main line we approach the annelidan type. Here we must put first the nemertean or proboscis worms, a group which, entirely without justification, was formerly included with the plathelminths. Now intervene the Gephyreans. Finally we reach the annelidan type, beginning with the Carchiannelides and leading off to the Chætopods and *Myzostoma*, the Echiurids and *Balanoglossus*, and to the Discophori.

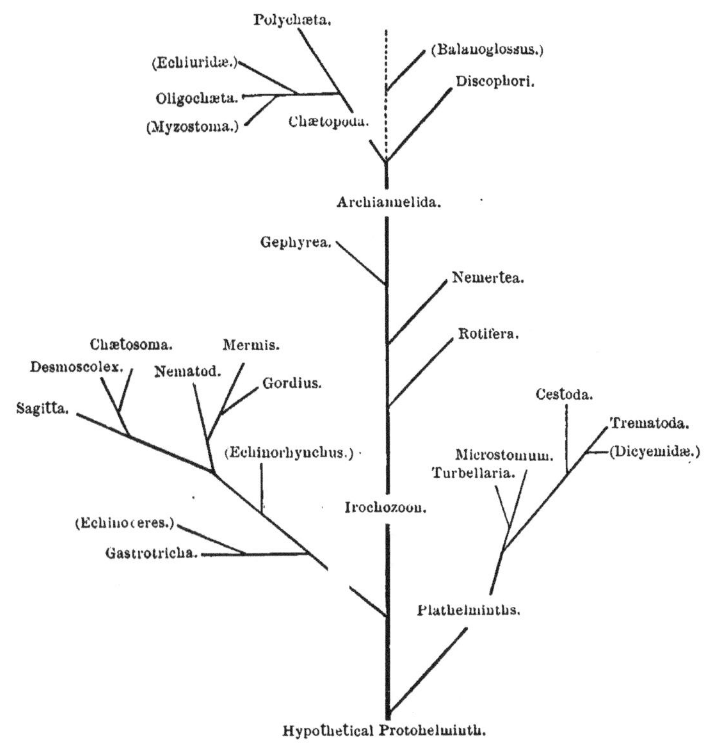

FIG. 163. — Diagram of the natural affinities of worms. The exact relationship of those forms which are enclosed in parentheses is doubtful.

CLASS I. — PLATHELMINTHIA.

In fresh and salt waters, and even on moist earth, are found the free living species; in nearly every animal are found some of the parasitic species of this huge and varied class of worms. The members of the group offer but little resemblance with one another as regards their external form; one would not be led by their appearance to place a planarian, a fluke, and a tape-worm in the same natural division, but the internal organization of them all most clearly demonstrates the closeness of their natural affinities. A great difference in the ways of life is corresponded to by an equal difference anatomically. On the one hand are the non-parasitic forms, roughly called the planarians, which require, and therefore have, well-developed organs of sense, good means of defence against attacking enemies, perfected means of locomotion, all the paraphernalia of independent existence; on the other hand are the parasites, which, being concealed in the body of their host, require little more than the apparatus for eating the victims that are also their dwellings, where they are well protected from external attack. Parasites are always degraded in structure. Within the present class we find a series of forms, beginning with those leading a free life and ending with those which are parasitic during their whole existence; with also intermediate links between the extremes, species that are parasitic through a longer or shorter term. In this series we find a progressive degradation, as the parasitism increases until it becomes the complete master of the worm's whole life. The free living forms are the unaltered types, the undegenerated patterns of the class, of which the parasites are the marred copies. Let us begin with the type, the planarians, or so-called Turbellaria.

Sub-Class I. — Turbellaria.

Order I. — DENDROCŒLA.

If we scoop up some mud and plants from the bottom of a ditch in which the water is tolerably clean, and let the collected mire settle in a basin of water, many strange and interesting animals will be discovered, — many insect larvæ, molluscs, crustacea, and worms. Among them, one usually discerns some short, very dark creatures, long in shape, quite broad and thin, which crawl about slowly, but almost incessantly, over the sides of the basin and the various objects in it, or indeed sometimes along the upper surface of the water itself; their soft, flexible bodies are highly contractile, so that the animal has hardly any definite shape, except indeed when crawling straight forward; for while thus progressing it always assumes a constant and characteristic form, as shown. These worms belong to the genus *Planaria*, and are typical turbellarians. Of the genus several species have been described, but it is difficult to determine the specific characteristics of a *Planaria*. The best known and most common form is the *P. torva*, found in Europe as well as America. The planarians do not have any means of locomotion visible to the naked eye, yet a close examination will disclose the existence of their motor organs, for it is possible to distinguish the passing and whirling of suspended particles in the water. These whirls are most characteristic; the very name of Turbellaria refers to them; they are produced by immense numbers of vibratile hairs or cilia which cover the body, especially on the ventral side. Even in proportion to the small size of the worm the cilia are tiny, but the united propulsions of the multitude of cilia are sufficient to achieve the worm's locomotion. The planarians, after the excitement produced by their transference has ended, subside into the inanimate sediment, amidst which they are well concealed by their dark brown color. At night time they are more active, for it is then that they gratify their carnivorous voracity, but they seem to me very uninteresting, except from an anatomical standpoint.

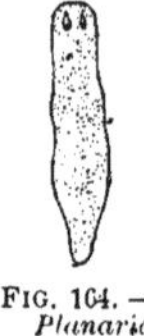

Fig. 164. — *Planaria torva.*

In the same basin we may find, beside the black planarians, many other allied worms, the largest among which will probably be the whitish *Dendrocœlum lacteum.* Some individuals measure over three-quarters of an inch in length. Its natural habitat is on the under-side of stones and leaves. It is white, with a shimmer of gray, and so translucent that the digestive canal shines through, and as it is usually gorged with dark-colored food it appears very distinctly. It is not a simple but a branched tube; there is a main central stem running through the front half of the body, and giving off many lateral ramifications; at the middle of the body the main stem forks, producing two branches, which run backwards, and give off many secondary branches. Where the three main stems unite, springs off the long, muscular proboscis, which is an extensive cylinder for seizing and swallowing food. The proboscis will swallow everything small enough; in fact, its deglutive propensities persist even after the death of the worm; for, sometimes, when anatomical research has destroyed all the tissues of the creature, the proboscis, being much tougher than the rest, still remains intact, and goes on swallowing everything it can seize, as if frenzied by hunger.

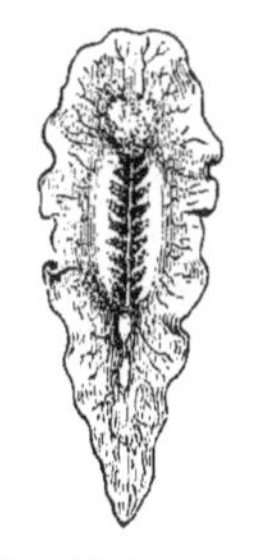

Fig. 165.—*Polycelis lævigata*; *a*, eyes.

There are many planarians having a branched digestive tract, and they are all classed together, under the common appellation of Dendrocœla, in opposition to the remaining Turbellaria, which have a simple straight tract, and are therefore called Rhabdocœla. It has been asserted by some writers that there are other species with no digestive canal, for which the term Acœla has been proposed. It seems, however, more probable that the authors alluded to have been careless in their observations and hasty in their conclusions than that there are any planarians really lacking a digestive tract.

Of the Dendrocœla, one of the commonest is *Polycelis lævigata*, of which we give a good figure, borrowed from Dr. Schmidt's great work on the natural history of the lower animals.

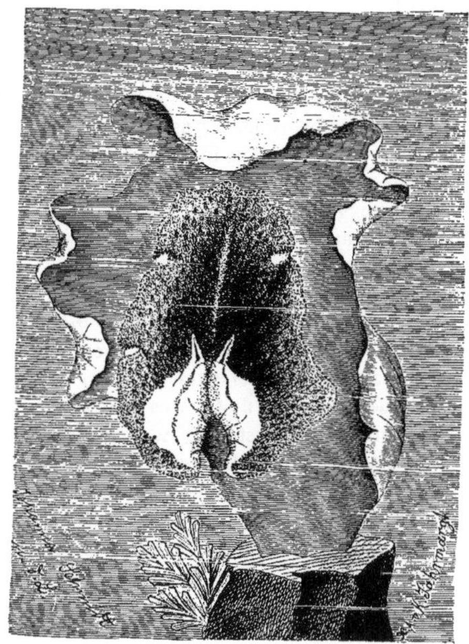

Fig. 166. — *Thysanozoon.*

Of the Dendrocœla, besides the three species already mentioned, the naturalist distinguishes many others. A few are found in moist earth, a goodly number in fresh water, but the majority are marine. The marine forms differ, for the most part considerably, from the genera above described. The reproductive organs, which are quite complicated in all Plathelminths, are especially different. The body is generally very broad and thin, often translucent and beautifully colored. In some cases a large size, a length of several inches, is attained. By way of illustration, we can mention only a single very beautiful form, very common in the Bay of Naples. The animal (*Thysanozoon*) may grow to nearly an inch in length. The back is covered with many rows of dark colored papillæ. On the head end is a pair of ear-shaped outgrowths, which project obliquely upwards, and appear to act as tentacles, with a most sensitive touch. The ventral surface is pure white. The artist has represented the animal clinging to a sea-weed, with the anterior end of the body raised as if in search of a new support. The genera occurring in America have hardly been studied yet. Let us hope that this gap in our knowledge of the American fauna will soon be filled.

The land planarians were first discovered by the celebrated Danish zoologist, Otto Friederich Müller, in moist earth under stones. A very few species have been found in Europe, but none as yet, to my knowledge, in the United States; but in South America Charles Darwin found numerous species inhabiting the moist earth of the primitive forests, where they attain a truly tropical luxuriance of size and color.

ORDER II. — RHABDOCŒLA.

The Rhabdocœla are planarians built on a smaller scale and simpler pattern. Some of them are sure to be found together with the true planarians in our ditches. There are two forms which I have found most abundantly in New England. The larger one, which I take to be identical with *Mesostomum ehrenbergi*, is a third of an inch or more long. It is whitish and translucent, and has a broad, dark streak in the middle of the body, an effect produced by the dark contents of the stomach, which the *Mesostomum* always keeps well filled as long as it can secure food. If one of these worms be kept in filtered or distilled water it finds nothing to eat, and the stomach is gradually emptied, and the worm appears of the same translucent white throughout. This species is admirably adapted to anatomical investigation because its transparency reveals all its internal organs. The two eye-specks are very conspicuous in front; the mouth is near the middle of the ventral surface, and is armed with a long proboscis; the stomach, as in all the Rhabdocœla, is a simple wide sac without branches. In summer time one can often distinguish several dark-brown, small spheres on each side of the body. These are the eggs, or more correctly the egg-capsules, which are deposited on aquatic plants. In reality each capsule contains several true eggs, which are very soft and delicate, and a certain quantity of nutritive material deposited around the eggs, to be gradually absorbed by the latter, and used as raw material to build up the structure of the embryo. The interesting manner by which the *Mesostomum* preys on *Daphnias* and Cyprids is thus described by Oskar Schmidt: — "It captures them as one might capture a fly with the hand, for it closes the hind extremity against the front, and by bending over the edges of the body forms a complete cavity; at first its captured prey rushes madly about, but soon the *Mesostomum* succeeds in fastening its powerful proboscis upon its prisoner. The struggles of the *Daphnia* gradually cease; its vampyre then stretches itself, and crawls away, having sucked the life-blood of its victim."

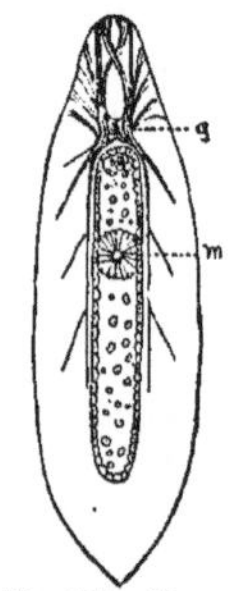

FIG. 167.—*Mesostomum ehrenbergi*; *g*, ganglion; *m*, mouth.

The second species is very common, being often found upon the well-known aquatic plant *Utricularia*. It measures scarcely an eighth of an inch in length, and is remarkable for its bright green color, so uncommon among animals; its anterior end is somewhat pointed or conical. This I believe to be the *Vortex viridis* of European naturalists. It is remarkable for its gregarious habits, large numbers being found together.

Although there are many genera and more species of this group known to naturalists, both from fresh and salt water, there is little in their habits to awaken general interest. We will therefore close our account by a brief mention of the genus *Con-*

voluta, which comprises small worms which have the thin lateral portions of their bodies curled over on to the ventral side.

In connection with the Rhabdocœla we may refer to a small aberrant group of worms, the MICROSTOMIDÆ, although they differ in two respects very strikingly from the true Turbellaria. They are mainly not hermaphrodite, and multiply not only by ova but also asexually by spontaneous division, like many annelids (*Nais*, *Autolytus*, etc.).

SUB-CLASS II. — TREMATODA.

The large class of animals to which we now turn our attention offers some of the most interesting life-histories known. The flukes or Trematoda are all parasitic upon other animals, and accomplish during their lives strange migrations and metamorphoses. According to their stage of development varies their habitat; usually the embryo swims about for a short time in the water; it then becomes a parasite by entering the body of its first host, where it changes its form, and by a singular process of asexual propagation it becomes the parent of several or many individuals belonging to a second generation. The members of the second generation in some cases multiply further, and the descendants mature to the final sexual stage, while in other instances they change directly into the adult form. In those species which pursue the more complex metamorphoses, the parasite may in successive stages infest as many as three different hosts. In general the adult fluke does not live in a host of the same species as the larval worm.

We cannot better gather a notion of the characteristics of the trematode worms than by following the history of one species as a concrete example of the habits of the class. For this purpose we choose the liver fluke, *Distomum hepaticum*. The adult worm infests the liver of mammals. It is an hermaphrodite, and every worm produces several hundreds of thousands of small eggs, which it discharges into the bile ducts. The eggs then pass into the intestines, and out with the droppings of the host, in which they may be found abundantly. The embryo is developed within the egg shell, and when mature bursts open the little cap or operculum; this occurs only when the egg is supplied with moisture. If it falls or is washed into some pool the embryo survives its birth, and immediately begins swimming freely in the water. Its form is an elongated cone, Fig. 168, with rounded apex, and measuring 0.13 mm. in length. The base of the cone is directed forwards, and in its centre is a short, retractile head papilla. The whole surface is covered by cilia, springing from large cells, which form the external envelope or so-called ectoderm of the embryo. In the interior are two eyes, and other structures, which we will not pause to describe. The embryo is exceedingly active, swimming about like an infusorian, though more rapidly. Now in England, where this worm has been most successfully studied, there lives in the ponds and ditches of the fields a snail known to zoologists by the name of *Lymnæus truncatulatus*. When the larval *Distomum* in the course of its gyrations happens to meet one of these unfortunate snails it attacks it. The worm presses its head-papilla against the surface of the snail, and begins spinning like a top around its own axis, and working its body until the tissues of the snail are forced apart, leaving a gap through which the embryo squeezes its way into its host. The embryo

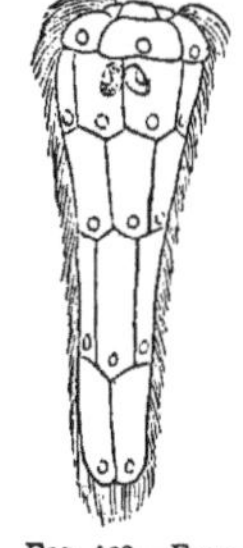

FIG. 168.—Free-swimming embryo of *Distoma*.

appears to have some means of instinctively recognizing the *trunculatus*, for it does not attack other species of snails. It cannot live much more than twelve hours in water, and it usually gets into a snail within eight hours. In its host the embryo changes into a new form, the nurse or sporocyst, within which arise the germs or spores producing new individuals. The outer ciliated cells swell up, and are finally cast off. The embryo then becomes a little bag or cyst, at one end of which the pigmented eye-spots of the embryo can still be recognized. The cyst or nurse grows and elongates. During warm summer weather it may reach its full size within a fortnight, but in autumn twice that time may be necessary. These cysts sometimes, but rarely, multiply by transverse division, but in other species this phenomenon is more frequent.

FIG. 169. — Cyst of *Distoma*.

The next larval forms, the rediæ, are developed within the sporocyst. The first clearly recognized appearance of the rediæ is a mulberry-like cluster of cells, over which a structureless membrane is soon formed, while the pharynx and other organs of the redia are produced in the cluster. There are usually several of these germs in each cyst. This is a very characteristic stage in the life history of the Trematoda; the embryo is converted into a bag, in which the germs of a new generation of individuals originate and are confined until far advanced in their development; the body of the parent is converted into a temporary prison-house for the progeny. The sporocysts of one species or another may be found in nearly every snail; many kinds are bizarre in shape, and all offer the curious spectacle of the living germs squirming about and nearly filling the whole of the cyst, their common parent. In the cysts of *Distomum hepaticum* there is usually one redia, less frequently two, nearly ready to leave the sporocyst, with two or three germs of medium size, and several small ones. When ready to leave the sporocyst, the redia by its own motion makes a forcible exit by rupturing the walls confining it. The free rediæ force their way through the tissues of the host, and are found especially in the liver. They increase in length, to 1.3 mm. or 1.6 mm.; a sort of collar is formed meanwhile a little behind the pharynx. In other respects, except that they have a digestive tract, which is wanting in the cysts, the rediæ resemble the sporocysts in structure; their most important new feature is the birth opening at the side of the body just behind the collar, which permits the exit of the new brood developed within the redia. The germs develop similarly to those of the sporocysts, but are more numerous. Sometimes they produce a second generation of rediæ, probably as long as the weather continues warm, but sooner or later, usually when cooler autumn weather begins, there come rediæ, which produce a new stage, the cercaria, in the series of metamorphoses. The cercaria is *the* typical larval fluke, and is easily recognized from its appearance, which reminds one of a tadpole, as is shown in Fig. 171; it has a large body nearly as broad as long, and flattened, with a long round tail. In the interior of the body one can

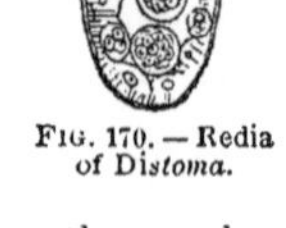
FIG. 170. — Redia of *Distoma*.

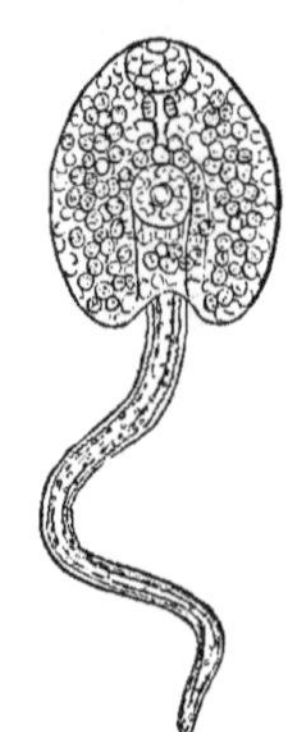
FIG. 171. — Cercaria of *Distoma*.

distinguish portions of the digestive tract and numerous cells, three of which with their granular contents are represented very much magnified in Fig. 172. These cells have a glandular character and serve to pour out a mucous secretion, the function of which will be immediately described. There may be as many as twenty-three spores in various stages of development in one redia; out of these spores there will be one, two, or three cercariæ approaching complete development.

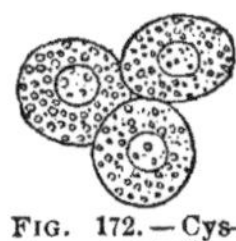

FIG. 172.—Cystogenous cells.

As soon as the cercaria has reached the limit of its development within the redia, it escapes from the parent by the birth opening. When free, the cercaria, Fig. 171, is very active, and constantly changes its form. By the aid of its suckers, the tadpole-shaped larva crawls or wriggles its way out of its host. When the infested cells are kept in an aquarium, the cercariæ may be found occasionally swimming about in the water; but not long, for on coming in contact with the side of the aquarium or with water-plants, it proceeds to encyst itself. The process can be readily observed under the microscope; for, on a glass slide, the cercaria soon comes to rest. It assumes a rounded form, whilst a mucous substance is poured forth over the body. together with the granules of the cystogenous cells, which we mentioned above. The tail is shaken off either before or during encystation, which is completed in a few minutes. These cysts are the means of infecting the final vertebrate host of the parasites, the infection being rendered possible by the habits of the intermediate host, *Limnæus truncatulus*, which might well be termed amphibious, so strongly is its habit of wandering on land developed. Indeed they can remain on land for long periods, and resist even prolonged droughts: hence when in the water, the snails become infested, and when on land, leave the cercariæ that crawl out of their first host scattered over the fields, where they encyst on the grass and are eaten by the sheep and other animals.

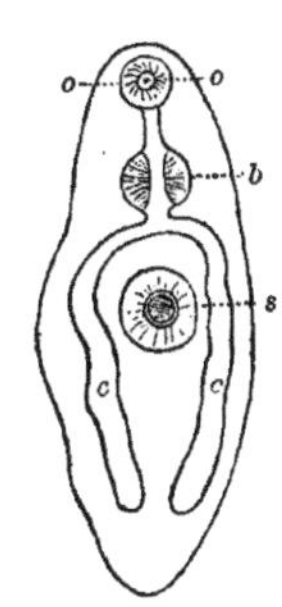

FIG. 173.—Young *Distomum*: *b*, pharynx; *c*, stomach; *o*, mouth; *s*, suckers.

In the stomach of the unlucky sheep the cyst is dissolved, leaving the worm free. The worm then makes its way into the liver and probably in about six weeks begins to produce eggs, growing itself meanwhile. During its growth its external form changes, the simple forked intestine develops numerous blind secondary branches, the posterior sucker is greatly enlarged and the sexual organs are matured. Thereafter the wondrous cycle of metamorphoses and emigration recommences with the new eggs. There are, perhaps, no instances more striking of the adaptation of animal species to particular conditions of existence than we encounter in the life histories of trematode worms, one of which we have narrated.

The adult worm, *Distomum hepaticum*, attains a length of three-quarters of an inch; it is broad and flat and at its anterior end has a small projecting lobe, which bears two ventral suckers, one in front at the very extremity surrounds the mouth, the second one considerably larger lies an eighth of an inch or more further back. It is found in most ruminants and in many other animals including man; it is commonly encountered in the biliary ducts, but is sometimes found in the intestine or venous system. Geographically it is very widely distributed not only throughout Europe and America, but also in Egypt, India, and even Australia and Van Diemen's Land. Although discovered by Gabucinus as long ago as

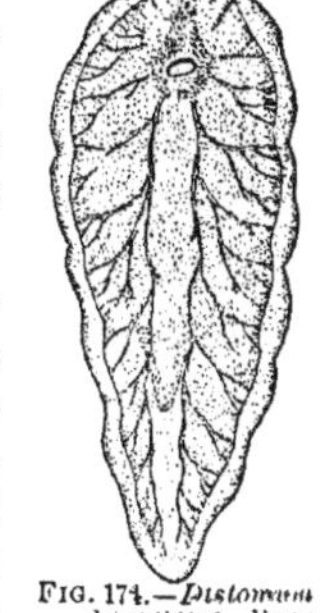

FIG. 174.—*Distomum hepaticum*, liver fluke.

the middle of the sixteenth century, it was not until 1882 that the complete life history was known, as we related above. They are very injurious to their hosts on account of the interference with the discharge of bile, which easily becomes serious and even fatal to their unfortunate entertainer.

The genus *Distomum* is a very extensive one, comprising a great many species, infesting an almost incredible variety of animals; indeed, it may be questioned if there is any other genus of living things privileged to be such a universal infliction. Man alone is exposed to the attacks of no less than five different species, while the poor frog is even more numerously endangered. Molluscs suffer especially from the attacks of the *Distomum* in its larval state, so that it is not at all uncommon to find the whole body, of a pond-snail for example, crowded with sporocysts or rediæ. The following are the species which attack man: *Distomum crassum*, occurring in China—it inhabits the intestine and grows to one or two inches in length; *D. lanceolatum*, which occurs together with the true liver fluke, but is much shorter, and, proportionally, much narrower, so that it can be very readily distinguished from the *hepaticum;* next a doubtful species which has been only once observed, *D. opthalmobium;* and, finally, the *D. heterophyes*, known only in Egypt. The genus may be readily recognized by the two ventral suckers, which lie near together at the anterior end of the body.

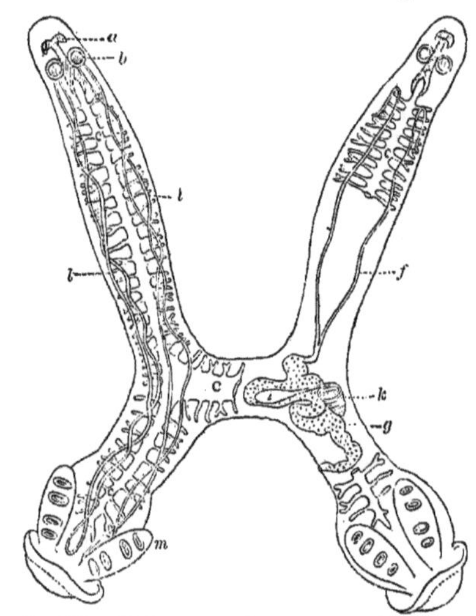

FIG. 175.—*Diplozoon paradoxum;* a, mouth; b, anterior suckers; c, stomach; f, oviducts; g, uterus; i, testis; k, vas deferens; l, Vascular canals; m, posterior suckers.

The Trematoda are divided into two orders, the Distomeæ and Polystomeæ. The former comprises those forms that are related to the genus *Distomum*, and have two or sometimes only one sucker, while the latter have two lateral small suckers at the anterior end of the body and one or several suckers posteriorly, while connected with the latter are often several hooklets. In the first order there reigns a considerable similarity of appearance, while in the second there is an unusual degree of diversity. Among the Polystomeæ, however, is one species which offers the wondering naturalist an unparalleled phenomenon—two separate and complete individuals united in one.

The *Diplozoon paradoxum*, to which the closing sentence of the previous paragraph referred, is indeed well named, for it is literally two animals united in one. There are two bodies precisely resembling each other in every particular, and united, like the Siamese twins, by a narrow communicating band, so as to form but one animal, the nutrient canals of one division communicating freely with those of the opposite half. One might easily regard this extraordinary arrangement as an accidental monstrosity, but observation has proved it to be common to all the individuals of the species. The animals, which are of very small size, being not more than two or three lines in length, are found attached to the gills of the bream, from which they derive nutriment. Siebold discovered that the *Diplozoon* arises by the union of two distinct worms, and the whole life history was subsequently worked out with great exactitude by Zeller. The larva was formerly supposed to be a distinct creature, and went by the name of Diporpa, Fig. 176. The *Diplozoon* is the mature sexual form, which produces the eggs, long oval capsules, with a long snailed thread running off from one end. The egg, Fig. 177, breaks open, and the larva swims about in search of its host, to which, when found, it attaches itself upon the gills, living there, in company with the adult, perhaps for months, but after a while they pair off; there is a little knob on the back of each Diporpa, and, of course, a ventral sucker; when two join they twist over so that each seizes with its sucker the dorsal knob of the other, and so they remain, and in due time actually grow together. They are, in truth, the most monogamous of animals, for each individual can have one mate only, from whom he can never be divorced. The union takes place in such wise that the animals form a cross. The left tail belongs to the right head. Each member of the *Diplozoon* has nine suckers, two in front, by the mouth, one near the middle, and six at the posterior extremity of the body.

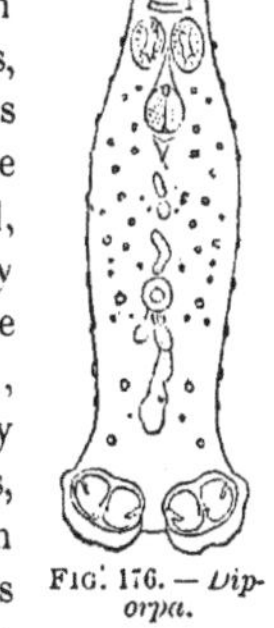

FIG. 176. — *Diporpa.*

FIG. 177. — Egg of *Diplozoon.*

Dr. Ernst Zeller has worked out very carefully the complicated history of a typical species of the second group of trematods, namely, the *Polystomum integerrimum*, parasitic in the bladder of frogs. The animal grows to a third of an inch in length, and is remarkable for having, unlike most Trematoda, a branching intestine; the posterior end of its body is expanded into a broad disc, with three pairs of suckers on its under side. The eggs are discharged by the parent in the bladder, and expelled into the water. The larva hatches out in from fourteen to forty days, according to the temperature. "The young worm," writes Dr. Zeller, "is an extremely lively, active animal, and swims about merrily in the water by means of its coat of cilia; contracting and stretching its body, bending and turning, and often, also, bending its head down, turns a somersault as quick as a flash." Under ordinary conditions the eggs are laid in the spring, when the frogs awake from hibernation, and the larvæ are hatched at a period when the tadpoles are in a somewhat advanced stage of evolution. From the water the active larvæ get into the branchial chambers of the tadpoles, where they take their abode for about two months; when the gills of the frog begin to disappear they migrate through the œsophagus and intestine to the bladder, and in three years attain sexual maturity. On the other hand, when the formation of the eggs and the evolution of the *Polystomum* larvæ are artificially accelerated by keeping the frogs in heated rooms, the larvæ are hatched at a period when the tadpoles are quite young and their gills very delicate. Their evolution is then very rapid. They become mature and produce eggs within five weeks; their life is at an end before the gills of their hosts are obliterated, and they never migrate into its internal organs. The remarkable conclusion of the varying life-history is a difference in the adult, for the gill-cavity *Polystoma* are very unlike the normal adults in form, appearance, and their whole anatomy. External circumstances here produce a maximum effect, for when they are charged in a certain manner the same eggs which would nominally produce the ordinary *Polystomum integerrimum*

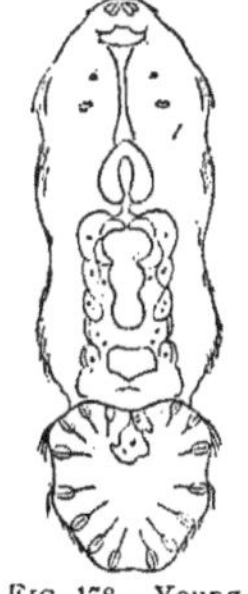

FIG. 178.—Young of *Polystomum.*

develop into a parasite, which, were its history not known, would not be supposed to have any connection with the species to which it really belongs.

The Polystomeæ include a great variety of strangely-shaped parasites, the majority of which infest fishes. Especially remarkable are the marine forms, many of which have been described by the elder Van Beneden, and illustrated by a series of beautiful plates, which record the bizarre outlines and colors of many species. Little, however, is yet known concerning the history of most of these trematods.

The following paragraphs, for which I am indebted to my able friend, Dr. C. O. Whitman, refer to a group of worms, the Dicyemidæ, which he has studied more profoundly than any other zoologist. Although their systematic position is doubtful, it seems to me most probable that they are larval trematods.

The DICYEMIDÆ are parasitic worms inhabiting the renal organ of the cuttle-fish, the poulp, and other cephalopods. Only ten species have thus far been described, and these, with one exception, belong to the fauna of the Mediterranean. These creatures are attached to the renal organ by the head, the body floating free in the fluid that fills the sack enclosing the renal organ. The different species vary in length from 1 to 7 mm., and all have the same habits and life history. The entire renal organ is often beset by these animals, which, to the naked eye, appears as short, white, undulating hairs. They have no mouth, no stomach, no muscle, no nerve, nor organ of any kind. The entire animal is made up of a few cells varying, according to the species, from twenty to thirty. There is one long axial cell (Fig. 179, *en*) stretching through the entire length of the parasite; the remaining cells form an envelope (*ec*) for the axial cell. The cells of the epithelial envelope are arranged in a single layer, and clothed with vibratile cilia. At the anterior fixed end, a certain number of these cells, which are elsewhere elongated and thin, are short and thick, and more closely ciliated, thus forming what may be called a head. The number, arrangement, and shape of the head cells, furnish, according to Whitman, the chief generic and specific distinctions. Seven species have each eight head cells, disposed in two sets, four propolar (*ap*), and four metapolar (*mp*), and thus form a generic group, to which the name *Dicyema* has been given (Fig. 179, A). The three remaining species have each nine head cells (Fig. 179, B), four propolars (*ap*), and five metapolars (*mp*), and have therefore received the generic name *Dicyemennea.*

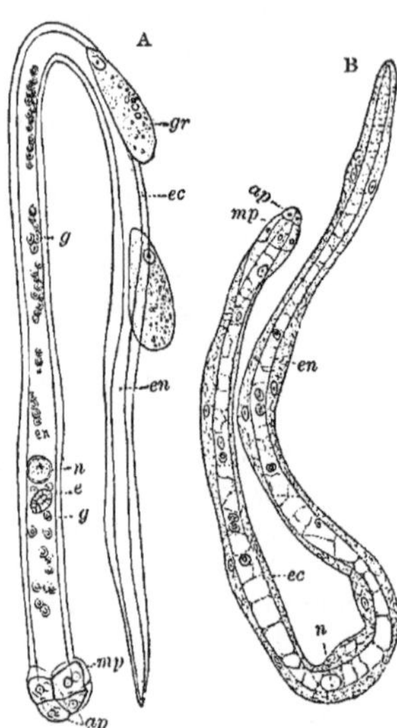

FIG. 179. — A, *Dicyema*. B, *Dicyemennea*; *ap*, propolar cells; *e*, embryo; *en*, axial cell; *ec*, outer cell; *mp*, metapolar cells; *n*, nucleus.

The structural simplicity of these animals is most probably due to degeneration resulting from their parasitic mode of life. "When we find an animal in the form of a simple sack, filled with reproductive elements, secured by position against enemies, supplied with food in abundance, and combining parasitism with immobility, we have strong reasons for believing that the simplicity of its structure is more or less the result of the luxurious conditions of life which it enjoys, even if its development furnishes no positive evidence of degeneration." Physiologically speaking, the axial cell is a uterus; for in it the germ cells, or ova, are lodged, and it is here that all the known stages of development are completed. The fully-formed embryos escape from the parent by pushing their way out between the cells of the epithelial envelope.

The reproductive cycle of these worms has not yet been fully ascertained; but some very interesting portions of it have been made known by the investigations of Van Beneden and Whitman. As Kölliker first pointed out, the Dicyemids produce two very distinct kinds of embryos, which he distinguished by the terms *vermiform* and *infusoriform*. The vermiform embryo (Fig. 180, F) develops directly into the parent form without metamorphosis. The fate of the infusoriform embryo (Fig. 181) still remains a puzzle; but it may be safely assumed that it either represents a male individual or a special form which serves to carry the species to new hosts. In the latter case, as suggested by Balfour, it is not improbable that in the course of a free existence, it may develop into a sexual form, the progeny of which are destined to complete the cycle of development by becoming again parasitic in the renal organ of a cephalopod.

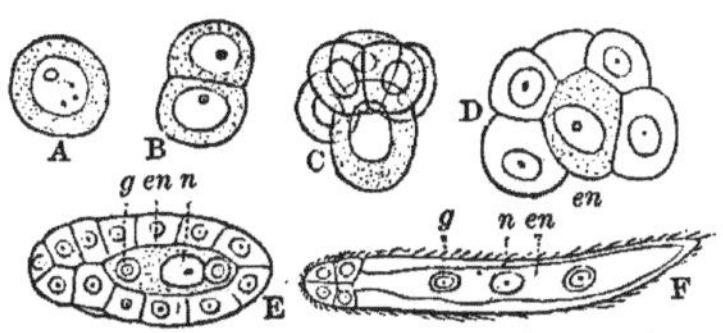

FIG. 180. — Development of the Vermiform embryo of *Dicyema*; *en*, axial cell; *g*, germ cell; *n*, nucleus.

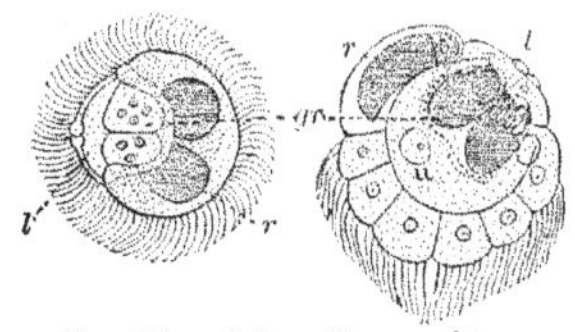

FIG. 181. — Infusoriform embryo.

Fig. 180, A to F, represents the more important phases in the development of the vermiform embryo. From the germ cell (A) arises, by division, a two-cell stage (B), then a four-cell stage. One of the four cells (*en*) now remains passive, while the other three go on dividing (C, D), and arrange themselves so as eventually to completely inclose the passive cell (E), which thus becomes the axial cell. The axial cell elongates as it becomes enclosed, and from its two ends two cells (*g*) are split off at an early date, which give rise by division to the germ cells. The embryo (F) attains nearly the adult form, and is clothed with cilia before escaping from the parent.

The infusoriform embryo, seen in Fig. 181, is somewhat pyriform, with the broader end directed forward in swimming. It has a complicated structure, the significance of which is entirely unknown. At the anterior end are seen two refractive bodies (*r*) which lie above an organ called the urn (*u*). The urn consists of a wall (*u*), and a lid (*l*), and contains four polynuclear cells (*gr*). The wall of the urn is hemispherical, and composed of two halves. The lid is made up of four cells.

The cells that form the posterior extremity of the embryo are ciliated. The germ cells that give rise to the infusoriform embryos are larger and much less numerous than those which develop into vermiform embryos.

As the two kinds of embryos are never found simultaneously in the same parent form, it has been supposed that there are two distinct adult forms, one of which produces exclusively vermiform embryos, the others exclusively infusoriform embryos. The former have been called Nematogens, the latter Rhombogens. It has now been ascertained that this distinction is not valid, the two forms being only two phases in the same individual cycle of life. There appears to be two kinds of Dicyemids, however, one of which, so far as known, produce only vermiform embryos, the other produces first infusoriform embryos, and subsequently, after the escape of all these embryos, vermiform embryos.

Considerable difference of opinion exists in regard to the systematic position of the dicyemids, some contending that they form an independent group, intermediate between the Protozoa and the higher animals, others, that they represent a degenerate branch of some division of the worms, perhaps of trematods.

SUB-CLASS III. — CESTODA.

The tape-worms or cestods are among the most dreaded enemies of mankind, and they inflict terrible destruction upon nearly all vertebrates, especially upon carnivorous species. A tape-worm is an elongated animal, usually so broad and thin as to suggest a ribbon or tape; one end is thickened and represents the head, having an imperfect or rudimentary brain, but lacking entirely organs of special sense, although possessing some specialized organs, hooks and suckers by which the worm anchors itself in the tissues of its unfortunate host. The long body is either a continuous band or else is divided up into parts called proglottids, forming a longer or shorter chain as the case may be. They have no trace of any digestive canal, but on the contrary have lost even that by the degeneration consequent upon their exclusively parasitic life; apparently their nutrition is effected by the absorption of the juices of the host through the skin of the parasites.

It is not difficult to trace out a series of genera by which one can pass gradually from the trematode type to the extreme of the cestode type, for there are certain flukes which approximate to the simpler tape-worms. In the genus *Caryophyllæus* we have in fact a tape-worm, which might well be described as a fluke that had lost its digestive tract by excessive degeneration. In *Ligula* the body is partially divided up, while in *Tænia* it is completely jointed, each joint or proglottis having its full set of reproductive organs. *Tænia* is in fact the extreme product of the changes which have occurred in the parasitic Plathelminths, and as it is also one of the best known as well as most feared of the whole class, we will consider their life history first. At least seven species of *Tænia* attack man; but of these only three are frequent, namely, *T. solium*, *T. mediocannellata*, and *T. echinococcus*. They all inhabit the intestines, where they burrow their heads into the walls. The two species named first grow to a very large size, *solium* reaching sometimes a length of three yards, *mediocannellata* a length of four yards, while *echinococcus* does not exceed four millimeters, scarce one one-hundredth of the length of *mediocannellata*.

The *Tænia solium* has a small head, about the size of that of a pin, within a long, thin neck, which gradually widens until the body is some six or seven millimeters broad. The whole body behind the head is divided up into proglottids or joints, which are so fine in the region of the neck as to be undistinguishable by the naked eye, but a couple of inches further back the narrow joints are plainly visible; the further down, the longer and larger the joints become. In each joint a set of eggs is formed, and in the last proglottid the eggs are mature; the last joint falls off and is discharged with the egesta; the new last joint, previously the penultimate, ripens and falls off in its turn; now the new joints are formed in front just behind the head, and each new joint as it shoves the others back acquires its place in the series, to be in its turn shoved back by fresh joints; thus the process continues, how long is not exactly known, but at least until hundreds of joints, each crowded with eggs, have been separated from the parent band. The head is remarkable for its armature; upon the crown is a circle of some six and twenty hooks, and four suckers arm the sides of the head.

The hooks, as seen under the microscope, form a regular circle, alternately big and little claws. The suckers are very muscular, the fibres being arranged in two systems, one equatorial and one meridional. Altogether the means are very ample to secure the hold of the worm upon its abiding place.

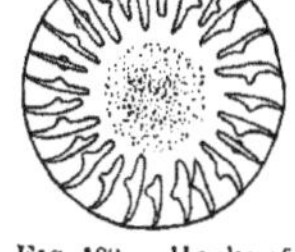

FIG. 182. — Hooks of *Tænia solium*.

The proglottids, after their expulsion from the body of the host, may be compared to the sporocysts of the flukes, for the eggs develop into embryos within the proglottids, so that they become cysts filled with larvæ. "The growth of the multitude of embryos within their interior causes the proglottis sooner or later to burst, and the embryos thus become dispersed; some are conveyed down drains and sewers, others are lodged by the roadsides in ditches and waste places, whilst great quantities are scattered far and wide, by winds or insects, in every conceivable direction. Each embryo within the egg is furnished with a special boring apparatus, having at its anterior end three pairs of hooks; after a while, as it were by accident, some animal, a pig perhaps, coming in the way of these embryos, or of the proglottids, swallows some of them along with matters taken in as food. The embryos, immediately on being transferred to the digestive canal of the pig, escape out of the egg-shells, and bore their way through the living tissues of the animal to lodge themselves in the fatty parts of the flesh, where they await their further destiny. The flesh of the animal thus infested constitutes the so-called measly pork. In this situation the embryos drop their hooks, or boring apparatus, and become transformed into the Cysticercus cellulosæ. A portion of this measled meat being eaten by ourselves transfers the Cysticercus to our own alimentary canal, to the walls of which the larva attaches itself" (Cobbold). The Cysticercus is the larva which infests the pig. Originally no helminthologist surmised the existence of a genetic connection between the parasites in swine and the human tape-worms. The great German zoologist von Siebold was the first to establish the exact metamorphoses, while Küchenmeister had the merit of clinching the proof by experiment.

The life history of *Tænia solium* illustrates the phases as they occur in nearly all tape-worms infesting carnivorous animals. The cestods of the herbivores have a simpler history, which we will relate directly. For the moment let us consider the more complicated type of development: The eggs or embryos are dropped about, to be swallowed by the first host, in which they assume the first or larval form; the host is preyed upon by some carnivorous animal, which, together with the flesh of its victim, swallows some of the larvæ, and these then undergo their final development. In order to reach their ultimate abode the parasite must twice be swallowed by a host; it accomplishes its migrations passively by the aid of the very animals it injures. The larva is always a vesicular structure, a membranous sack with a little appendix, which becomes the head of the adult worm. Imagine a glove of which the hand makes a closed bag, and which possesses only a single finger; imagine also the tip of the finger armed with a circlet of hooks and four suckers; finally turn the finger in so that it forms an inverted tube extending into the hand of the glove; thus we can conceive a good model of a Cysticercus; the vesicle itself is about a quarter of an inch long, and of a pale flesh color. Ordinarily the head is found turned in, but sometimes it is everted, and eversion always takes place after the larva enters the intestine of its second host. Now when the head is protruded the vesicle becomes the posterior portion of the body, and the proglottids being developed behind the head, the tape-worm is gradually formed in about three months by the continued interpolation of new joints in front, and the steady enlargement of those once formed, until the hindmost ones, being fully matured, drop off the chain.

Tænia mediocannellata is distinguished from *solium* by the want of hooks on the head, and by the fact that the broadest proglottids are those half mature at the middle of the ribbon. The Cysticercus stage is passed in the flesh of cattle.

In *Tænia echinococcus* the history is reversed, and offers many extraordinary peculiarities besides: reversed because the larval stage infests man and other animals, while the adult is found in one of man's domestic animals, the dog; peculiar chiefly because one larva produces several adults, and also because the larva differs very much when found in man from its form otherwise. The larva in the pig bears the name of *echinococcus;* it is a large round vesicle usually the size of a walnut, but sometimes growing to be as big as an orange. Usually it is found in the liver, sometimes in the lungs, and even in other positions. The thin-walled bag has several ingrowths, as shown at A in the accompanying wood-cut, each ingrowth is an invaginated head, and becomes a distinct adult individual. These are everted and broken off, and are found in the intestine of the dog shortly after it has eaten the infested flesh of a pig. The crown of numerous small hooks around the head enables the young worm to anchor itself. It forms only a few proglottids at a time, Fig. B, so that the adult is not more than four millimeters long, the terminal joint soon matures, breaks off, and is replaced. In man, however, the larva is still more complicated, for from the main vesicle grow out secondary sacks, and from the walls of these latter the heads are suspended; the heads thus lie in distinct capsules outside the main vesicle. The illustration, copied from Leuckart, shows the edge of the central sack with one "Brutkapsel" projecting from it. The Icelanders are very extensively afflicted by this parasite, and the fact is attributed to their want of cleanliness and the number of dogs that they keep around them. The dogs scatter about the proglottids with their dung, leaving the eggs directly or indirectly upon the plants which the Icelanders eat; for they gather for food certain mosses, sorrel, dandelion, and so forth, from the midst of the plains, in which live flocks of sheep guarded by dogs. The vesicles are found in all parts of the body, for when the larvæ are set free in the intestine of man they wander about everywhere, until they finally come to rest, and change to the many-headed vesicles. As the vesicles, or hydatids, as they are often called, enlarge, they produce very serious disturbances, and are often fatal.

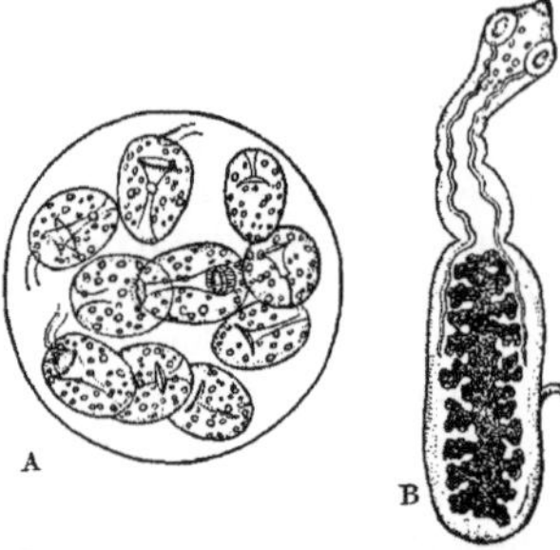

FIG. 183. — *Tænia echinococcus* from pig and dog.

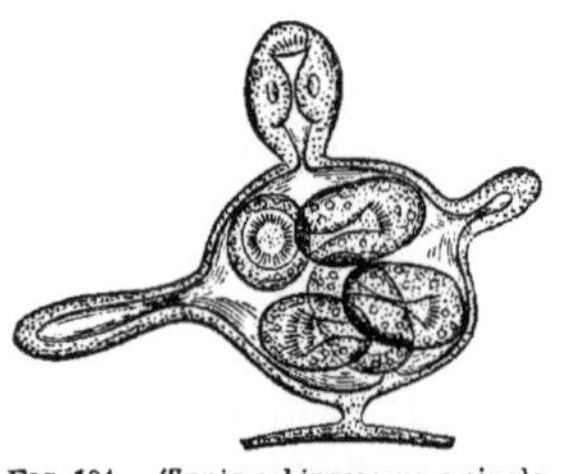
FIG. 184. — *Tænia echinococcus*, a single "Brutkapsel," from man.

Besides the *echinococcus*, dogs are infested by *Tænia serrata*, which lives as the *Cysticercus pisiformis* in the peritonæum of rabbits and hares, and also by *T. cucumerinæ.* "Some years ago," says van Beneden the elder, "while making a post mortem examination, at the Museum of Paris, of some young dogs which I had previously infected with *Tænia serrata* at Louvain, there were found by the side of these some *Tænia cucumerinæ.* These dogs had taken nothing but milk and Cysticerci!

Whence came these *Tænia cucumerinæ?* I knew not, and I frankly owned it to members of the commission who proposed the question to me. This, however, did not prevent my being greatly puzzled by the presence of this worm, of whose origin I had no idea. Now we know whence they came. An acaris, the *Trichodectes*, lives in the hair of young dogs, and harbors the scolex (larva) of this cestode. The dog, by licking its own hair, grows infested, like the horse, which in a similar manner introduces the gad-fly, and, though it has taken no other nourishment, harbors its own epizoaria."

Sheep are afflicted by a disease known as the "gid," or "staggers." The animal goes round and round; its power to walk straight ahead is lost. This curious effect is produced by the presence of a hydatid, a many-headed larva of a tape-worm; the larva has long been known under the name of *Cænurus cerebralis*, and is the cause of a mortal disease but too well known to the farmer. The pressure and irritation caused by the hydatid produces inflammation and degeneration in the surrounding tissues. For a long time the further metamorphoses of this species remained undiscovered, but it has since been ascertained that the mature stage is reached in the dog.

The second family of Cestodes, the Tæniadæ being the first, are the BOTHRIOCEPHALIDÆ, distinguished by having only two weak and shallow suckers on the head. *Bothriocephalus latus* is the largest of internal human parasites, specimens occurring twenty-five and thirty feet in length. Its history has not yet been satisfactorily worked out, but it is known that the embryo first enters the water, and probably passes its larva life in a fish.

We will not attempt to describe all the types which mark off the families of tape-worms, but we can at least indicate the great variety of appearances among the species. This is especially noticeable in the tape-worms of birds and reptiles, whose predaceous habits lead them to engulph many aquatic animals laden with larval plathelminths, which reach their full development in the bird or lizard. The accompanying wood-cut shows two forms, A, *Tænia filum*, which was described long ago by Goeze; the head is scarce over a millimeter in width, and is furnished with a crown of ten small hooks; the cysticercus stage is unknown, but the adult is found in the intestine of *Scolopax*, *Totanus*, and *Tringa*. B represents the very singular *Ophryocotyle proteus*, originally discovered by Dr. Friis; the head ends in a large fan-shaped cupula, and bears six distinct festoon-like suckers; it grows to a length ordinarily of four inches; it is found in the intestine of certain sandpipers and plovers. There is a rich field of interesting study open to a naturalist who will turn his attention to the parasitic plathelminths of our water and shore birds. Another very remarkable genus has been found in *Varanus*, a genus of lizards, and recently described by Prof. Perrier of Paris, and were obtained by M. Vallée from the lizards kept at the Jardin des Plantes. The tape-worm in question has been named *Duthiersia*, after Prof. Lacaze-Duthiers; it is distinguished by the enormous development of the lateral suckers, which form two large cups with

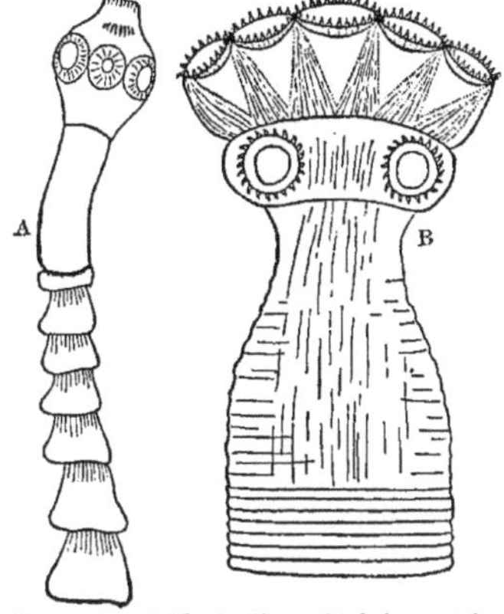

FIG. 185.—*A*, *Tænia filum*; *B*, *Ophryocotyle proteus*.

crenullated edges; the cups are soldered together in the median line; a similar arrangement, but on a much smaller scale, exists in *Solenophorus*, and indeed there can be no doubt that both the genera alluded to belong to the family of Bothryocephalidæ.

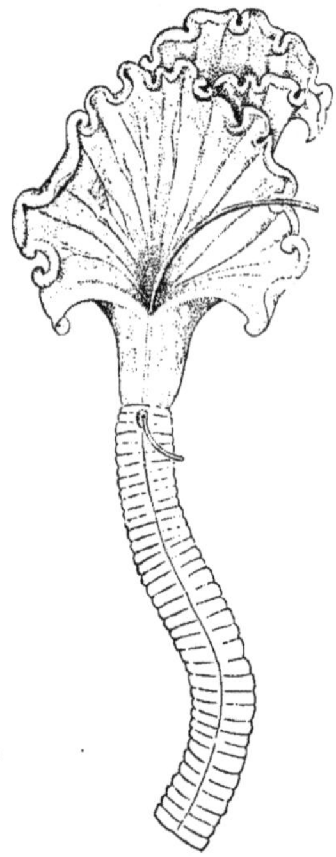

Fig. 186. — *Duthiersia expansa.*

Our herbivorous domestic animals are much infested by tape-worms, the horse, sheep, goat, and cattle, each have distinct species. The horse is especially subject to their attacks, and has three species peculiar to itself, namely, *Tænia plicata*, *T. mamillana*, and *T. perfoliata*. These worms sometimes occur in large numbers in a single host; Chabert counted ninety-one from one horse. It is supposed that these species have free-swimming aquatic larvæ, which the horses swallow in drinking; but of none of them is the history satisfactorily known.

Thus we terminate our brief account of the parasitic plathelminths. The subject is a repulsive one, but is full of instruction to the thoughtful naturalist. To trace out the life histories, so baffling to the student, has required the greatest perseverance and acumen; and our present knowledge is a very remarkable monument to the patience and skill of scientific naturalists.

Class II.—ROTIFERA.

The Rotifera or wheel-animalcules are small creatures found in marine and fresh waters, but most abundant in stagnant pools, and often in places where water has stood for a few weeks only. They equal a pin's head in size, and are very transparent, so that an entire animal may be forced to display its complicated anatomy at one view to the inquisitive microscopist. They are all, except the sessile forms, agile and restless, and dart about eagerly and rapidly, so that they are hard to follow with the eye; but, fortunately, they have a liking for occasional repose, and will sometimes keep delightfully still, long enough for the keen observer to discover some of the secrets of their organization and of their physiological processes.

One of the most familiar forms is the little wheel-bearer, *Rotifer vulgaris*, which may be collected during the warm season from almost every ditch. The body of this animal, when fully extended, possesses greater length, in proportion to its breadth, than most others of its class. The tail commonly has three joints or segments which are capable of being drawn one within the other. This animalcule may be considered typical of its class. There are no legs; the anterior portion of the body is furnished with a retractile lobed disc, of which the margin is covered with vibratile cilia, while at the opposite end of the body is a cylindrical process forked at its ex-

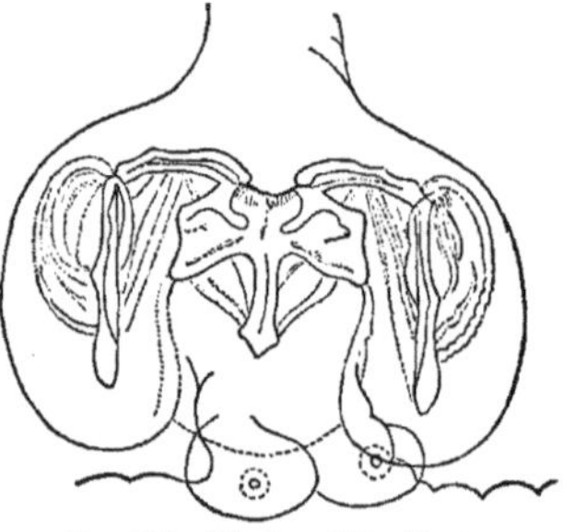

Fig. 187.—Mastax of *Euchlanis*.

tremity. This false foot or tail is jointed, and can be contracted and extended like a telescope. It does not form a direct prolongation of the end of the body, but arises from and is situated upon the ventral aspect. In most wheel animalcules there are two eye specks, which are usually reddish in color. The œsophagus is provided with a complicated masticating apparatus, the so-called "mastax," which is very interesting to the anatomist. Such are the general characteristics of the group.

The class was formerly confounded with the chaotic assemblage of minute creatures, to which the name of infusorial animalcules was applied rather for convenience than from discrimination; but, says Rymer Jones, "the information at present in our possession concerning their internal structure and general economy, while it exhibits, in a striking manner, the assiduity of modern observers and the perfection of our means of exploring microscopic objects, enables us satisfactorily to define the limits of this interesting group of beings, and assign to them the elevated rank in the scale of zoological classification to which, from their superior organization, they are entitled." The ciliated lobes at the anterior end of the body have given the class its name; yet there are forms known in which the cilia are wanting and the lobes are excessively modified in shape. This is notably the case in the parasitic genera *Albertia*, *Balatro*, *Seison*, etc., which form the family of the Atrocha (wheelless), and also in certain species not parasitic. Of the latter, the best known is a rotifer, described in 1857 by the distinguished veteran among zoological investigators, Professor Leidy of Philadelphia, under the name of *Dictyophora vorax.* This rotifer is now united with several other allied species in one genus, *Apsilus. Apsilus vorax* is spheroidal, has no jointed tail, but is sessile. Instead of the ordinary rotary discs it possesses a large protractile cup or disc. The animal has the power of turning upon its point of attachment, but does not appear to have the power of letting go its hold. The animal is about one twenty-fifth of an inch long; in shape ovoid, the narrower pole being truncated and bearing the mouth. The body is transparent, colorless, and even, and exhibits no signs of annulation, nor does it become transversely wrinkled by contraction. The interior exhibits the digestive apparatus and other organs, mostly more or less obscured by an accumulation of eggs in various stages of development. From the truncated extremity of the body, the animal projects a delicate membranous cup, forming more than half a sphere and more than half the size of the body. At will the cup is entirely withdrawn into the body, but, when protruded, expands outwardly like an opening umbrella, and is the substitute for the ordinary trochal discs of rotifers, and appears a most efficient means in catching the animalcules which serve as its food.

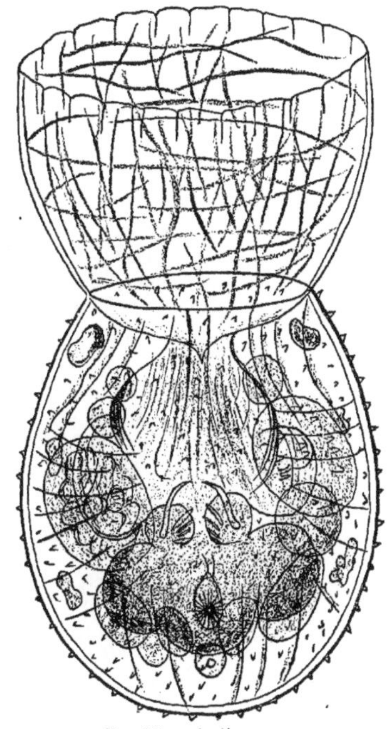

Fig. 188.—*Apsilus vorax.*

There are also many other rotifers, which normally remain attached permanently to some water plant or submerged stone. Among the attached forms we may call attention especially to the Flosculariaus. They are commonly found attached to the stems and leaves of aquatic plants. The foot-stalk, bearing the bell-shaped body, is very long. Dr. Carpenter describes this one of the most beautiful species, *Stephanocera eichornii.* The body has five long tentacles, beset with tufts of short bristly cilia, reminding one of the ciliated tentacles of the Bryozoa. The body and foot-stalk are enclosed in a cylindrical cell resembling that of the Hydrozoa and Bryozoa. A comparison of this with other forms, however, shows that these tentacles are only extensions of the ciliated lobes which are common to all the members of these families. The so-called cell is not formed by a thickening and separation of the outer integument, but by a glutinous secretion from it, so that, as the rest of the organization is essentially conformable to the rotiferous type, no such passage is really established by this animal towards other groups as it is commonly supposed to form. In the allied genus *Melicerta* the body is protected by an artificial cell, constructed of little pellets. Mr. Gosse, who fortunately had an opportunity of watching the animal build its cell, describes how the work is done. The ciliary wheels sweep particles into the upper end of the digestive canal, where they are fastened together, probably by a glutinous secretion, and moulded into a rounded pellet, which is then disgorged, and the animal, bending over, deposits the pellet upon the edge of the wall it is building, thus imitating the art of the bricklayer.

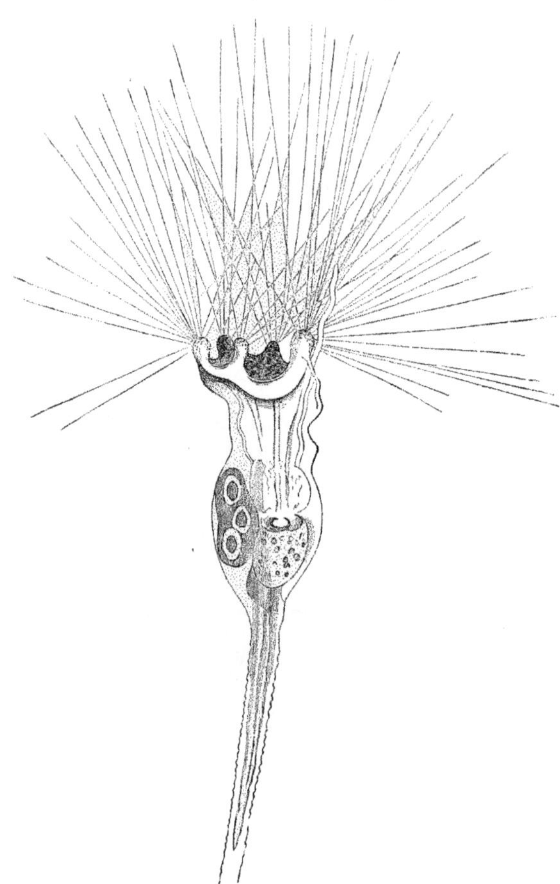

FIG. 189. — *Floscularia ornata.*

The Rotifera include many odd forms, among them the extraordinary spherical creature discovered by Professor Semper in the Philippines, and the group of the queer ASPLANCHNIDÆ, remarkable for having one opening of the digestive canal; but the majority of the wheel animalcules, which the collector obtains, belong among the relatives of the *Rotifer vulgaris*, yet present to our inquisition a great variety of shapes and diversity of habits. They have been divided into three families: First, the Philodinidæ, of which the genus *Rotifer* is the type; second, Brachionidæ (*Brachionis*, *Noteus*, *Mastigocerca*, *Stephanops*, etc.); third, Hydatinidæ (*Hydatina*, *Monocerca*, *Notommata*, *Diglena*, etc.). We shall give an account of a typical species of the second and third family each, having already described the type of the first.

In the BRACHIONIDÆ the body is broad, almost shield-like sometimes, the foot is short and jointed, and the integument is, in part, so hardened as to create a so-called carapax. *Noteus quadricornis* is perhaps the best example of the family. The broad rump is covered by the shell, which has many nodules upon its surface, and is prolonged into four horns, — two by the head and two near the tail. The wheel forms a notched cup, and serves to sweep food down to the mouth at the bottom of the cup, as well as to act as an organ of expeditious locomotion.

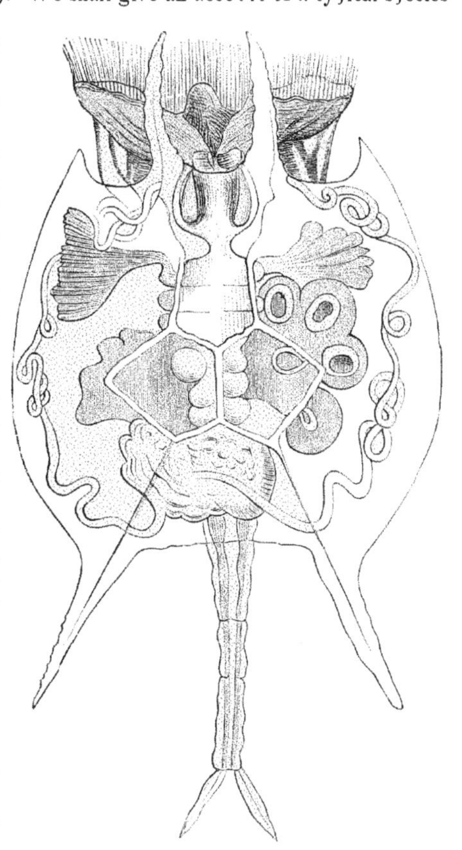

FIG. 190. — *Noteus quadricornis.*

Hydatina senta is a classical animal, because it was principally on this species that the illustrious Ehrenberg studied the anatomy of this group of animalcules. The broad body has only a very short foot-stalk, which is forked behind. The mouth is armed with two jaws and many teeth. There are no eye-specks whatsoever. The cuticle is delicate and soft. *Hydatina* illustrates the extreme rapidity with which the development of the young is accomplished. The egg is extruded within a few hours after the rudiment of it becomes distinctly visible. Within twelve hours more the shell bursts and the young animal comes forth. In *Rotifer* and several other genera the young are extruded alive, incubation being completed within the parental body. In *Brachionus* still a third method prevails, the eggs remaining after their extrusion attached to the posterior end of the body, until the young are set free. Ehrenberg has made the astounding estimate that *Hydatina* may multiply so rapidly that one hundred millions may be produced within ten days from a single individual! I confess that this appears to me a totally erroneous calculation, resting upon an utter misapplication of the multiplication table.

The rotifers, like a few others of the lower animals, have an almost incredible power of withstanding desiccation. Apparently, in the most unfavorable conditions of deficient moisture, evaporation from their bodies does not proceed further than to interrupt their activity without destroying their vitality. Apparently they may exist in this dried condition for an indefinite period, and resume their course of life upon being again supplied with water. "This fact, taken in connection with the extraordinary rate of increase mentioned in the preceding chapter, removes all difficulty in accounting for the extent of the diffusion of these animals, and for their occurrence in incalculable numbers in situations where, a few days previously, none were known to exist; for their entire bodies may be wafted in a dried state by the atmosphere from place to place," and they return to active life whenever they encounter necessary conditions of moisture and warmth.

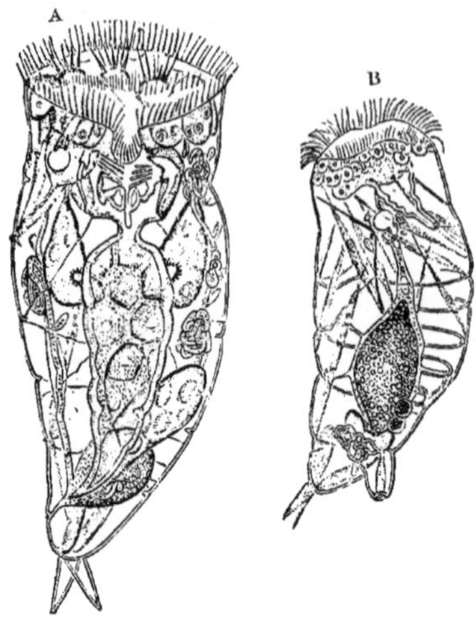

FIG. 191. — *Hydatina senta*; *A*, female; *B*, male.

CLASS III. — GASTROTRICHA.

These minute worms, barely more than microscopic in size, are found in quiet waters in company with other animalcules. They owe their name to the presence of locomotive cilia upon the under surface of their bodies. The number of species is limited, and there have been thus far only seven genera established, of which *Icthydium* and *Chætonotus* are the most common. The body is flask-shaped, and terminates in a fork formed by two short processes. The mouth lies at the front extremity, and leads into the very noticeable muscular œsophagus. There is usually a pair of pigmented eye-specks in front. All the species are probably hermaphroditic. Beyond these few anatomical details we have little knowledge about the group, — nevertheless it is particularly interesting because the type of organization is extremely simple, indeed, I think the simplest of all the animals with an organization essentially bilaterally symmetrical. We hope that some patient investigator will soon work out the development and entire life history of these worms.

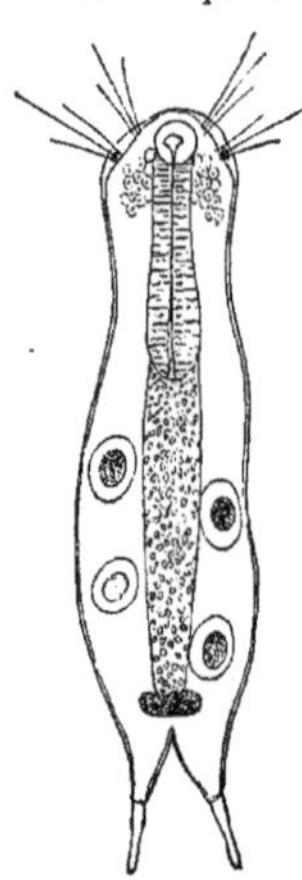

FIG. 192. — *Ichthydium*.

Closely related to the Gastrotricha, and yet offering striking peculiarities of its own, is a small marine worm, *Echinoderes*, which was first discovered at Saint Malo, in 1841, by Dujardin. The body is from a third to a half of a millimetre long. The shape of the animal,

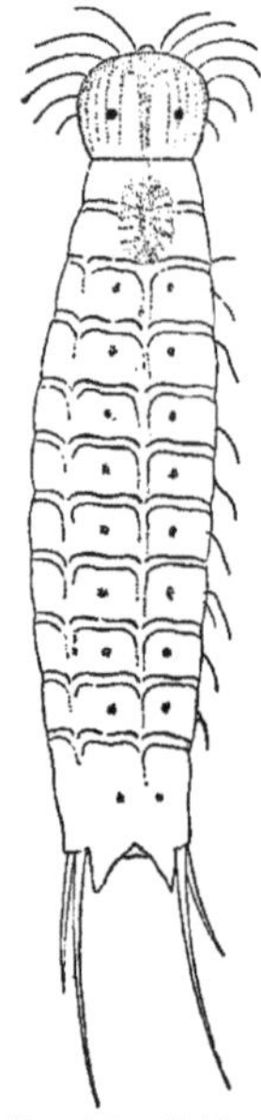

FIG. 193. — *Echinoderes*.

as well as the bristles and red eye-specks on the head, and also the caudal fork, reminds one strongly of the Gastrotricha. A false appearance of segmentation or jointing of the body is produced by ten rings of spines. The head is quite distinctly marked off, and can be withdrawn into the body. The sexes are said to be distinct.

Class IV. — NEMATODA.

The well-known "vinegar eels" are typical Nematods. The true name of vinegar eel is *Leptodera oxophila*, but most authors still call them *Anguillula aceti*. The older writers describe them as very abundant in vinegar, but nowadays they are not so common. Formerly vinegar was prepared so that a good deal of sugar was left in solution, but at present the processes of manufacture used eliminate nearly all the sugar, and, moreover, the vinegar is too often adulterated with sulphuric acid, so the *Leptodera* has a poor chance. The vinegar eels, however, do not live on the vinegar, but on the fungi which grow in it, and which depend on the sugar for nutriment. If any one wishes to observe these microscopic worms, it is only necessary to add a little sugar and mucilage to some vinegar and allow the mixture to stand in an open dish for a few days. A fungous growth appears, in which the eels develop, and they may be seen readily by scooping up some of the fungous mass, spreading it upon a glass slide, and examining it under the microscope. The same worm, apparently, appears in fermenting starch paste, although the paste worm has received a different specific name, *L. glutinis*. The worm never exceeds a couple of millimetres in length; in English measure it is always less than the twelfth of an inch. It is long, cylindrical, transparent, and has a digestive canal which occupies nearly the whole length of its body. At the anterior end is the long, muscular œsophagus; in the middle lie the elongated reproductive organs, alongside the intestines. In the female the eggs are quite conspicuous, shining through the body wall. The external surface of the body has a tough crust, yet a very elastic one, else the animal could not wiggle and twist as it does. Now all these features are common to the nematods in general, and they owe their very name to their thread-like shape. Indeed, despite the great number of species, they present a remarkable uniformity, not only in appearance, but also in organization. The most noticeable variations are in the outline of the body, the armature of the mouth, and the form of the caudal extremity. Now in all three of these respects the *Leptodera* is as simple as possible; the mouth is a simple opening, the body a simple cylinder, and the tail simply tapers off. It is organized like one of Mother Goose's repetitive melodies.

Fig. 194. — *Leptodera oxophila*, Vinegar eel.

Nearly all the species of *Leptodera* and the allied *Pelodera* live in moist earth and putrefying substances. Our knowledge of their curious habits is mainly due to the experiments of Anton Schneider, whose account we reproduce in abstract. To obtain the species it is only necessary to have a pot with some earth in it, it matters little of what soil; put a small piece of meat or pour a little blood or milk into the earth, and keep it moist, not wet. The earth contains great numbers of the larvæ, which attracted, perhaps, by the smell, crawl to the centre of putrefaction, where they soon swarm. They become sexually mature, and the young they produce develop on the spot to mature worms. But after a time they are seized with a migratory instinct, and crawl off in all directions, the young worms with the rest. In the drier earth the young slough their skin, but do not leave it; on the contrary, it forms a protective case, which the larva inhabits until it again encounters moisture and putrefaction, when it bursts from its imprisonment and renews the cycle of life. According to Schneider also, some other species of these genera occasionally exchange their free life for parasitism upon the large black field snail of Europe and upon the common earth-worm, but in favorable external surroundings leave their host again. It is but a step to change from this habit to the true parasitic life led by most Nematoda.

The *Rhabditis nigrovenosa* is a parasitic member of the family of Anguillulidæ, and has a most remarkable life history. There are two generations of sexually mature animals. The members of one generation are very tiny, perhaps two hundredths of an inch long, lead a free life in moist earth or mud, and resemble anatomically the *Leptodera*. The members of the second generation grow to half an inch long, inhabit the lungs of frogs, and were formerly supposed to belong to the Ascaridæ, a very distinct family. Evidently something is wrong, — either our system of classification or else nature itself. Let us trust to the future for the resolution of our dilemma. The first generation is developed from the eggs of the first, and *vice versa*. The eggs from

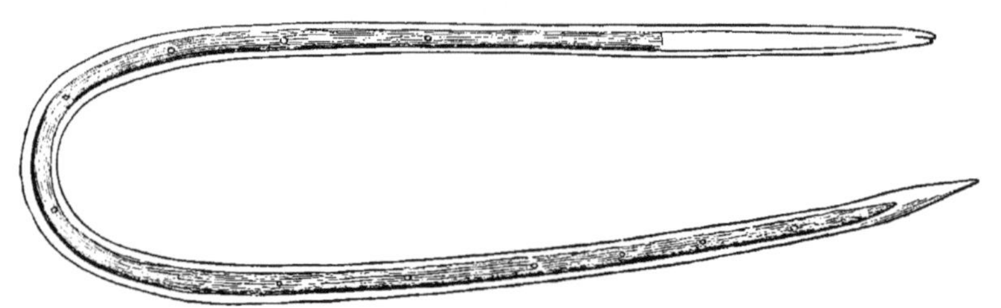

FIG. 195. — *Anguillula tritici*, wheat worm.

the parasitic form are discharged through the intestines of the host, and develop directly into the free form; the female of the latter, however, retains the eggs, and the young develop within the parent. "She exceeds the pelican, for the mother nourishes her offspring not only with her blood, but all her internal organs break up until nothing remains but the skin, to form a lifeless case around the brood of squirming young worms." This stage of life lasts some time, and then the young jump out and remain perhaps for weeks in the moist earth, until they have an opportunity, by way of a frog's mouth, of getting into his lung and there growing up into the *Ascari nigrovenosæ*.

Other members of the Anguillulidæ are very injurious to plants. The most notorious of these enemies to husbandry is the *Anguillula tritici* (known also under the name of *Tylenchus tritici*), for it inhabits and destroys the ears of wheat, producing the dreaded disease known as smut. These worms were discovered about the middle of the eighteenth century by the English microscopist Needham, but our present knowledge of them is due mainly to Monsieur Davaine. In the diseased ears the grains are malformed, small, and black; they have a hard crust, which encloses a core of white powdery substance, which upon being moistened falls into little bodies, which the microscope shows to be Anguillulæ, but still immature. In nature the kernel of wheat falls to the ground, gets moist and rotten, the wall breaks open, and the worm is set free, to crawl about in the earth until the new grain shoots forth. The worm then climbs up, resting motionless wherever it gets dried, but resuming its climb as soon as moisture restores its activity, and finally reaches the top while the ear is young, and there entering the parts of the flower, the worms cause a gall-like growth to the plant. There they dwell, and, having meanwhile become mature, the eggs are developed, deposited, and hatched out. The parents dying off, the embryos become the only inhabitants of the gall and form the dust-like substance which we mentioned at the beginning of the paragraph. The power of this species, and of its immediate congeners, to withstand desiccation and to recover their full activity upon the return of moisture, is almost incredible. This wonderful capacity is most developed in the larvæ, the adults being endowed with less resistance. It is known that one of the galls containing larvæ may be kept dry for over twenty years and at the end of that period be revived by moisture. Spallanzani's experiments showed that the loss of moisture must be gradual, for the larvæ need apparently to make some preparation for their long confinement. It is not to be supposed that the worm dries up, but rather that it is able to prevent its own loss of moisture for an indefinite period. The excessive development of this strange ability to stop the vital processes is evidently a means by which the animal is enabled to escape what were elsewise the fatal dangers of its life cycle.

The primitive type of the Nematoda is probably more nearly preserved in the family of **Enoplidæ**; they are not parasitic, but lead a free life; for the most part they are marine animals. Very little is known of their habits or metamorphoses. They are found among plants, oftentimes in snarled bunches. Some species live in fresh water, but rarely, except in pure running streams, and others again in moist earth. The most common genera are *Enoplus* and *Dorylaimus*. Many of the species have a peculiar spinning gland at the posterior end of the body and opening on the under side of the tail. "So soon," writes Professor Schneider, "as the animal has fixed its tail upon some support, it moves along and draws out the secretion of its gland to a vitreous thread several lines long. One end of the thread is glued fast, on the other floats the animal in the water." Most of the Enoplidæ avoid the neighborhood of putrefaction, but delight in pure soils and waters, in which they often abound.

The remaining families of the thread-worms are parasitic; arranging them as nearly as possible according to the extent they depart from the type of the free forms of the class, we have, beginning with those the least changed, Filariadæ, Trichotrachelidæ, Strongylidæ, Ascaridæ, Mermithidæ, and Gordiidæ. Each of the first four of these families includes parasites very dangerous to man and the domestic animals.

Of the **Filariadæ** the *Filaria sanguinis-hominis* is said to be the cause of the elephantiasis, so familiar to physicians in the oriental tropics. The larvæ are found in the blood vessels and lymphatics of man, and by clogging the passages impede the circulation and produce, it is asserted, the enormous enlargement or hypertrophy of parts known as elephantiasis. It is supposed that the mosquitos suck into their own bodies the larvæ in the human blood; that when the mosquitos go to the water to lay their eggs, they soon die, and the worms escape into the water, these become mature and produce their young, which enter the human system when the water is drunk. *Filaria* (*Dracunculus*) *medinensis*, the so-called Guinea-worm, occurs in the tropical districts of the old world, and is found parasitic in the subcutaneous tissue of man. It is very thin, but may attain a length of several feet; only the female is known. The parasite lies coiled up in the soft tissue, in which it produces an ulceration, and the discharge of the ulcer serves to set free the brood of young, but the history of the species has not yet been further elucidated.

The TRICHOTRACHELIDÆ include the most dangerous by far of human parasites, the one which most often and most rapidly proves fatal to the life of its host, the much-feared trichina. The worm occurs in little distinct capsules in the muscles of the body of many animals; now when some of the infested muscles of one animal is eaten by another, the capsule is dissolved by the action of the digestive juices, and the larval worms are liberated in the intestines of their new host, where in the course of a week they become mature and then produce millions of eggs, which soon hatch out minute larvæ. The larvæ shortly penetrate the walls of the intestine and wander about through all parts of the body, injuring and irritating all the tissues they traverse. Finally they settle down in the muscles, penetrate the muscular fibres, feed on the substance thereof, and grow in a few weeks to several times their original size. While wandering and growing they produce the most serious consequences in the health of the host, and being often present literally in millions, the accumulated effects of all the slight injuries produced by each trichina cause the greatest suffering to their victim, and often lead to his death within a few weeks. Still the host may be restored to good health if he can survive the first few weeks of danger; for after the trichinæ are established in the muscle fibres and have finished their growth, there they remain for years and years, enclosed in tough capsules, in which they lie coiled in a spiral. In this state they produce no serious injury, and, as after the first attacks are over, and the inflammation they cause has subsided, the muscles and other injured tissues of the host are regenerated, there may be a complete recovery, after passing through the first period of danger.

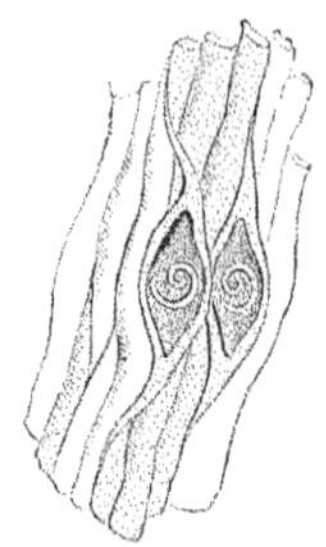

FIG. 196. — *Trichina spiralis* encysted in muscle.

Pigs, from their omnivorous habits, are peculiarly exposed to the attacks of trichinæ. In some German towns it was found that one pig in every three hundred or even less was infested; now such a pig slaughtered and made into sausages and uncooked ham has often served to feed nearly a whole German village, and thus to cause a veritable epidemic of trichinosis. In 1884, shortly after the exclusion of American pork from Germany, ostensibly because of the danger of trichinæ, there occurred one of the most frightful epidemics just in the manner told above. From one pig the parasites were conveyed to three hundred and sixty-one known persons, of whom no less than fifty-seven died within four weeks. This was in the villages around Emersleben, Saxony, having only about sixteen hundred inhabitants. Such outbreaks have been by no means uncommon in Germany, owing to the habit universal there of eating raw or imperfectly cooked meat. To every prudent person it has become an inviolable rule, never to eat any sausage, pork, or ham, which has not been very thoroughly cooked, so that any encapsuled trichinæ in the meat may have been killed by the heat.

FIG. 197. — *Trichina spiralis.*

The trichinæ are found in the muscles in immense numbers, and often closely crowded together, each worm looks like a coiled hair, whence the name *Trichina spiralis.* The oval capsules are not over a fortieth of an inch in length, while the worm of course is very much smaller, and always coiled. They were first found by Hilton and first named by Richard Owen, who published a very inaccurate description of them in 1835, but the origin and dangerousness of these parasites was never suspected until 1860, when the investigations of Zenker, Leuckart, Pagenstecher, and Virchow suddenly revealed the danger by ascertaining the history of the trichina and its pathological action. The whole civilized world was horrified at the discovery of the great and unthought-of risk which our daily meals involve. Everywhere the information was spread, and never, perhaps, has any subject occasioned such universal discussion as this new knowledge. In the intestine the trichina grows rapidly, the female becomes about an eighth of an inch long, but the male does not measure over a sixteenth. The eggs in the female have no shell, and develop to tiny embryos within the body of the parent. The adult worm is thin and tapering towards the head. A number of related species and genera are known, but none are so dangerous as the *Trichina spiralis*, although one form, the *Trichocephalus dispar* is not rare in the human colon.

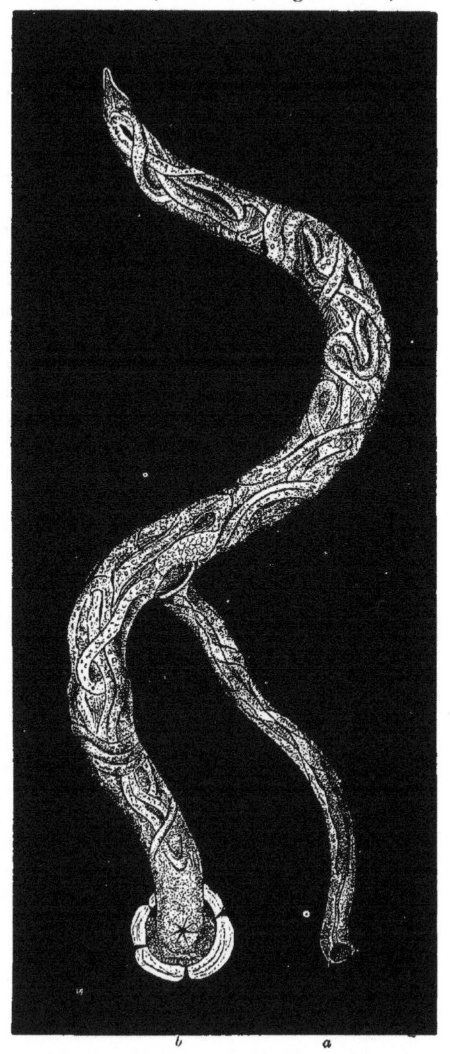

FIG. 198. — *Syngamus trachealis*; *a*, male; *b*, female.

Of the STRONGYLIDÆ several species attack man, for example *Eustrongylus gigas*, *Dochmius duodenalis*, etc., but the species we select for a special biography is the one so detested by poultry raisers as the cause of the "gapes." The parasite in question goes by the name of *Syngamus trachealis*, and inhabits the respiratory passages, tracheæ and bronchi of birds, living in bunches, which soon enlarge so much that breathing becomes difficult to the unlucky bird, which gasps for air, — the name "gapes" refers to the characteristic symptom of the disease. The generic name *Syngamus* refers to the peculiarity of the small males of attaching themselves to the large females to form indissoluble pairs. The constant effort to dislodge the parasites by coughing serves to expel the eggs laid by the female; the eggs are thus scattered about, and are swallowed again by other birds, and thus the disease is spread. In a hennery the malady readily becomes an epidemic; the sufferers should be isolated from the rest and kept carefully cleaned; it is said the parasites may be dislodged by brushing out the trachea with a feather.

The ASCARIDÆ also include several species long known as human parasites; two of these, the pin-worm, *Oxyuris vermicularis*, and the large round-worm, *Ascaris lumbricoides*, belong to the earliest and most familiarly known of parasites. The pin-worm is extremely common, and very generally distributed over the world. The eggs are so light that they are easily scattered about, and when swallowed along with some other object, they develop in the intestines to the adult worm. The habits of civilized life diminish the danger from these parasites, especially as modern systems of sewage constantly remove the eggs, which pass out with the natural evacuations. The *A. lumbricoides* grows to a large size, six or seven inches; is cylindrical, but tapers towards the head and somewhat towards the tail; the color is reddish brown. The animal is a parasite of the small intestine. The eggs pass out, and in water or moist earth await the completion of embryonic growth, but how the larva reaches the place of its final development has not yet been ascertained.

The MERMITHIDÆ and GORDIIDÆ were long placed near to, but apart from, the Nematoda proper, but the best opinion now groups them with this class in spite of certain peculiarities of their anatomy and development. Both *Mermis* and *Gordius* are commonly known as hair-worms, and are found while sexually immature in the body cavities of various insects; but they become mature only in the free state, *Mermis* living in the earth, *Gordius* in the water. The habits of *Mermis* have given rise to the belief in Europe in a rain of worms; for often in summer after a warm rain at night they swarm to the surface, and appear to have been indeed rained down. The larvæ are parasitic in caterpillars, but exactly how they gain entrance to the body of their insect host is not known. The metamorphoses of *Gordius* are still under discussion, for M. Villot, who has published a long memoir on the subject, does not agree with earlier observers. "The eggs of *Gordius variabilis*," writes Dr. Leidy, "are extruded in a delicate cord resembling a thread of sewing cotton; the eggs are very minute, and as the parent may be a foot long, it is able to produce an enormous number of young; Leidy estimates over six million. The development of the young is readily observed from day to day, and it takes about a month before the process is completed. . . . In about four weeks the *Gordius* escapes from the egg, totally different in appearance from the parent. The newly developed *Gordius* is about $\frac{1}{250}$ of an inch long. The body is constricted just posterior to the middle, so as to appear divided into two portions. The anterior thicker portion of the body is cylindrical, distinctly annulated, and contains a complex apparatus, which the animal is capable of protruding and withdrawing." Meissner observed the larvæ of another species penetrate the larvæ of May flies and caddis flies. Villot maintains that there always follow several successive migrations before the worm reaches its last host, but his observations are not convincing. The adult *Gordius* is sometimes found in the water, and country superstition affirms that a horse hair has fallen in and been changed to a worm. *We* know better!

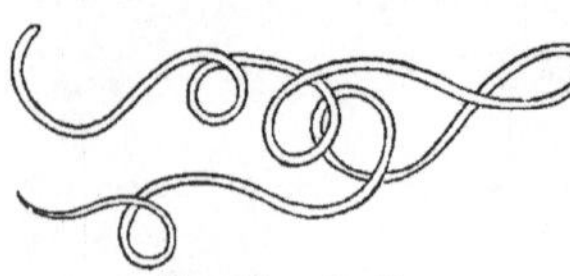
FIG. 199. — *Gordius.*

FIG. 200. — Larva of *Gordius.*

CLASS V. — ACANTHOCEPHALI.

We again enter the dominion of parasitism. The Acanthocephali, or barbed-headed worms, comprise the single genus *Echinorhynchus*, a genus rich in parasites which infest the higher animals. Of them all the best known is the *Echinorhynchus gigas*, which is especially inimical to the pig, but does not hesitate to make its home in the intestines of many other species of mammals, and even, it is said, in man. It grows to over a foot in length, and is, towards the middle of its body at least, thicker than a lead pencil. Other species abound in fishes, big and little. All *Echinorhynchi* (there are over one hundred species known) have an elongated sack-like body, without any digestive canal. Their essential external characteristic is the short, more or less nearly cylindrical proboscis, which is armed with several rows of hooks or barbs, which, like those of the tape-worms, serve to anchor the parasite in the tissues of its host. The eggs are generally spindle-shaped, and are retained in the body of the mother until the embryo is developed; the embryo is enclosed in several membranes, and has a bilateral armature of thorns around the anterior extremity. The embryos are supposed to pass the larval state in some aquatic host; but the life history of these animals is still very imperfectly elucidated, therefore it does not appear desirable to attempt a fuller account of the migrations and metamorphoses of these parasites, because such an endeavor would lead only to an unsatisfactory compilation of fragments.

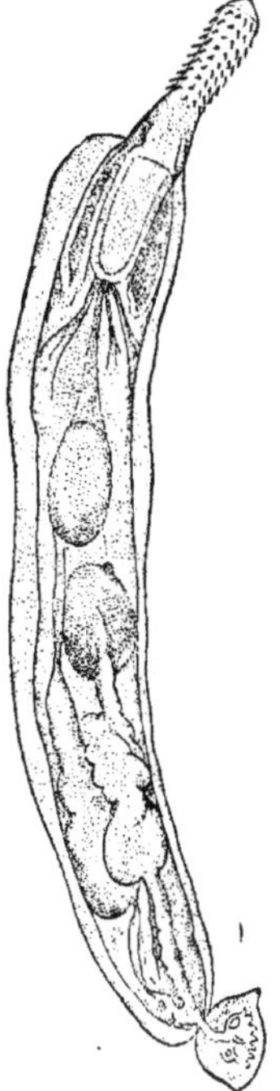

FIG. 201. — *Echinorhynchus anthuris.*

The internal anatomy of the Acanthocephali has been quite thoroughly studied, but we have thereby only gained a knowledge of the extraordinary singularities of their organization, without obtaining any real clew as to their affinities with other worms. The most we can safely assert is that they present some resemblances with the nematods, after which we have accordingly placed them in the system here adopted.

CLASS VI. — CHÆTOGNATHI.

Along the Atlantic seaboard the skimming net may gather from the ocean's surface considerable numbers of little elongated animals, perhaps half an inch or more in length, and somewhat javelin-like in shape. They move with quick jerks, which are highly characteristic. Examined more carefully they are seen to be almost pearly white in color, and quite translucent. Their form is nearly rod-like, with lateral fins; the anterior end is enlarged to make the head, which is further characterized by two dark specks, the pigmented eyes, and two groups of curving hooks, acting as jaws. On each side of the body is a horizontal membrane or projecting fin; this lateral fin varies considerably in size and outline in the different species, but is always most developed on the posterior half of the body, but is often subdivided, and in that case forms an anterior and posterior fin. A similar fin projects around the tail, and is probably merely a specialized part of the horizontal fin, which we may assume to have extended in its undifferentiated form along both sides of the body and around the tail as a continuous flap. The posterior portion of the body is divided off from the anterior by a transverse septum; the anterior compartment is much the longer, and contains the simple, straight intestine, which terminates just behind the septum and the ovaries; the smaller caudal compartment contains the male genitalia.

These worms are probably related to the nematodes or thread-worms, but their anatomical singularities are so numerous that it is well to keep them apart. They were first described about the middle of the last century by a Dutch author, Martin Slabber; for nearly one hundred years little was added to our knowledge of these worms, but, during the last forty years, their structure and development have been quite thoroughly worked up, principally by German zoologists. But even the complete information which we now possess has not yet enabled us to form any satisfactory or definite opinion as to the affinities of the group.

The number of species known is only thirteen, which have been separated into two genera, *Sagitta* and *Spadella*. Some of the species, at least, are cosmopolitan. The spawn may be obtained by confining the pregnant animals at the proper season, in most cases the spring or early summer. The spawn drops to the bottom of the vessel and lies there unattached. The ova are enveloped in a gelatinous mass, which does not surround each ovum separately, but belongs to the whole mass of eggs in common. In the species studied by Gegenbaur the process of segmentation occupies some seven to nine days, and it is probable that maturity is reached within a year. A singular interest attaches to the literature upon the development of *Sagitta*, in that it includes a publication by Charles Darwin, which contains one of the few serious blunders committed by the great English naturalist. He observed certain fish-eggs which he mistook for, and described in 1844 as, the eggs of this worm, and it was not until many years later that the error was corrected. There is no metamorphosis, but the embryo gradually assumes the adult form. The animals are predaceous, feeding upon small crustacea and other pelagic animals.

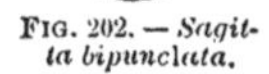

FIG. 202. — *Sagitta bipunctata.*

We ought to mention here, rather than elsewhere, two other aberrant types, which have as yet been only very incompletely studied. The little information we possess indicates that they are related to *Sagitta*, and also to the Nematodes. The first type is known under the name of the CHÆTOSOMIDÆ, and is probably the connecting link between the Chætognathi and thread-worms proper. There are two genera, *Rhabdogaster* and *Chætosoma*, both marine, and found crawling about over the surface of algæ. The former genus has been described by Metschinkoff. The mature female measures only 0.36 millimeters. The anterior

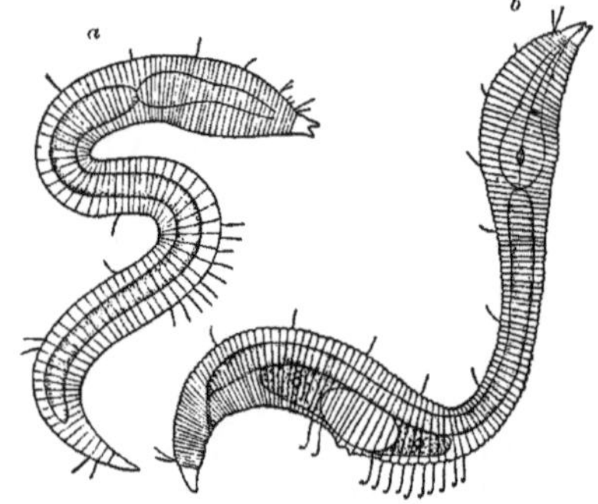

FIG. 203. — *Rhabdogaster cygnoides*: *a*, immature male; *b*, mature female.

end of the body, particularly the region of the œsophagus, is very much thickened; in *Chætosoma* this thickening ends abruptly posteriorly. The body shows fine transverse lines, and has hairs or bristles upon the back. The mouth is surrounded by three lips. The male has two hooks around the genital aperture, which are very similar to the so-called spiculæ of nematod worms. The most singular and characteristic feature of *Chætosoma*, however, is a double row of short rods on the ventral side, just in front of the anal opening; each rod has a knob at the free end. Nothing is known of the life history of these interesting forms. Hitherto they have been observed in Europe only, but it seems probable that they will yet be discovered in America.

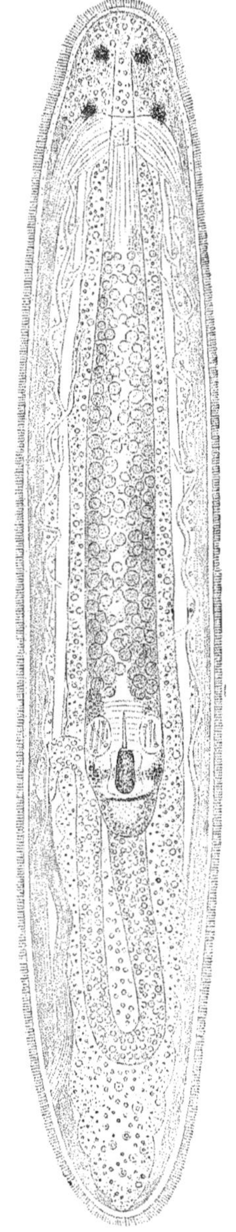

Fig. 204.—*Tetrastemma obscura.*

The second type, the Desmoscolecidæ, is also imperfectly known. It presents some indications of segmentation of the body, but whether it is really jointed like a true annelid is very doubtful. We cannot do more than give a brief description of a typical species, *Desmoscolex minutus.* The animal has a head-like enlargement of the anterior end of the body, and, behind that, a series of protuberant rings, producing an appearance of segmentation. The head carries two pairs, and each ring (save the eleventh and fifteenth) a single pair of bristles, which Greef states are used as locomotive organs. The sexes are distinct, and may be recognized by secondary external sexual characters. Three or four eggs are laid at a time, and borne about by the female for some while. The animal is marine.

Class VII.—NEMERTEA.

The nemertean worms are free-living animals, usually very much elongated in shape, but especially characterized by an enormously long slender proboscis, which when at rest, is withdrawn into the body, but is thrown out upon slight provocation. When one of these animals is captured, the proboscis is often broken off and might readily be mistaken for a second worm, for it is a long filament, which maintains its autonomic contortions for a considerable period. The Nemertea were long classed with the Turbellaria, but the union with the latter was based upon gross misconceptions, and we now know that they must stand as an independent class, having more affinity with the true annelids than with the plathelminths.

We may take the well known genus *Tetrastemma* as typical of the class, although most of the forms are very much larger. The species represented in our illustration is only two millimetres long, while some nemerteans exceed a foot in length. In *Tetrastemma* the opening of the mouth, and also that of the canal, in which the retracted proboscis lies, are situated in front on the under side; at the front also, but upon the upper side, are placed four groups of eye-specks, but in most forms there are only two such groups. The proboscis occupies the centre of the body, and is the most conspicuous of the internal organs; in it may be distinguished the stylet, which in the exserted proboscis projects from the tip. The presence or absence of this stylet serves to distinguish the two sub-classes into which this group of worms is divided; for the Enopala, to which the *Tetrastemma* belongs, are furnished with a stylet, while the Anopla are without. The second sub-class includes the majority of the better known and larger nemertean worms.

FIG. 205. — *Polia crucigera.*

Muddy bottoms, both in deep water and between the tides, yield to the digger many species of this class. Perhaps the most remarkable of these thus obtainable is the giant *Cerebratulus ingens*, which is met with under stones where there is sand. It is an enormous, smooth, flattened worm, yellowish, whitish, or flesh-colored, and sometimes grows to be ten or twelve feet long and over an inch in width. *C. rosea* also occurs in similar places; in form it resembles its larger relative, but is less flattened, smaller, and decidedly red in color; it is often covered by adherent grains of sand. Another species belonging to the nemerteans is also found on the New England coast in great abundance under stones from mid-tide to high-water mark; it is the *Nemertes socialis*, a gregarious species, many of them coiling together into large clusters; it is filiform, measuring five or six inches in length; its color is dark ash brown or blackish, a little lighter underneath, and it has three or four eyes in a longitudinal group on each side of the head. (Verrill.)

A very beautiful and common European species is the *Polia crucigera*, so called because its dark green body is marked by crosses, made by white stripes and transverse rings. Its favorite haunts are calcareous rocks bored out and excavated by other creatures, or else it seeks refuge among the branches of coral, as in our figure, which represents a very perfect specimen drawn life size.

There is, perhaps, nothing more interesting in the history of the nemerteans than the curious metamorphoses which some of the forms accomplish, although in other cases the development is direct. In the former instance the larva passes through the remarkable pilidium stage, so called, because the larva was originally described as a new animal under the name of Pilidium; in that condition the embryo is found swimming about in the ocean; it is almost microscopic in size, transparent and somewhat like an inverted bowl in shape, with two flaps hanging down from the edges, one on each side; on top a thick mobile hair, the flagellum; the edges of the bowl and the two flaps are fringed with delicate cilia to serve, together with the flagellum, as locomotive organs.

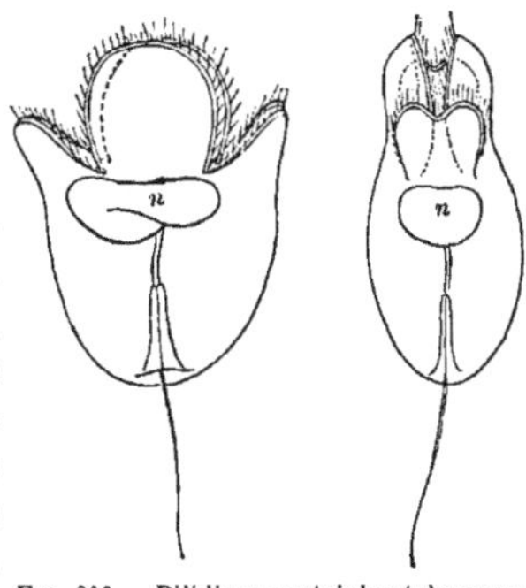

FIG. 206. — Pilidium containing (*n*) a young nemertean.

CLASS VIII. — GEPHYREA.

The Gephyrea are an illy-defined group of forms intermediate in many ways between the lower worms, particularly the nemertean type, and the higher worms or annelids proper. Until very recently the Echiuridæ, which Hatschek's brilliant researches have proved to be degraded annelids, were also included in this group, — and indeed this important rectification has not yet found its way into zoological text-books. The only true gephyreans are those known hitherto as the Inermes, while the Gephyrei Chætiferi (Echiuridæ) belong to the chætopods (see Annelida).

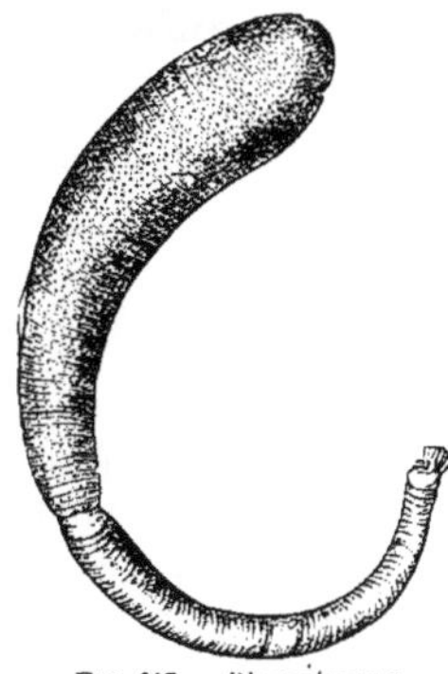

FIG. 207. — *Phascolosoma.*

The Gephyrea are not hermaphroditic; they are all marine, usually have a retractile proboscis, and always have a ring of nervous matter round the œsophagus and a ventral nerve cord; but their bodies are not segmented. The ventral nerve cord distinctly marks an approach to a higher type; in the lower forms the nervous system is far less concentrated. The true gephyreans comprise two families, Sipunculidæ, having tentacles around the mouth at the end of the proboscis, and Priapulidæ without tentacles. Of the SIPUNCULIDÆ we may consider *Phascolosoma* typical; an undetermined European species of this genus is represented in the accompanying figure. *Phascolosoma cœmentarium* is very common upon the sandy or shelly bottoms in deep water along all the northern coasts of New England. This worm takes possession of a dead shell of some small gastropod, like the hermit crabs, but as the aperture is always too large for its body, it builds out the rim of the aperture until only a small round opening is left, through which the worm can stretch forth the anterior extremity of its body, or withdraw into the shell it has appropriated for its dwelling. It lives permanently in its borrowed home, dragging it about by the powerful contractions of its body. The material with which the *Phascolosoma* patches out the shell so as to constrict the mouth thereof, is a hard and durable composition of sand and mud, cemented together by a secretion of the animal. When fully extended the forward part of the body, or so-called proboscis, is long and slender, and furnished close to the end with a circle of small slender tentacles, which surround the mouth. There is a band of minute spinules just behind the mouth. When the worm grows too large for its habitation, instead of exchanging it for a larger shell, as do the hermit crabs, it gradually extends its tube outward from the mouth of the shell, a labor sometimes rudely interrupted by a fish swallowing the worm, shell and all.

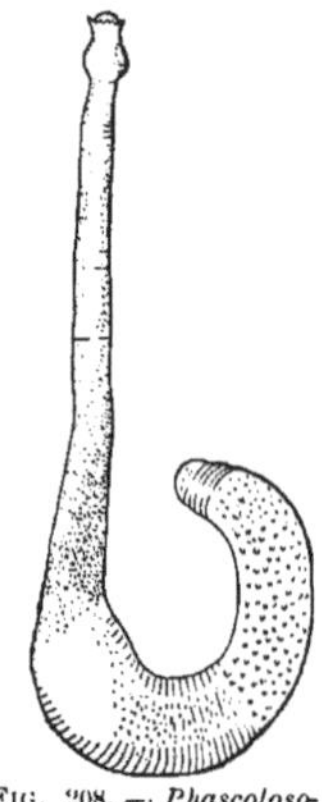

FIG. 208. — *Phascolosoma cœmentarium.*

Priapulus, of which we give an admirable figure, is the type of the second family, the PRIAPULIDÆ. Even its external form indicates the necessity of placing it apart from Sipunculids. The anterior club-shaped portion of the body is the proboscis with numerous dentate ridges. The tail is covered with numerous round papillæ. The genus is common in deep water along the shores of northern Europe, and burrows a hole in sandy bottoms, where it resides as happily as the *Phascolosoma* in its stolen shell, and leads an equally uninteresting life.

FIG. 209. — *Priapulus.*

The genus *Phoronis* stands quite apart, and in fact has been classed with the annelids by some writers, but our present knowledge indicates rather affinities with the gephyreans. It lives in tubes on the sea bottom, and has a crown of tentacles around the mouth. Its larva, known as Actinotrocha, is a remarkable creature, with long ciliated arms about its body, and changing into the adult worm by turning itself apparently inside out.

CLASS IX. — ANNELIDA.

We pass now to the last of the worms in the ascending series, the large and varied class of segmented or jointed worms, which includes the leeches, the common earth worms, and an immense number of marine, fresh water and terrestrial species, which attract the naturalist by the wonders of their organization, development and habits. Every sort of life which is open to crawling and swimming creatures is led by one or another of the annelids; some grovel in the earth or live in mud; some take refuge amid rocks and corals; some wander freely over the ocean, and are gaily decorated; others are mean parasites, feeding and dwelling on larger creatures, and leading greedy, selfish lazy lives, like human criminals, stealing the property of others.

The annelids attain the highest anatomical grade known within the vermian type. By this I mean that their organs are more complicated, or as one says, specialized. This advance is recognizable in all parts; the sensory apparatus is much perfected, especially the eye and ear have become capable of better functions than in lower worms. The nervous system has distinct centres, or ganglia, of which the largest lies in the head and is comparable with a true brain, and exhibits, upon proper microscopical examination, considerable complexity of structure. The peculiarity, however, which is most striking, and which gives the name to the class, is the division of the body into a succession of more or less similar short parts, known as rings, joints, or, more correctly, as segments. In the common earth-worm these joints are readily seen, being marked on the outside. The essential part of a true segment is a distinct division of the muscles, but besides we also find a separate nerve-centre for each segment, a separate excretory organ, separate markings and separate external appendages, all repeated for each segment. In the group known as the Polychæta, the external appendages are numerous and conspicuous, and serve to accentuate, in the eyes of even a hasty observer, the serial repetition of parts, which is the most obvious result of the segmentation of the body.

The annelids fall naturally into three well marked sub-classes, the Archiannelida, having no bristles and no suckers; the Chætopoda, having bristles upon their sides, and the leeches, having two suckers, one around the mouth and the other on the ventral side of the posterior extremity.

SUB-CLASS I. — ARCHIANNELIDA.

This sub-class includes a small number of worms, whose affinities and importance as the nearest living representatives of the archetype of the segmented worms was first pointed out by Hatschek, to whom science is also indebted for the recognition of the group, which is only just beginning to find its way into text books (1884). The best-known form is *Polygordius*, a long, slender worm, without bristles or appendages nor external joints, although its body is segmented. Its larva was long familiar to naturalists as a free swimming pelagic embryo, very minute indeed, and always designated as Lovèn's larva. Many an attempt was made to trace out the metamorphoses of this larva, but always unsuccessfully until within a few years, when it was discovered to be the young of *Polygordius*.

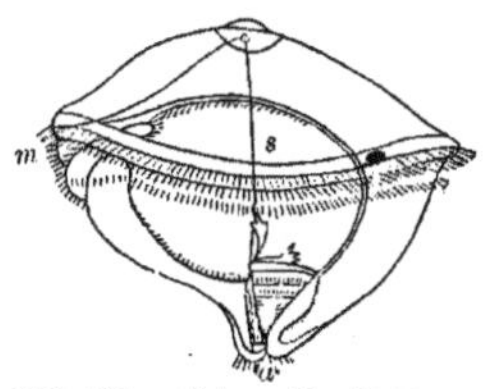

FIG. 210. — *Polygordius* larva; *a*, anus; *m*, mouth; *s*, stomach.

SUB-CLASS II. — CHÆTOPODA.

These worms may be easily recognized by their jointed bodies, and by the presence of two rows of spines on each side thereof. The places of insertion for the spines are often protuberant, and sometimes project so far as to form false limbs — (parapodia). We then have on each segment four of these outgrowths, each bearing a single spine or a group of spines. When upon each protuberance there is but a single, or very few spines, the worm belongs to the Oligochæta, the lower order of the sub-class; in the Oligochæta the protuberances are but slightly if at all marked; in the other and higher order, the Polychæta, there are usually true parapodia, movable like a limb and each having at the end a cluster of many bristles, often of bizarre shapes and brilliant shades of color.

Order I. — OLIGOCHÆTA.

The Oligochæta are all hermaphrodites, are without cephalic appendages, and have no armature around the mouth. The best known of the order, and indeed of the whole class of Annelida, is the familiar earth-worm, the favorite victim of an art which serves to lure boys to brooks, and to betray fishes into the frying-pan. We owe to Darwin a series of most interesting observations on the habits of earth-worms, and the role they play in the economy of nature, and we are almost equally indebted to the acute studies of Von Hensen. The characteristics of earth-worms are, the absence of appendages, the large number of small segments, the small size and number of the bristles, the want of sense organs, and the development of a belt or clitellus; this is a thickened portion of the body, not far behind the head, and having a smooth, glistening surface.

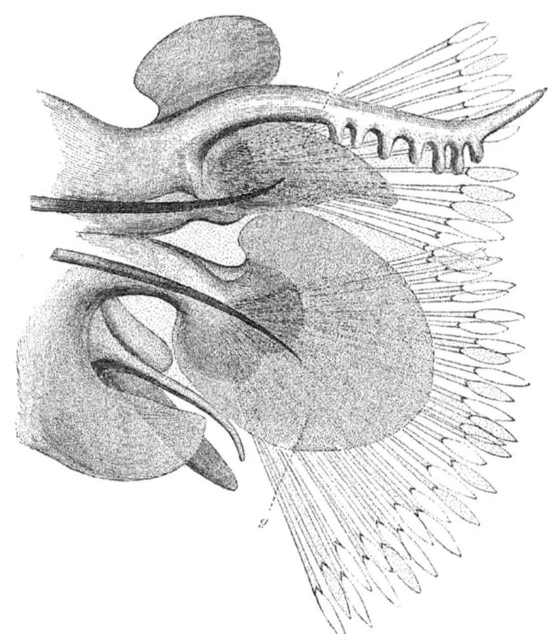

FIG. 211. — Parapodia of *Heteronereis*; *c*, upper, *g*, lower divisions.

Lumbricus terrestris is common both in Europe and America. I quote Linnæus' description as given in Turton's quaint translation: "Inhabits decayed wood and the common soil, which by perforating it renders fit to receive the rain; devours the cotyledons of plants and wanders about by night; is the food of moles, hedgehogs, and various birds. *Body* with about one hundred and forty rings, each of which contains four pair of prickles, not visible to the eye but discoverable by the touch; when expanded is convex each side, and when contracted is flattish beneath, with a red canal down the whole body; the belt is wrinkled and porous; mouth placed beneath the proboscis." Now this description is altogether inadequate for the discrimination of the species, nevertheless we will let it pass; still less satisfactory are the remarks on the habits, which we must amend and extend. First of all let us add that "belt" means the clitellus, which in this species includes six segments, so fused that the external demarcations are obliterated.

Earth-worms are essentially burrowing animals, nocturnal in their habits, although they sometimes leave their holes and crawl over the ground to a new locality, and also are occasionally active during the daytime. They are exceedingly dependent on moisture, for a single day in the dry air kills them, while on the other hand they will survive in water for a long period; hence, wherever there comes a "dry spell," they all retreat into the lower stratum of soil not yet parched by the heat upon the surface. I have known them to retire to a depth of four feet in a period of prolonged drought, which completely exhausted the moisture to that depth. In winter, too, they always go down below the frost and make a little hollow or chamber at the bottom of their burrow, in which they coil up to hibernate, often several of them getting together. They usually carry down with them a few small stones, for what purpose is not known. In summer they live close to the surface, if it is not too dry, shutting up the mouths of their tubes with little pellets gathered from about, or with their own castings. Keeping quiet during the day, they emerge at night, stretching forth the anterior end of their bodies and exploring the neighborhood; keeping, however, most of their long selves within doors and retreating entirely upon the least alarm; a jar of the soil, or light falling upon them, is sufficient to awaken their timidity and cause an instantaneous retraction of the protruded part of their bodies. But their habit of remaining so near the surface renders their timidity, even, an insufficient protection, for they are often discovered and dragged forth by robins and other birds, which, unlike Luther, esteem the diet of worms. The *Lumbrici* are omnivorous; beefsteak, cabbage, fruit, green leaves and dead, dirt, stones, broken glass are all swallowed with an impartiality that would do credit to Aristides. But, although it has its preferences as to what it will eat, *Lumbricus* is not content without dirt and small stones, or other hard, indigestible objects, together with more nutritious fare. Apparently from the dirt it is able to extract some matter, perhaps to assimilate the microscopic organisms it contains; the stones probably act as grinders, serving to crush the food proper and mix it thoroughly with the digestive juices. The earth-worm, then, passes through its intestine pretty much everything in and on the ground, which can possibly get through; but it discharges its castings upon the surface, a manure which is universally known as vegetable mould, but would be more correctly termed if called animal mould. Now, as the worms burrow in every direction, they constantly bring up from below and deposit on the surface, so that the superficial layer grows slowly but steadily. Thus it happens that if ashes are strewn on a field the earth-worm castings are deposited over them, gradually burying them until they finally disappear. In his book on the earth-worm, Mr. Darwin gives many instances of this apparent subsidence which, under the most favorable circumstances, goes on at the rate of two or three tenths of an inch per annum. We quote the following account from his pages: "Near Maer Hill in Staffordshire, quick-lime had been spread about the year 1827, thickly, over a field of good pasture land, which had not since been ploughed. Some square holes were dug in this field in the beginning of October 1837, and the sections showed a layer of turf, formed by the matted roots of the grasses, one half inch in thickness, beneath which, at a depth of two and a half inches (or three inches from the surface), a layer of lime in powder or in small lumps could be distinctly seen running all round the vertical sides of the holes." Even large objects, big stones and extensive pavements, are gradually buried by the worms, because their burrows extend underneath, and by their collapse let the overlying object sink, while their castings raise the surface around it. "When we behold," writes Mr. Darwin, "a wide, turf-covered expanse, we should remember that its smoothness, on which so much of its beauty depends, is mainly due to all the inequalities having been slowly levelled by worms. It is a marvellous reflection that the whole of the superficial mould over any rich expanse has passed, and will again pass, every few years, through the bodies of worms. The plough is one of the most ancient and most valuable of man's inventions, but long before he existed the land was in fact regularly ploughed, and still continues to be thus ploughed, by earth-worms. It may be doubted whether there are many other animals which have played so important a part in the history of the world as have these lowly organized creatures."

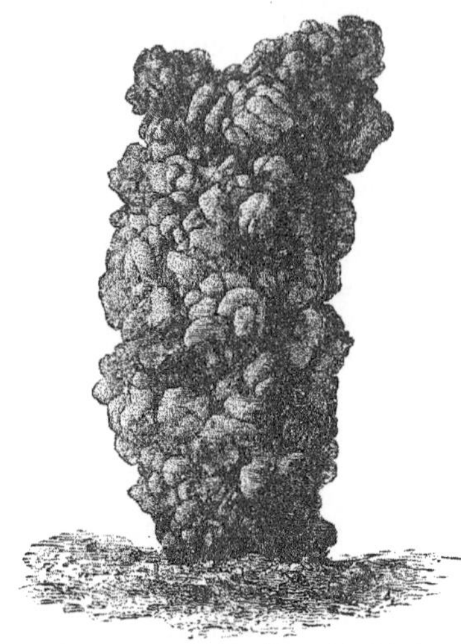

FIG. 212.—Tower of castings of *Perichæta* near Nice.

The amount of the castings is strikingly shown by those earth-worms which belong to the genus *Perichæta*, for these animals deposit their ejections in remarkable towers, which rise like turrets, with their summits often broader than their bases, to a height of two and a half and even three inches. Near Nice, in France, these towers abound in extraordinary numbers, and are probably formed by a species naturalized from the east. Mr. Scott complains of the trouble they cause in the botanic garden at Bombay: "Some of the finest of our lawns can be kept in anything like order only by being almost daily rolled; if left undisturbed for a few days they become studded with large castings." The period during which worms near Calcutta display such extraordinary activity lasts for only a little over two months, namely, during the cool season after the rains. Hensen believes that the importance of the earth-worm is not so much in the preparation of humus as in making passages for the roots of plants, and he describes the manner in which the burrows are utilized by plants. Indeed we owe to him the demonstration of the relation of the worms to the fertility of the soil, increasing it as just mentioned.

During the mating season the earth-worm leaves his burrow, seeking a mate. The eggs are laid in the ground, and are two or three lines in length. Our figure delineates one of them with the enclosed mature embryo, and its top closed by a valve-like structure adapted to facilitate the escape of the worm. The shell generally contains several yolks, but only one of them usually develops. It was once erroneously believed that *Lumbricus* might be multiplied by mechanical section, but although the front part of a divided worm survives, the back part dies, unless, indeed, when the front includes only the head and a few segments, for then the survival is reversed, the posterior division living on and manufacturing a new head for itself.

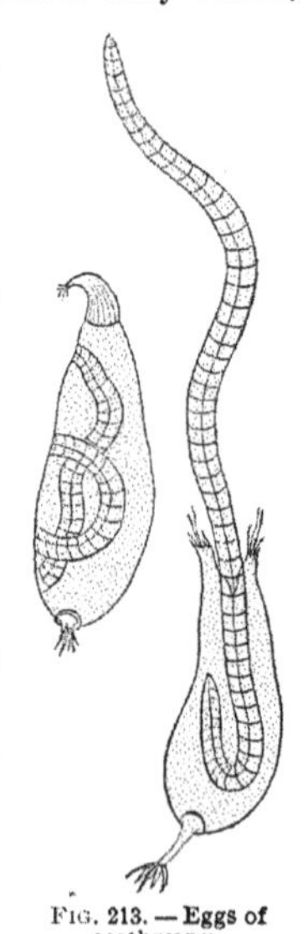

FIG. 213.—Eggs of earthworm.

The aquatic Oligochæta (Limicolæ) are very numerous both in species and individuals, and have been separated into four families. 1. PHREORYCTIDÆ, of which the type is a very long and slender worm, discovered by Professor Leydig, which displays a marked predilection for deep wells. As far as I am aware it has hitherto been found only in Germany. 2. TUBIFICIDÆ. The common and graceful *Tubifex* makes an interesting inhabitant of a small fresh-water aquarium. It is easily obtained by digging up some dark mud from the bottom of almost any meadow brook, and then placing it with water in a jar; when the settling is completed the worms will soon reconstruct their long tubes which run down from the surface of the mud; they will then stretch forth their long slender bodies, which undulate incessantly until some disturbance causes the frightened worm to jerk back into its domicile. The animal is long and slender as a thread, somewhat reddish in color and transparent. With a lens the segments of its body may be readily distinguished, and the little lateral bristles; those of the lower row are forked and hooked: similar bristles, together with simple hair-shaped ones, form the upper row. Along the New England coast the allied genus *Clitellio* is common, being, unlike *Tubifex*, an inhabitant of the ocean; it is found, in

FIG. 214.—*Phreoryctes menkeanus.*

company with two or three other cognate genera, under stones and decaying sea-weeds near high-water mark.

The members of the third family, ENCHYTRÆIDÆ, live in the earth, rotten woods, waters of swamps and the ocean. The red-blooded *Pachydrilus* may be taken as typical; it being a common marine genus. The last family, the NAIDÆ, comprises the best-known and most interesting members of the order, the two chief genera, *Nais* and *Chætogaster*, having been studied again and again by naturalists during the last century and a half, their wonderful reproduction by transverse division always possessing a vivid interest. The Naidæ are small, transparent worms, which may be readily captured by scooping blindly through the plants growing in fresh water, among which these creatures swim about. Many of them have a long snout or horn growing out from the head. Very common is the *Nais proboscidea*, which has a relatively immense appendage "poking out before its eyes." This long trunk is used to feel the way. Another member of the same genus, however, has a simple rounded head. The genus itself is easily recognized by the fact that the upper row of bristles on each side are hair-like, while the lower row are hooked; *Chætogaster* is characterized by having no dorsal, but only the ventral row of bristles. Both forms lay large eggs singly, enclosing them in protective capsules. It is, however, the asexual reproduction of these worms which is so interesting. There appears in the midst of the body a little zone of tissue, occupying at first less space than one of the segments between which it is interpolated. The microscope shows that this tissue is of a very elementary character, consisting of so-called embryonic or germinal cells. Gradually the tissue of this interpolated zone transforms itself into muscles, nerves, etc., and, growing meanwhile, it forms in front a new tail piece to patch out the anterior half of the worm, and behind it forms a new head for the posterior half of the original body. The zone then breaks, and there are now two worms, one with a new tail, the other with a new head. In *Nais* it is very common to see several "budding zones" at once in various stages of development. But the process of division does not always proceed in precisely the same way. The genus *Lumbriculus*, one of the Tubificidæ, also multiplies by transverse division, but it breaks in two first, and then develops the germinating tissue out of which the missing parts are redeveloped for each individual, a new tail for the front, and a new head for the hinder of the two. The new hind end arises as a little bud, consisting of new cells and ciliated over its surface, and subsequently forms the requisite number of segments. If the head is cut off it is reformed — a most conve-

FIG. 215. — *Nais proboscidea* reproducing by fission, enlarged.

nient arrangement, which, were it feasible with man, would essentially diminish the inconveniences of capital punishment.

The delicate fresh-water annelids are much preyed upon by carnivorous insect larvæ, and it is not uncommon to see a *Dytiscus* larva, for instance, seize one in its jaws and snip the poor worm in two, one half escaping. This mishap, which would be fatal to most animals, is only a temporary inconvenience to a *Nais* or *Lumbriculus*. It is evident that their extraordinary reproductive endowments must be one of the most important factors in the preservation of the species. Bonnet, one of the most accurate of observers, found that the process of regeneration is completed within a week, and proved that one worm, divided into several pieces, might produce, under favorable circumstances, an equal number of new complete individuals, so that sometimes the very act of destruction, as in the fabled hydra, multiplies instead of annihilating the victim.

ORDER II.—POLYCHÆTA.

The members of this order are generally diœcious, and pass through striking metamorphoses in the course of their development. The head is conspicuous on account of the feelers, cirri, and gills, which are often very prominent. They are nearly all marine. They far exceed all other worms in the variety of species and the diversity of their lives: indeed within our limits it is utterly impossible to refer even to all the families of the Polychæta, unless we should content ourselves with a bare catalogue. Roughly speaking, a polychætous annelid may be recognized by its jointed body, the false feet with numerous bristles, and the possession of cephalic tentacles. The order has two main divisions: 1, Sedentaria or Tubicolæ; 2, Errantia,—the former with fifteen, the latter with twelve families.

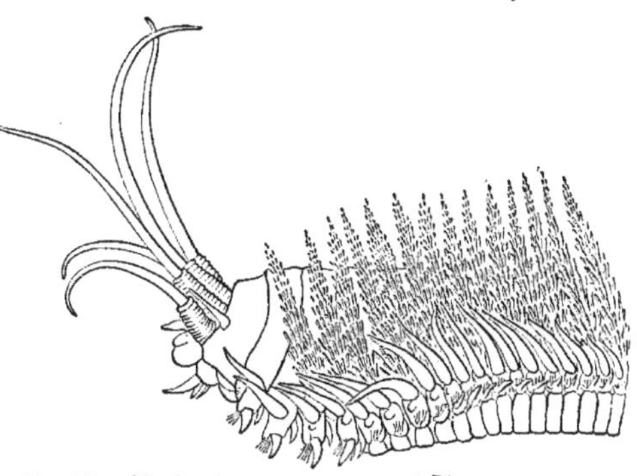

FIG. 216.—Head and anterior segments of *Diopatra cuprea.*

SUB-ORDER I.—TUBICOLÆ.

This sub-order owes its name to the general habit of building a tube in which the worm lives. The dwelling is constructed, now of one kind, now of another, of foreign particles, according to the tastes and habits of the builder, who cements his materials together by a secretion of his own body, or sometimes the secretion itself hardens, making a tube without extraneous adjuncts. From the fact that their lives are spent in this armor, the anterior end of the body becomes highly specialized, and is usually very different from the more posterior segments.

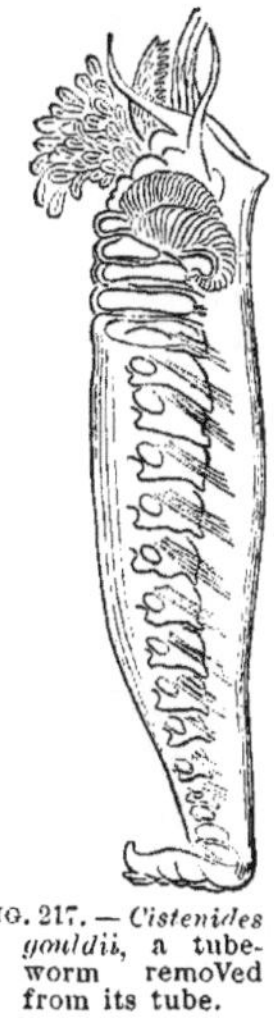

FIG. 217.—*Cistenides gouldii*, a tube-worm removed from its tube.

The handsome *Terebella* (*Amphitrite*) *ornata* of our North Atlantic coast, a large and interesting worm, is common both among and under rocks, and on muddy shores. It constructs firm tubes out of the consolidated mud and sand in which it resides, casting cylinders of mud out of the orifice. It grows to be twelve or fifteen inches in length, and is usually flesh-colored,

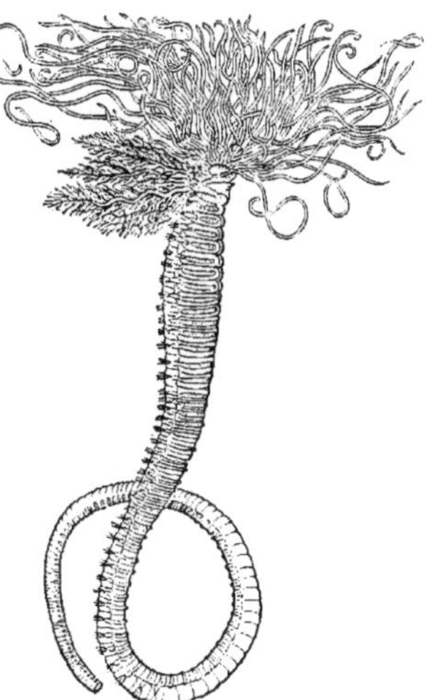

FIG. 218.—*Amphitrite ornata.*

although variable in hue from reddish to orange and dark brown. From the head spring very numerous flesh-colored tentacles, and three pairs of large feathered gills, which are bright red from the blood showing through them. The tentacles are capable of great extension, and may be stretched out to a length of eight or ten inches. They are incessantly in motion, apparently to gather food and materials for tube-building. This species may be considered a type of the large family of TEREBELLIDÆ, and possesses the following features characteristic of the family: The body is thicker in front, the thin posterior extremity bears no bristles; the tentacles are filiform; the head is not marked off from the body; the gills are confined to a few anterior segments; the bristles are short, those of the upper row hair-like, of the lower, hooked.

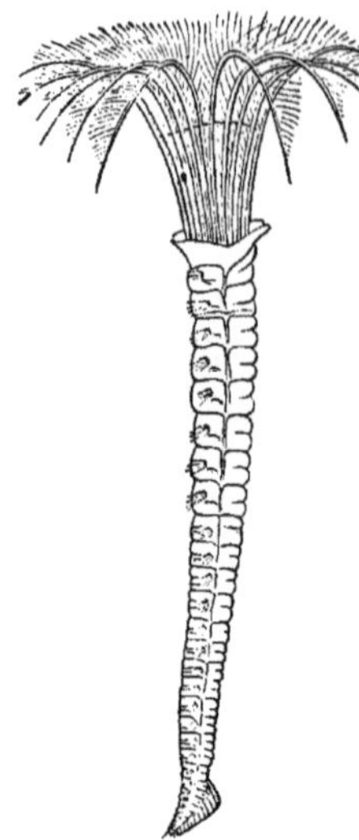

FIG. 219. — *Euchone elegans.*

Euchone elegans is a beautiful species found on the New England coast. When expanded, the pale yellow or greenish branchiæ are recurved and connected by a broad thin membrane. The body anteriorly is reddish, changing into flesh color and then into a darker green or brown as we proceed posteriorly. The species lives in water from five to thirty fathoms in depth, and makes slender tubes covered with fine sand.

In the large family of the SERPULIDÆ also, the gills are confined to the anterior end of the body, and are covered with cilia, which maintain a stream of water, which sweeps food down towards the mouth, which is placed at the base of the gills. The head is usually set off from the body by a collar; all the bristles are hair-like, except those on the anterior half of the lower row. Their larval life is free-swimming, but when they settle down they excrete a calcareous shell which the worms enlarge subsequently to meet the necessities of the growing inhabitants. The secretion and shaping of the tube are performed principally by the basal portion of the gills. The round, crooked tubes made by the American *Serpula dianthus* are often found on the under surfaces and sides of stones, and even in more exposed situations, — always near low-water mark. When disturbed, the worm suddenly withdraws its beautiful wreath of gills, and closes the aperture of its tube with a curious plug, called the operculum, — the portcullis of its castle. The branchiæ, when fully displayed, reveal their elegance of form and color; they are a round cluster, parted into symmetrical halves, some eighteen delicate feathered filaments on each side. The colors are extremely variable, but always brilliant; usually the branchiæ are purplish at the base, with narrow bands of light red or yellowish green; further from the centre the filaments are transversely banded with purplish-brown, which alternates with yellowish-green; in another variety they are all citron-yellow, and in yet another all whitish, banded with brown.

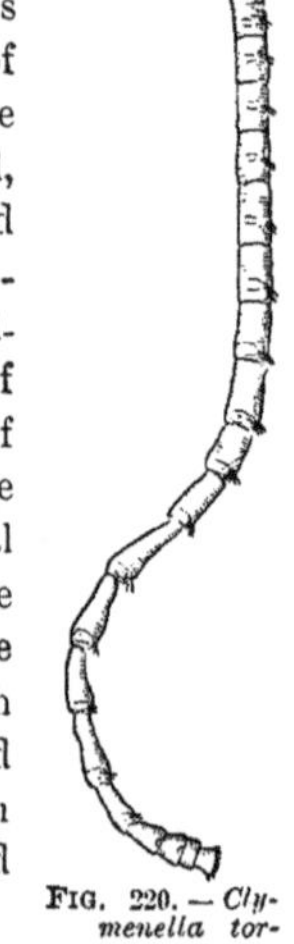

FIG. 220. — *Clymenella torquata.*

Very different is the abundant *Clymenella torquata*, which is plain in shape though pretty in color. It belongs to the MALDANIDÆ, Polychæta in which all the external appendages are very much reduced. *Clymenella* constructs long tubes of agglutinated sand, rather avoiding muddy bottoms. It loves quiet, and often seeks a home among the roots of eel grass. It is usually pale red, with bands of bright red around the swollen parts of the segments, but it is most readily recognized by the collar on the fifth ring and the peculiar funnel appended to the tail. There still remain a host of curious genera, *Sternaspis*, *Manayunkia*, *Polydora*, *Cirratulus*, *Capitella*, and many others, which we would fain describe, were it not for the painful conviction that the general reader's interest in worms, even in those that are polychætous, is exhaustible. We content ourselves, therefore, with a trio of brief allusions; first, to the lug-worm, *Arenicola marina*, which is eagerly sought after as bait by English fishermen, who dig it from the holes it excavates in the sands. On our coast it occurs north of Cape Cod, but is not used in fishing. The branchiæ are confined to the central portion of the body, where they form on each side a series of small tufts, remarkable, during the life of the creature, for their brilliant red color. It is a type of the family closely related to the *Clymenella*, above described. The second form is the *Spirorbis*, one of the Serpulidæ, whose white, coiled tubes might easily be mistaken

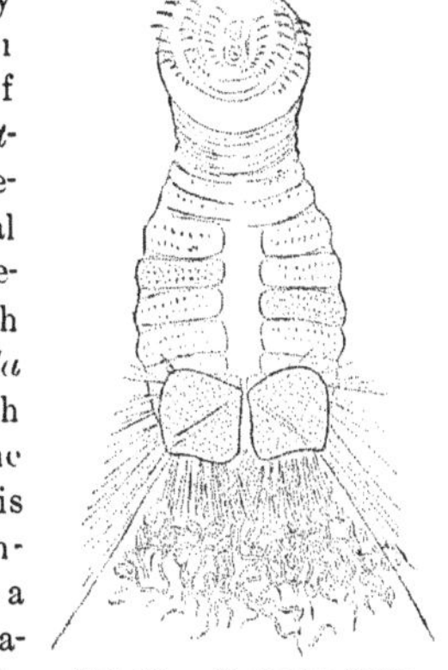

FIG. 221. — *Sternaspis fossor.*

FIG. 222. — *Serpula contortuplicata.*

for a snail shell. They occur on rocks, shells, etc., but are most numerous on bits of rock-weed (*Fucus*) thrown up from shallow water. Each individual worm is as pretty and delicate as any species of *Serpula*, and, like the members of that genus, when it retracts it closes its calcareous tube with an operculum. Over half a dozen species are found on our northern coasts. The last form is *Cistenides*, or, as it is called in the older works, *Pectinaria*. Our common species, *C. gouldii*, is light red or flesh color, handsomely mottled with dark red and blue. This species forms long conical tubes of sand, which are remarkable in the fact that the grains of which they are composed are built up, a single layer in thickness, "like miniature masonry, and bound together by a waterproof cement." The animal is shown in Fig. 217.

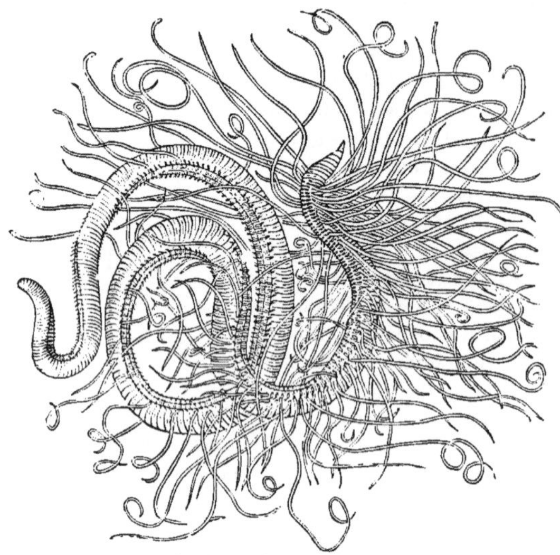

FIG. 223. — *Cirratulus grandis.*

SUB-ORDER II. — ERRANTIA.

The Errantia are active and fierce beasts of prey, of which the NEREIDÆ and NEPTHYDÆ may be regarded as the central type. The parapodia are large movable limbs, and bear numerous bristles varied in shape and color. The tentacles on the head are often of several sorts, and the segments of the body may vary in different regions, so that an anterior portion of the body looks very different from another posterior, as is so strikingly exemplified in *Heteronereis* and its allies; the segments of the bodies may also have gills of various kinds; in some cases these gills are long and delicate filaments which entirely change the physiognomy of the worm.

The term *Nereis* was given by Linnæus to a group of annelids, which he characterized as having an elongated vermiform body, furnished with soft, well-developed appendages, and a head bearing eyes and tentacles. He thus included nearly all the Errantia under one genus, and the tradition of the Linnæan name still lingers in the habit of naturalists, who use the term Nereis or Nereid as a vague designation. The genus has now been very much reduced, by cutting off hundreds of its original members to establish them under numerous new genera, so that the genus *Nereis* of to-day is but a very small fragment of the original one. This has been the general fate of all Linné's names, so that while we have kept the form, we have rejected the substance, because the essence of his system of nomenclature was to give two names, one to indicate the general place of the form in the animal kingdom, the other to designate the species; but the modern genera, unlike the Linnæan, are no longer general but special, and one of the chief merits of binomial nomenclature has been done away with.

Nereis pelagica is common on both sides of the North Atlantic, and is among the longest and best known of vermes, but on the New England coast the much larger *Nereis virens* is more readily found, for it lives in muddy and shelly sand between tides, preferring, however, to be near low water. *N. virens* grows to a length of eighteen inches or more. The color is dull bluish-green, with an iridescent tinge of red and other brilliant hues; the large lamellæ or gills along the sides are greenish anteriorly, but further back often become bright red, owing to the numerous blood-vessels which they contain. It is a very active and voracious worm, terrible to smaller animals, upon which it preys, capturing them by its large proboscis, which it suddenly thrusts out, seizing its victim with the two large jaws which arm the tip of its efficient weapon of attack; the proboscis is then withdrawn, and the food torn and masticated at leisure. These large worms, called "clam-worms" by the fishermen, are frequently dug out of their burrows and eagerly devoured by tautog, scup and other fishes, — in nature it is even thus, the eaters are in the end themselves eaten. In *Nereis* the proboscidean armature emulates in strength and sharpness the jaws of the ant-lion or some of the more formidable carnivorous beetles; nearly all the free-swimming Polychæta are similarly weaponed, and sometimes even more formidably. Thus, in *Phyllodoce maxillosa*, the fangs of a tiger seem to have been conjoined with the cutting teeth of a shark, to perfect such a model of carnivorous dentition as can find no rival in the animal creation. The *Nereis* does not always confine itself to its burrow, but, like all its relatives, frequently goes a-journeying. It is a nocturnal traveller, and at certain times swims about in vast numbers near the surface of the ocean; probably this habit has some connection with the reproduction. The life history of *Nereis* is still very obscure, for in some cases it produces sexually young which become *Nereis;* in other cases there intervenes what is known as the *Heteronereis* stage; *Heteronereis* again is capable of reproduction, and apparently the same species may assume different forms; moreover *Nereis* is found as a hermaphrodite as well as a unisexual animal. Now since the connection of these forms with one another has not yet been satisfactorily determined, the whole history of the manifold possible changes is in confusion.

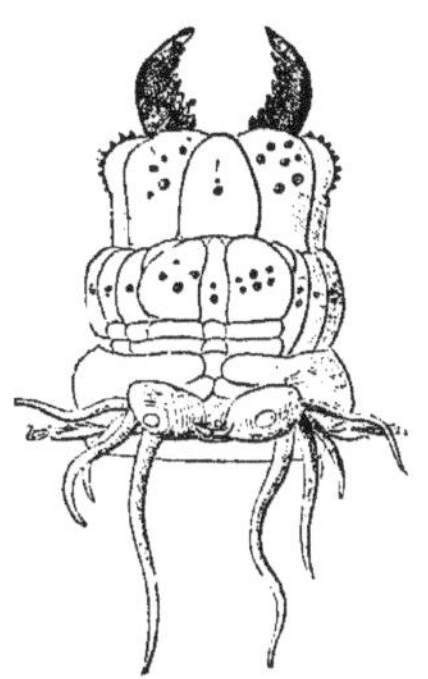

FIG. 224. — Head and jaws of *Nereis virens*.

FIG. 225. — *Phyllodoce.*

Very different is it in regard to *Autolytus*, whose vital career, at least for the present, is more comprehensible. The genus may serve also as the type of the SYLLIDÆ, one of the chief families of the order, and remarkable for the great length of the dorsal cirri of the body-segments. The eggs of *Autolytus* produce an asexual individual, which multiplies by division, the anterior end remaining the asexual worm, while the posterior individual is divided off and becomes male or female as the case may be. We have here a pure example of alternation of generations, a phenomenon first recognized about half a century ago. The individual born from the egg has neither the form nor value — that is to say, the physiological significance — of a sexual adult, but propagates itself by budding, division, or internal gemmation. Of this we have already given various instances in the sections referring to parasitic species among the lower worms. In *Tomopteris*, a pretty, transparent, pelagic creature, the false feet are very long and thick, while two long cirri spring from the first segment so that the outline of the animal is bizarre.

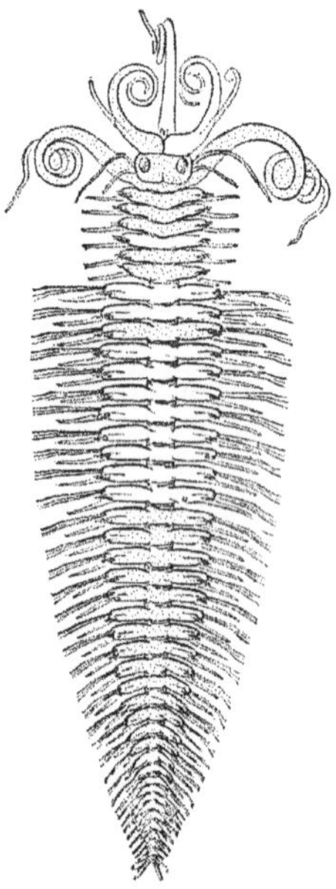

Another type altogether is shown by the scale-bearing annelids, APHRODITIDÆ; the upper parapodia, or false feet, carry large scales, which lie over the back of the animal and form an imbricated covering, serving the double purpose of protection and respiration. The most common of our species in New England is probably the *Lepidonotus squamatus*, which inhabits the rocky shores of bays and sounds, where they may be found hiding in crevices or on the under side of stones. It has twelve pairs of rough scales on its back, while its cousin *L. sublevis* has the same number, but smooth, and is found, though less abundantly, in the same localities as the first-mentioned species. In many members of this family, however, the bristles are greatly developed, in *Hermione* so as to partially, in *Aphrodite* so as to completely, hide the scales under a felting of hair. Nothing can exceed the splendor of the colors that ornament some of these hairs; "they yield, indeed, in no respect to the most gorgeous tints of tropical birds, or to the brilliant decorations of insects: green, yellow, and orange, blue, purple, and scarlet — all the hues of Iris play upon them with the changing light." In *Aphrodite* the respiratory function of the scales is evident; they exhibit periodical movements of elevation and depression; as they are overspread by a coating of felting, readily permeable to the water, the space between the scales is filled with filtered water while they are elevated; when they are again depressed the water is forcibly emitted at the posterior end of the body, and the back, which serves as an organ of respiration, is thus washed by fresh water. Although these animals are not active, yet they are highly organized, and are to be considered the tip-top of the world of worms, no slight dignity. They are rather inactive compared with many of their relatives, and are usually very dirty, so that repeated washings are necessary to uncover their natural beauties. Among worms, also, high rank does not ensure personal cleanliness.

FIG. 226. — *Autolytus cornutus*, male.

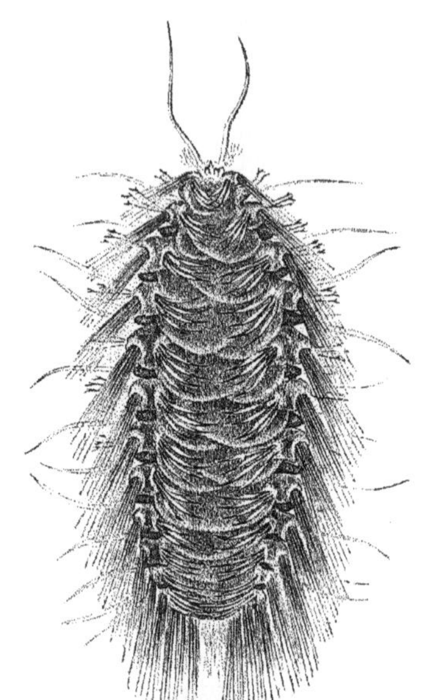

FIG. 227. — *Hermione hystrix*.

The ECHIURIDÆ, which were formerly classed with the Gephyreans, are now known to be true annelids, but their precise affinities are uncertain, so we will slip them in by appendix. They are easily recognized by the pair of hooked bristles on the ventral surface, and by the two crowns of bristles which occur around the caudal extremity of some forms. The three principal genera are *Echiuris*, *Thalassema* and *Bonellia*. The last-mentioned is very striking in appearance, as will be seen by the figure of *Bonellia viridis*. Oskar Schmidt, referring to his visit to the Dalmatian island Sesina, writes: "I noticed about a foot under water, beneath a large stone, an intensely green worm-like moving creature; I quickly lifted the stone, and my supposed worm revealed itself as the two-pronged proboscis of *Bonellia*. We kept it alive in a basin for a day, and never tired of watching its movements." The body is covered with little warts, and, like the proboscis, is vivid green; it is capable of manifold contractions and constrictions, and the proboscis is an even greater proteus, and may stretch out in large specimens to half a yard in length.

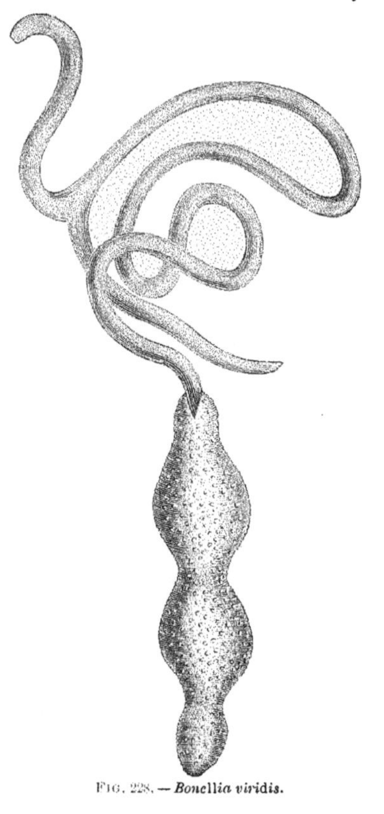
FIG. 228. — *Bonellia viridis.*

Myzostoma is another puzzle to zoologists, but is best guessed to be a degenerated parasitic annelid. The genus includes a considerable number of species which are all external parasites of the *Comatulas;* they are small, disc-shaped, have four pairs of lateral suckers on the ventral surface, and a retractile papillated proboscis, and there are five pairs of cirrus-bearing false feet.

SUB-CLASS III. — ENTEROPNEUSTI.

There now remains only one aberrant type for us to consider, namely, the whale's tongue. The singular and little known animals we have studied as isolated forms do not fall readily into any of the great classes, and this very fact of their standing so much apart renders them so much the more interesting to the thoughtful naturalist. Many of these species are rare, and it becomes therefore the more desirable to call general attention to them, in order that they may be sought and found by those who might otherwise let precious opportunities go by unutilized. Very few of them have yet been recorded from America, but there cannot be much question that many of them, together with others equally singular but yet unknown forms, will, in the future, be discovered in our fauna.

The whale's tongue, *Balanoglossus*, so named from a fancied resemblance, is a very interesting animal to the scientific zoologist. The adult worm was originally discovered at Naples; the free-swimming embryo was subsequently named Tornaria, and was long considered to be the larva of a starfish, until Metschnikoff established its real affinities by tracing out its metamorphosis into the adult. The shape of the transparent larva is well shown in the magnified drawing; it is just large enough to be recognized by the naked eye by those familiar with it. It may be caught in July and August by skimming the ocean surface with a fine net. Much as it differs from the adult worm, it yet passes by a series of gradual changes into the mature animal, which inhabits muddy bottoms between the tides. *Balanoglossus* is, I am convinced, a modified annelid, although its precise relationships are obscure, especially on account of the singularities of the nervous system. It has a long, tapering, fragile body, the anterior half somewhat flattened, the posterior rounded; at the anterior extremity is a large, pedunculated, top-shaped proboscis, followed by a thickened ring or collar; behind the collar follow the gills, a complicated set of branchial openings, recalling somewhat the branchiæ of the tunicates. Now gills mark an advance in organization from the vermian towards the vertebrate type, and *Balanoglossus* interests us scientifically just for this reason, that it appears in some respects a connecting link between widely separated divisions of the animal kingdom.

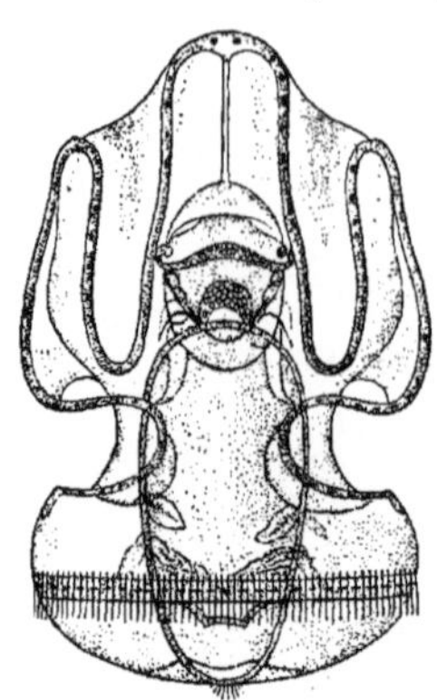
FIG. 229. — Tornaria.

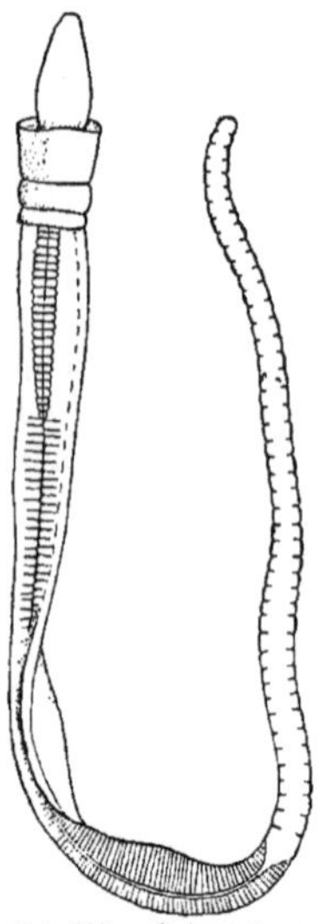

FIG. 230. — *Balanoglossus minutus.*

SUB-CLASS IV. — DISCOPHORI.

The sucker-bearing annelids, or Hirudinei, or leeches, are segmented worms adapted to a parasitic or semi-parasitic existence. They are all blood-suckers, veritable vampires, and to most persons the mere thought of their habit is revolting; but the antipathy they excite is not an altogether well-founded emotion, for they have their role to perform, and man has converted them into his servants, and given them a medical office, the duties of which they discharge with praiseworthy alacrity. The leeches have been the theme of one of the most singular of zoological delusions, in that they have been considered to be related to the Tremadota. I know no reason whatsoever for this queer conjunction, which is worthy of the time when a whale was called a fish. It is true that both leeches and flukes have suckers, but there all anatomical resemblance ceases. The leeches are true segmented worms, but the transverse lines visible on the external surface of the body do not mark off the segments, but, on the contrary, hide the true joints; hence their internal anatomy must be studied before the true plan of their organization can be recognized. In general shape we find that they are somewhat flattened; that they are usually broadest posteriorly, tapering off rapidly towards the tail, slowly towards the head; there is a sucker

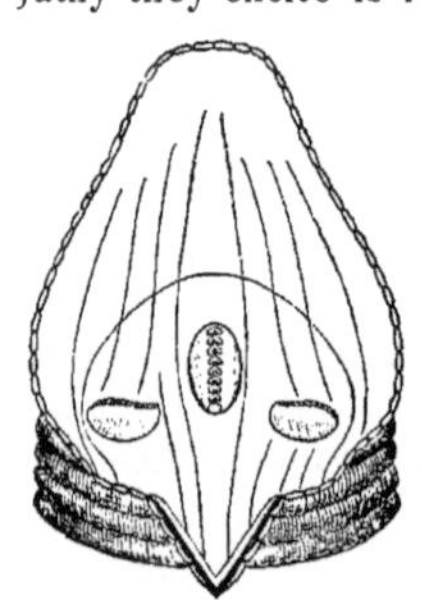
FIG. 231. — Sucker and jaws of leech.

around the mouth, which is armed with jaws, and a larger one at the tail; they are generally dark colored, very much mottled, often having fine lines and dots of bright hues; they are properly aquatic, and occur in both fresh and salt water.

There are three families, of the first of which, the GNATHOBDELLIDÆ, the medical leech, *Hirudo medicinalis*, is typical. The variety known as *officinalis*, measures, when at rest, some three or four inches. The color of the species is greenish or olive green, with six rust-red, thread-like longitudinal bands, speckled with black; ventral surface, greenish yellow, spotted with black. The mouth has three radiating jaws, with saw-like edges; when the animals bite through the skin, the wound made consists of three cuts radiating from a common centre, each jaw making a separate slit. The head is furnished with ten eyes and other special sense organs. The natural habitat of the common leech is in swamps and brooks where the water flows slowly; stagnant pools are unsuited to it; it preys on all vertebrates, both fishes and amphibia, and on mammals which come to the water to drink or bathe. It fastens itself upon its victim by means of its suckers, then cuts the skin, fastens its oral sucker over the wound, and pumps away until it has completely gorged itself with blood, distending enormously its elastic body, when it loosens its hold and drops off. Its attacks cause very little pain; boys in bathing are often feasted upon without being aware of it until they see the dark foe against the light skin. It is this power of extracting blood almost painlessly which has induced physicians to put them in requisition. They are generally kept by apothecaries, and supplying the market has become a considerable industry in western France. In that country it is said that they have leech plantations, swampy territories, which are carefully freed from all animals that might destroy the leeches, such as large frogs and certain fishes. To nourish them, worn out horses and cattle are purchased and driven into the leech enclosures, in which they are left to perish, death coming soon from loss of blood, and preceded by probably very little pain. The custom is not so humane as one would be glad to demand, but, on the other hand, there is no reason to suppose that it inflicts great suffering on the horses, etc. The leeches are collected in the fall, and, to a less extent, in the spring; for if gathered in summer they do not bear transportation well. In autumn, however, they are in the best condition, and, if captured then, will survive months without food if properly cared for. The animal casts its skin very frequently, and requires something to rub and scrape against, to remove the old slough, hence they must be supplied with plants and soil through which to crawl, as well as water. They move about either by crawling with the aid of their suckers somewhat in the fashion of an inch-worm, or else by swimming, at which they are adepts, but they evidently prefer to have a firm hold. Sometimes they plant themselves by the posterior sucker, and, stretching out their body, throw it into undulations, which pass from the head backward, and are sustained, an uninterrupted succession of waves, for long periods. When reposing, they assume all sorts of attitudes, to look at some of which makes one's back ache. The common way to catch them is for a man to stand in the water with bare feet and legs, then stir up the mud around him and pluck off those leeches which fasten upon him; some of these collectors become quite bloodless and sickly. Other modes of capture have been tried, but the belief that leeches must be captured upon the human skin still prevails, and leech culture therefore still goes on according to the ancient and semi-barbaric rules.

The eggs are laid in the ground; in the spring the leeches burrow into the moist earth, a little above the water level. Towards the end of June they form their co-

coons or egg-capsules, in each of which are several yolks, so that, when the young hatch out at the end of five or six weeks, several are born at once, and immediately make their way to the water. They grow very slowly. It is said that five years elapse before they attain their full size; they may live for twenty years.

Our waters contain many leeches similar to *Hirudo;* most of them belong to the genera *Nephelis* and *Hæmopis*. A giant among them is the big, spotted *Macrostomum* of our ponds. In southern Asia there live also terrestrial forms.

"Of all the plagues," writes Sir J. Emerson Tennent in his charming book on Ceylon, "which beset the traveller in the rising grounds of Ceylon, the most detested are the land leeches (*Hæmadipsa ceylonica*). They are not frequent in the plains, which are too hot and dry for them, but amongst the rank vegetation in the lower ranges of the hill country, which is kept damp by frequent showers, they are found in tormenting profusion. They are terrestrial, never visiting ponds or streams. In size they are about an inch in length and as fine as a common knitting needle; but they are capable of distension till they equal a quill in thickness, and attain a length of nearly two inches. Their structure is so flexible that they can insinuate themselves through the meshes of the finest stocking, not only seizing on the feet and ankles, but ascending to the back and throat, and fastening on the tenderest parts of the body. In order to exclude them, the coffee planters, who live among these pests, are obliged to envelop their legs in 'leech-gaiters' made of closely woven cloth. The natives smear their bodies with oil, tobacco, ashes, or lemon juice, the latter serving not only to stop the flow of blood, but also to expedite the healing of the wounds. In moving, the land leeches have the power of planting one extremity on the earth and raising the other perpendicularly to watch for their victim. Such is their vigilance and instinct, that, on the approach of a passer-by to a spot which they infest, they may be seen amongst the grass and fallen leaves on the edge of a native path, poised erect, and prepared for their attack on man and horse. . . . Their size is so insignificant, and the wound they make is so skilfully punctured, that both are generally imperceptible, and the first intimation of their onslaught is the trickling of the blood, or a chill feeling of the leech when it begins to hang heavily on the skin from being distended with its repast. Horses are driven wild by them, and stamp the ground in fury to shake them from their fetlocks, to which they hang in bloody tassels. The bare legs of the palankin bearers and coolies are a favorite resort; and as their hands are too much engaged to be spared to pull them off, the leeches hang like bunches of grapes round their ankles. . . . Both Marshall and Davy mention that during the march of troops in the mountains, when the Kandyans were in rebellion, in 1818, the soldiers, and especially the Madras Sepoys, with the pioneers and coolies, suffered so severely from this cause that numbers perished."

"One circumstance regarding these land-leeches is remarkable and unexplained; they are helpless without moisture, and in the hills, where they abound at all other times, they entirely disappear during long droughts; yet reappear instantaneously at the very first fall of rain, and in spots previously parched, where not one was visible an hour before, a single shower is sufficient to reproduce them in thousands. Whence do they re-appear? May they, like the rotifers, be dried up and preserved for an indefinite period, resuming their vital activity on the mere recurrence of moisture?"

The second family is distinguished by having a proboscis, and has, therefore, been named RHYNCHOBDELLIDÆ, of which the very common fresh-water *Clepsine* is a good illustration. *Clepsine* is remarkable because it carries its young about for some time attached to its belly. *Pontobdella* is a marine representative of the family, noticeable on account of the large size of the anterior sucker and the warts over its body. This greenish-gray leech lives on rays, and is apparently a lazy creature with dull senses. Its powerful muscles enable it to fasten itself upon a rock and sustain its body in a horizontal position for a long time, but it prefers to hang down, with the head rolled up.

Leeches are related to fisheries in three ways. Some of the large, blood-sucking forms, such as *Macrobdella* and *Hirudo*, attack many fishes directly, even when of considerable size, and destroy them very quickly by sucking their blood; some genera, like *Icthyobdella* and *Cystobranchus*, are true parasites, and often, when numerous, do the fish great injury. Others, among which belong *Clepsine* and *Nephelis*, destroy small molluscs and worms, which might otherwise become the food of fishes. On the other hand, the leeches are in their turn fed upon by the white fish of the lakes, and probably other fishes (Verril).

Here we must close, although we would gladly narrate the biographies of *Malacobdella*, *Branchiobdella*, *Histriobdella* and *Acanthobdella* — but we must leave the four 'Bdella's in the stables; we have driven far and rapidly through a large province of nature's realm, pausing to catch glimpses of a few of her "sights;" let us hope not to be of those travellers who "do" a country only to forget its appearance and character.

CHARLES S. MINOT.

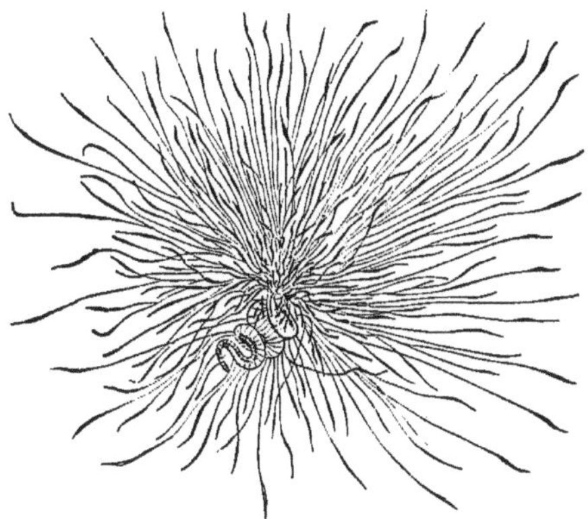

Polycirrus eximius, a tube worm with extended tentacles.

BRANCH VI. — MOLLUSCOIDEA.

Among the forms of disputed position occur two well-marked groups of aquatic animals, the Brachiopoda and the Polyzoa. In the older works the former of these was included among the Mollusca, while the second was accorded a place along with the hydroids in the heterogeneous group of Zoophytes. The next step was to recognize the affinities of the two, and, as a consequence, the Polyzoa were placed alongside the Brachiopoda, as members of the great group Mollusca. Then embryology was invoked, and, led by certain resemblances, many naturalists separated the brachiopods from the molluscs and gave them a place among the worms, the Polyzoa being dragged along with them. At present the tendency seems to be to recognize the affinity of the two groups to both the worms and the molluscs, while assigning them a place intermediate between the two.

At first sight the Polyzoa and the Brachiopoda seem widely different, but a deeper knowledge reveals many and important points of contact, especially in their early stages. They are all, with few exceptions, attached to some sub-aquatic object. In the adult all traces of metameric segmentation are lost. The tentacular apparatus is ciliated, and is borne upon a circular disc, or a two-armed process arising from the oral region. The alimentary canal forms a single loop, the mouth and anus (when the latter is present) being near each other. The principal nervous centre is a ganglion just beneath the œsophagus.

The larvæ of each, though presenting themselves in various forms, can be reduced to a common type, consisting of a body divided into two regions by a ring of cilia, in the anterior of which is the mouth, and not infrequently the anus as well. This larva can be but little removed from the trochozoon, the hypothetical ancestor of the worms and molluscs. Some recent investigations tend to show that the relationships supposed to exist between the two groups are really those of analogy, and not of homology, and that the Polyzoa have some connection with the rotifers, a group here treated of among the worms.

CLASS I. — POLYZOA.

The term Polyzoa, which means many animals, is highly appropriate for the group of aquatic forms which we now take up, for they, with a very few exceptions, form colonies composed of many individuals united in a common stock. Another name, which was applied but a year later, is Bryozoa, or moss-animals, a term no less apt, but debarred from use by the law of priority which governs scientific nomenclature. In general appearance many of the group closely resemble the hydroids, and especially the sertularians. Like them, they have a compound structure, and are enveloped by a cuticular sheath, while the circle of tentacles, which in life projects from the openings of the little cups, still further strengthens the similarity, and affords a justification for their association with those lower forms by the older naturalists. But when we come to study the anatomy and the development of these animals, important differences at once show themselves, and the resemblances which at first struck us so forcibly are seen to be of a merely superficial character.

As we have just said, most of the Polyzoa form colonies, the size of which is increased by budding, exactly as with the sertularians. In form they vary greatly; some, as *Gemellaria*, forming branching tree-like colonies, some, like *Membranipora*, spreading in flat sheets over the surface of submarine objects, while others, of which we may mention *Alcyonidium*, form soft and moss-like sheaths upon the rock-weed between tide marks. In *Rhabdopleura*, *Laguncula*, etc., a creeping root-stalk is formed, from which arise the cells, in a manner which strikingly resembles that of some of the campanularian hydroids; while in *Loxosoma* the individuals are separate, and no colonies occur.

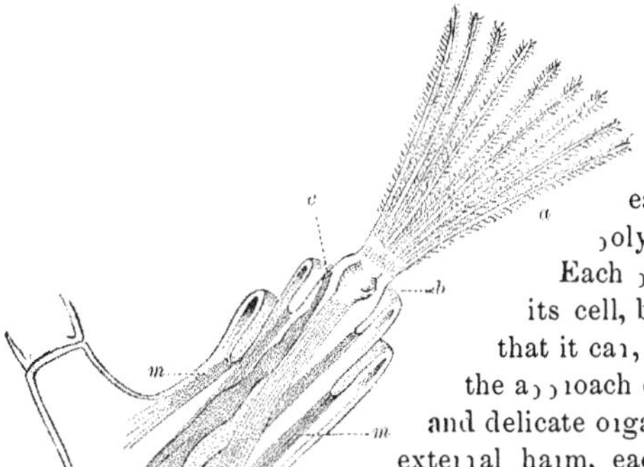

The chitinous or calcareous skeleton is composed of a series of cups, each of which contains one of the polypides, or individuals of the colony. Each polypide is fastened to the interior of its cell, but the mode of attachment is such that it can, at will, partially extend itself or, at the approach of danger, it can withdraw all its soft and delicate organs. To better afford protection from external harm, each cell of the colony is frequently armed with strong teeth or long spines, or there may even be an operculum developed, a little lid, which, when the animal is retracted, closes the opening through which the body extends itself at other times.

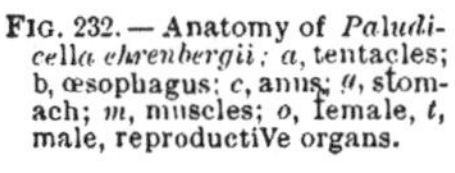

Fig. 232.—Anatomy of *Paludicella ehrenbergii*; *a*, tentacles; b, œsophagus; *c*, anus; *g*, stomach; *m*, muscles; *o*, female, *t*, male, reproductive organs.

When the polypide is extended, the most prominent feature is a disc, known as the lophophore, from which arises a more or less circular row of tentacles. Each of these tentacles is ciliated, and the constant motion of these small organs produces in the surrounding water currents which flow to the mouth, which in some is situated within, in others without, the circle of tentacles. The mouth communicates with a large pharynx, which in turn empties into the œsophagus, the distinction between these two being frequently emphasized by the presence of a valve. In several forms the œsophagus terminates posteriorly in a muscular gizzard, the function of which is to thoroughly triturate the food before it enters the stomach, the next division of the alimentary tract. The stomach is lined with small follicles, which are regarded as hepatic in function, while its upper portion bears numerous cilia, which, by their constant motion, keep the food in a state of agitation. The stomach is flexed upon itself, and after the food is digested, the excrementa are passed to the intestine, and thence out at the vent, which is placed close to the mouth.

No heart or circulatory organs exists in the Polyzoa, but the products of digestion pass through the walls of the stomach into the body cavity, where they bathe the various portions of the body. The nervous system is chiefly composed of a central ganglion placed between the mouth and the anus. In some forms, nervous cords have been described, connecting the various individuals of the colony, and although the nervous nature of these cords has been disputed, it is evident that some means of inter-communication exists, for there is frequently such a unison in the movements of the various members of a stock that no other explanation is possible. Nothing definite is known of the organs of sensation. The muscular system is well developed, the most prominent portions being the retractors and protractors of the lophophore.

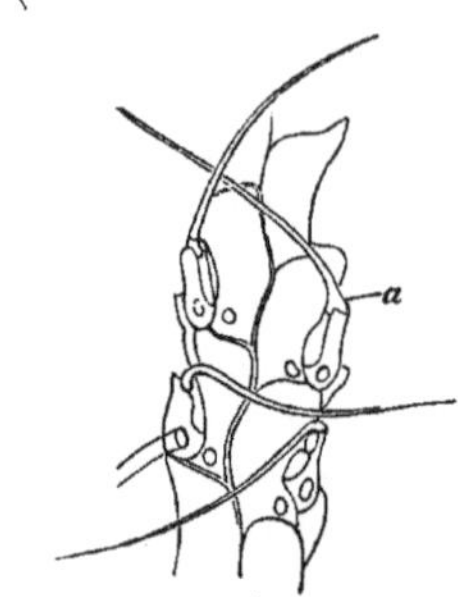

FIG. 233. — A portion of *Scrupocellaria ferox*, with (*a*) vibracula.

Possibly the structures known as avicularia and vibracula are the most interesting to the layman, on account of their motions and problematical functions. These organs are not found in all forms. The vibracula are long, whip-like appendages, which are attached to the cells of the colony by a single joint, and which, moved by appropriate muscles at the base, keep up a constant lashing motion. The avicularia, as is partially indicated by their name, are shaped like the head of a bird, with fixed upper and movable lower mandibles. These avicularia are either directly attached to the cell, or are elevated on a short stalk, and, in life, keep in constant motion, opening and closing the mandibles, thus rendering a colony of some such form as *Bugula* a most interesting object under the microscope. The purposes of these organs are as yet uncertain. It has been suggested that the constant lashing of the vibracula serves to clear foreign matter from the colony. The avicularia are frequently seen to seize small aquatic objects, but as they cannot carry the prey thus caught to the mouth, the part which they play in the nutrition of the polypide is at least indirect. Mr. Gosse, the entertaining English writer on natural history, has suggested, with considerable plausibility, that the decay of the objects caught by the avicularia attracts other organisms to the vicinity, thus bringing them within the influence of the currents produced by the cilia on the tentacles, and thus to the mouth.

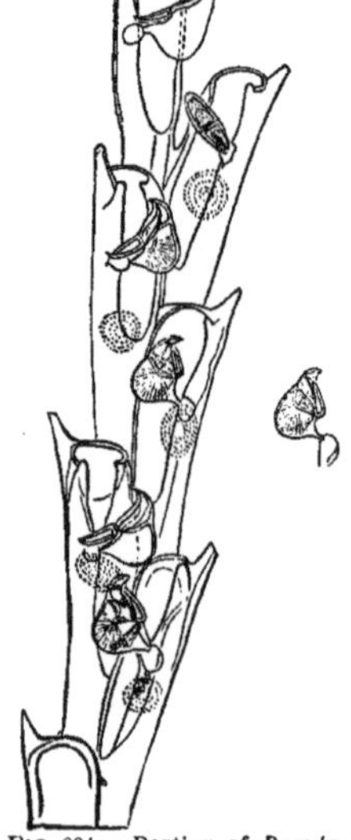

FIG. 234. — Portion of *Bugula*, showing birds' heads or avicularia.

The Polyzoa reproduce both by budding and by eggs. Usually the buds remain attached to the parent stock, thus causing it to increase in size; but in one or two forms the buds become separated from the parent, and form distinct individuals. Closely allied to this budding process is the formation of statoblasts. These are modified buds (and not true eggs), which are produced agamogenetically, the purpose of which, like the statoblasts of sponges previously described, is to perpetuate the species during the winter, or through a period of dry weather. The statoblasts arise from the funiculus, or cord which connects the stomach with the cell. As they increase in size, they become invested with a thick, horny brown envelope, which in many forms is ornamented by slender spines terminating with hooks, which more or less vividly recall the flukes of an anchor. Late in the autumn the fresh-water polyzoan dies, and then the statoblasts are set free to perpetuate the colony in the following

FIG. 235. — *Statoblast.*

spring. Then, under the influence of the warmth, the statoblast hatches with its organs already well developed. At first it swims freely through the water, but soon it becomes attached to some submerged object, with which hencefort h its fortunes are inseparably united. It now develops its capsule, and soon a bud is seen upon one side, which eventually grows into an individual, undistinguishable from the parent. This process is again and again repeated, until a large colony is formed, either extending its branches like a tree, or incrusting some submerged object with a gelatinous or calcareous envelope, forming in some instances clusters several feet in diameter and eight or more instances in thickness. This same process of budding takes place in the marine genera.

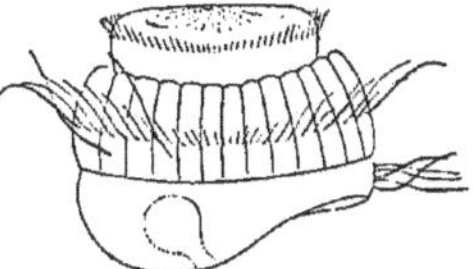
FIG. 236. — Larva of *Alcyonidium.*

We do not yet know enough about the development of the eggs of the Polyzoa to reconcile all the widely different features of the embryology. Still, all the various forms of larvæ may be reduced to a body surrounded by a ring of cilia (possibly corresponding to the lophophore), which divides it into two faces. On one of these is the mouth, and in some the anus also. On the other is a ciliated disc, by which, it may be, the animal attaches itself.

The Polyzoa first appear in time in the silurian rocks, and have persisted to the present day. The oldest forms known are referable to groups now living. At one time it was thought that the graptolites might belong here, but now the best authorities are inclined to place them among the hydroids

Sub-Class I. — Entoprocta.

The primary feature characterizing this group is the position of the vent, which is placed within the circle of tentacles, thus indicating that the group is the lowest of the sub-classes, a feature which is characteristic of the larvæ of some of the higher groups here persisting in the adult, as can be seen by comparing the figure of Cyphonautes (fig. 246), which is the larva of *Membranipora*, with that illustrating the anatomy of the adult *Pedicellina* (fig. 237). In this group the tentacles are not retractile, but can be rolled up.

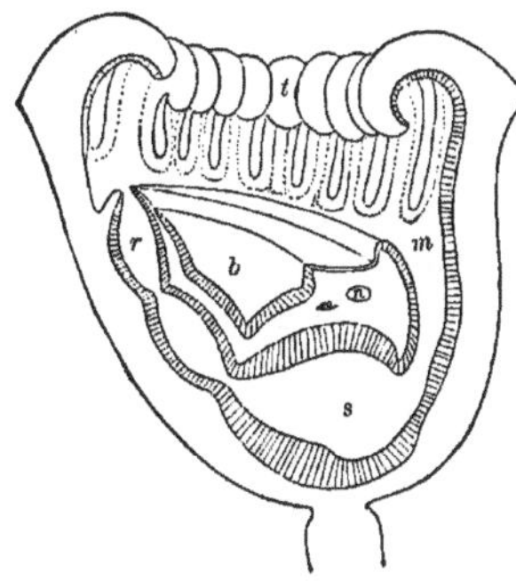

FIG. 237. — Anatomy of *Pedicellina*; b, blood pouch; *m*, mouth; *n*, ganglion; *r*, rectum; *s*, stomach; *t*, tentacles.

In the Pedicellinidæ there is a creeping rootstock, from which at intervals the long-stalked individuals arise and the colony increases by budding. On our coasts is found *Pedicellina americana*, which creeps over other Polyzoa, hydroids, etc., forming small white branching stems, the stalked individuals resembling so many clubs. *Urnatella* is a fresh-water genus, represented, so far as is at present known, by only a single species, *Urnatella gracilis*, from the Schuylkill River at Philadelphia. In this brightly colored form the cells are borne on the extremities of long noded branching stalks. In former years the species was very abundant just below the dam which supplies the city of Philadelphia with water, but now, doubtless owing to the pollution of the river by sewage, specimens are but rarely found.

The Loxosomidæ, which were first made known in 1863, are long-stalked, solitary entoproctous Polyzoa, without a partition between the cell and the stalk, and with a cement gland on the end of the stalk. The genus *Loxosoma* is represented in European waters by several species, distinguished, among other peculiarities, by the number of tentacles. These forms attach themselves to sertularians and other Polyzoa, and reproduce by budding. These buds, instead of remaining attached to the parent as in other Polyzoa, become separated, and settle down to begin life for themselves.

SUB-CLASS II. — ECTOPROCTA.

This division, which contains by far the greater proportion of the Polyzoa, is far more complicated in its structure than the Entoprocta. A most important distinction is found in the fact that the anus is placed outside the circular or horseshoe-shaped ring of tentacles. There is further a tentacular sheath. Other characters will be noticed in our subsequent account of the two orders into which the sub-class is divided.

ORDER I. — GYMNOLÆMATA.

The forms embraced in this order are almost wholly marine. They agree in having the ring of tentacles in a complete circle and in the absence of a lophophore, a structure which will be mentioned when treating of the other order. Statoblasts are but rarely present (as in the fresh-water genus *Paludicella.*) The larvæ leave the eggs as ciliated embryos, which swim freely for a time, and then settle down to spend the remainder of their lives attached to some submerged object, forming a colony by the process of budding. In the shape and constitution of the external skeleton the greatest diversity exists, it being sometimes calcareous, sometimes chitinous, and at others gelatinous. The order is divided into three sub-orders, founded upon the shape and ornamentation of the mouth of the cell containing the polypide.

SUB-ORDER I. — CYCLOSTOMATA.

The cyclostomatous Polyzoa, as is indicated by the name, embraces those forms in which the mouth of the cell is round and unarmed by spines, and in which, when the animal is retracted, the opening is not closed by an operculum. Most of the genera and species are extinct, yet many are found living in the colder seas, the sub-order reaching its highest development in Arctic waters. The living forms are arranged in six families, three of which (CRISIADÆ, DIASTOPORIDÆ, and TUBULIPORIDÆ) are represented on the New England coast. *Crisia eburnea*, of which we give enlarged figures, is an ivory white, calcareous species, frequently

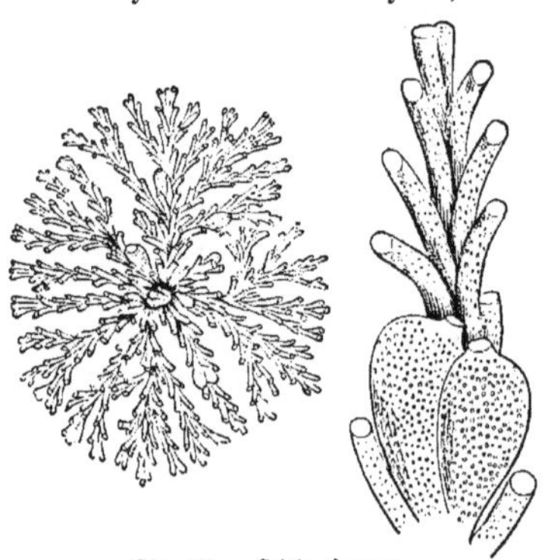

FIG. 238. — *Crisia eburnea.*

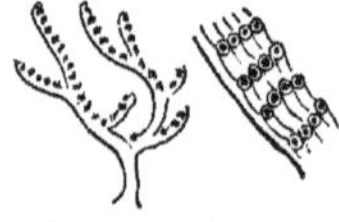

FIG. 239. — *Idmonea.*

found attached to seaweeds in tide-pools and in deeper water. In the Tubuliporidæ the cells are placed in rows, arranged transversely to the branches. Species of *Tubu-*

lipora and *Idmonea* are common in the shallow waters north of Cape Cod, one species of the former genus extending to the south of that barrier.

Sub-Order II. — Ctenostomata.

Here the cell is closed, after the retraction of the polypide, by processes of the tentacular sheath, or by bristle-like projections. Two families, represented by several species, occur on our coasts. In the Alcyonidiidæ the colony forms a fleshy or membraneous mass very irregular in form. Our most common species is *Alcyonidium hispidum*, and scarcely less frequent is the closely related *A. hirsutum*. These forms are found most abundant surrounding the stems of the rock-weed (*Fucus*), between tide-marks. The former is thicker, and may be readily recognized by the slender reddish bristles which surround the mouth of the cells. In the second species each cell forms a small soft papilla, from the centre of which the polypide protrudes itself. *A. ramosum* is a large branching species, which not infrequently forms colonies over a foot in length, the branches sometimes being nearly half an inch in diameter. In the Vesicularidæ the general form of the colony is a creeping or upright branching mass, from which the cells arise as free sheaths. In *Vesicularia* these sheaths are sessile upon the stock, while in *Farella* they are seated upon short peduncles. Several species of the former are common upon our coasts, while *Farella familiaris*, which extends from Long Island Sound to Europe, is found on rocks and sea-weed. "When it surrounds the stems of small algæ, the whitish pedicels project outwards, in all directions, and thus produce the appearance of a delicate chenille cord." The members of the family Paludicellidæ are inhabitants of fresh water.

Sub-Order III. — Chilostomata.

The Chilostomata are characterized by having the mouth of the horny or calcareous cell capable of being closed by a lid, while the oral area is usually membranous, rather than horny. It is in this group alone that we meet with the vibracula and avicularia, which we have already described, but their presence is not universal. It is divided into four super-families, the lowest being the CELLULARINA. Here the horny or slightly calcareous cells are tubular, funnel-shaped, the lower attached extremity being tubular or conical. In *Æta anguinea*, which may be taken as the type of the family Ætidæ, the tubular cells, with the mouths at the apex, arise from a creeping root-stalk, while in *Eucrate chelata*, representing the Eucratidæ, the mouth of the cell is one side of the extremity, and the cells are arranged in a single row. In *Gemellaria loricata*, a large form common in shallow water north of Cape Cod, we have a close similarity to the last species, except that the cells are arranged in pairs, back to back. In the Eucratidæ no vibracula or avicularia are present, while in the Cellularidæ these structures are usually found. The colony branches dichotomously, and the cells are arranged in two or more rows. It is represented in our waters by species of *Cellularia* and *Caberea*.

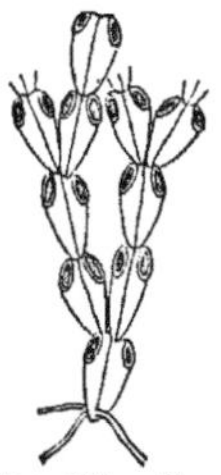

Fig. 240. — *Gemellaria.*

In the Bicellaridæ the cells are conical or quadrangular, and the large, laterally placed mouth is placed near the median axis. Our most prominent species of this family belong to the genus *Bugula*. Here the branches are arranged in a spiral, giving the colony a very graceful appearance. Occasionally the species are found in the greatest abundance, flumes leading to tide-mills being especially favorable localities. These forms are most favorable for studying the motions of the avicularia.

Fig. 241.—*Bugula turrita.*

The second super-family, the **FLUSTRINA**, embraces flattened forms with quadrate cells and an even surface. Very frequently the colony is reduced to a mere incrusting scale upon stones or sea-weeds. Our most common forms belong to the genus *Membranipora*, which is separated from *Lepralia* in having the anterior cell-wall membranous instead of calcareous, as in that genus. The edges of the cells are ornamented by long and slender spines, the numbers and shape varying according to the species. The species of *Flustra* assume a branching form, the branches being broad and flat.

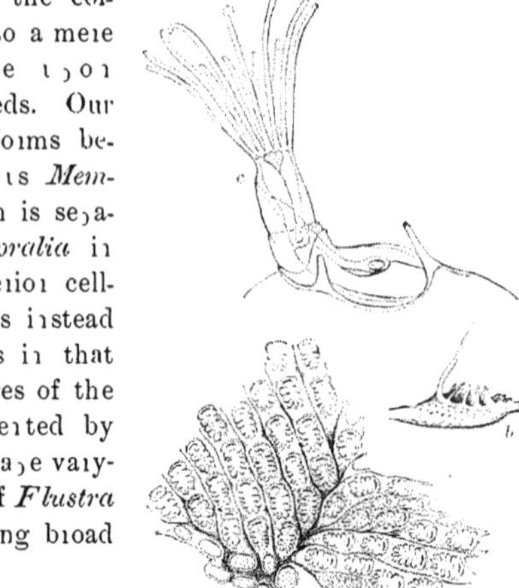

Fig. 242.—*Membranipora pilosa*; *a*, portion of a colony; *b*, side view of a single cell; *c*, a single polypidea; all enlarged.

The **ESCHARINA** are Polyzoa with a lateral opening to the quadrate or half-oval cell. The first family, the Eschariporidæ, has the cells rhomboid or cylindrical, while the opening is semicircular, with the anterior margin split or perforated with a median pore. In the Myriozoidæ we have erect forms, with more or less cylindrical branches, the posterior margin of the mouth of each cell being excavated. *Myriozoum subgracile*, which we figure, is found north of Cape Cod. The Escharidæ have the principal mouth of the cell semicircular or round, the secondary being reduced to accommodate the occasional avicularia. The colony may be either in the form of round branches or of broad, flat divisions, the cells occupying the opposite sides. The Discoporidæ have oval or rhomboid cells, with semicircular mouths, the posterior margins of which are armed with one or more spines.

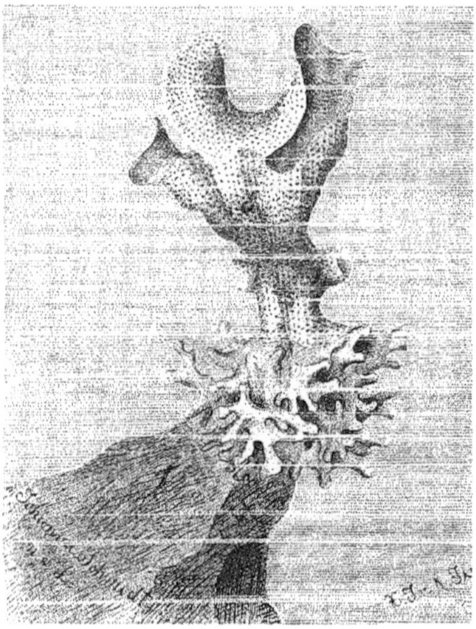

Fig. 243.—*Lepralia.*

In the **CELLEPORINA** the colony is calcareous, the cells being rhombical or oval, and the mouth is terminal. Two well-marked families exist. The first, Celle-

FORIDÆ, has the colony lamellar and irregular, incrusting the surfaces of sub-marine objects, or upright and branching. In *Cellepora* an avicularium exists in the median line, just behind the posterior margin of the mouth of the cell, while in *Celleporaria* it is absent. The RETEPORIDÆ are graceful forms, in which the cells unite to form a flat, leaf-like colony, which is perforated with numerous oval openings.

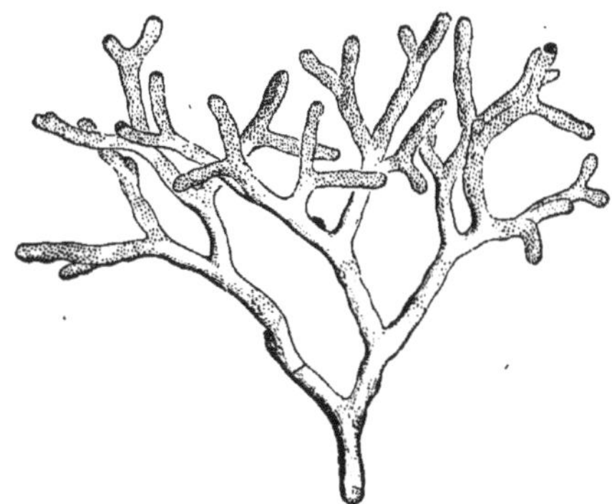

FIG. 244. — *Myriozoum subgracile.*

ORDER II. — PHYLACTOLÆMATA.

This order embraces the fresh-water Polyzoa proper, none of its members being found in the sea. They have the tentacles arranged on a horseshoe-shaped lophophore, while the mouth may be closed by a tongue-shaped lid, known as the epistome, which is placed just above the rudimentary brain. In size these forms

FIG. 245. — *Retepora cellulosa.*

are much larger than their marine relatives, while their general appearance is much more uniform. They are found beneath stones in running brooks, attached to submerged logs in lakes and ponds, and in the ease of *Cristatella* the whole colony is free, and has the power of motion as a whole. The number of species known, from the whole world, is relatively small, not over fifty, of which North America possesses about a dozen. The names which the different genera have received are more musical than in some other groups, owing to the diminutive "ella," with which most of them terminate.

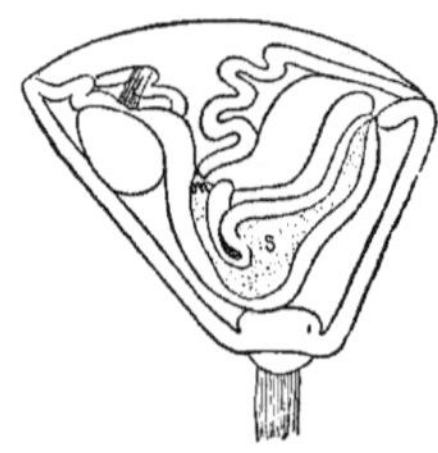

FIG. 246. — *Cyphonautes*, young of *Membranipora*; *m*, mouth; *s*, stomach.

The CRISTATELLIDÆ are readily distinguished by their free condition. They form large colonies, the individuals being arranged in concentric circles or ovals, on the upper surface, while the lower is modified into a contractile fleshy foot, of use to the colony in its slow, creeping motion. Two species are known in this country, and one in Europe, both belonging to the genus *Cristatella*.

FIG. 247. — *Fredricella.*

The family PLUMATELLIDÆ, which is much larger, embraces sessile forms, the various genera of which are distinguished by the gelatinous or parchment-like nature of the cells, the structure of the statoblasts (with or without spines), and the massive or branching nature of the colony. Four genera, *Fredricella*, *Plumatella*, *Lophopus*, and *Pectinatella*, are represented in the waters of eastern North America.

SUB-CLASS III. — PODOSTOMATA.

The genus *Rhabdopleura* so differs from the other Polyzoa as to warrant the erection of a sub-class for its reception. It approaches most closely of all the class to the Mollusca. It consists of a creeping root-stalk, of a chitinous nature, from which arise the tubular branching cells. Each cell has a round terminal mouth, and the walls of the cell are annulated for some distance below the mouth. The various cells of the colony are separated by transverse partitions. The long arms of the lophophore bear two series of tentacles, and resemble somewhat those of the Phylactolæmata, but much more closely those of the brachiopods. The animal is fastened to its cell by a long, contractile filament, by which it draws its body down out of the way of harm. When the danger is past, according to Sars it literally climbs out of its tube, by means of a disc between the arms, which appears to represent the epistome of the fresh-water forms. *Cephalodiscus* is the only other genus of the sub-class. It was dredged by the 'Challenger' expedition.

CLASS II. — BRACHIOPODA.

The shells of the brachiopods, at first sight, closely resemble those of the lamellibranch molluscs, and hence it is not strange that these forms were for so long a time associated together. Indeed, even at the present time most geologists fail to recognize the important differences, a fact probably due to their ignorance of the anatomy and embryology of the living forms.

The brachiopods, which are all marine, are provided with a bivalve shell, but the two valves of the shell are always dissimilar, while the two sides of each valve are alike, just the reverse of what obtains among the lamellibranchs. In the chemical structure of the shell, also, an important fact is to be noted, that phosphate of lime is present in much larger proportion than in that of any true mollusc. Most of the brachiopods are fastened to some marine object by a fleshy peduncle, which passes out between the valves in the centre of the hinge line, or in a corresponding position in those forms where no true hinge is present.

The inside of the shell is lined by a membrane, which is called the mantle, from its resemblance to a similar structure among the molluscs. Close up to the hinge line is the visceral mass, which is small in proportion to the size of the shell. The mouth is situated in the centre of the visceral mass. The œsophagus communicates with the stomach, into which open the ducts from the liver, while the intestine, in most forms, is short, and ends without any external opening, but in others is longer, and terminates in a vent on the right side of the mouth. The alimentary tract is supported in the spacious body cavity by a membrane analogous to the mesentery of the vertebrates. The body cavity is lined with cilia, which keep the contained fluids in constant motion, while prolongations of the cavity extend into the lobes of the mantle, thus forming a rudimentary circulatory system. The body cavity communi-

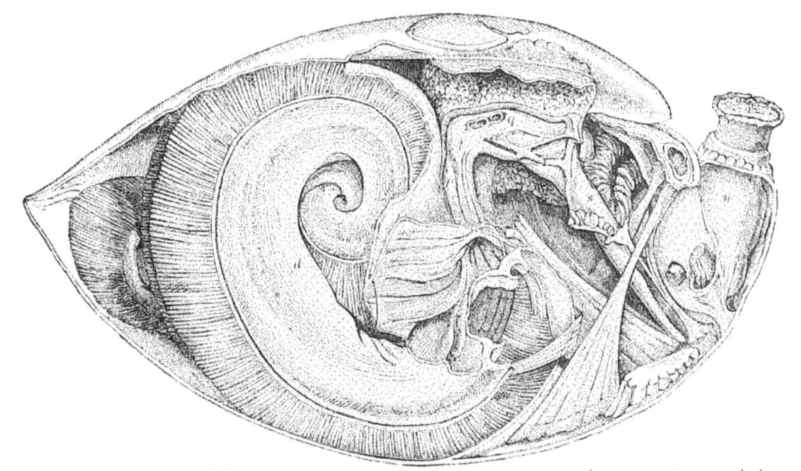

FIG. 248.—Anatomy of *Waldheimia*; *d*, arms; *m*, *n*, peduncle; *p*, œsophagus; *q*, stomach; *r*, liver; *s*, intestine.

cates with the exterior by two or four ducts, which in the older works were described as hearts, but it is now known that they are urogenital in function, and should be compared with the segmental organs of worms.

The nervous system is much better developed than in the Polyzoa, and consists of an œsophageal ring and, in the lower forms, two lateral cords; in the higher, of a more complex structure. No sense organs are known. The muscles which open and close the shell and control the other movements of the animal are well developed.

There now remains to be described, in this hasty sketch of the anatomy of the group, the arms, which in almost all forms are large.

FIG. 249.—Development of *Terebratulina*; *a*, three segment stage; *b*, attached; *c*, middle segment folding up to envelop the first; *d*, *e*, later stages.

These arms, from which the group receives its name (brachiopoda, arm-footed), arise on either side of the mouth, corresponding to the lophophore of the Polyzoan, and, like that structure, support a greater or less number of tentacles, like a fringe. These arms are long, and, in order to be accommodated within the shell they are folded, or coiled in a spiral. In some species they can be slightly protruded from the shell, but the extent of motion, in most forms, is small, since they are frequently supported upon a calcareous process of the shell itself, a structure frequently preserved in fossil forms. The Brachiopoda are divided into two groups, accordingly as the two valves of the shell are hinged or not.

ORDER I. — INARTICULATA.

In the Inarticulata, or Ecardinia as it is sometimes called, the hinge of the shell is wanting, as is also any calcareous support to the arms. The alimentary canal is complete, the anus emptying into one side of the chamber of the mantle. The borders of the mantle are completely separate.

The family LINGULIDÆ embraces forms which have lived on the earth since almost the earliest geological times. The genus *Lingula* appeared in the rocks of the Potsdam group, at the very base of the lower Silurian, and to-day species of *Lingula* are found in various warmer seas. In these the thin, horny valves of the shell are nearly equal and similar, while from near the point of attachment of the valves proceeds a long, fleshy stalk, or peduncle. The best-known species of *Lingula* to-day is *L. pyramidata*, occurring on the sandy shores of Virginia and North Carolina. In this form the stem is about two inches in length. The animal lives with its peduncle buried in the sand, in water of from one to ten fathoms, while the shells, in the centre of which is the mouth, project above the bottom. Not only is the genus a long-lived one, but the individuals themselves are able to withstand very adverse circumstances. Specimens can readily be carried to all parts of the country, and Professor Morse relates that individuals which he obtained survived after being several hours loose in his pocket. While our species of *Lingula* is small, those found in the eastern seas reach a length of nearly a foot. The development of our species has been studied, and Dr. Brooks says, "that the recent and fossil shells of the various species of *Crania, Discina, Lingula, Lingulella, Obolus*, and other hingeless brachiopods, furnish a series of adult forms representing all the changes through which the outline of the shell of *Lingula pyramidata* passed during its development." Of *Lingula* there are now seventeen living species, and a large number fossil. All the other genera of the family, *Obolus, Lingulella*, etc., are extinct.

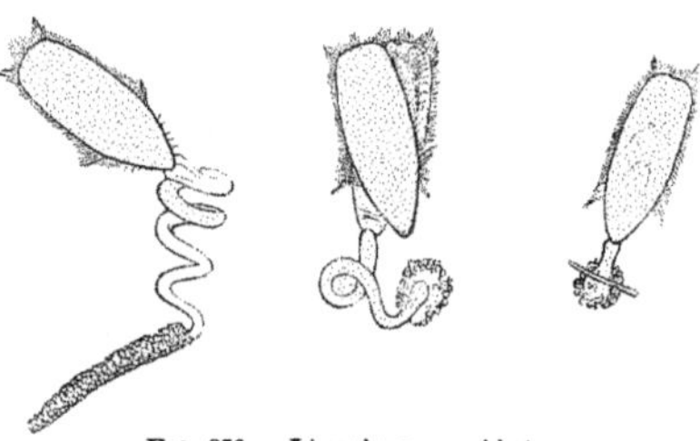
FIG. 250. — *Lingula pyramidata.*

The DISCINIDÆ, in which the shell is nearly circular and the peduncle passes through the flat lower valve, have only a single existing genus, *Discina*. These forms externally closely resemble the genus *Anomia*, a true mollusc. The CRANIIDÆ, represented in the seas of Europe by the genus *Crania*, have no peduncle. Fourteen species of the two families are found in the existing seas.

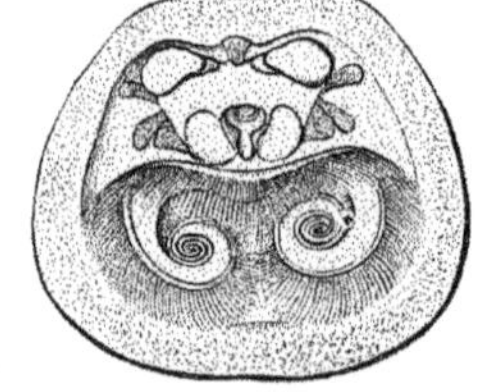
FIG. 251. — Upper valve of *Crania anomala*, with the animal.

Order II.—ARTICULATA.

The Articulata, or Testicardinia, have the valves articulated by a hinge, usually formed by teeth on the lower valve, fitting into sockets in the upper one. The intestine ends blindly. On the inner surface of the upper valve a more or less complicated calcareous loop, the object of which is to support the arms. In the existing forms this loop is usually quite simple, but in some of the fossils it is very complicated, portions being coiled in a spiral, which evidently supported all parts of the arms, so that their extension from between the valves was impossible. In the living forms a slight protrusion may be occasionally seen.

Passing by the three extinct families, Productidæ, Calceolidæ, and Orthidæ, we reach first the family Rhynchonellidæ, of which forms are represented in the northern seas. In these the arms are coiled in a spiral; the shell is either free or anchored by a peduncle, which passes through an opening in the beak of the larger valve. The hinge line is either curved or straight, and the outer surface of the shell is impunctate. *Rhynchonella psittacea* is a common form in the colder waters of the northern hemisphere, from the Gulf of Maine to Europe. Other species are found in Japan, New Zealand, Fijis, etc.

The Spiriferidæ attained its greatest development in the paleozoic rocks, disappearing in the jurassic. In these forms the shells are unequal, have a straight hinge line, while the support for the arms is coiled in two spirals, much like a watch-spring. Occasionally these spirals bear hardened supports for the tentacles, thus indicating that these parts could have but the slightest motion.

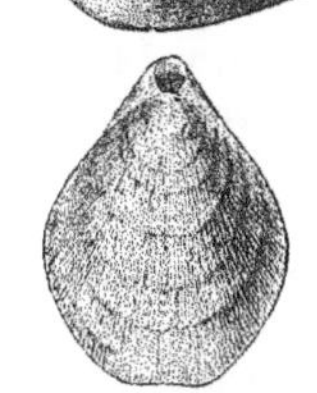

Fig. 252.—*Terebratulina septentrionalis.*

The Terebratulidæ is the largest of the recent families. In these forms the arms are not coiled in a spiral. The shell is punctate and ventricose, the lower valve is perforated for the passage of the peduncle, and the two valves are hinged together by two teeth. On our New England coasts, *Terebratulina septentrionalis* is the most abundant, being brought up by the dredge from a depth of only a few fathoms. Usually the specimens are encrusted with a yellow sponge. In life the animal has considerable powers of movement, raising itself at times so that it stands upright upon its peduncle, or twisting itself around upon the same support. In the more northern waters of America the genus *Waldhamia* is found, while the genus *Thecidium* is found in the Mediterranean and the West Indies.

These forms are popularly known as lamp shells, their rounded shell, with its perforated beak, presenting no inconsiderable resemblance to the lamps used by the ancients. The existing species possess no inconsiderable vitality, and Professor Morse has called attention to the striking fact that the power of the recent species to withstand adverse circumstances has a curious parallel in the history of the group, the *Lingula* of the Potsdam sandstone being congeneric with the forms living in Japan and the Carolinas to-day.

J. S. Kingsley.

BRANCH VII.—MOLLUSCA.

With the possible exceptions of the insects and the birds, there is no group in the animal kingdom which is such an universal favorite among all classes as the one now under discussion. This is very natural; for the hard armor which they bear, and the bright colors with which many of them are ornamented, renders them attractive, while the comparative indestructibility of the same shells renders the care of a collection an easy task. But while the collectors of the shells are many, the real students of the animals are few, and even now, although these forms have been collected and studied by conchologists for many years, a satisfactory classification is still desired.

The word Mollusca means soft, and it was applied by Linné to a group of animals embracing of the true molluscs only the naked forms, together with the hydroids, echinoderms and annelids, while the shell-bearing molluscs were arranged as Testacea in a section of his group of Vermes. Cuvier was the first to introduce order into the group. His studies during the seven years spent as tutor on the Normandy coast resulted in a classification of the Mollusca upon truly scientific grounds. The group, as recognized by him, embraced, besides the forms now admitted, the barnacles, the ascidians, and the brachiopods, truly a heterogeneous assemblage. In after years the Polyzoa were drawn in. The first of these groups to be separated were the barnacles, which were shown by Thompson to be Crustaceans in 1831. Then Kowalewsky, in 1865, described the embryology of the ascidians, from which it was apparent that they had no relationships with the molluscs, but were rather to be classed with the vertebrates, and lastly, the brachiopods were absolutely divorced from the group, taking the Polyzoa with them.

A concise definition of the Mollusca is impossible. Here, as elsewhere, nature refuses to be bound by strict rules, and the best we can do is to form a general conception which shall be true of the majority of forms, and which will, at the same time, be loose enough to admit all. A mollusc, then, is a bilaterally symmetrical, unsegmented animal, usually covered with a univalve or bivalve shell. It has a ventral, muscular portion (the foot) well developed; a symmetrical nervous system, consisting of a brain or supra-oesophageal ganglia, an oesophageal commissure, and a secondary brain beneath the throat. Most forms, in their development, pass through a trochozoon stage.

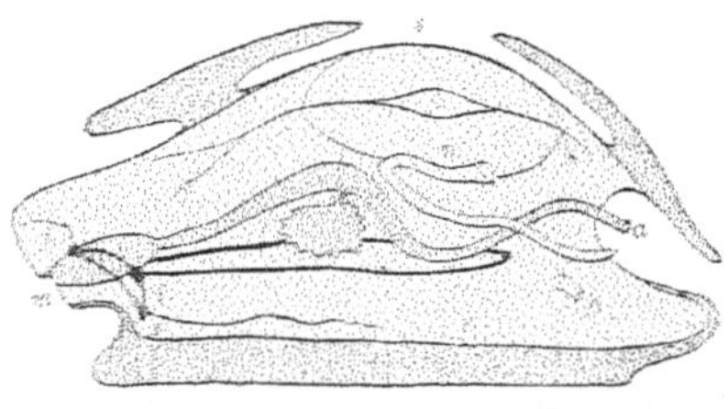

FIG. 253.—Diagram of mollusc: *a*, anus; *b*, brain, cerebral ganglion; *f*, foot; *g*, genital opening; *h*, heart; *i*, pleural ganglion; *l*, liver; *m*, mouth; *n*, kidney; *p*, pedal ganglion; *r*, visceral ganglion.

In some forms, especially in the gasteropods, the bilateral symmetry of the body is more or less obscured, owing to what may be called a torsion of the body, but, nevertheless, if we make allowance for this twisting, it can readily be traced. In the young, the segmentation of the body is frequently evident, but it entirely disappears in the adults, except among the chitons, where the elements of the shell and the gills are metamerically repeated.

The foot is a muscular process on the lower surface of the body, which is highly distinctive of most molluscs. In it one can frequently find three distinct portions in serial order, known respectively as the propodium, (in front) mesopodium, and metapodium. Occasionally lateral portions, epipodia, are developed. From the dorsal portion of the body arises a fold of the body wall, the pallium, or mantle, which partially or completely envelops the body. In some the two halves of the mantle may be distinct, while in others they are connected. This mantle plays no inconsiderable part in the economy of the animal, for from it is developed the shell so characteristic of most molluscs, and which deserves more than a passing mention.

The shell is largely composed of carbonate of lime, together with more or less animal matter, the whole being secreted by the outer layer of the mantle. This shell is entirely without blood-vessels, and is absolutely incapable of interstitial growth. Such being the case, it is an interesting question to decide how it increases in size. This is readily settled if we burn a bit of some shell like that of the clam, to destroy the animal matter, and then break it across from the hinge to the margin It will then be found that the shell is built up of a series of layers, each of which, as we proceed inward, is larger than its predecessor. The way in which the shell is formed by the mantle explains this structure. When the animal is very small it secretes a layer on the underside of the embryonic shell. With an increase of growth another layer is laid down, but since the mantle is now larger than it was before, this layer extends beyond the preceding one. Other similar depositions follow, the result being that the shell is thicker at the hinge than at the edge, while the outer surface is marked with parallel lines, the edges of the successive layers.

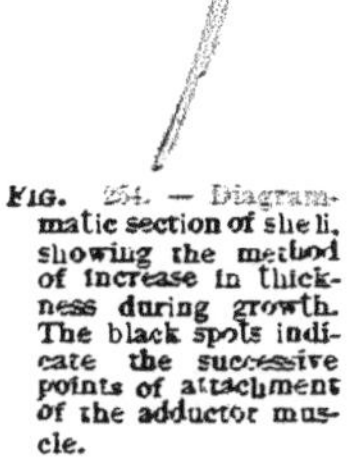

FIG. 254. — Diagrammatic section of shell, showing the method of increase in thickness during growth. The black spots indicate the successive points of attachment of the adductor muscle.

The structure of the shell presents many interesting points. It may be hard and opaque, like porcelain, fibrous, glassy, horny, or pearly, or nacreous, giving beautiful iridescent colors. On microscopic examination it is seen that these latter owe their hues to minute undulations of the layers, and that they are diffraction spectra similar to those now produced for physical researches by fine rulings. The external color of shells is due to pigment deposited by the edge of the mantle, which frequently bears the same pattern of ornamentation as does the shell. Usually shells are covered with a horny external layer, the so-called epidermis, which is likewise a product of the edge of the mantle. Its purpose is to protect the shell from the corroding power of the water in which they live, or from other external injury.

At some stage of growth almost all molluscs bear a shell, but with some it disappears with growth. The shell may be univalve or bivalve, or in the case of that aberrant group, the chitons, it may be composed of eight pieces serially arranged. In the first case it is usually coiled in a spiral, although a conical form is not rare. Among the bivalves the two halves of the shell are nearly alike, though in some the similarity is largely lost.

Turning now to the internal structure, we have first to take up the digestive tract. This is always separated from the body cavity by proper walls. It begins with a median mouth at the anterior end of the body, and terminates at the anus, which is also primitively in the median line at the posterior end of the animal. The torsion which brings the vent in another position will be discussed further on. The three divisions of the digestive tract, stomodeum, mesenteron and proctodeum are well developed, the middle region being characterized by a very large liver. Salivary glands are frequently present, emptying into the stomodeum, and the same region frequently bears a lingual ribbon, armed with teeth, for the comminution of food. This organ is employed to characterize the Cephalophora, one of the two great divisions of the Mollusca, and will be described when treating of that group.

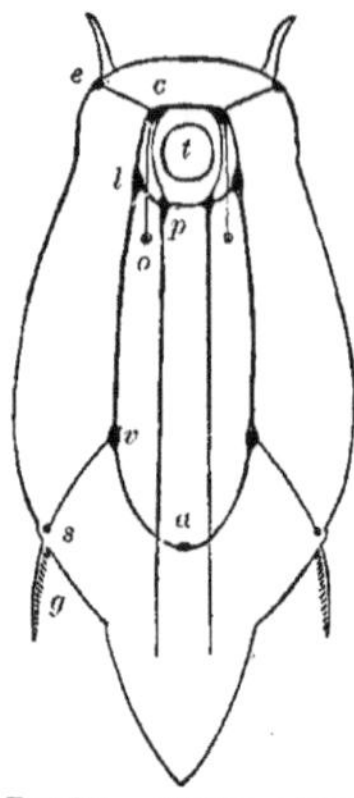

FIG. 255.—Diagram of nervous anatomy of mollusc; *a*, abdominal ganglion; *c*, cerebral ganglion; *e*, eye; *g*, gill; *l*, pleural ganglion; *o*, ear; *p*, pedal ganglion; *s*, organ of smell; *t*, throat; *v*, visceral ganglion.

The nervous system typically consists of two ganglia above the œsophagus (cerebral); two at its sides (pleural); and two beneath (pedal). These are connected by a ring of nervous tissue. From each of the pedal ganglia arises a nerve cord which traverses the length of the foot (the pedal nerve) while from the pleural ganglia two similar cords arise, which also pass backward, but at a higher level (the pleural commissures). These terminate in a ganglion on either side, known indifferently as the visceral or parieto-splanchnic ganglion. These two visceral ganglia are connected with each other by a cord known as the visceral loop, in the middle of which is the abdominal ganglion. From the cerebral ganglia nerves go to the eyes, and primitively to the auditory organ. An additional commissure on either side connects the cerebral with the pedal ganglion.

The heart, which is situated dorsally, consists of a ventricle and one or two auricles. It is always arterial, receiving the blood from the respiratory organs and forcing it to all parts of the body. The circulation is not completely closed, the blood for a portion of its course flowing through channels without proper walls.

Though the whole surface of the body has respiratory functions, special organs for the aëration of the blood exist in the shape of gills, or, less frequently, so-called lungs. The gills are ciliated outgrowths from the body, usually placed in the cavity of the mantle between that envelope and the foot or body wall. Each gill may be reduced to a type called by Lankester a ctenidium. This, as its name indicates, is like a comb, the back of the comb being the rhachis or stalk, while the gill lamellæ correspond to the teeth of the comb. In the rhachis are two canals, one carrying the blood to the gill plates, there to be brought in contact with the water, the other returning it to the heart. From this type most of the forms of gills can be derived.

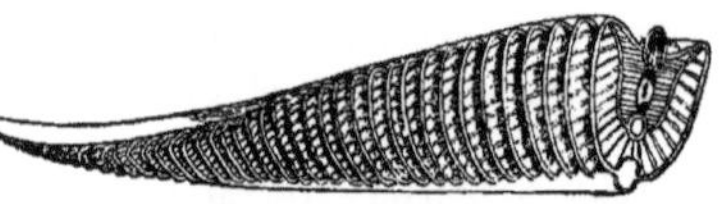
FIG. 256.—Gill of *Sepia*.

All of the gills of Mollusca are not homologous, a fact first pointed out by Spengel. This anatomist has shown that in the true or typical gills are normally paired organs, one or more being found on either side of the body. These true gills receive their nerves from the visceral loop of the nervous system, and he has also pointed out that at the base of each gill is a sense organ, the purpose of which is to test by smell the quality of the water supplied to each gill. This olfactory organ is also innervated from the same part of the nervous system as are the true gills. Other respiratory organs exist in some forms, but we have a sure test of their homology in their relations to the nervous system and to the organs of smell.

Lungs, which are cavities of the mantle lined with respiratory folds, occur only in the pulmonate gasteropods, where they will be described at length.

The renal organs, nephridia, or organs of Bojanus as they are frequently called from the celebrated anatomist who discovered them, are always present. They are usually symmetrically disposed, there being one on each side of the body. Each nephridium consists of a tube, the inner portion of which communicates with a portion of the body cavity, while the other opens externally. In the interior portion are well-developed glands, which excrete uric acid, while the outer, non-glandular portion is merely an afferent duct. That these nephridia are homologous with the segmental organs of worms is more than possible, and the probability is strengthened by the fact that their internal openings are ciliated, and that in many forms they serve for the extrusion of the seminal, as well as for excretory, products.

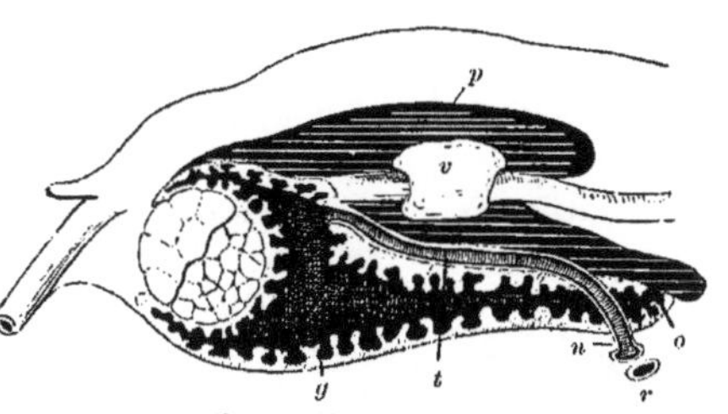

FIG. 257.—Nephridium of Unio; *g*, glandular portion; *n*, external opening; *o*, opening between (*p*) pericardium and glandular portion of nephridium; r, reproductive orifice; t, non-glandular portion of nephridium; *v*, Ventricle.

Reproduction is here always a sexual operation, fission and budding being unknown. As a rule the two sexes are combined in the same individual, but numerous marine gasteropods, and all cephalopods, are diœcious. The sexual glands are placed on either side of the body, and either open through ducts of their own, or by means of the nephridia, as mentioned above.

In all except the cephalopods there is a more or less complicated metamorphosis in passing from the egg to the adult. According to the amount of food-yolk, the segmentation is regular or irregular, the result being a morula or mulberry-like mass. Soon a portion invaginates, just as we may push in one side of a rubber ball, or, owing to the presence of a great quantity of food-yolk, this process may be obscured. The result, however, is in both cases the formation of a two-layered sac, the gastrula. The mouth of the gastrula, the blastopore, soon closes more or less completely, and from the middle portion is developed the foot, while the two ends correspond respectively with the mouth and vent. Occasionally one of these openings persists, but not infrequently a new invagination takes place to form the openings, the inpushing of the integument being always within the limits of the blastopore. From the outer layer of the gastrula is developed the epidermal structures of the body, while the inner gives rise to the middle division of the digestive tract. From this inner division cells are also budded off between the two layers, forming the mesoblastic tissues, and later one or more spaces appear in this mesoblast, the body cavity. Further details of the internal development may be found in special works, but for our purposes we need to follow the changes in external form a little further.

At about the time of the invagination, a portion of the outer surface develops a circle of long hairs or cilia. This circle, which is known as the velum, embraces only a small portion of the exterior, and since both mouth and anus, when formed, are behind it, it follows that the area so circumscribed is pre-oral. Not infrequently a single longer hair or flagellum occupies the centre of the velar area, marking the differentiation of the ectodermal layer into nervous tissue, the future supra-œsophageal ganglia. This stage is the trochosphere, and presents a close resemblance to the larva of many worms, and especially the rotifers. At this stage, or even earlier, another important feature appears, the shell gland. This is at first an invagination on the side of the body opposite the mouth, but still outside the velar area. The gland soon flattens and begins to secrete the shell, which at first appears as a single delicate plate. Following the trochosphere comes the stage known as the veliger. The velar area is now a flattened plate, fringed with cilia, and frequently expanded into lobes, while the rest of the body is greatly enlarged in proportion. The foot is also more prominent. With subsequent development the disproportion between the velar area and the rest of the body increases in all except a very few cases, as the pond-snail, *Limnæa.*

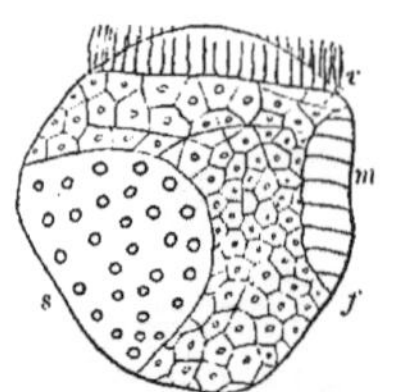

FIG. 258. — Trochosphere of oyster; *f*, foot; *m*, mouth; *s*, shell gland; *v*, velum.

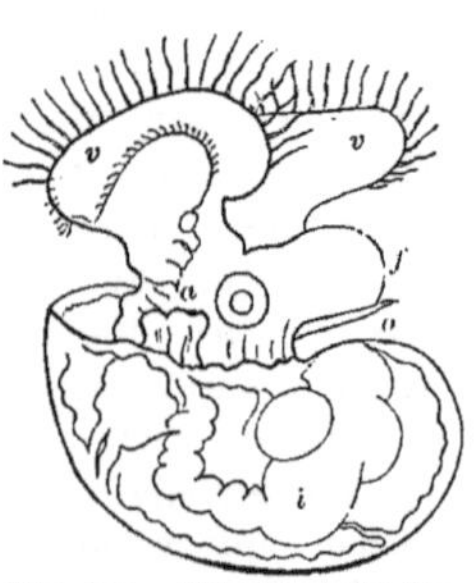

FIG. 259. — Veliger of Opisthobranch; *a*, anus; *f*, foot; *i*, intestine; *o*, operculum; *v*, velum.

Shells, more than any other objects of natural history, have played a part as objects of merchandise, and for the rarities, conchologists have in times past paid the most fabulous prices. The following, copied from Tryon, may prove of interest to those who have not yet caught the fever of shell-collecting: —

" *Scalaria pretiosa*, which can now be had for one or two dollars, was worth $100 in 1735, and $200 in 1701. *Phasianella bulimoides*, which also brought $100, can now be purchased at from one to two dollars, or even less. In 1865 a great English collection, that of Dennison, was sold by auction in London, and some extravagant prices realized. *Cypræa guttata* brought $200; *Cypræa princeps*, the same; *Conus gloria-maris*, also $200; *Conus cervus* nearly $90; *Conus cedonulli* (not a very rare shell), $90 and $110; *Conus omaicus*, (also not rare), $60; *Voluta festiva*, $80; *Oniscia dennisonii*, $90; *Pholadomyia candida*, $65; *Carinaria vitrea*, (which Montfort stated to be worth $600), brought $50. The very rare *Pleurotoma quoyana* brought in London, in 1872, $125. In 1876 the Röters van Lennep collection was sold, including: *Voluta junonia*, $50; *Mitra belcheri*, $40; *Spondylus regius*, $36, etc. For this same *Spondylus regius* Professor Richard had previously paid several thousand francs. *Voluta junonia* has always been considered a rare species, and dealers have obtained as much as forty pounds sterling for it. . . . *Cypræa umbilicata* has been sold for thirty pounds, and may now be had for one pound. The Boston Society of Natural History possesses an *Argonauta argo*, or paper nautilus shell, which is said to have been purchased by the gentleman who presented it to that society, for $500. It is a common species, and the only reason of the greater valuation of this specimen is that its diameter is about two or three inches greater than any other individual known to naturalists."

CLASS I. — ACEPHALA.

This group of the molluscs has been burdened with a large number of names. Among them we find Conchifera, Endocephala, Lipocephala, Lamellibranchiata, and Pelecypoda, as well as the older, and consequently preferable, designation adopted here. The group will readily be recognized by all under the popular designation of bivalve molluscs. In this more familiar name is embodied one of the most characteristic features of these forms, — a shell divided into halves, one on either side of the body. This bilateral symmetry pervades the whole organism, and frequently one side is almost an exact repetition of the other. Just inside the shell is found the fleshy mantle, which, like the shell it secretes, forms a flap on either side of the body. In these bivalves this pallium, or mantle, acquires a great development, and not infrequently its edges are joined together, so that the rest of the animal is enveloped, as it were, in a bag. Still the bag is never completely closed; at the front end a small

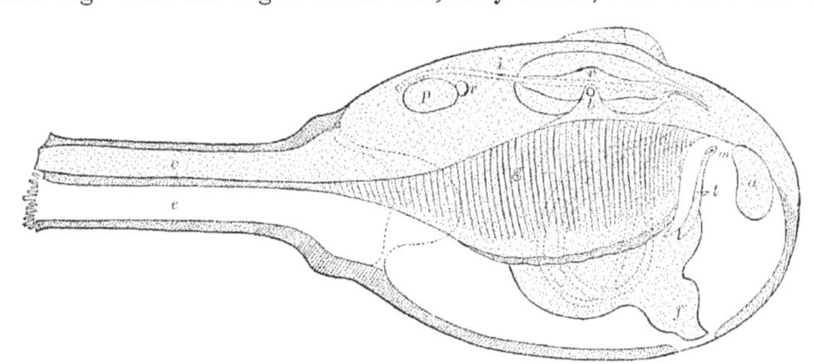

FIG. 260. — Diagram of anatomy of a clam (*Mya*); a, anterior adductor; *b*, auricle; *c*, excurrent siphonal tube; *e*, incurrent siphonal tube; *f*, foot; *g*, gills; *i*, intestine; *m*, mouth; *p*, posterior retractor; r, retractor of foot; *t*, labial palpi; v, Ventricle of heart.

hole is left for the protrusion of the foot, while, at the opposite extremity, means is afforded for the entrance of water, bringing food and oxygen to the animal, and also for the escape of the same fluid, bearing away the waste products of respiration and digestion. Not infrequently this posterior opening becomes divided into two tubes, which sometimes can be extended a long distance from the shell. This is known as the siphon, and will readily be recognized by most people in the 'head' of the clam. Head it certainly is not, for it is at exactly the opposite end of the body from where the head should be. These tubes, which are, in reality, but expansions of the mantle, are very contractile, and each tube has its own function. The lower one (the one furthest from the hinge of the shell) is for the incurrent stream, while from the other the water which has played its part in the economy of the animal is discharged. In other forms there is no siphon, and in still others the two halves of the mantle are entirely free from each other.

The mantle joins the body near the hinge line, and between the two hang down the gills, to which we shall again recur. From the lower side of the body proceeds the foot, which in some forms is well developed, even proving an organ of locomotion of no mean capacities, while in others, like the oyster, the foot has nearly, or even entirely, disappeared. Near the extremity of the foot in the adults of some species, and the young of others, is a gland, the function of which is to secrete the byssus. This is a bundle of fibres, more or less closely united, by which the animal attaches itself. The byssus can be cast off by the animal when desired, and a new one formed at pleasure.

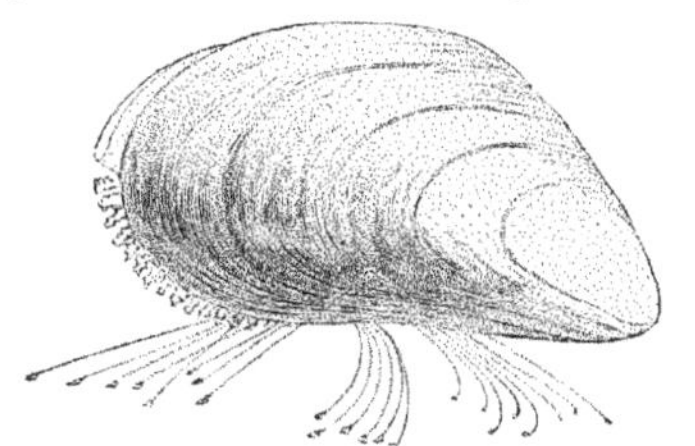

FIG. 261. — Mussel with byssal threads.

The mouth, which is at the opposite end of the body from that at which the siphon arises, bears, at the sides, a pair of leaf-like or tentacular folds, the labial palpi, the function of which is to direct and conduct currents of water to the mouth, the cilia with which they are covered aiding greatly in this respect. It has been suggested that these palpi represent the velum of the larva, but no known facts of embryology confirm this view. They are, in reality, the greatly expanded upper and lower lips.

The alimentary canal always traverses the whole length of the body, terminating in a vent at the posterior end. Usually its course is much contorted, the intestine, in some forms, passing through the ventricle of the heart. The œsophagus is short, and communicates with a more or less spherical stomach, into which a voluminous paired liver pours its secretion. The intestine is very long and convoluted. No organs of mastication are present, and, if we make one possible exception, nothing that can be compared to the lingual ribbon of other molluscs. This exception is the crystalline style. This is a transparent elastic rod, of unknown functions, which lies in a pouch arising from the stomach. Whether it be a representative of the odontophore is very uncertain.

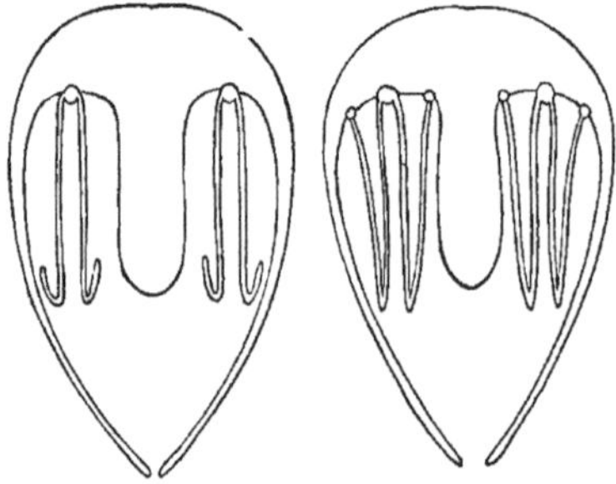
FIG. 262. — Diagram showing the development of the gills of a lamellibranch.

The heart always consists of a median ventricle, which forces the blood to all parts of the body, and two auricles, one on either side, which receive the blood from the gills and pour it into the ventricle. The gills possess a very complicated structure, but one which can without much difficulty be reduced to a simple type. Of these organs there are usually two on either side. Embryology shows us that each of these gills is primitively made up of a series of little tubes running down from the body wall. These tubes then turn and grow back until they reach the inner surface of the mantle, as shown in the adjacent figures. The filamentary condition persists in some acephals, but in others the adjacent filaments become united so that a broad lamellar gill (whence the name Lamellibranchiata) is the result. In some forms this union is produced by bunches of hooked cilia on the sides of the tube, while in others the walls of the branchial filaments become solidly grown together. The blood from all parts of the body gathers in a large tube at the base of the gills; thence it passes down through one half of the little tubes, and up in the other, to another vessel, whence it is conveyed to the heart. During this passage it is brought in contact with the water, discharging its carbonic acid, and taking a new supply of oxygen.

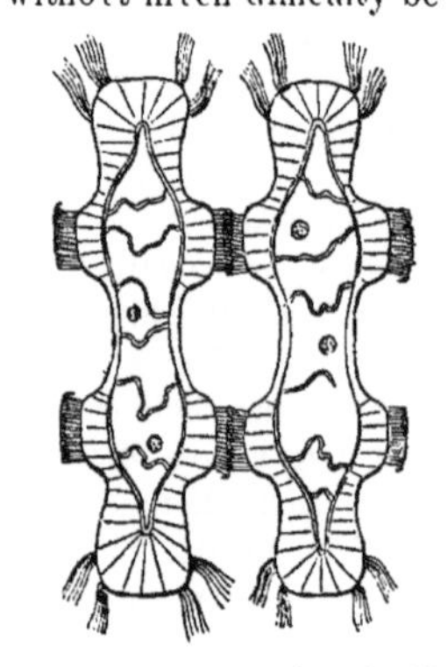
FIG. 263. — Section through gill filaments of *Mytilus*, showing the hooked cilia which fasten them together.

The way in which the water is brought into the cavity of the mantle, and in contact with the gill, is very interesting. The gills, and for that matter the whole inner surface of the mantle cavity, are covered by innumerable little hairs, or cilia, which, by their constant motion (always in one direction) create currents in the water, drawing it in through the incurrent siphon, passing it over the gills, around to the mouth, and then out through the excurrent opening. Singly these cilia are very weak, but together they exercise a great deal of force. Many experiments have been tried by cutting out a piece of the gill, placing it on a flat surface, and covering it with a weight. The amount which will be moved by these minute lashes, under these circumstances, is almost beyond belief, the motion in one instance being six millimetres a minute.

The excretory organs are paired, and communicate internally with the cavity (pericardium) surrounding the heart. The nervous system consists of three pairs of ganglia, a cerebral or supraœsophageal, a pedal, and a pariete-splanchnic pair. The arrangement of these shows many minor variations. Normally, the first pair is situate above the œsophagus, but they may be brought beneath that tube, occupying a position just outside the pedal ganglia, which are, as their name implies, situate in the foot. The last pair are placed just beneath the posterior adductors. The two latter pairs are connected with the first by double cords. Besides the sense of touch, organs of smell, hearing, and sight are developed in most of the group. The olfactory organs are situated upon the parieto-splanchnic ganglia, the auditory organs near the ganglia in the foot, while the eyes are very variable in position. The organs of smell are merely patches of elongated epidermal cells, strictly homologous with similar organs in other molluscs. The ears are small sacs lined with cilia, each containing a single otolith, which, by its vibration against the cilia, conveys the sensation to the nervous system. The eyes may be found either upon the edges of the mantle or upon the tip of the siphon. In some forms like *Spondylus*, *Pecten*, *Mactra*, etc., these eyes on the edge of the mantle are well developed, and, like those of *Onchidium*, which will be mentioned in a succeeding page, are similar to those of vertebrates, in that the nerve fibres penetrate the retinal body, and distribute themselves

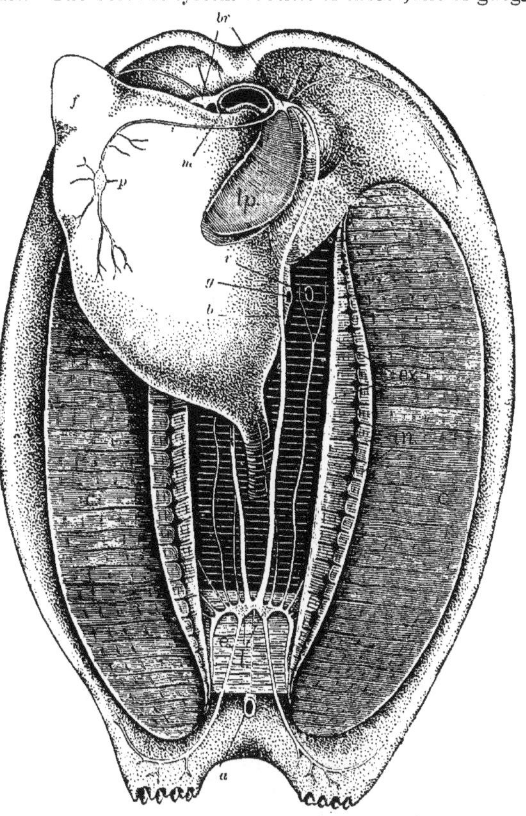

FIG. 264.—Diagram of *Anodonta*; *a*, anus; *b*, cerebro-Visceral connective; *br*, cerebro-pleural ganglia; *c*, gill; *d*, mantle; *e*, posterior accuctor; *ex*, external lamella of inner gill; *f*, foot; *g*, genital opening; *in*, inner lamella of inner gill; I, Visceral (parieto-splanchnic) ganglia; *lp*, labial palpi; *m*, mouth; *p*, pecal ganglia; r, renal opening.

on the outer ends of the rods and cones. In the siphonal eyes found in *Solen*, etc., we have the merest apology for a visual organ.

The sexes of the acephals are usually separate, though in rare instances they are united in the same individual. The genital glands are on either side of the body, and empty by paired ducts. The eggs are either cast free in the water or are retained for a time between the lamellæ of the gills of the parent. The veliger presents a prominent difference from that of gasteropods, in that the primitively simple shell soon becomes bivalve. The peculiar larval form known as glochidium will be mentioned in connection with the Unionidæ farther on.

Lastly, in our general account, comes the shell, which occupies so important a place in existing schemes of classification, and with it may be mentioned some of the features of anatomy which have been neglected in the preceding page. On examining the outer surface of any bivalve shell one notes the lines of growth concentrically arranged. These have as a centre an elevated portion of the shell known as the umbo, the position of which marks the dorsal border. Usually this umbo points towards one end of the shell, which may thus be recognized as the anterior. Having these landmarks, we can readily decide the question of right and left. At the dorsal margin, where the two valves join, is the hinge line, and just in front of the umbo is frequently a distinct area, half on each shell, the lunule.

Now for the mechanism which opens and closes the valves. The closure is effected by one or two transverse muscles (adductors they are called) which pass from one shell to the other, and by their contraction the two valves are approximated. No divaricators exist, but instead the valves are separated, the moment the muscles are relaxed, by means of an elastic ligament. This ligament may be either external or internal. In the former case, as shown in Fig. 265 A., the ligament connects the two valves, and by its contraction spreads them. In the other (Fig. 265 B.,) the internal ligament is placed between two portions of the shell, so that when closed it is compressed, but upon relaxation of the muscles the elasticity and expansion of the ligament forces the valves apart.

On the internal surface of the shell we also find certain features which are made prominent in systematic work. Just where the two valves join together is the hinge, usually provided with projections and depressions, forming what are known as teeth. Those in the centre are known as the cardinal, and those frequently present at the sides as the lateral teeth. Near the hinge line, but on the inner, concave surface, will be found one or two approximately oval marks, the impressions produced by the attachment of the adductor muscles. When only one muscle is present, it is morphologically the posterior one. Usually close to these may be seen other similar but smaller scars, marking the spots where the muscles of the foot had their origin. Going around the margin of the shell is a line, more or less distinct, called the pallial line, which

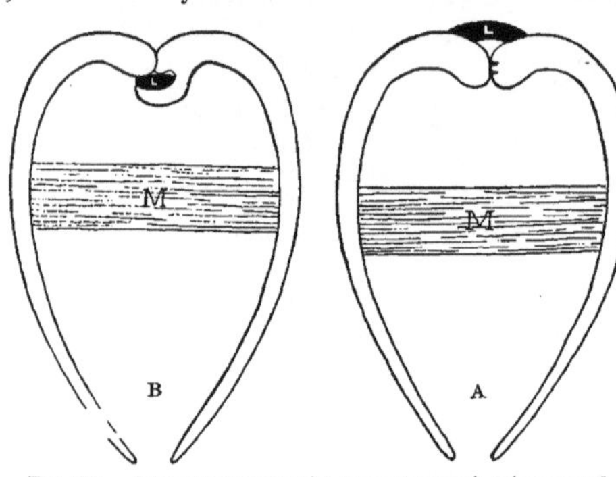

FIG. 265.—Diagram showing the hinge ligament, internal and external; *l*, ligament; *m*, muscle.

marks the limit of the thickened edge of the mantle, and in one large group of shells a portion of the pallial line makes a re-entrant angle. This is the pallial sinus and is found only in bivalves with a siphon, where it marks the place of attachment of the muscles of that organ.

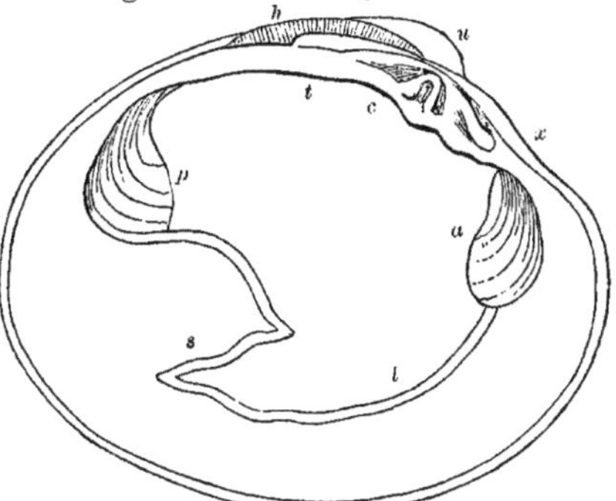

FIG. 266. — Inner surface of left valve of *Cytherea*: *a*, impression of anterior adductor; *c*, cardinal tooth of hinge; *h*, hinge ligament; *l*, pallial line; *p*, posterior adductor; *s*, pallial sinus; *t*, lateral teeth of hinge; *u*, umbo; *x*, lunule.

The classification of the acephalous mollusca is still in a very unsatisfactory condition. In the system of Lamarck the group was divided into two subclasses, based upon the number of adductor muscles, those in which only one of these muscles was present forming the Monomyaria, while those with two were called Dimyaria. Woodward, who wrote one of the most valuable manuals of conchology which has as yet appeared, used the presence or absence of siphons as a means of division, those where no siphon was formed being the Asiphonida, the others forming the Siphonida. The Siphonida in turn were subdivided, according to the presence or absence of a pallial sinus, into the Sinupallialia and the Integropallialia, respectively. With each of these systems many grave faults may be found, and hence, for our purpose we will divide the Acephals directly into families, without the intervention of sub-classes and orders, and other intermediate divisions.

In economic importance the OSTREIDÆ, the oyster family, stands pre-eminently first. The characters of the family, taken in its older and broader sense, are as follows: The two valves of the shell are unequal, the hinge is without teeth, and, as a rule, the single adductor is nearly median in its position. The two halves of the mantle are free from each other, and the borders are fringed with small tentacles, and the foot is rudimentary or even entirely absent.

The genus *Ostrea* has a shell so irregular that specific limits are very poorly defined. The left valve, which is attached to some submerged object, is hollowed out to receive the body, while the upper right valve is nearly flat. With us Americans, *Ostrea virginiana* is the important form, and to it most of the succeeding account applies. In Europe the species which form the bulk of those eaten are *O. edulis* and *O. angulata.*

Ostrea virginiana extends, on our Atlantic coast, from the Gulf of St. Lawrence to the Gulf of Mexico. In the former region the oysters are found from the Bay of Chaleur to Prince Edward's Island. The beds are small and in many places seem to be decreasing. The oysters found here are usually large. These beds are separated from the nearest natural living beds to the south, by a thousand miles of coast line, but between these points evidence is abundant that in former times the gap was far less, for remains of extinct beds are found all along the coast, from Mount Desert Island to Cape Cod. At Damariscotta, Maine, was once a large bed which furnished the Indians with the means of many a feast. Here, year after year they came, and with the refuse shells they formed huge heaps which to-day are the delight of the archæologist. The oysters in these "kjökkenmöddings" (a Danish term meaning heaps of kitchen refuse) were of enormous size, one having been found at Damariscotta which measured fifteen inches in length. Remains of other beds, and indications of still others in the shape of kjökkenmöddings, are abundant along the coast, at Portland, Harpswell, and points in Massachusetts.

It seems that most of these beds, north of Cape Cod, have become extinct since the first settlement of the country. The old records tell of beds in the Mystic and the Charles Rivers, near Boston, which were well known to the early settlers, but all trace of them is now lost. The same is true of many other places. What was the cause of this extinction is not easy to decide. Professor Verrill, who has taken into consideration the facts afforded by the past and present distribution of other molluses, is inclined to attribute it to a climatic change; others to over-fishing, pollution of the waters, etc.

South of Cape Cod the oyster still flourishes in its native vigor, and an enumeration of the places with extensive beds would be about equivalent to giving a catalogue of the shore towns from Buzzard's Bay to Texas. In Long Island Sound, and on the Jersey coast, besides the supply of natives, large numbers are transported from the Chesapeake. These latter are brought in the shape of small (seed) oysters, and are scattered in suitable situations, where they increase in size. When large they are taken up and sent to the market. The profits on this operation are said to be large. It is in Chesapeake Bay, and its various creeks and sounds, that the oyster business reaches its greatest development, and where the study of the various economic and scientific problems connected therewith, have received the greatest attention, an account of which will not be out of place here.

In our American oyster the sexes are separate, and, thanks to the labors of Dr. W. K. Brooks and Mr. J. A. Ryder, we now have a pretty complete knowledge of the life history of this valuable form. The eggs are fertilized after leaving the parent, and undergo their development in the water. The development is normal and direct. After the formation of the velum the young oyster swims freely through the water, during which time the shell and the internal organs are being gradually developed. In a few days this free life ceases, and the young oyster attaches itself to some submerged object. The way in which attachment is effected is at first by the rudimentary shell of the left valve, and subsequently, as lime salts are deposited in the shell, they serve to unite the young oyster more firmly to the point of support. When first attached, the young oysters are termed 'spat;' when large enough for transplanting, 'seed.'

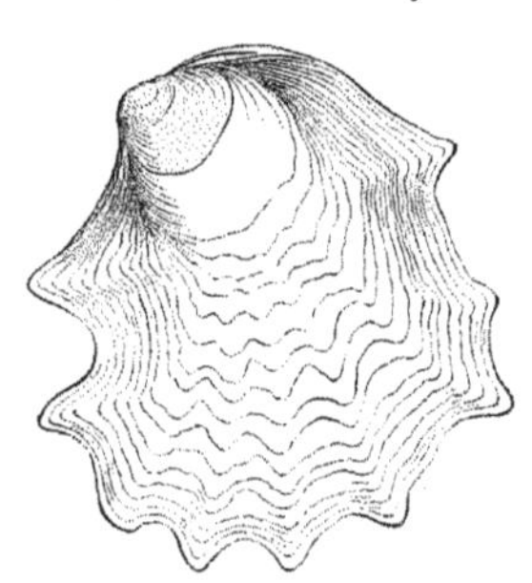

Fig. 267. — Young oyster.

The life of the oyster depends upon many things, the most prominent being the location. At the time the embryos are to be converted into spat they need some solid object to which to attach themselves. On the natural beds this is found on the shells of other oysters and on rocks, but where artificial beds are formed it is customary to throw down old shells. In artificial propagation 'collectors' of earthenware, slate, etc., are used. The young oyster needs plenty of water and food, and also needs protection from mud, which would soon smother them. The food consists almost wholly of microscopic animals and plants.

Many experiments have been tried to raise oysters in confinement, but until recently they have not been successful. The great trouble has been in keeping the young alive.

Mr. J. A. Ryder solved the problem in the following manner. An artificial pond was formed, and filled with salt water, which was filtered through sand, and all connection with the ocean was by means of a ditch interrupted by a bank of sand. This permitted a slight change of water with each tide, but prevented the entrance of injurious forms and the exit of the young before their transformation into spat. In the pond were placed numerous collectors, to which the spat could attach itself; and after the young shells had attained sufficient size to take care of themselves, these collectors, with their molluscan load, could be transferred to the beds in the adjacent sounds. By this process the period of greatest mortality is passed in comparative safety, and the result of Mr. Ryder's labors will doubtless be to greatly increase the supply of this delicious bivalve.

The oysters are taken from the beds by rakes, tongs, and dredges; the names of which indicate their general appearance. When brought to the shore, some are sent to market, while others are 'shocked,' and sold as solid meats. The extent of the oyster industry in the United States can be seen from the following figures, extracted from the census of 1880. The business gives employment to over fifty thousand persons and over four thousand vessels, and involves an investment of over ten million dollars. The number of bushels of oysters produced is over twenty millions, and, at first hand these sell for over thirteen millions of dollars. Each State has its own laws and regulations regarding its shell-fisheries, and in Maryland and Virginia an oyster-police is maintained, to prevent irregular fishing and depredations upon the beds.

Our east-coast oyster has been described under four specific names, *virginiana*, *virginica*, *borealis*, and *canadensis*, but all have been shown to be varieties of one and the same form. The average length of those brought to market is, perhaps, four to six inches, but larger ones are not uncommon; but how much credence is to be given to the following quotation from the "Mobile [Alabama] Register," of April, 1840, taken from "Ingersoll's Monograph of the Oyster Industry," is a question. The quotation runs: "The large oyster taken by Xavier François, while oystering on Monday last, was brought up from the wharf, on a dray, last evening. An oyster measuring three feet one inch in length, and twenty-three and a half inches across the widest part of it, is a curiosity."

On our western coast two species of oyster are eaten. Of these *O. conchophila*, of California, is small, while the *O. lurida*, from Shoalwater Bay, Washington Territory, is much larger and better. Still a large proportion of the trade is supplied by oysters shipped from the Atlantic coast. In Europe the oysters most eaten are *O. angulata* and *O. edulis*. Of these the former has the sexes separate, while the latter is hermaphrodite. Space will not allow a detailed account of the many and successful trials that have been made in the propagation of these forms, and we can only refer to the green oysters so highly prized by the Parisian epicures. Many theories have been proposed to account for the color, — such as the presence of copper, etc., — but it now seems probable that it is due to the effect of the food, the chlorophyl of the vegetable food being transferred to the blood, or some peculiarity in other objects eaten affecting the liver. The British market is now largely supplied from the United States, the American oysters being much larger, and better flavored than those of European seas. The trade amounts to about half a million dollars yearly.

Edible oysters are found at the Cape of Good Hope, in Australia, Japan, and elsewhere. The *O. talienwanensis*, of Japan, occasionally measures three feet in length. Other forms which deserve mention are the species of the sub-genus *Alectryonia*, which are found attached to the roots of mangrove and other trees, in the warmer seas of the world. The oysters appeared in the carboniferous age, and have persisted until the present time. Among the fossil forms closely allied to *Ostrea* may be mentioned the genera *Gryphæa* and *Exogyra*, in which the shell attained great thickness and weight.

FIG. 268.—Very young *Anomia*, showing the byssal notch not yet converted into a foramen.

The genus *Anomia*, though possessing but little economic importance, is especially interesting from the fact that for a long time it was supposed to form a connecting link between the brachiopods and the molluscs. Like many of those forms it lived attached to rocks or shells by means of a peduncle, which perforated one of the valves of the shell. Could the correspondence be stronger? When the mode of growth was studied it was found that the two were entirely different. *Anomia*, in its early stages, spins a byssus, by which it attaches itself to some foreign object. As the shell increases in size, the byssus interferes with the growth of one of the valves, producing a notch in its margin. As growth continues, the two edges of the notch close around the byssus, resulting in the perforation. As this closure occurs at an early stage of development, the foramen produced is close to the hinge, thus strengthening the resemblance to the similar opening in the brachiopods, though, as will be readily seen, there is no homology between them.

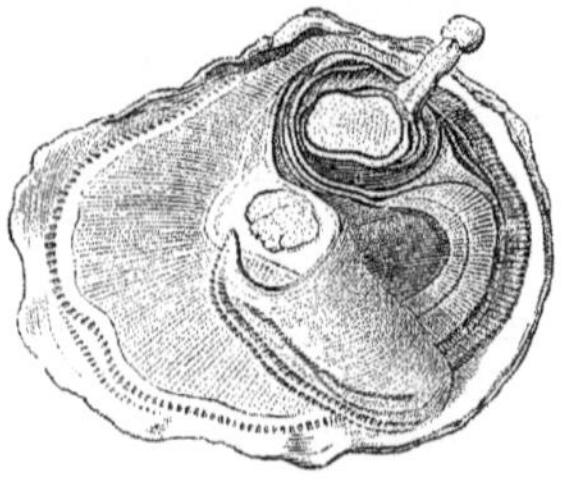

FIG. 269.—*Anomia ephippium*, showing byssus.

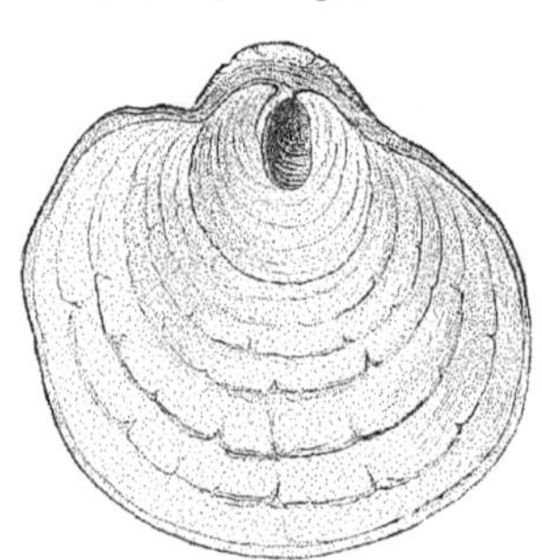

FIG. 270.— *Anomia glabra*.

Besides the perforation of one valve for the byssus, the genus *Anomia* is characterized by having the two valves unequal, one being larger and convex, while the other (the perforated one) is flattened, or even concave, according to the location on which it settled. The hinge is without teeth. The species vary greatly, so that it is very difficult to define them. On our coasts two forms are commonly found, *A. glabra* and *A. aculeata*, the former being essentially a southern, the latter a northern form. *Anomia glabra* is especially common on oysters. In color it is glistening white or yellow, and hence has received the names silver shell or gold shell. *A. aculeata* is usually spiny or scaly. It is more varied in its places of attachment, rocks, shells, and large algæ forming favorite localities. This is very closely allied to the *Anomia ephippium* of Europe. On the coasts of California occurs *A. pernoides*.

Placuna is a genus with thin shells, in which no byssus is formed. The valves are nearly equal, and the species inhabit sandy shores. The species are inhabitants of the Indo-Pacific Ocean.

The family PECTINIDÆ, or scallops, comes next in order. Here the valves may be similar or different, while the anterior and posterior halves of each valve are nearly alike. The hinge is prolonged on each side of the umbo into an ear-like process. The borders of the mantle are free, and frequently they bear a number of brightly colored eyes, which have been made a subject of study by Mr. Holman Peck. A single adductor is present, and the gill filaments are free, showing as a permanent structure, a condition characteristic of the young of the higher forms. The foot is small, and frequently spins a byssus, by which the animal attaches itself.

The typical genus of the family is *Pecten*, in which the regular shell is usually ribbed, the lines radiating from the umbo. The anterior ears of the shell are the larger. In older times species of this genus were known as "pilgrims' shells," from the fact that for some unknown reason the pilgrims of the middle ages were wont to ornament their clothing with these shells. So prevalent was the practice, that when, in the early days of science, it was adduced as a proof of the biblical record of the flood that fossil shells were found in the Alps; the reply was made by sceptics, that these were merely shells dropped by pilgrims returning from Palestine.

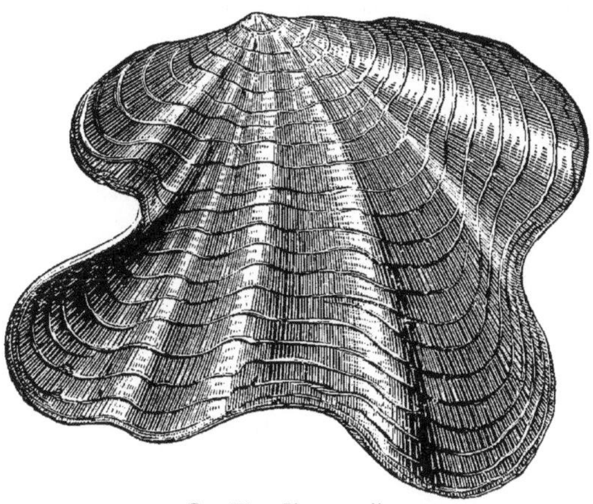

FIG. 271. — *Placuna sella.*

The common scallop of the southern shores of New England is known in scientific terminology as *Pecten irradians*. It lives in shallow places, among the eel-grass, and swims away at the slightest alarm. This swimming, which is somewhat rare among bivalve molluscs, is effected by rapidly opening and closing the valves of the shell, the result being a sub-aquatic flight in a backward direction. In color this species varies considerably, but the flat valve is always lighter than the other, being often white. The other valve may be reddish, orange, purplish, or mottled with two of these colors. The eyes, upon the edge of the mantle, are silver or bluish, and are thirty or more in number. This is the scallop of the markets, and is highly prized by some, though its sweetish taste makes it unpleasant to others, while some find it actually unhealthy, and productive of nausea. Only the adductor muscle is eaten.

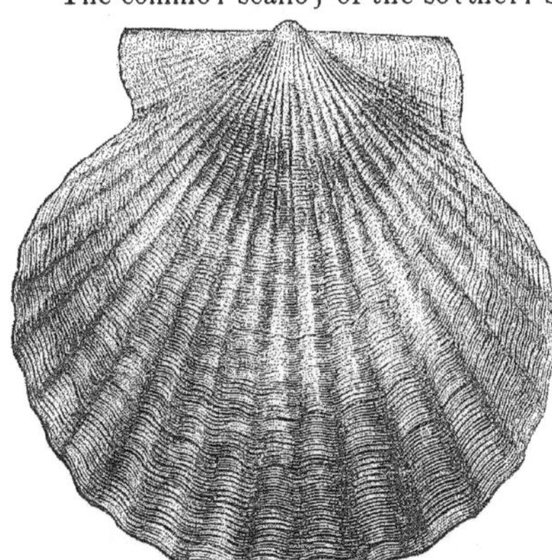

FIG. 272. — *Pecten irradians*, scallop.

North of Cape Cod this species is replaced by two larger ones, *P. islandicus* and *P. tenuicostatus*, neither of which are of much economic importance. In Europe the scallop (*P. maximus*) and the quin (*P. opercularis*) are extensively eaten. Other species, nearly two hundred in number, are found in all the seas of the world.

In *Lima* the shell is obliquely oval, and gapes anteriorly. The hinge is straight, toothless, and the ears are small. The border of the mantle is fringed with long cirri, but the eyes have not yet been discovered. Whether or no they are really absent has not yet been decided, but the recent investigations of Dr. Benjamin Sharp show that light-perceiving organs exist in many forms where their presence was not previously suspected, and so it may be here. *Lima hians*, the species figured, swims with great ease, and in the same manner as do the scallops. It also spins a byssus, and the adults not infrequently build a rude nest or burrow, by cementing together bits of coralline, shells, and sand. "The species of *Lima* usually live quietly at the bottom, with the valves widely extended, and thrown flat back, like the wings of certain butterflies when basking in the sun; but when disturbed, they start up, flap their light valves, and move through the water by a succession of sudden jerks. The cause of alarm over, they bring themselves to an anchor by means of their provisional byssus, which they seem to fix with much care and attention, previously exploring every part of the surface with their singular, leech-like foot."

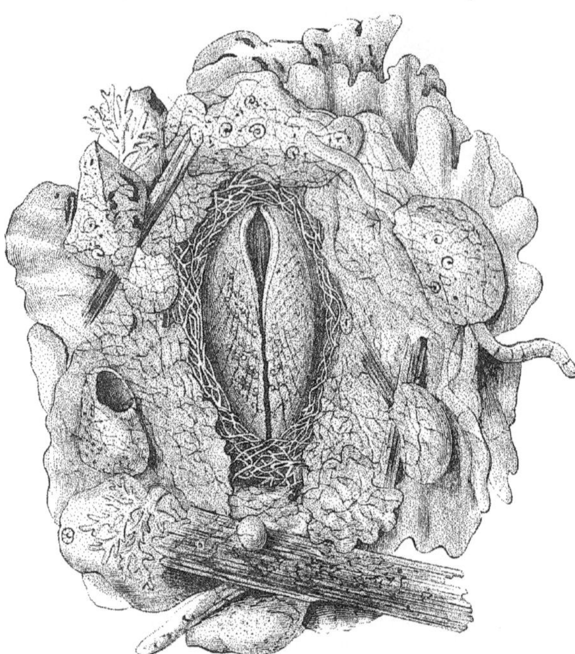

FIG. 273.—*Lima hians* in its nest.

The species of *Spondylus* are known as thorny oysters. The unequal valves are usually armed with spines, which not infrequently are very long and flat. The right valve is the largest, and is attached at the beak, the hinge ligament is internal, and the hinge is provided with two teeth in either valve. By the process of growth, the hinge area of the lower valve becomes converted into a triangular space furrowed down the centre by the groove for the hinge ligament. Slight ears are present at the hinge line. The ocelli on the margin of the mantle are numerous. The species are all inhabitants of tropical and sub-tropical seas. In the West Indies occurs *S. americana*. *S. gœderopus*, of the Mediterranean, is said to produce pearls. Most noted of all is the *Spondylus regius* of the East Indies, which is classed among the rare shells. In times past perfect specimens have brought immense prices, and no longer ago than 1876 a specimen sold for thirty-six dollars. The long and delicate spines are so easily broken that perfeet specimens are comparatively rare.

The AVICULIDÆ is the family of the true pearl oysters, and although many other molluscs produce pearls valued as ornaments, it is to this family that the world owes the largest proportion of these so-called precious stones. The family is characterized by having the hinge line straight, and produced on either side into wing-like ears. The valves, which are very oblique (their axis being at a considerable angle with the hinge line) have a foliaceous texture, and are lined on the interior with mother-of-pearl, giving them an iridescent appearance. These rainbow hues are due to the fact that the surface is covered by minute lines, which produce diffraction spectra. Soon after the fact was discovered that fine lines produced this appearance in the mother-of-pearl, some ingenious person applied the same method in the arts, and at one time buttons, etc., were made from steel, which had this same iridescent appearance produced by engraving microscopic lines upon the surface. Lately this phenomenon of diffraction has been turned to a scientific use, and to-day glass or speculum metal,

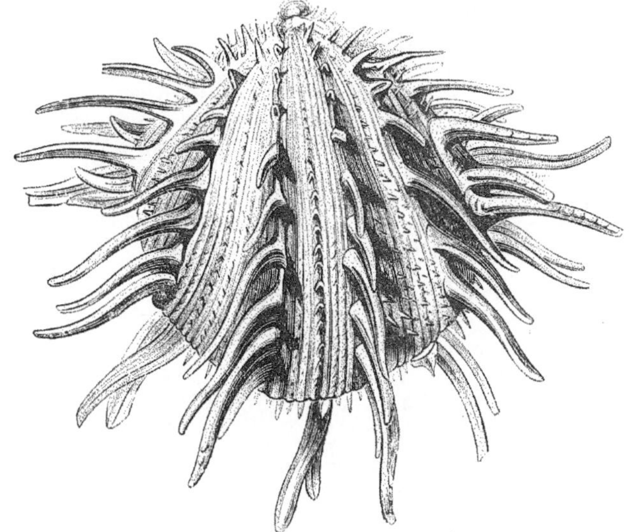

FIG. 274. — *Spondylus regius*, thorny oyster.

ruled with very fine lines, is used to produce the spectrum studied in spectroscopic analysis.

The hinge of the shell is without teeth, or with these elements obscure, while the ligament is partly internal. The shells gape in front, but are closed behind. The small foot spins a byssus; the mantle margins are free throughout their extent, and two adductor muscles are present. Of these the posterior is large, the anterior small and placed under the beaks of the shell, producing an almost imperceptible scar upon the inner surface. All the species are from the warmer waters of the globe.

In *Avicula* there is a single cartilage pit, and the hinge is furnished with two teeth; the right valve has a notch near the anterior ear for the passage of the byssus. *Meleagrina* lacks the hinge teeth, and the ears of the hinge line are small. The most prominent species is *M. margaratifera*, the true pearl oyster, which has an extensive distribution, being found in Madagascar, the Persian Gulf, Ceylon, Australia, Philippine Islands, the South Sea Islands, Panama, West Indies, etc. The pearl fishery is carried on at many points, but the finest pearls are said to come from the islands of Bahrein, Karak, and Congo in the Persian Gulf. The chief fisheries are those of

Ceylon. The oriental pearl oyster is much larger than the American form, the average diameter being, perhaps, nine inches, while specimens a foot across are not very rare. The principal locality of the Ceylon fishery is on a bank about ten or twelve miles off the north shore of the island.

Each night, at about ten o'clock, during February, March and April, a fleet of small vessels starts from Condatchy and Arippo, bound for the oyster banks, which are about twenty miles long. Arrived on the ground, the fishing begins. Each boat has a crew of twenty-three, ten of whom are divers, and these last are divided into two gangs of five each, one lot resting while the others are below. The average depth of

FIG. 275.—*Meleagrina margaratifera*, pearl oyster.

the bed is between nine and ten fathoms, and it nowhere exceeds thirteen. Each diver has a rope weighted at the lower end by a stone weighing about thirty pounds, and just above this is a loop for the foot of the diver, while a large net-work basket is fastened above. The diver, placing his foot in the loop, is rapidly lowered to the bottom, and there, working as fast as possible, he fills his basket with the oysters, and then, giving the signal, he, together with the weight and the basket of oysters, is quickly hauled to the surface. Incredible tales are told of the length of time that these divers can remain beneath the surface, but no well-authenticated case exists where one remained longer than eighty seconds, and but few can remain longer than a full minute.

When the boat is filled (which requires from fifteen to thirty thousand oysters) it returns to the shore, and the cargo is placed in an earthen bin, with walls about two feet high, and then left to die and decompose. When the flesh is pretty thoroughly disintegrated, it is washed away with water, great care being taken that none of the pearls loose in the flesh are lost. When the washing is concluded, the shells themselves are examined for pearls, which may be attached to the interior of the valves. The loose pearls are the most valuable, as they are round and more apt to be free from defects. Those attached to the shells have to be removed by clipping, and as one side is thus defective, they can only be used in settings. For over two thousand years this pearl fishery has been carried on in this place, and the result is that the shell heaps are perfectly enormous, miles of territory being buried to an average depth of about four feet.

Concerning the fishery in other localities but little has been written, although Panama, the island of Margarita, and the Sulu islands produce considerable numbers. The best pearls are usually about the size of a pea, but the largest known was two inches in length and four in circumference, and weighed three and three fourths ounces troy weight.

The pearl oyster is valued not only for the pearls which it produces, but for the mother-of-pearl as well. Of this there are three varieties recognized in the trade, the best of which are the silver-lipped, from the South Seas; next come the black-lipped, from Manilla and Ceylon; and lastly those known as bullock shells, from Panama, etc. These last are smaller and thicker than the others. Reliable statistics of the amount of the trade are difficult to obtain, but its extent may be seen from the fact that Great Britain uses annually about three thousand tons, valued at half a million dollars. Mother-of-pearl is used for inlaying, knife handles, etc., but the greatest consumption is in the manufacture of pearl buttons.

Mother-of-pearl is but the nacreous shell of the pearl oyster, which has an iridescent appearance, due to the fine striæ caused by the undulating layers of which it is composed. The true pearls are, like the shell itself, produced by the mantle, and owe their beauty to the same cause. They are, however, abnormal products, caused by the deposition of the nacre around some foreign object. This nucleus may be a bit of sand, a parasite, or some similar object, but it is said that usually it is an egg which has failed to develop properly. Other forms than the pearl oyster (*Meleagrina*) form pearls of value, while almost all bivalves occasionally secrete similar bodies; but, owing to the fact that these partake of the nature of the shell, they have not the beauty of those produced by molluscs with nacreous interiors. Of some of these other pearls we shall have occasion to speak further along when treating of some of the other families.

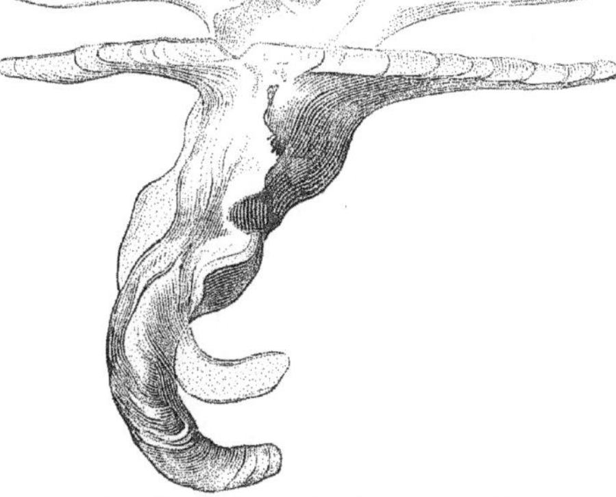

FIG. 276. — *Malleus vulgaris*, hammer shell.

The genus *Malleus* includes the hammer shells. With their long, winged hinge, and their still longer valves at right angles to the hinge, they well deserve both their common and their scientific names.

In the young stage they closely resemble the genus *Avicula*, even to having the notch in the right valve for the byssus; but, as they grow older, the body of the shell becomes more ribbon-like, its edges grow wavy, and at last the shell takes on the adult characters. Several species are known, all inhabitants of the eastern seas.

In *Perna*, which in general appearance resembles *Avicula*, the cartilage grooves are several in number, arranged at right angles to the line of the hinge. In some of the fossils of the tertiary age, the pearly layer lining the shell is an inch in thickness.

The family MYTILIDÆ embraces the mussels, in which the two valves of the shell are equal, convex, and covered with a thick epidermis. The hinge is weak, without teeth, and with the ligament internal; the posterior muscle is large, the anterior small; the foot is cylindrical and grooved, and secretes a byssus. The mantle is mostly free, but at the posterior end the margins unite to form a rudimentary syphon with fringed margins. Most of the species are marine, but a few live in fresh water.

Mytilus, the typical genus, has a world-wide distribution, and is represented on the northern shores of both continents by the common mussel, *Mytilus edulis*. On our east coast this extends as far south as the Carolinas, to San Francisco on the west coast, while on the eastern continent it is found in Great Britain and the Mediterranean and in China and Japan. In color the specimens from exposed situations are dark-brown or bluish-black, while in more sheltered localities one frequently finds specimens of a light, pellucid, olive-green, striped with darker, or occasionally all banding may be absent. These mussels grow in immense quantities in certain situations, rocks, piles, etc., being covered with a thick matting, each individual of which is anchored by its silken, yellow byssus. In Europe these mussels, as the specific name implies, are eaten in large quantities, but with us they form a very inconsiderable portion of the diet of people living near the shore. The cause of this neglect may lie in the fact that they are said to be poisonous to some people. In some regions they are gathered in immense quantities and used as manure. In France the natural growth is far from sufficient to supply the demand for table purposes, and hence large tracts near the shore are used for their cultivation. Numerous sticks are driven firmly into the bottom, and the ends which project above the surface of the mud are interwoven with a wicker-work which affords an anchorage for large quantities of mussels. At high tide these are covered, but at low water they are exposed, and it is at this time that they are gathered for the market.

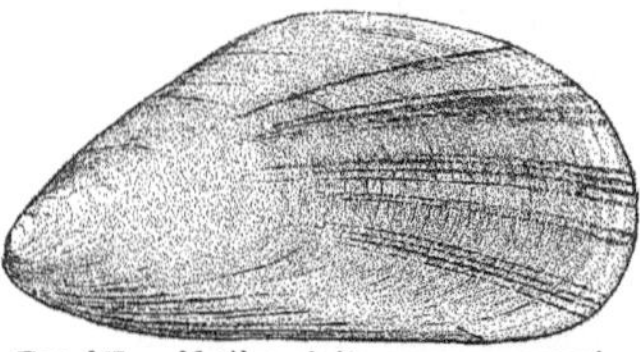

FIG. 277.—*Mytilus edulis*, common mussel.

Although the mussels are anchored by a byssus, they are not compelled to live sedentary lives, for at will they can drop the byssus and move about by the aid of their slender foot. They can even climb, and their method of accomplishing this is interesting. The foot is moved about in the direction in which they wish to go, and a byssal thread is attached. This supports the animal while the foot is again extended, and another thread applied to a more distant point. By continued repetitions of this operation the heavy shell is gradually lifted to the desired situation. *Mytilus edulis* flourishes best in the zone between high and low water marks, and a little below, although specimens are frequently dredged in much deeper water.

In *Mytilus* the umbones of the shell are terminal, and the hinge is either toothless or furnished with minute teeth, while in *Modiolus* the umbones are a little behind the end of the shell, a distinction which is well shown in our figures. *Modiolus plicatulus* is a shore-inhabiting species, varying in color from nearly clear yellow to dark bronze green, and ornamented with a series of radiating ridges. It likes especially shores where a slight admixture of fresh water renders the sea brackish. Another species, *Modiolus modiolus*, is larger, and lives at extreme low water mark, and below. The surface of the shell is not ribbed, but specimens from sheltered localities have the epidermis of the external surface produced into bristles and hairs. It occurs on our shores as well as those of Europe.

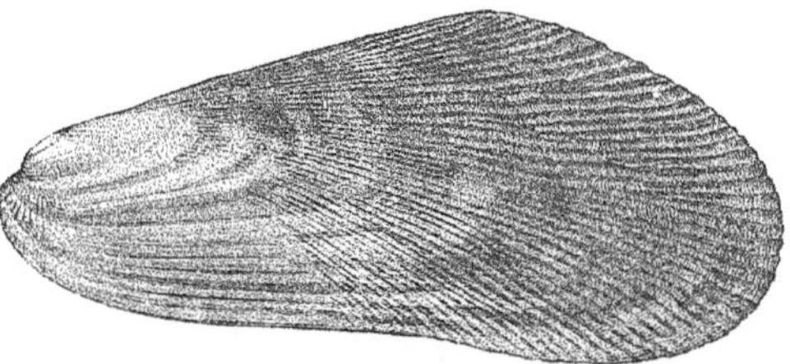

FIG. 278. — *Modiolus plicatulus.*

In *Lithodomus* the shell is small, long, and nearly cylindrical, resembling somewhat a date; and is covered with a thick, dark epidermis. In the young a byssus is spun, but not by the adult, which excavates a hole in some soft rock, in which it subsequently lives.

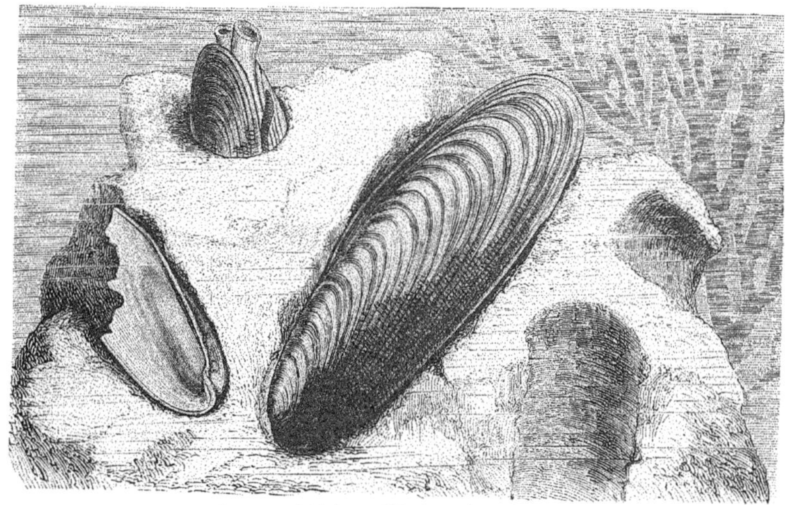

FIG. 279. — *Lithodomus lithophagus* in its burrows.

Like all rock-excavating forms, it is not known how it bores its holes, a question which will be mentioned again when treating of the family Pholadidæ on a subsequent page. Three species of *Crenella* (small thin shells with one tooth in each hinge, and straight beaks) are found in our northern shores, one extending to the south of Cape Cod.

The species of *Pinna* have long, triangular shells, tapering to an acute angle at the umbones. The shells are very thin and delicate, and are usually ornamented exteriorly by large or small scale-like projections. The animals spin a very large and strong byssus, which, as a curiosity, has been used in textile arts, the product somewhat resembling silk.

The last number of this family which needs mention is the form which science has at last decided shall be called *Dreissena polymorpha*. Its specific name is indicative of its many vagaries of shapes, a fact which has led to its description under a large number of generic and specific names. The shell is much like that of *Mytilus*, but in the animal some important differences may be observed. The mantle is closed, leaving only a small opening for the byssus, while at the posterior end a short siphon is formed. The most interesting fact in connection with *Dreissena* is its distribution. It is a native of the Aral and Caspian Seas, and was discovered by Pallas in 1769 at the mouth of the Volga. Later it was found in the rivers flowing into the Black Sea, and from these it is supposed to have been transported into the rivers of Germany by the pontoon trains during the Napoleonic wars. Its introduction into England in 1824 is supposed to have been effected by means of foreign timber, and now it is a recognized member of the fauna of almost all parts of Europe. In London it has proved itself something of a nuisance, as it has obtained entrance to the pipes which supply the city with water.

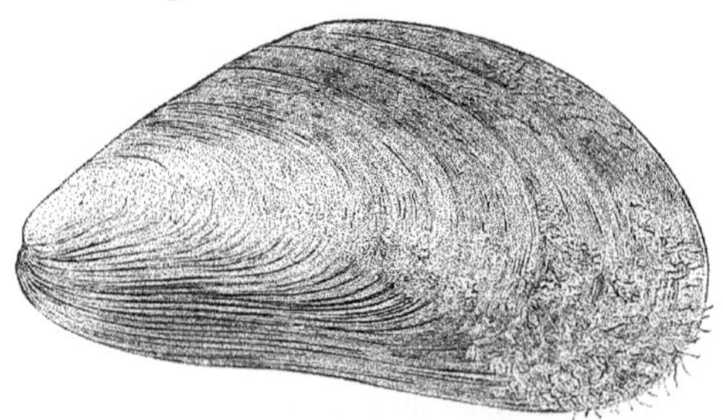

FIG. 280. — *Modiolus modiolus.*

The family ARCADÆ embraces thick equivalve shells covered with a thick, often hairy, epidermis. The hinge ligament is external, and the hinge itself is very characteristic, being formed of a large number of teeth arranged in a single row like those of a comb. Both adductor muscles are present, and equally large, producing corresponding impressions on the shell. The edges of the mantle are free, and the gills terminate in free filaments. The foot is large, but very variously shaped.

Of the genus *Arca* numerous subgenera have been made, and many species commonly parade under the names *Byssoarca*, *Scapharca*, *Argina*, *Parallelipedum*, etc. For our purpose this refinement of classification is not necessary, and we here accept the genus with the Lamarckian signification. With this classification we may define *Arca* as a genus in which the hinge teeth are nearly equal, and form a straight line, the shell is ventricose, the umbones are widely separated, and the free edges of the shell are frequently gaping. *Arca noæ*, Noah's ark, received its name from some queer fancy existing in the mind of Linnæus. It secretes itself under stones at low water in the Mediterranean, and closes the gape of its shell with a byssal structure shaped like a cone, and composed of numerous thin plates. Occasionally violet-colored pearls are found in this species. On our eastern coast three species occur; a very thick and heavy form known as *Arca ponderosa*, a more closely ribbed species, *A. pexata*, and the longer and smaller, *A. transversa*. Further south other species occur. *Arca pexata* is known as the 'bloody clam,' on account of the red gills and the red fluid with which the tissues are filled. It is covered with a thick, hairy epidermis, and has from thirty-two to thirty-six radiating ribs on the outer surface of the shell. In *Arca*

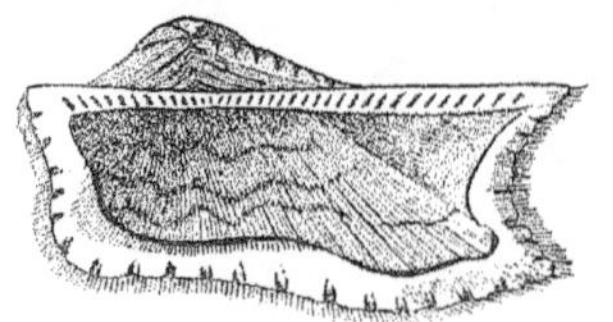

FIG. 281. — *Arca noæ.*

FIG. 282. — *Arca transversa.*

transversa the ribs are about the same in number, but the greater length of the shell readily distinguishes it. Both these species occur under stones near low-water mark. *Arca tortuosa* of the Chinese seas is remarkable for the way in which the valves are twisted. It is very common in collections. The species of *Pectunculus* have a nearly circular outline, and the row of comb-like hinge teeth are arranged in a circular arc. As the shell grows, the number of teeth increases by additions to either end of the series.

The following genera are frequently separated under a family, NUCULIDÆ, but for our purposes they can be retained, as in the older works, as members of the Arcadæ. *Nucula* embraces small trigonal forms covered with an olive epidermis and a simple pallial line. The species are mostly inhabitants of the colder waters of the northern hemisphere. *Leda* is closely similar to *Nucula*, but has a small pallial sinus, and the shell is much longer. *Yoldia*, like the other genera, is boreal in its distribution; in shape and pallial sinus it is much like *Leda*, but the siphons are long and slender. Of the several species we need mention but two. *Yoldia limatula* is a very long species, which, according to Dr. Mighels, has the power of leaping "to an astonishing height, exceeding, in this faculty, the scallop-shells." *Yoldia thracæformis* is larger and comparatively shorter. For a long time it was among the rarest of our New England shells, and the only source of supply was the examination of fish stomachs. More recently the various dredging expeditions have found them in large numbers, and the writer remembers that on one occasion the dredge brought up about two bushels of nothing but dead specimens of this species.

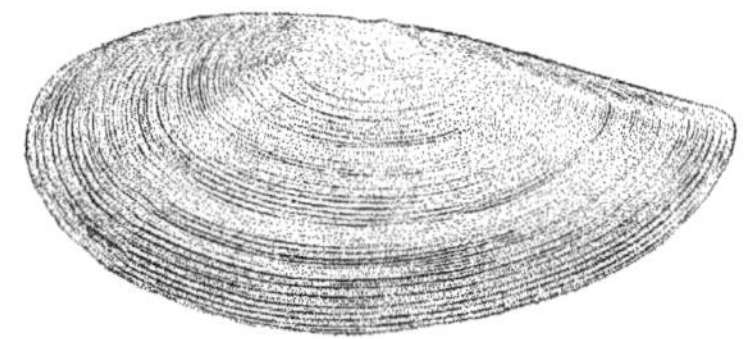

FIG. 283. — *Yoldia limatula.*

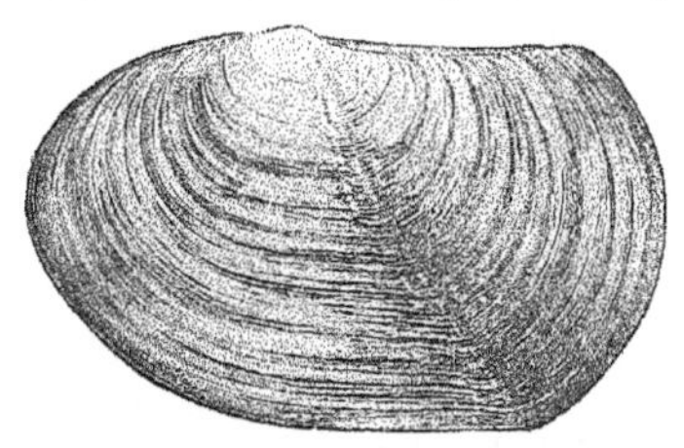

FIG. 284. — *Yoldia thraciæformis.*

Of the TRIGONIDÆ, a small and nearly extinct family, but little need be said. The two valves of the shell are equal, somewhat triangular in outline, pearly inside, and are marked by a simple pallial line. The hinge ligament is external, and the few hinge teeth are diverging. A somewhat unusual feature is found in the posteriorly directed umbones. The margins of the mantle are free, the foot is large and long, and the labial palpi small and pointed. The few existing species belong to the genus *Trigonia*, and are found only in the Australian seas. They are very active and are supposed to wander about on the sea bottom.

Far more prominent is the next family, which embraces the fresh-water mussels, the UNIONIDÆ or NAIADÆ of science. Nearly fifteen hundred nominal species exist in the fresh waters of the world, a large proportion being found in the streams and ponds of the United States. It seems probable that further investigations will relegate a large number of these so-called species into synonymy, as many

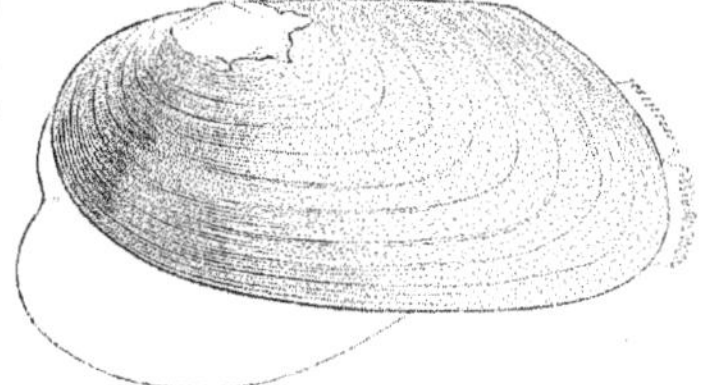

FIG. 285. — *Unio complanatus*, with foot and siphons extended.

seem to be founded on individual variation. The shells are usually long, equivalve, and are covered externally with a smooth, thick epidermis, which is of various shades of brown, olive, and green. Internally the shell is pearly; sometimes white, at others dark pink or almost purple. The hinge ligament is large and external, and the anterior hinge teeth are large and thick, the posterior compressed and laminar; or all may be wanting. Both adductor muscles are present. The mantle edges are united posteriorly so as to form a rudimentary siphon. The foot is very large and tongue-shaped, and in the adult occasionally secretes a byssus.

One of the most interesting features connected with the fresh-water mussels is found in their development. The eggs are carried in brood pouches between the lamellæ of the outer gills, and there undergo their early development, the details of which need not be described here. Eventually a shell is formed, which soon divides into two valves united by a straight hinge. Soon at the fore extremities of each valve a strong beak-like process is developed, and a little later a byssus is secreted. Next, peculiar sense organs are formed on the inner surface of the mantle, the function of which is, doubtless, to ascertain the presence of fishes. Now the young is in the condition known as a Glochidium, and when the mother is under natural conditions or placed in a tank with fish, the young are expelled from the brood pouch and almost immediately attach themselves to some submerged object by means of the byssus. If no fishes be present, the mother will retain the young for a long time after the glochidium stage has been reached. As soon as an opportunity is afforded, the glochidia attach themselves to the gills, fins, or other parts of a fish, by closing the valves of the shell and holding on by the spiny beaks mentioned above. Here they undergo an extensive metamorphosis, in which the single adductor muscle disappears and is replaced by the two of the adult, the gills appear, the sense organs atrophy, and, according to one author, the shell and even the mantle lobes are formed anew. Finally, when all the organs except those concerned in reproduction are formed, the embryo quits its host and sinks to the bottom, where by a regular growth it attains the adult condition. This parasitic or semi-parasitic life is very unusual among the Mollusca.

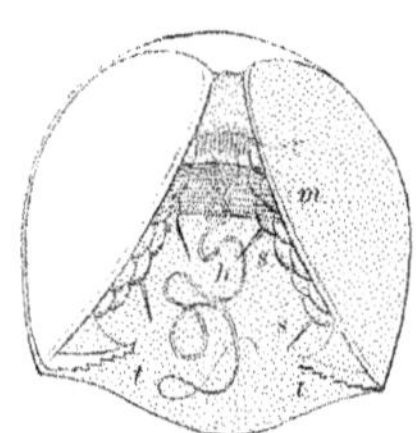

Fig. 286. — Glochidium of *Unio*, still within the egg membrane; *b*, byssus; *m*, adductor muscle; *s*, sense organs; *t* teeth; *v*, velum.

The Unionidæ live in still water or in running streams, usually about half buried in the mud. They do not, however, lead sedentary lives, but burrow or plough along the bottom, the part of the shell which contains the siphons being above the surface of the gravel or mud. According to Dr. Lea one species occurring in the United States (*Margaritana dehiscens*) buries itself completely in the mud and draws its water and food through a tube, as do so many of the siphonated marine forms.

In the typical genus *Unio*, the hinge is provided with one or two anterior and one posterior teeth in one valve, and two anterior and two posterior in the other. The shell is smooth or ribbed or even spined, and the variations in shape are extraordinary. In some of the species the valves become soldered together at the hinge, so that motion would be impossible were it not for the fact that a fracture takes place near the line of junction, so that one valve bears two wings while the other has

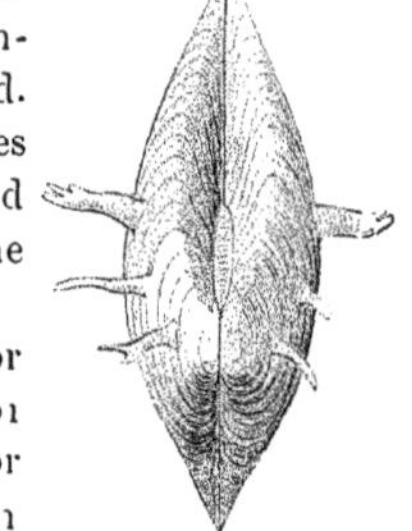
Fig. 287.—*Unio spinosus.*

none. This fact has been used by Dr. Lea to divide the numerous species of *Unio* into two groups, those with soldered hinge being called symphynote, and those with the normal structure asymphynote forms. Although we have a vast number of species of *Unio*, there is but little of popular interest to be said concerning them.

Margaritana (or *Alasmodon*) is closely similar to *Unio*, the only differences being in the details of the hinge teeth. *M. margaritifera*, which occurs both in Europe and in the northern United States, is especially noticeable from the fact that it produces pearls which are frequently valuable, equalling those of the true pearl oyster described on a preceding page. During the Roman occupation of the British Isles, these pearls were famous, and in more modern times the search has been continued, especially in Scotland, Bavaria, and Bohemia. These pearls not unfrequently have a slight pinkish hue, which is permanent. In the United States the search for these has never been prosecuted with any great vigor, although the writer has seen several fine pearls found in New York State. This branch of the pearl fishery can be conducted with more ease and less danger than that of the true pearl oyster, since the species can be found in brooks and rivers which can be waded, and which are so clear as to render the collection of the mussels an easy task. All the pearls collected have a value, no matter how imperfect, since pearls can be properly polished only by the use of pearl powder, and for this purpose inferior pearls are as good as those of the best quality.

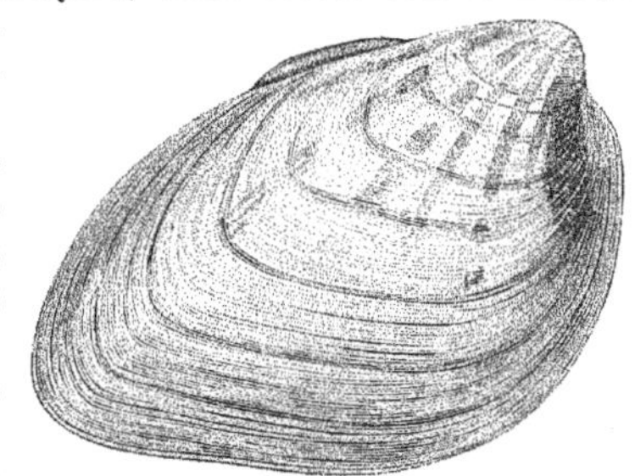
FIG. 288.—*Unio clarus.*

The only other genus which needs mention is *Anodonta*, in which as the name indicates, the hinge teeth are lacking. The species have a light, thin, smooth shell, and, like the preceding forms, are found both in streams and ponds. Regarding the names to be applied to our North American Unionidæ, the utmost confusion exists. Many forms were described almost simultaneously by Rafinesque, Say, Hildreth, Barnes, Lea, and Conrad. The descriptions of the latter author, like all his work, were decidedly poor, and it is said that here, as elsewhere, he described, from accounts furnished him by people other than naturalists, shells which he had never seen. To make confusion worse confounded, Rafinesque, in his later years, when his intellect was clouded, named a set of shells which now figure as his 'types.' Besides this, almost innumerable questions of priority exist; the slightest variation of outline has been seized upon to form new species, and in many ways the whole subject has been tortured and twisted so that the person who in the future endeavors to unravel the skein will have a task of no small difficulty.

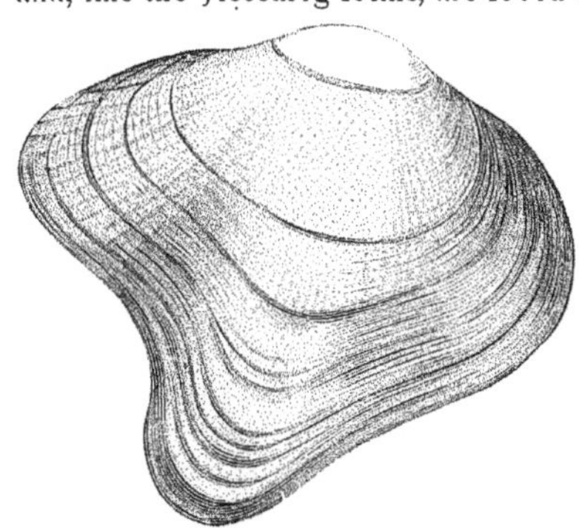
FIG. 289.—*Unio flexuosus.*

In the preceding families of Acephals we have found considerable variation in the free or united edges of the mantle, and in the presence or absence of siphonal tubes.

In all the remaining families the siphons are well developed, elongate and tubular, and the edges of the mantle are united to a greater or less extent. In the Chamidæ, Tridacnidæ, Hippuritidæ, Cardiidæ, Lucinidæ, Cycladidæ, and Cyprinidæ, the pallial line produced by the edge of the mantle on the shell is simple, while in all the remaining families at the posterior end of the shell it makes a re-entrant angle known as the pallial sinus. This is produced by the great development of the siphonal muscles.

The members of the CHAMIDÆ are thick, irregular, inequivalve shells, with external hinge ligament and strongly developed hinge teeth, one in one valve and two in the other. The mantle margins are united, leaving only the small siphons and a small opening for the foot. The living species are found in the tropical seas of both hemispheres, and all, both recent and fossil, are usually attached to some rock or shell by one of the umbones. In *Chama* the shell is lamellar in texture and the umbones are coiled in a spiral; but this peculiarity reaches its greatest development in the fossil genus *Diceras*, either valve of which, seen alone, would be taken for the shell of a gasteropod until a closer examination showed the hinge, muscular impressions, etc.

The HIPPURITIDÆ (ranked as an order, Rudistes, by Lamarck) is usually placed here, but its position is far from certain. The family is characteristic of the rocks of the cretaceous age and in some parts of Europe have given to certain beds the name of Hippurite limestone. They are also found on our continent. The shells are *sui generis;* the two valves are very unequal, one being frequently very flat, while the other is immensely drawn out, in one form resembling a cow's horn, and giving rise to the name *Hippurites cornu-vaccarum*. The shells were attached, and frequently the larger valve was chambered so that in forms like *Caprinella* it bore no inconsiderable resemblance to the chambered shells of the cephalopods. The position which these forms should occupy in the animal kingdom has been very differently regarded by different authors. Besides being placed in varying relationships among the Acephals, they have been considered as corals, barnacles, worms, and various combinations of these distinct groups. The principal features which place them among the bivalve molluscs are the structure of the shells, which are hinged together and furnished with an internal cartilage, the presence of two adductor muscles and a well-marked pallial line. One noticeable peculiarity is found in *Hippurites*, where the free valve "is perforated by radiating canals which open around its inner margin, and communicate with the upper surface by numerous pores, as if to supply the interior with filtered water." In other genera there is no trace of these canals.

The TRIDACNIDÆ, which are all inhabitants of the eastern seas, contain the largest bivalves known. The shell is regular and equivalve, the valves being ornamented by strong ribs radiating from the umbones. The shell is very hard, almost every trace of organic matter being removed. The hinge ligament is external, and there are either one or two cardinal teeth in each valve, and one posterior tooth in one, and two in the other. The foot is comparatively small, finger-like, and provided with a byssal groove. Some of the species spin a byssus, but others do not. The adductor muscle is single, and there are two gills on either side, the outer ones composed of but a single lamina.

The genus *Tridacna* embraces some very large forms, and the generic name is given in allusion to this fact. The ancients used to tell of an Indian oyster which was so large that it required three bites (*tri* and *dakno*) to devour one of them, and so, when a more modern science came to name these forms, the old terms was resurrected and applied as a generic name. The appropriateness of it will readily be seen when it is stated that the soft parts of *Tridacna gigas* will sometimes weigh twenty pounds. The shells of this same species will sometimes, together, weigh five hundred pounds, and in some of the churches of France they are employed to hold the holy-water, a use which well accords with the beautiful white of the inner surface of the shell. This species and the smaller *T. squamosa* are very common in collections, the first having the ribs of the valves nearly smooth, the other having them ornamented by large foliaceous processes. Both these species are of considerable importance to the South Sea Islanders, as they afford an important part of the food supply. The giant clam is also useful from a mechanical point of view. In many of these islands, stones are unknown, but, as a substitute, the natives make their knives and axes from the fragments of this shell. In one species the gills are bright blue.

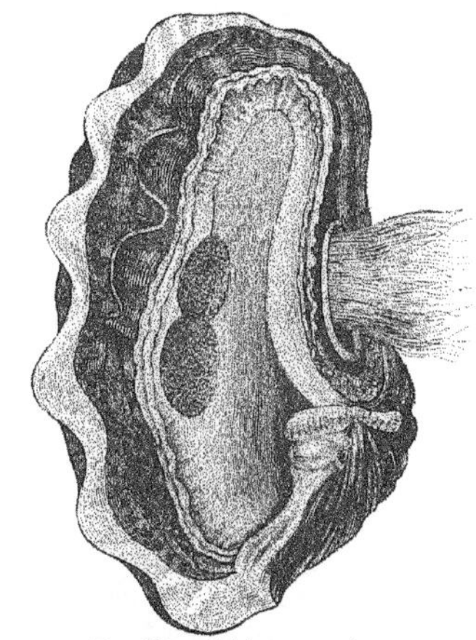

FIG. 290.—*Tridacna mutica.*

Hippopus, which differs but slightly from *Tridacna*, has a similar distribution, and the common species, *H. maculatus*, is found in almost every collection. The ribs on the outside are much smaller and more numerous than in the other genus, and scattered over the outside are numerous little, scale-like projections. The color, both inside and out, is a delicate, creamy white, ornamented on the exterior with small blotches of lake. The generic name means horse-foot, but the common name is the bear's-paw clam. The applicability of either is not very evident.

The name CARDIIDÆ is applied to the next family which we have to consider on account of the heart-shaped outline which the species present when viewed from the end. The valves are equal and swollen out, and the umbones are rolled inward, so that the appearance is like the conventional heart, which, it may be remarked, resembles no heart existing in nature. The hinge ligament is external, and the hinge teeth are two in each valve beneath the umbones, and one in each on either side of the central ones. The external surface of the shell is usually radiately ribbed, the ornamentation on the posterior part differing from that in front. The edges of the mantle are united, so that a small slit is left for the protrusion of the strong, sickle-shaped foot, which plays a very important part in the locomotion of these molluscs. The siphons are usually very small, but occasionally they are better developed, and at such times there is a slight indication of a sinus in the pallial line.

Cardium, with its various subdivisions, embracing about a hundred and fifty species, is found in all the seas of the world, living in shallow water, from low tide down to a hundred and fifty fathoms. The shell is strongly ribbed and very swollen. Our largest species is *Cardium islandicum* which is found north of Cape Cod. Other smaller species are found on our coasts, one of which (*C. pinnulatum*) occurs as far south as Long Island Sound. These shells are known as cockles, and one species is eaten in England. They prefer sandy bays near low water, and are usually found in large numbers together.

In *Lævicardium*, as the name indicates, the surface is nearly smooth, and our single species ranges from the West Indies to Nova Scotia, being common in the waters south of Cape Cod. *Serripes* is noticeable from the fact that the cardinal hinge teeth have disappeared; the species *S. grönlandica* occurs in the polar seas, extending south on our coasts to Massachusetts Bay. *Hemicardium* is a tropical genus, in which the posterior part of the shell is separated from the anterior by an abrupt angle. The cardinal teeth are distinct.

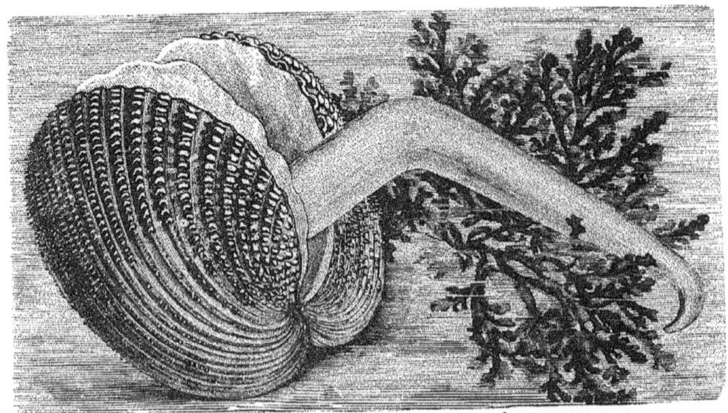

FIG. 291.— *Cardium echinatum*, cockle, with the foot extended.

The LUCINIDÆ embrace forms with a circular outline, one or two cardinal hinge teeth, and the laterals one or obsolete. The hinge ligament is external or partially internal. The mantle is open in front, but behind is united into one or two short siphonal tubes. The foot is long, cylindrical, or worm-shaped. *Lucina* is represented on our coast by two species, one of which, *L. dentata* (sometimes included in a subgenus *Cyclas*) is noticeable for the peculiar ornamentation of the valves. The shell is pure white, and exhibits the regular lines of growth, but over these is laid a second series of lines in a manner difficult to describe in a few words, but readily understood from the figure. One or two other genera (*Strigilla*, *Choristodon*) belonging to the Tellinidæ are ornamented in a similar manner.

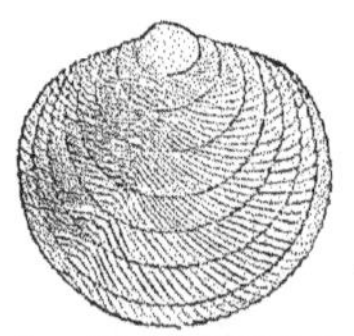

FIG. 292.— *Lucina dentata*.

In *Lucina filosa*, which is not very common, these striations are absent, and the shell is roughened by a number of concentric lamellar ridges with smaller, thread-like ridges between.

In *Cryptodon* the hinge teeth are wanting. *Ungulina*, a genus of the eastern seas, is noticeable from the fact that it excavates winding galleries in the coral on the reefs where it is found. *Kellia planulata* is found on the northern coast of America, under stones, at low-water mark, and other species in various parts of the world live above the reach of ordinary tides, or burrow in sandstone. They creep about freely, and anchor themselves by a byssus at pleasure. One species of *Montacuta* occurring in Great Britain (*M. substriata*) is to be mentioned because it has never been found except attached to the spines of the sea-urchin, *Spatangus purpureus*. It cannot be regarded as a true parasite, for not only has it no organs by which to feed upon the urchin, but it is never attached to any of the soft parts of its host. It is rather to be regarded as a mess-mate or commensal, obtaining its food from the same source as does the urchin.

In the fresh waters the world over occurs a group of usually small bivalve shells, covered with an amber or brown epidermis, while in the brackish waters of warmer countries occur some larger forms. The family under which these are assembled is variously known as CYCLADIDÆ or CYRENIDÆ, the latter name being preferable. In all, the shell is nearly circular in outline, the ligament is external, the hinge is provided with several teeth. Usually there are apparent indications of a pallial sinus, most marked in the American species.

Cyrena is the typical genus and embraces over a hundred nominal species, which live in brackish water in the warmer parts of the globe. They are frequently found buried in the mud of mangrove swamps, where the tide rises and falls slightly, but where the admixture of rain renders the water less dense than that of the ocean. *Cyrena carolinensis* occurs in the rivers and swamps of some of our southern states. In our northern states the family is represented by the genera *Sphærium* and *Pisidium*, our fauna containing about fifty species of both genera. *Sphærium* (known in some of the older works as *Cyclas*) has the shell nearly equilateral, the hinge teeth minute and rather weak, and two nearly separate siphons. In *Pisidium* the part of the shell in front of the umbones is larger than that behind, the teeth are stronger, and the two siphons are united the whole length. The species abound in the still water of some of our ponds, and are very active.

FIG. 293. — *Cyprina islandica.*

The CYPRINIDÆ is a much larger family than the last, and its members are inhabitants of salt water. The shell is regular, oval, and equivalve, and is covered with a thick, strong epidermis. The hinge ligament is usually external, and the hinge is provided with from one to three cardinal teeth, and usually one posterior lateral in each valve. The margins of the mantle are fringed, the pallial line simple, and the two siphonal tubes are short.

Cyprina islandica is a large boreal shell, common in sandy bottoms north of Cape Cod, but is less frequently met with south of that barrier. The hinge has three unequal diverging cardinal teeth and one lateral. The shell is thick and heavy, and the color in the young is very light brown, but old specimens are very dark. With age, the epidermis, near the umbones, usually disappears, and the shell itself is frequently eroded. Large specimens measure four inches across. Several species of *Astarte* are found on our northern coasts, all of which can be recognized by the smooth or concentrically furrowed surface, and the two hinge teeth in each valve. The shell is covered with a strong epidermis.

FIG. 294. — *Astarte undata.*

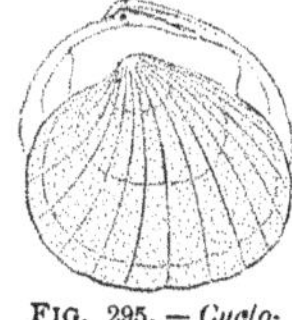

FIG. 295. — *Cyclocardia novangliæ.*

In *Cyclocardia* the shell is radiately ribbed, and nearly circular in outline. Cardinal, but no lateral teeth are present. *Isocardia* is a genus with very ventricose shell, and with the beaks regularly inrolled. *Isocardia cor* is found in the Mediterranean.

With the VENERIDÆ we enter upon a group embracing all the remaining bivalves, in which the siphons are well developed and the pallial sinus well marked. In the Veneridæ the shell varies in outline between nearly spherical to oblong, the ligament is external, and usually there are three diverging teeth in each valve. The siphonal tubes are unequally developed, and are united at the base. The foot is tongue-shaped, and compressed, and the triangular labial palpi are very large. The forms embraced in this family include some of the handsomest of the bivalve molluscs, the distribution of color being frequently very striking, chevron-shaped lines being most frequent.

In *Venus*, the typical genus, the shell is oval and thick, and the pallial sinus is small and angular. Most important to us is the quahog, round clam, or hard-shell clam (*Venus mercenaria*), which forms a considerable article of diet in those regions where the long clam, or soft-shell clam (*Mya arenaria*), is not to be had. In its range it extends from Texas to Cape Cod, but north of that cape it is comparatively rare and local. It is "common on sandy shores, living chiefly on the sandy and muddy flats, just beyond low-water mark, but is often found on the portion laid bare at low-water of spring tides. It also inhabits the estuaries, where it most abounds. It burrows a short distance below the surface, but is frequently found crawling at the surface, with the shell partly exposed." The mantle is widely open in front, allowing the large foot to be placed in almost any position. The siphonal tubes are united for quite a distance. "This species is taken in large quantities for food, and may almost always be seen of various sizes in our markets. The small or moderate-sized ones are generally preferred to the full-grown clams. Most of those sold come from the muddy estuaries, in shallow water, and are fished up chiefly by means of long rakes and tongs, such as are often used for obtaining oysters. Sometimes they are dredged, and occasionally they can be obtained by hand at or just below low-water mark. These estuary specimens usually have rough, thick, dull-white or mud-stained shells, but those from the sandy shores outside have thinner and more delicate shells, often with high, thin ribs, especially when young; and, in some varieties, the shell is handsomely marked with angular or zigzag lines or streaks of red or brown."

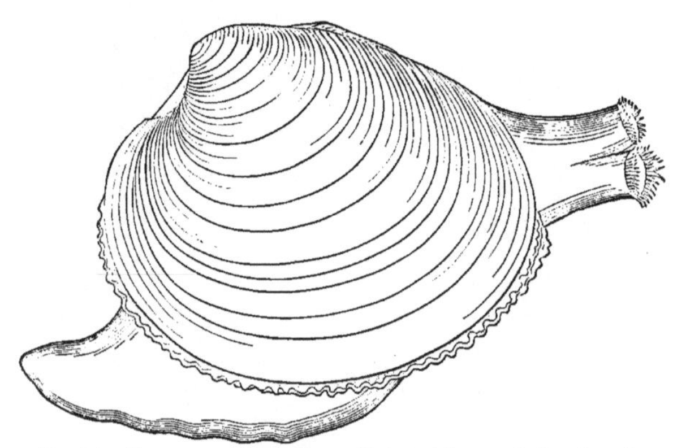

FIG. 296.—*Venus mercenaria*, quahog, with foot, siphons, and edge of mantle extended.

In most of the shells, when nearly full-grown, the borders of the interior of the valves is colored purple to the distance of about half an inch from the margin, and it was by breaking up this portion and converting it into beads that the Indians of New

England made their purple wampum, or *suckauhock*, which was regarded as twice as valuable as the white money, or *wompom* proper. This latter was made from various shells, but mostly from *Busycon*.

Many other species of *Venus*, in its broader sense, are found in the warmer seas of the world, the west coast of America being better supplied than the east.

Cytherea and its sub-genus *Callista* are readily distinguished from *Venus* by the presence of an anterior lateral tooth in the left valve, which fits into a corresponding depression in the other. Like the last genus it is rich in species, especially in the warmer seas of the world. Our northern *C. convexa* has an outline much like that of the quahog, but its dead white surface does not render it as attractive as its southern relative, *Callista gigantea*, which is found on our southern coasts. *Cytherea lusoria* is a Chinese species, which derives its specific name from the fact that the inhabitants of the celestial empire paint certain figures on the inner surface of the valves, and then employ them in some of their many games of chance. *Cytherea scripta* has a ground of white or yellow, over which is laid a series of zig-zag reddish-brown lines, which require a rather vivid imagination to be regarded as resembling writing. It comes from the Indian Ocean, as does the *C. erycina*. *C. dione*, from the west coast of America, is a remarkable species, from the fact that it is ornamented by a series of long slender spines, running in a row down the posterior side of the shell, from the umbones to the margin. The color is a rosy purple, varying considerably in depth. In the more recent systems of classification it is made the type of the genus *Dione*.

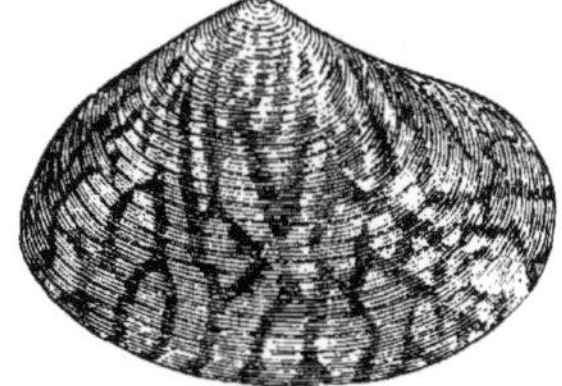

FIG. 297. — *Meroë*.

Meroë embraces a few species of oval shells, with three cardinal teeth and a long anterior tooth. The general shape can be seen from our figure, but there is nothing of popular interest to be said concerning the species. *Dosinia* is represented on our shores by a species (*D. discus*), the specific name of which is very apt. The shell is flat, and nearly circular in outline; the siphons are united, and the foot is large.

Gemma embraces only a single species, found on our coast, and known under the repetitive name of *Gemma gemma*. In size it is minute, scarcely more than an eighth of an inch in length, and in color it is a yellowish white, or rosy, tipped posteriorly with an amethystine purple, so that the name is very appropriate. It would appear that it was known to the early settlers of this country, and that they sent specimens of it, along with other curiosities, to the old world; and yet it was unknown to naturalists until the year 1834, when the eminent engineer, General Totten, who was a good naturalist withal, published a description of it. It is an active species, found on sandy shores, where it burrows quickly. One of the most interesting facts known in connection with it is that it retains the young inside its valves until the shells are fully formed, sometimes thirty young being found inside of the parent shell.

No species of the genus *Tapes* occur on our coasts, but the seas of other parts of the globe contain nearly a hundred species. The shells are long, the siphons separate at their extremities, and the long, slender foot spins a byssus. Many of the species are ornamented with zig-zag lines, of darker color, and in Europe, especially on the Mediterranean coast, *T. geographica* is used as an article of food.

The family PETRICOLIDÆ is a small one, and its only member which requires mention is the form known as *Petricola pholadiformis*. This is a thin, long shell, ornamented with radiating ribs, arranged much like those in *Pholas*, whence the specific name. The shell cannot be completely closed, but gapes, while the mantle is almost entirely united in front, leaving but a slight opening for the small, pointed foot. It has a rather extensive distribution, reaching from Nova Scotia to Texas; and Professor Verrill writes that he has received specimens from the Gulf of California which are scarcely to be distinguished from this. It is sometimes found among rocks and stones, below low-water mark, but more frequently it makes deep burrows in stiff mud or clay bottoms, climbing up and down in them by means of its slender and flexible foot. The siphons are very long, and are united for only a short distance. Other species of *Petricola* excavate and occupy burrows in the softer limestone rocks, and from this fact the generic name has been given. I am not aware that this habit has been noticed in our species. The shell of *P. pholadiformis* reaches a length of two and a half inches, but specimens over two inches long are uncommon. In color it is a chalky white.

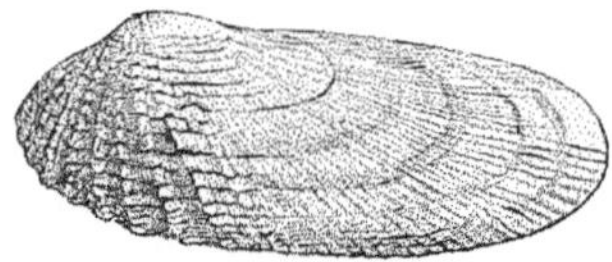

Fig. 298. — *Petricola pholadiformis.*

The Mactridæ embraces a considerable number of trigonal equivalve shells, which can be completely closed, or which gape slightly. The ligament is usually internal, and contained in a deep pit, but occasionally it is external; the hinge has two diverging cardinal teeth, and laterals are frequently present. The outside of the shell is covered with a thick epidermis. The pallial sinus is short, and rounded or angular, and the siphonal tubes are united and have their extremities fringed with small tentacles.

Fig. 299. — *Mactra ovalis.*

Mactra solidissima and the closely allied *M. ovalis* are known along our northern coasts as hen-clam, sea-clam, and surf-clam, these names properly belonging to the more common and more littoral species first mentioned. The first is distributed from the Carolinas to Labrador, while the second is only found north of Cape Cod. On every sandy beach *M. solidissima* occurs in large numbers, and after gales the other species is thrown up on the shore. They are used in a limited way for food, and when properly prepared they make a chowder no less palatable than the *Mya arenaria*, so extensively used. The toughness of the older individuals prevents their more extensive use, and their preparation calls for the use of the chopping-knife. The species are very active, and, instead of leading sedentary lives in burrows, they plough through the sand by means of their well-developed foot. They have a considerable leaping ability, and many a time have I known a basket full at night, and left in a 'trap,' to be half emptied in the morning, the clams leaping out by means of their foot. The species of *Mactra* are valuable for the student of molluscan anatomy, from the fact that their ganglia and nerves are colored a reddish hue, and are thus more easily distinguished, and dissected out, than in many other forms. *Mactra solidissima* occasionally reaches a length of over six inches, has the cardinal teeth delicate, and has a shallow pallial sinus. *M. ovalis* but rarely exceeds four inches in length, and has the teeth short and the sinus deep. Separated from *Mactra* proper by an angular pallial sinus is the sub-genus *Mulinia*, of which one species, *Mulinia lateralis*, occurs nearly the whole length of our eastern coast.

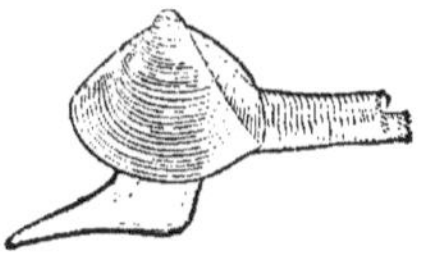

FIG. 300. — *Mulinia lateralis.*

Rangia (also known as *Gnathodon*) is a brackish-water genus, represented in the southern United States by the species *R. cyrenoides*. In the gulf states this species occurs in vast numbers. Banks of dead shells, three or four feet in thickness, occur in many places, some of them twenty miles inland. The city of Mobile is built on one of these banks, while the road over which the inhabitants of New Orleans travel in order to reach Lake Pontchartrain "is made of these shells, procured from the east end of the lake, where there is a mound of them a mile long, fifteen feet high, and from twenty to sixty yards wide; in some places it is twenty feet above the level of the lake."

FIG. 301. — *Rangia cyrenoides.*

The family TELLENIDÆ vies with the Veneridæ in its beautifully-colored species. Like the members of that family, they are mostly inhabitants of the warmer seas of the world, the small number of species which stray into the more northern waters being dull colored, and far less attractive than their tropical relatives. The shells are long, compressed, and usually closed and equi-valve; and one marked feature that is generally found is that the half of the shell in front of the umbones is longer than that behind them. The cardinal teeth are at most two in number; the lateral teeth are occasionally obsolete. The hinge ligament is usually external and posterior. The mantle is widely open in front, and the two very long and slender siphonal tubes are completely separated. The foot is triangular and compressed. In most species the shell is very dense, and highly polished, and not infrequently is white, enlivened by bands or stripes of delicate shades of red or yellow, while in others the effect is heightened by the finely sculptured lines.

In *Tellina*, which reaches its highest development in the Indian Ocean, the shell is rounded in front and angular posteriorly, a fold running from the angle to the umbo. The siphons are very long, and can be extended to at least twice the length of the shell, and, as a necessary sequence of such extensibility, the pallial sinus is very wide and deep. In *Strigilla* the surface is ornamented with divaricating lines, like those of *Cyclas*, described on a preceding page.

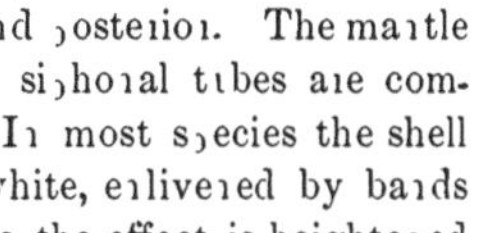

Our northern forms belonging to this family are dull colored and unattractive. *Tellina tenera*, which occurs from Florida to the Gulf of St. Lawrence, is an exception, for the delicate rose-color which tinges the otherwise white shell makes it an object of beauty. The siphons are very long, several times the length of the body, and hence

FIG. 302. — *Tellina tenera.*

this species can burrow beneath the sand, and thence extend its long siphons to the surface. In *Macoma* the thin white shell is covered with a dusky epidermis. *M. sabulosa* is a more northern form, extending as far south as Long Island Sound, while the more common *M. fusca* reaches the coast of Georgia. In muddy places this latter species possesses but little beauty, but when living in clean sand the epidermis becomes thinner, and the shells frequently have a delicate rose or lemon color.

FIG. 303. — *Macoma sabulosa.*

FIG. 304. — *Macoma fusca.*

Donax is an easily recognizable genus, in which the posterior end of the shell appears as if cut off at a more or less oblique angle. The species occur in tropical and semi-tropical regions, where they bury an inch or two beneath the sandy shores. Our two species, *D. fossor* and *D. variabilis*, occur abundantly in some places in the southern states, the former reaching as far north as New Jersey and Long Island.

Space will allow but a mere mention of the other prominent genera of this family; *Paphia*, *Semele*, *Scrobicularia*, *Psammobia*, *Sanguinolaria*, etc., which embrace many beautiful forms, but only few of which anything of popular interest can be said. *Scrobicularia piperata* of the Mediterranean is occasionally eaten, and receives its specific name from its peppery taste.

The MYIDÆ has an importance among the molluscs of the northern seas far out of proportion to the number of species, for the long clam, or soft-shell clam, is its most prominent member. In all the family the shell is thick and strong, and when closed as far as possible, it still gapes at one or both ends. The hinge cartilage is internal, and rests in a peculiar process of the left valve, as diagrammatically shown in Fig. 265, on a preceding page. The foot is small, the mantle margins have but a slight pedal gape, while the very extensible siphons are united their whole extent.

Mya arenaria, the clam *par excellence*, which figures so largely in the celebrated New England clam-bake, is found in all the northern seas of the world, and is too well known to need any technical description. All along the coasts of the Eastern States, every sandy shore, every mud flat, is full of them, and from every village and hamlet the clam-digger goes forth at low tide to dig these esculent bivalves. The clams live in deep burrows in the firm mud or sand, the shells sometimes being a foot or fifteen inches beneath the surface. When the flats are covered with water, his clamship extends his long siphons up through the burrow to the surface of the sand, and through one of these tubes the water and its myriads of animalcules, is drawn down into the shell, furnishing the gills with oxygen and the mouth with food, and then the water, charged with carbonic acid and fœcal refuse, is forced out of the other siphon. When the tide ebbs, the siphons are closed and partly withdrawn. The clam begins its burrow at a very early stage, and keeps enlarging and deepening it as it grows older. An old clam dug up and left on the surface has hard work to make a new burrow.

The white settlers were not the first to find out the palatable qualities of the clam, the Indians knew it and loved it long ages before. On the shores of every promontory and bay along the New England coast are old shell heaps, 'kjökkenmöddings' the archæologist calls them, which tell of many a feast by the red man. The careful explorer, by raking them over, finds not only the shells of the common clam, but those of many other molluscs, and bones of various animals as well, together with the tools and implements of the former inhabitants. In these heaps on the coast of Maine, the bones of the now extinct great auk have been found.

However eaten, whether raw, or in a chowder, or fried, the clam is good, but best of all is the clam-bake, where a long-continued fire heats the rocks red hot. Then the cinders and ashes are swept away, a thin layer of rock-weed is laid on the hot stones, the clams placed in the centre and covered with more of the rock-weed. The steam generated cooks the clams in the most perfect manner.

A second species of *Mya*, *M. truncata*, occurs on our coast, and but one other is known in the world. *Mya truncata*, instead of being rounded at both extremities, is truncated behind, while in the living specimen the epidermis is extended in the shape of a tube, some six or seven inches in length, from the posterior edges of the shell. This species lives below low-water mark, and is not common south of Eastport. On the English shores it is more common than *Mya arenaria*.

The species of *Saxicava* bore into hard mud, stones, etc., and have very irregular and greatly distorted shells, so that specific limits are far from certain. Probably our only species is that figured, *Saxicava arctica*, but several others have been described upon the variations of this form. Indeed, Woodward states that no less than five genera and fifteen species have been made upon the form known as *S. rugosa*, and since this and *S. arctica* are probably the same, the synonymy of this protean species is something awful to contemplate. In places where limestone is abundant, *Saxicava* bores holes in it to the depth of about six inches, and as it is not at all careful where it goes, it not infrequently cuts across the burrow of another individual. When a specimen dies the soft parts decay, but the valves remain in the burrow, and another individual occupies the same burrow, seating itself "between the valves of its predecessor. In this way four or five pairs of shells may be frequently seen nested one within the other."

FIG. 305. — *Saxicava rugosa.*

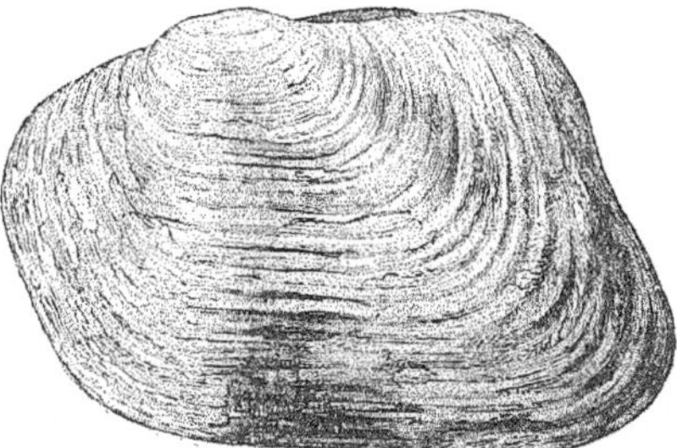

FIG. 306. — *Panopæa norvegica.*

Panopæa is a genus of large shells, the animal of which closely resembles *Saxicava*. *Panopæa norvegica* is a boreal species, specimens of which are occasionally found in the stomachs of fishes; the shell is covered with a thick epidermis.

FIG. 307. — *Glycimeris siliqua.*

Glycimeris siliqua is the only species of the genus known, and presents several noticeable peculiarities. It is covered with a thick, glossy-black epidermis, while the interior of the thick valves is rendered very irregular by the great deposition of calcareous matter seemingly without rhyme or reason. The animal is much larger than the shell, and the soft parts cannot be drawn entirely within the valves. Like the last, it is boreal in its distribution, but it is not uncommon in Massachusetts Bay.

The ANATINIDÆ is an almost extinct family. To-day it is represented by a few species, but in geological time they played a much more important part than now, their remains being found in the oldest palæozoic rocks. The internal surface of the shell is pearly, the external granular, and the hinge is usually toothless. About half a dozen species are found on our coasts, representing the genera *Pandora*, *Thracia*, *Periploma* and *Lyonsia*.

The SOLENIDÆ, the family of the razor-clams, is next in order, and the typical forms, like *Solen*, well deserve the common name which has been given them. In all the family the shell is long, and in some of the genera immensely so; at each end it gapes, and the hinge teeth are usually two in one valve and three in the other. The foot is very large, and more or less cylindrical, the siphons are short or moderate. The old genus *Solen* has recently been broken up, those with straight shells and one tooth in each valve being retained in that genus, while those with the typical number of teeth, and usually slightly curved shells, belong to *Ensis*.

In American waters *Ensis americana* is the razor-fish, or razor-clam. The shell is long and sub-cylindrical, and bears no slight resemblance to the familiar tonsorial instrument. It is a common inhabitant of the sandy shores. These clams excavate large elliptical holes, which penetrate downward, usually in a nearly vertical direction, to a depth of two or three feet. Up and down this hole they go. When the tide is in, and no danger is near, one end of the shell usually projects above the surrounding surface of the sand for an inch or so, but a sudden jar startles them, and down they go with great rapidity. It is useless to attempt to dig them out, for they can burrow as fast as a man can shovel out the sand, and besides they have two or three feet the start. The process of burrowing is interesting. The foot is bevelled off to a point, and this is readily pushed down into the sand. Then the animal inflates the foot with water so that it becomes bulbous at the extremity; this at the same time crowds aside the sand and gives the animal a hold whereby it can draw itself down. By a repetition of the process it still further increases its distance from the surface. The razor-clam can start this burrow when lying on the surface; at first its progress is slow, but as soon as it gets the shell in a vertical position, it goes much more rapidly.

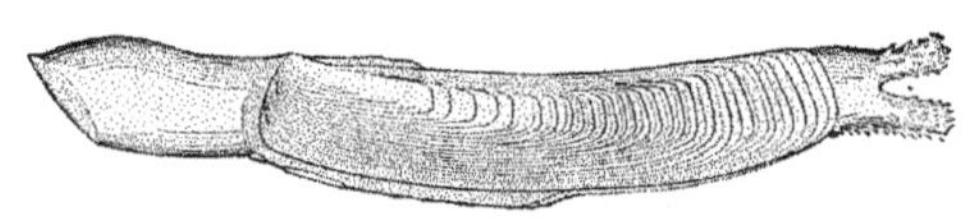

FIG. 308. — *Ensis americana*, razor-clam.

The razor-clams are used for food, but to a far less extent than many other bivalves. For this there are several reasons. First, it is not so common as the quahog or the soft-shell clam, and, again, it is not so easily procured. We have already alluded to the rapidity with which it disappears when alarmed, and the recent investigations of Dr. Sharp show that it does not depend, for its warning, upon the senses of sight and feeling alone for its warnings of impending danger. On the ends of the siphons Dr. Sharp finds organs of vision, very rudimentary, it is true, but still sufficient to recognize various degrees of light and shade. His attention was called to this subject by the fact that a shadow cast upon razor-clams exposed for sale caused them to immediately withdraw their siphons. Histological investigation followed, which showed that the essential parts of optic organs were present.

The fishermen, in procuring them, walk quickly up to the animal, and grasp it before it has time to retreat. In Europe the clam-digger pours a little salt down the hole; this brings the clam to the surface, when it is quickly grasped. If not successful, no subsequent salting will arouse the clam. Where the water is still, another method is adopted. At low tide the fisherman goes over the flats and puts a little oil near each hole. When the tide rises and the clams come to the surface, this oil marks the spots where they are, and thus the fisherman is readily guided to them. These clams are said to be very good, but as to their merits compared with *Mactra*, *Mya Venus*, and *Ostrea*, the writer can from his own experience say nothing.

FIG. 309. — *Siliqua costata.*

Siliqua costata is another common species on our northeastern coasts. The shells are covered with a greenish epidermis, which is enlivened by one or two rays of more or less vivid violet. On the inside of the shell is a thickened rib running outward from the umbo. This is one of the most common shells thrown up on sandy beaches, and in life it is found buried in the sand a little below low-water mark. *Solemyia velum* is a very pretty shell, covered with a light brown radiated epidermis which projects far beyond the edges of the shell, its margins being slit into numerous lobes. The species is active, leaping about with its foot, and swimming by opening and closing its valves. A larger species is *S. borealis*, in which the lobes of the epidermis are proportionately much longer and narrower.

FIG. 310. — *Solemyia velum.*

The family GASTROCHÆNIDÆ, or TUBICOLIDÆ, has a very heterogeneous appearance, some of its members bearing close resemblance to the next family. The shell is equi-valve and the hinge is toothless, and not infrequently, in the adult, the shells are imbedded in a calcareous tube, so that the whole has but little resemblance to an ordinary bivalve. In *Gastrochæna* the valves are distinct from the tube, in *Clavagella* one valve is fixed, while in *Aspergillum* both valves are united with the tube, of which they form a very inconsiderable portion, as is shown in our figure. All of the members of the family are borers, penetrating hard mud, shells, coral, or rock. The most noticeable species is the 'watering pot,' *Aspergillum vaginiferum*. Here the valves are very small, while the lower end of the tube is closed below by a disc, which is perforated by numerous holes and short tubules. The other part of the tube is much longer, and at its distal portion is surrounded by one or more calcareous ruffles, so that the whole has a very bizarre appearance. This species comes from the Red Sea, and all other members of the genus belong to the Indo-Pacific region.

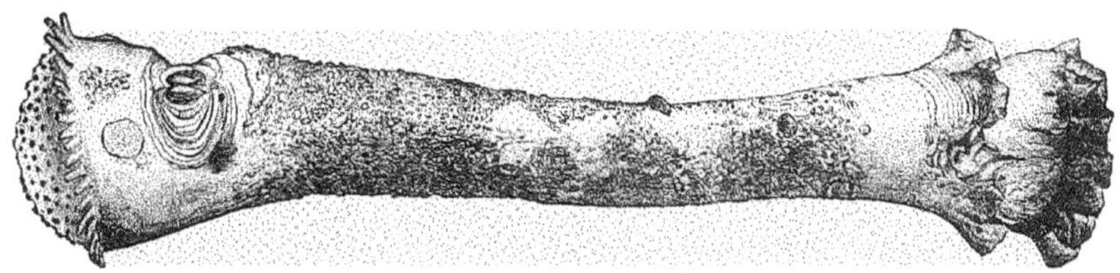
FIG. 311. — *Aspergillum vaginiferum*, watering-pot shell.

The PHOLADIDÆ, taken in its widest sense, is a family characterized by the absence of hinge teeth and of hinge ligaments. Instead, we usually find one or two accessory pieces (pallets they are called) which, in the Pholadinæ, lie between the valves at the hinge line, or, as in the Teredinæ, are borne at the extremity of the long, calcareous tube formed by these animals. The margins of the mantle are almost completely united, leaving only a small opening in front for the protrusion of the short and truncated foot. At the other end it is prolonged into a very large siphon, which, in the *Teredos*, has the power of secreting a calcareous tube. The gills are long and narrow, and, posteriorly, are drawn out into a point which extends some distance into the excurrent siphon.

They are all boring animals, and make their burrows, some in mud or sand, some in submerged wood, while others bore into rocks, shells, or corals, at times doing considerable damage to human interests. The distinctions between the two sub-families is sufficiently emphasized in the foregoing paragraph.

The genus *Teredo*, with about twenty-five so-called species, has gained a somewhat extensive notoriety under the popular name of ship-worm, and hence deserves some little attention at our hands. The *Teredo* is a long, worm-like animal, bearing at the larger end a comparatively small bivalve shell, while near the other are the two accessory pieces, the so-called pallets, beyond which extend the separate extremities of the siphonal tubes. The development of the *Teredo* has been made the subject of exhaustive papers by Quatrefages and Hatschek, from which we learn that, like other molluscs, it passes through a veliger stage. Soon after this the young larva comes across a piece of submerged wood, or, in case it does not, it dies. At first it creeps over the surface of the timber, but soon it settles down and begins the excavations which are to result in that prison, which it never leaves until death. Exactly how it excavates is still a matter of dispute, but it seems probable that it is partly by means of the edges of the pallets. Another theory of action is that the foot, with its thick coriaceous epidermis, cuts away the wood. The hole made at first by the young *Teredo* is minute, about as large as a pin's head, but, once within the wood, it grows rapidly, and its burrow is enlarged in the same proportion. As it excavates farther and farther into the wood, it lines its channel by a calcareous deposit, thus forming a shelly tube, which, on our coasts rarely exceeds ten inches in length and a quarter of an inch in diameter, but in favored localities some species attain a length of two feet and a half. In this tube the animal lives, its only means of communication with the external world being through the small hole by which it entered the timber.

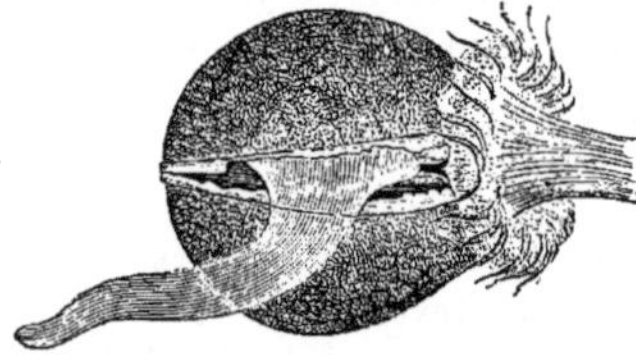

FIG. 312. — Young *Teredo* before it begins its burrow.

The *Teredo* does not feed upon the wood, the small particles which it erodes being passed out through the excurrent siphon. The food which nourishes the animal is, here as elsewhere, brought in through the incurrent siphon, and consists of microscopic animals and plants. Notwithstanding the fact that the *Teredo* does not eat the wood, the damage it does is very great. It was first brought prominently into notice at the beginning of the eighteenth century, when, by its ravages in the piles and other submerged wood which supported the dikes and sea-walls of Holland, it seriously threatened the safety of that country. Hundreds of individuals will obtain entrance to the same bit of timber, and, boring either with or across the grain, they soon convert it into a mere shell, ready to break down at the slightest strain. The rapidity with which they work is well illustrated by a fact recorded by Quatrefages. In the early spring an accident sank a coasting boat near St. Sebastian in Spain. Four months later some fishermen raised the vessel, hoping to turn the materials to advantage, if not to repair the vessel itself, but in the short space of time that had elapsed, the planks and timbers were so completely riddled by the *Teredos* that they were valueless. There is a curious fact noticed in connection with their burrowing. No matter how many of these molluscs gain entrance to the same piece of wood, their tubes never interfere with one another, but there is always left at least a thin partition between two adjacent burrows.

Since their appearance in Holland so long ago, these ship-worms have done an incredible amount of damage to wharves, ships, etc., and many devices have been suggested for checking their ravages. The use of chemicals, creosote, etc., has but comparatively slight effect, for since these animals do not eat the wood, the chemicals do not poison them. While kyanising (soaking the wood with creosote) is an effectual check against that injurious crustacean, the gribble (*Limnoria lignorum*), it is but a slight defence against the *Teredo*. In Norway, timbers which were saturated with creosote under a pressure of ten pounds to the square foot were found two years later to be filled with the molluscs. To iron rust they have a decided aversion, and piles and other timbers which are driven full of broad-headed nails escape their ravages. Our modern vessels also escape their injurious action, thanks to the copper sheathing with which their hulls are covered. On our coasts south of Cape Cod, it is customary to coat all spars and buoys with verdigris paint, and to take them up every six months for cleaning and a new coat of this poisonous paint. Notwithstanding this, the average life of a buoy is only about twelve years, but half of which is spent in the water.

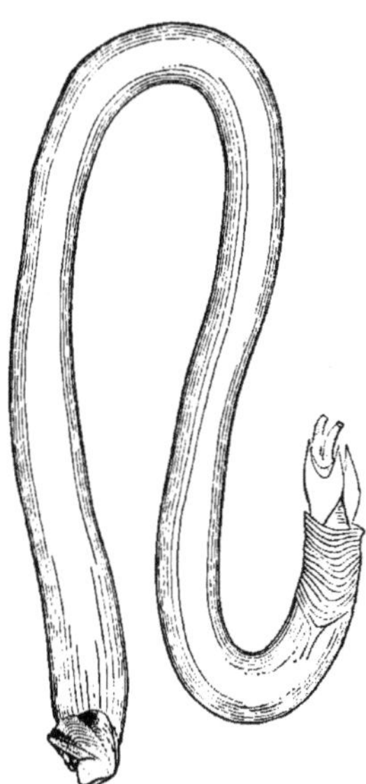

FIG. 314. — *Teredo navalis.*

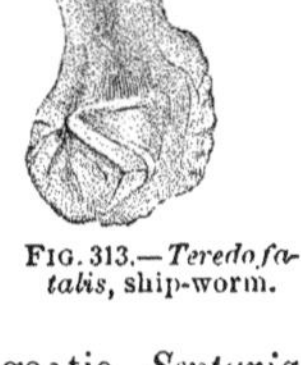

FIG. 313.—*Teredo fatalis*, ship-worm.

On our coast, *Teredo navalis* is the most common and most injurious species, but three other species of *Teredo* and one of *Xylotrya*, an allied genus, occur in larger or smaller numbers. In tropical waters many other forms occur, of which we need only mention *T. corniformis*, which burrows in the husks of cocoa-nuts and other woody fruits floating on the sea, and the gigantic *Septaria arenaria*, of the Philippine Islands, which burrows in the sand, sometimes attaining a diameter of two inches and a length of nearly six feet.

Pholas and its allies are also burrowing forms, but, unlike those just mentioned, they do not form tubes, nor do they, except *Xylophaga*, live in wood. In *Pholas* the valves are large and the shells are never completely closed in front. The shell is long and cylindrical, and the pallial sinus reaches to the middle of the valves. The common name for these molluses in England is piddock, but no appellative has gained much currency here. The species bore in sand, clay, limestone, and even gneiss. Doubtless here the instrument of boring is the foot with its hardened dermal armor. When we consider the hardness of some of the rocks perforated, we can scarcely realize that this organ is sufficient to produce the effects, but time is a matter of small importance to the pholads. As they increase in size they increase the size of the burrows, which are always just a little larger than the shell. These burrows are always nearly vertical, and but rarely encroach upon each other.

FIG. 315. — *Pholas* in its burrow.

In Europe the piddock are esteemed a delicacy, and on the coast of Normandy their capture furnishes employment to a good many women and children, who pull them from their burrows with an iron hook. They are usually cooked, but are said to be very palatable raw. One remarkable peculiarity of the pholads is their phosphorescence or capacity of shining in the dark, which is here better developed than anywhere else among the Mollusca, unless it be in *Phillirhoe*. The common European species is *Pholas dactylus*, while on our coasts are found *P. costata* and *P. truncata*. A closely allied genus, *Zirphæa*, has a species, *Z. crispata*, common both to New and Old England and the northern waters between. In this genus the shell is short and the accessory valves are absent.

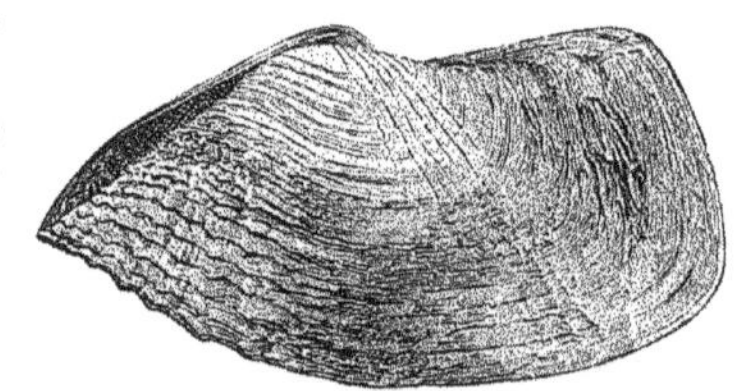

FIG. 316. — *Zirphæa crispata*.

In another section of the family we find pholads in which the anterior gape of the shell is closed by a callous plate. Here belong the genera *Pholadidea* and *Martesia*. *Martesia smithii*, on our coast, burrows in oyster shells, while other forms have different habitats. While most of the Pholadidæ are marine, Adams reports finding a species in the fresh water of Borneo, living in dead trunks of trees.

Class II. — CEPHALOPHORA.

The remainder of the Mollusca differ greatly from the group which we have just left, and the fact that nearly all of them possess a lingual ribbon (an organ to be described further on), while there is a head distinct from the rest of the body, has caused them to be included in one large group, variously termed Glossophora, Odontophora, and Cephalophora. In the present work, however, the term Cephalophora is restricted to indicate the forms between the squids and cuttle-fishes (Cephalopoda) and the Acephala.

The anterior end of the body is more or less distinctly marked off as a head, and this differentiation is the more marked in many forms from the fact that it usually bears tentacles and eyes, and thus is seen to be the locality of the senses, increasing its claim to the term head. The body possesses a bilateral symmetry, but, owing to the fact that most of the forms live in spiral shells, this resemblance between the two sides is somewhat obscured. The body is enveloped (at least in the young) in a mantle comparable to that of the Acephals, which in most forms secretes the shell, which is usually calcareous, but not infrequently, as in our common snails, is more or less horny. As these shells are very important from a systematic point of view, and, indeed, are the only portions usually preserved, they demand far more attention than they otherwise would.

The shells of the Cephalophora are always, except in the chitons, univalvular; that is, composed of a single piece, which, though presenting the most various forms, can in reality be reduced to a simple type. This type is a cone. The cone may be broad and low, as in the limpets, or it may be greatly drawn out and very slender, as in the tooth-shells. It may be coiled in a nearly flat spiral, or it may be curled in a conical spiral, the form found in most of the shells of the group. In some few forms the shell is internal, being enveloped in a fold of the mantle, while in a large number of these animals no shell is present. In the chitons the shell consists of a number of pieces, never more than eight, arranged in a linear series on the back.

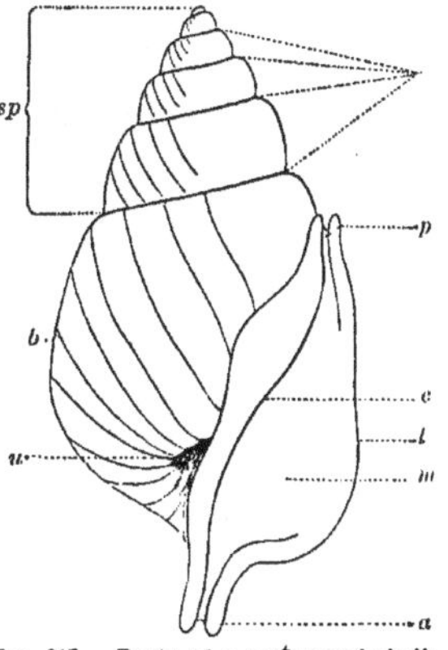

Fig. 317. — Parts of a gasteropod shell; *a*, anterior canal; *b*, body whorl; *c*, columella; *l*, outer lip; *m*, mouth or aperture; *p*, posterior canal; *s*, sutures; *sp*, spire; *u*, umbilicus.

In systematic works each part of the shell has its name. The upper spiral portion is known as the spire, and the curved portions of which it is composed are known as the whorls, the last and largest being the body whorl. The whorls are separated from each other by the sutures. The opening is known as the mouth or aperture, the outer edge of which is the lip, the inner the columella. Sometimes the lip is prolonged into one or two grooves or canals which are always approximately parallel with the axis of the shell. The one nearest the spire is the posterior, the other the anterior canal. Frequently an opening is left in the axis of the shell, which is known as the umbilicus.

Returning to the animal itself, the next thing we have to notice is the foot, which is usually large and muscular, and is used as an organ of locomotion. It may bear on either side a lateral appendage (epipodium), while frequently on the dorsal surface of the foot is a corneous or calcareous structure, the operculum, which is employed to close the aperture of the shell when the animal retracts itself. The operculum may be either horny or calcareous, and frequently shows a spiral structure. Some of the calcareous opercula of the smaller top shells are in common use as 'eye-stones.' By the older conchologists it was sometimes held that the Cephalophora possessed bivalve shells like the Acephals, the true shell being regarded as one valve, and the operculum as the other. This view has been shown to be erroneous, and now it is usually thought that the operculum corresponds to the byssus of the other group.

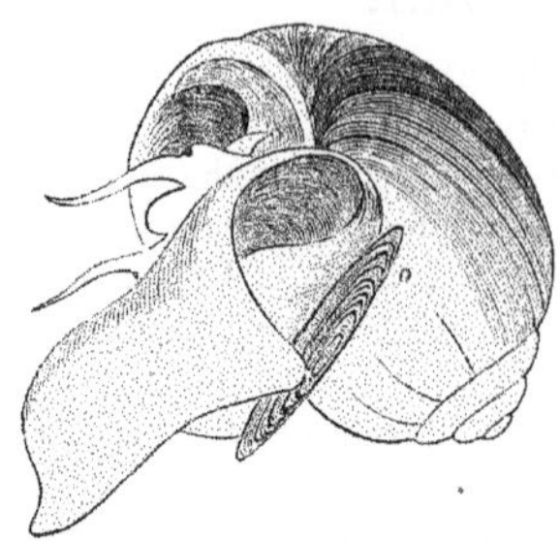

FIG. 318. — *Melantho ponderosa* partially extended, showing the operculum (*o*) on the upper surface of the foot.

In most of the forms there is a chamber formed on either side of the body by the free edge of the mantle and the body itself. Usually the pallial chamber on one side (usually the right) is larger than the other, and contains the gills when these organs are present. It also contains the outlet of the alimentary canal.

The mouth is situated on the under side of the head, and is armed with variously arranged jaws or plates of a hard chitinous or calcareous nature. Besides these, there is found in all except a very few forms an odontophore, or, as it is occasionally called, a tongue or lingual ribbon. This consists of a ribbon-like band of chitin, attached at one end and free at the other, and bearing on its upper surface numbers of hard, tooth-like processes. The odontophore is attached to the floor of the mouth, and is moved by appropriate muscles. When in use it is drawn over some supporting cartilages, and the teeth, acting like a file, rasp away the substance to which the mouth is applied. The action can be partly seen by watching pond snails feeding upon the green slime which frequently collects on the sides of an aquarium. The size of the odontophore varies greatly, as does also the number of teeth. In some it is very long, in others it is more oval. The teeth themselves are arranged in transverse series, there being in some species about two hundred in a single transverse row, while in others there are but three. By use this ribbon is constantly wearing away at the tip, but the loss is compensated by a continuous growth at the other end.

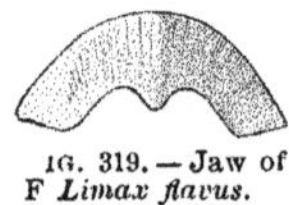
FIG. 319. — Jaw of *Limax flavus*.

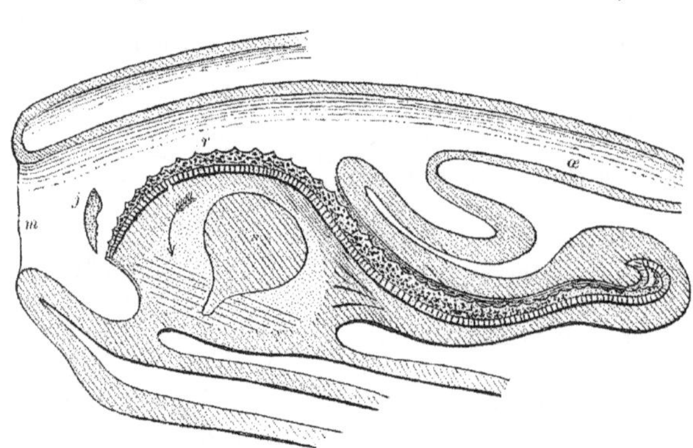

FIG. 320. — Diagram of the mouth and lingual ribbon of a gasteropod; *j*, jaw; *m*, mouth; *œ*, œsophagus; *r*, lingual ribbon; *s*, support of ribbon.

Within recent years the characters derived from the lingual ribbon have been regarded as very important in the arrangement of molluscs, but like all other good things this means of classification has been carried to an extreme; forms which in every other respect agree closely with each other being occasionally widely separated. The characters derived should be compared with those obtained from other structures and thus all such extremes would be avoided.

In most forms the body is distorted to fit the spiral shell, and even where this is not the case, the alimentary canal usually follows a tortuous course, doubling on itself and terminating usually on the right side of the body, frequently in front of the middle. The cavity of the mouth communicates with an œsophagus which sometimes dilates into a sort of crop, and eventually empties into the stomach, from which arises the intestine either opposite to or beside the cardiac opening. In a few forms the stomach is armed with plates or horny teeth. Salivary glands are almost always present, usually two in number, but occasionally four are found, and it is interesting to note, in passing, that in *Dolium* and some other forms these glands pour out a saliva containing sulphuric acid. The liver is well developed.

FIG. 321. — A row of teeth from the lingual ribbon of *Paludina intertexta*.

The circulatory system is usually well developed, though in *Dentalium* a heart is wanting, while other forms show a correspondence to the Acephals. Thus in the ear and top shells, the alimentary canal perforates the heart, while in the first, as in the chitons, there are two auricles. Usually, however, there is one auricle and one ventricle, which propel a blood containing colorless, nucleated corpuscles. As in the Acephals, the heart receives the blood from the gills and forces it to all parts of the body.

Respiration is effected by means of gills or by pulmonary organs. The gills, which are usually contained in a cavity of the mantle, may consist either of lamellar organs or of plume-like branchiæ. Lankester, who has recently investigated the structure of the gills, is of the opinion that the primitive type was what he calls a ctenidium, consisting of a central stalk to which lamellar respiratory processes were attached, a view which seems open to some objections. The variations occurring in the respiratory organs are of great value in systematic work, and will be referred to again in connection with the different groups. The pulmonary cavity of the Pulmonifera is formed by a cavity of the mantle, and is richly supplied with blood vessels which extract the oxygen necessary for respiration from the air. In a few forms special respiratory organs are lacking.

Closely connected with the organs of circulation are the renal organs. There are one or more sacs near the heart and opening to the exterior, which extract their secretion from the blood going to the heart, and convey it outside the body.

The nervous systém acquires a different development in the various groups. In the tooth shells it most nearly approaches that of the lamellibranchs, consisting of two ganglia above the œsophagus, connected by two nervous cords with a pair of pedal ganglia, while two more cords connect the brain with two ganglia near the vent. These two last are evidently comparable to the visceral ganglia of the bivalve molluses. In the other forms the visceral ganglia may be increased in size and number, and become closely connected with the other two pairs. The details of structure should be sought in more technical works.

The auditory organs are usually seated near the ganglia of the foot. Eyes also are generally present, and usually are two in number, situated upon the head or its appendages. The peculiar eyes of *Onchidium* and the chitons will be mentioned further on. At the first glance the eyes of the gasteropods (and also those of the cephalopods) seem strikingly like those of the vertebrates. They have an external cornea behind which is a lens, a vitreous humor, and lastly a retina containing rods and cones. A great difference exists in the fact that these rods and cones are on the front side of the retina, and between them and the optic nerve is the pigment layer. Frequently these organs are seated upon stalks or tentacles which are capable of retraction. In our common garden snail may be found an example of this structure, and the retraction is accomplished by the drawing in of the end of the tentacle in the same way that one inverts the finger of a glove.

The sexes of the gasteropods may be separate or combined in one individual, and copulatory organs are frequently present. Most forms lay eggs, a few, however, bring forth living young, the eggs undergoing development within the parent. The eggs, which are numerous, are frequently enveloped in capsules, differing greatly in form and ornamentation, each capsule containing a number of eggs. We shall note the appearance of some of the more interesting of these egg-cases in connection with the forms to which they belong.

In the development of the eggs of the Cephalophora there is much diversity in the early stages, according as the amount of food-yolk present is large or small. This also influences the character of the gastrula which is formed, and the entoderm may either be solid or in the form of a sac. Soon after the invagination, the outer surface almost always becomes covered with cilia in certain regions, the most prominent being a ring around the body in advance of the mouth. This is the velum, so characteristic of molluscs. At about the same time a small portion of the outer layer sinks in to form the gland which eventually secretes the shell, and the foot begins to appear.

These changes all take place within the egg, and upon the development of the cilia, the larva begins to turn round and round, thus rendering it difficult for the student to obtain a distinct view of what is going on inside the embryo, or even to draw the external appearance. The shell gland begins to secrete the shell, which arises first as a small plate, but soon takes the form of a cap enveloping the posterior part of the body and then gradually acquires a spiral form. The region of the velum also exhibits a change. Instead of being a ring around the body it becomes a two-lobed plate fringed with cilia, which serve as locomotory organs after the young has hatched.

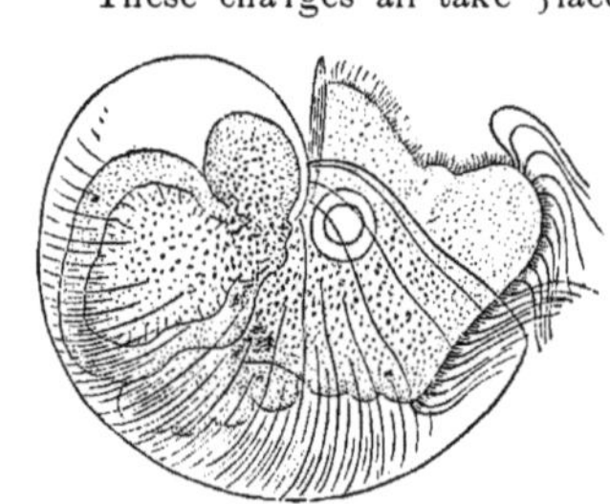
FIG. 322. — Veliger of *Eolis diversa.*

From this point the development is usually direct, no startling metamorphoses being introduced. The velum almost always disappears, and the body and shell gradually acquire their adult structure. Still the variations and the changes undergone throw considerable light on the classification and arrangement of the different forms.

The classification of the odontophorous molluscs is still in an uncertain condition, notwithstanding the fact that they have been so extensively studied. In fact, there are scarcely two authors who agree as to the rank and relationship of the different associations of forms. This difference of opinion is partly due to the varying importance accorded to the different characters, and partly to the fact that a linear arrangement does not exist in nature, there being numerous inter-relations between the different groups. With this uncertainty it matters but little what classification we adopt, though that which follows seems to the writer to best represent the present state of our knowledge. All authors admit that the Scaphopoda are the lowest, while the position of the Pteropoda is very uncertain, one of the latest writers including them among the Cephalopods.

SUB-CLASS I. — SCAPHOPODA.

The tooth shells, as they are commonly called, are few in number, but their peculiarities have caused them to be regarded with considerable interest. They are of all the Odontophora the most closely related to the Acephals. The shell is very long, tapers slightly, and is either straight or curved like the tusk of an elephant, and is open at both ends. The animal is attached to the shell near the smaller end, while from the larger it protrudes a large number of long and slender tentacles which are used in obtaining food and as prehensible organs. These tentacles arise far within the shell, from a muscular ring surrounding the body. In advance of this ridge, but still within the shell, is a cup, at the bottom of which lies the mouth. There are no eyes, no heart, while the head is very rudimentary. The foot is large, three-lobed, and protrudes from the larger end of the shell. The mantle is relatively very large and, like the shell, is open at both ends.

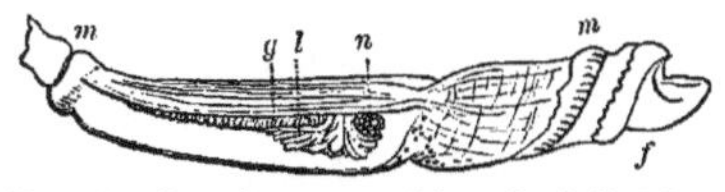

FIG. 323.—*Dentalium* removed from the shell; *f*, foot; *g*, genital organ; *l*, liver; *m*, mantle; *n*, kidney.

The mouth is armed with a lingual ribbon, and the alimentary canal, which is convoluted, terminates behind the origin of the foot. The liver is large and two-branched, and the sexes are distinct. There are no special respiratory organs. We have already spoken of the nervous system.

The development of *Dentalium*, which has been studied by Lacaze-Duthiers, is very peculiar. After the formation of a morula, the body of the embryo becomes surrounded with a number of rings of cilia, while at the anterior end is a tuft of longer cilia. Gradually the anterior end becomes flattened into a disc, the edges of which are ciliated, while the posterior part develops the mantle. This mantle secretes the shell, which at first is open, but finally the edges of the mantle and the shell unite, producing the tubular form found in the adult.

The tooth shells are divided into two families (by Sars accorded a higher rank), founded on characters of the shell and foot. In the first (DENTALIIDÆ) the margin of the smaller end is entire, or has a medium ventral slit, while the foot is three-lobed; in the SIPHONOPODA the foot bears a circular disc, the edges of which are armed with papillæ, while the posterior end of the shell is either entire or with numerous notches.

The species are found in all seas and live buried in the sand, the smaller end of the shell protruding, and through this the water necessary for respiratory purposes is drawn. The large foot is employed in burrowing, while the tentacles, which are ciliated, are employed in capturing food, which consists of Foraminifera and other minute animals. About one hundred species are known, represented on our coasts by *Dentalium occidentale*, *Entalis striolata* and three or four other species. The largest

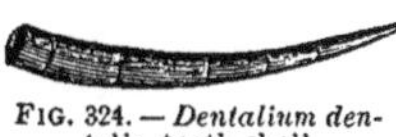
FIG. 324. — *Dentalium dentalis*, tooth-shell.

species is about four inches in length. The group of tooth shells first appears in rocks of Devonian age.

Sub-Class II. — Gasteropoda.

This group, which contains by far the largest number of forms (about twenty-five thousand species being known), embraces the molluscs commonly known as snails, slugs, sea-slugs, whelks, cowries, limpets, and the like. In all the head is well developed and bears one or two pairs of tentacles. The body is usually asymmetrical, owing to the presence of a spiral shell, though this is far from being invariably the case. The alimentary tract is straight or doubled on itself, and usually terminates on one side of the body. A heart is always present, except in the problematical form *Entoconcha*. Respiration is effected either by gills, by a pulmonary cavity, or by the general surface of the body. The sexes are separate in the majority of the forms, but are combined in the same individual in the Pulmonata.

The classification here adopted is based on that of Lankester, which, while it varies greatly from that in common use, has the merit of agreeing well with our knowledge of the anatomy and embryology of the group. The basis of Prof. Lankester's primary divisions is found in the symmetry or torsion of the body.

Super-Order I. — ISOPLEURA.

The name given to this division means equal-sided, which emphasizes the most important feature of their structure. They retain in the adult the primitive bilateral symmetry. The alimentary canal traverses the entire length of the body, and terminates posteriorly in a median vent. Renal organs, gills, circulatory organs, and genitals, are paired and symmetrical. The pedal and visceral nerve cords are straight and parallel, extending the length of the body.

Order I. — CHÆTODERMÆ.

This group contains but a single genus, *Chætoderma*, which was originally placed among the Gephyrean worms. *C. nitidulum* is a small, worm-like body, with an enlarged head at one end, while the cavity of the mantle is found at the other. In this small cavity are a pair of small gills. The external integument is roughened by minute calcareous spines, which give the body a hairy appearance. The foot is obsolete, and the lingual apparatus is greatly reduced, the lingual ribbon being represented by but a single tooth. Nothing is known of the embryology.

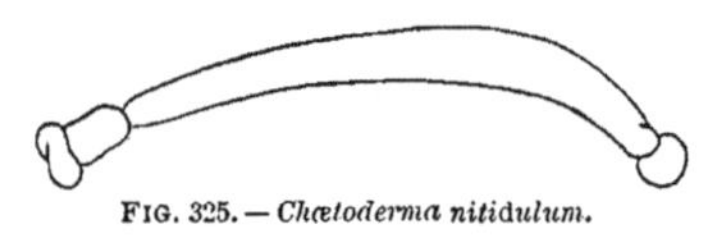

Fig. 325. — *Chætoderma nitidulum.*

Order II. — NEOMENOIDEA.

Neomenia is a peculiar genus found on the western coast of Sweden. *N. carinata* reaches the length of nearly an inch, grayish in color, with a shade of rosy red at the posterior end of the body. The outer surface is covered with minute spines, giving it a velvety appearance. In shape the body somewhat resembles a pea-pod, a dorsal ridge giving rise to the specific name. The mantle is reduced to a small ring around the vent, enclosing the paired gills. The lingual ribbon is poorly developed, but bears many teeth. The eggs pass out by the renal ducts. The mouth and pharynx can be retracted or extended at will. The second genus of the order is *Proneomenia* which has been found in the North Sea and in the Mediterranean. It is more elongate and worm-like than *Neomenia*. Nothing is known of the embryology of either form.

FIG. 326. — *Neomenia carinata*, Ventral and side views; *a*, anterior, *b*, posterior extremity; *c*, furrow in which the foot is concealed.

ORDER III. — POLYPLACOPHORA.

The chitons are a group which have made no little trouble for zoologists. In the early days of science they, together with the barnacles (which are really Crustacea) were united in a group characterized by the possession of multivalve shells. More recently they have been assigned a place among the gasteropods, but here they have not been allowed to remain in quiet, almost every author assigning them positions of varying rank and relationship, and one, influenced by their peculiar development and the structure of the nervous system, has actually placed them among the worms.

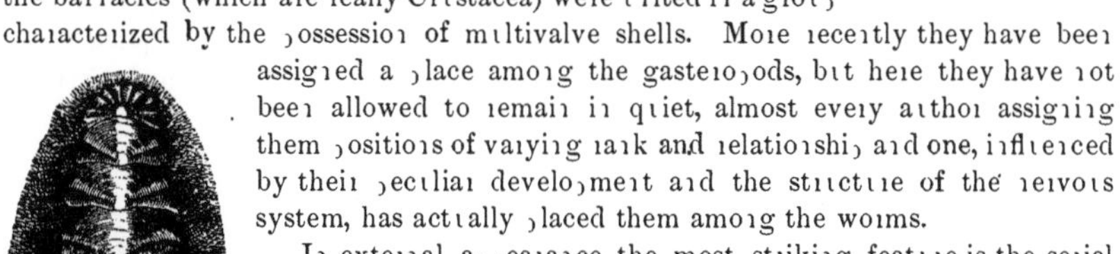

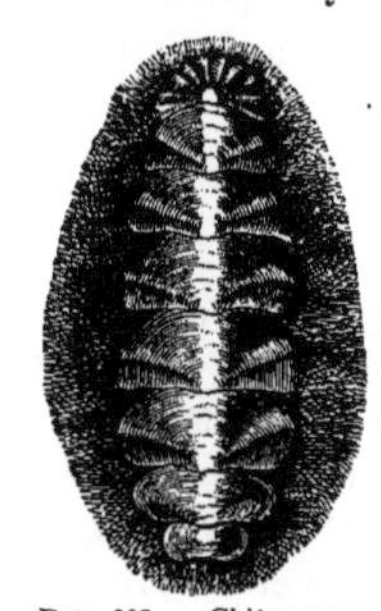

FIG. 327. — *Chiton wossnessenskii.*

In external appearance the most striking feature is the serial arrangement of eight calcareous shells upon the back, indicating a segmentation far from common among the Mollusca. This segmentation is carried still farther, and we find the gills similarly arranged on either side of the body (Fig. 328), to the number of sixteen or more, each accompanied by an olfactory organ. Around the margin of the dorsal surface frequently occur calcareous spines, or other forms of ornamentation useful in classification.

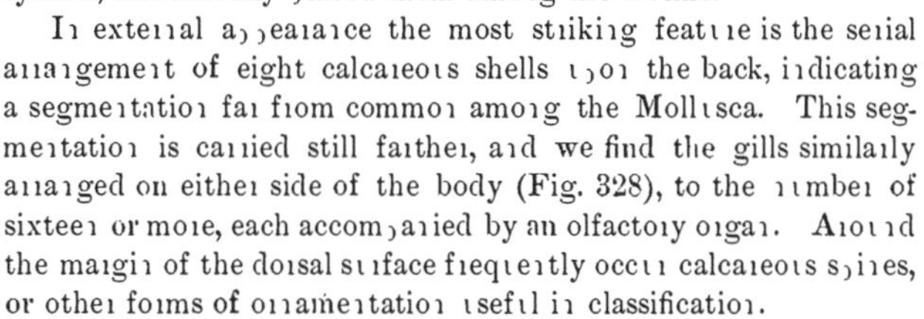

One of the most interesting of recent discoveries is that the chitons, which have been so long studied and so long regarded as blind, are (in most genera) really very well provided with visual organs, the whole dorsal surface of some forms being studded with eyes of which not less than eight thousand occasionally exist on a single specimen. These eyes are unlike the dorsal eyes of *Onchidium*, and like those on the tentacles of *Helix*, in that the retina is between the nerve and the exterior. These eyes are further shown by Professor Moseley to be developed from peculiar sense organs covering the dorsal surface. No trace of these eyes has yet been found in the fossil chitons.

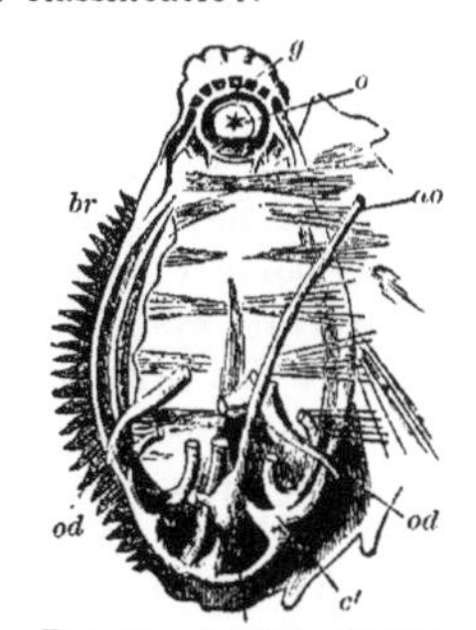

FIG. 328. — Anatomy of *Chiton*; *ao*, aorta; *br*, gills; *c*, ventricle; *c'*, one auricle; *g*, nervous ring; *o*, mouth; *od*, oviducts.

The mouth is armed with a well-developed lingual ribbon, in which the teeth are arranged in the following manner, — 5. 1. 1. 1. 5; the laterals being large and hooked. The intestine is coiled in a loose spiral and terminates in a median vent at the posterior end of the body.

Little is known of the development of the chitons, but that little is very interesting. Segmentation gives rise to a true gastrula and at the same time the velum is produced. At the anterior end a single flagellum is produced, which is soon replaced by a tuft of cilia. Shortly a constriction appears behind the velum, and on the dorsal surface appear six or seven transverse plates which may represent the shell glands.

Two large eyes are also formed, which are remarkable in being behind the velum. The details of the closure of the blastopore, the formation of the pedal nerves (of too technical a character for recital here), the bilateral symmetry and the segmentation of the body, all point to the fact that the chitons branched off from the gastropodous stem at an early date.

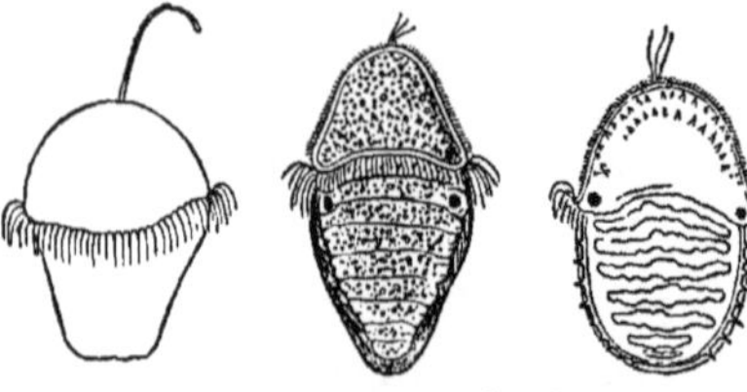

FIG. 329. Development of *Chiton*.

FIG. 330.—*Trachydermon ruber*, red chiton.

The chitons are mostly littoral forms living in the shallow waters of the ocean. Over three hundred species are known; but until the manuscripts of the late Dr. P. P. Carpenter are edited and published, we shall have no adequate review of the group. A large number of genera have been made, but with these we need not concern ourselves.

SUPER-ORDER II. — ANISOPLEURA.

In this, by far the largest division of the Gasteropoda, the symmetry so marked in the preceding group is greatly obscured. The head and foot, indeed, retain the primitive bilaterality; but here the resemblance usually ends. The cause of this lack of symmetry in other parts of the organism is to be explained on mechanical grounds. On the back there is usually developed a large shell, which, with its included viscera, acquires a very great proportional weight. This shell naturally falls over to one side, and by thus doing twists the various organs so that the primitively median anus occupies a position at the anterior portion of the body, usually upon the right side, or may even be placed in the median line above the head. Not only is the alimentary tract affected by this torsion, but the openings of the kidneys, the gills, and other organs are transposed, so that the gill, for instance, of the normal right side is in reality borne upon the left. Part of the nervous system may or may not share in this twisting, accordingly as the visceral loop is above or below the anus. The effect of this twisting is to coil the nerves in the shape of the figure 8, and an illustration of the stages of the process may be seen in the adjacent diagrams copied from Lankester who was first to point out the systematic importance of these facts. Coincident with this torsion frequently occurs an atrophy of parts, and, from the fact that the twisting usually occurs in one way, it is the gills, kidneys, etc., of the left side which usually suffer or even entirely disappear.

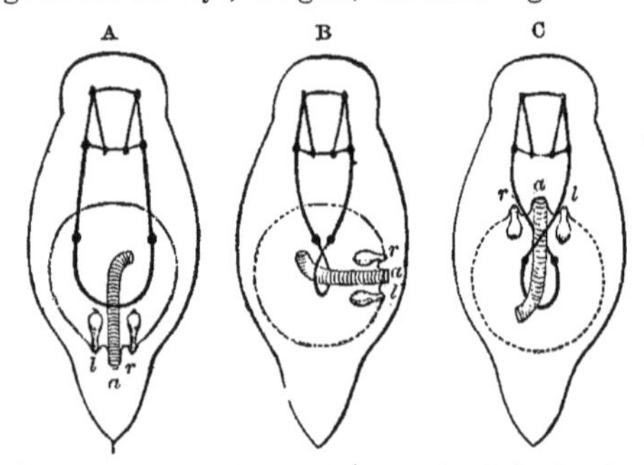

FIG. 331. — Diagram showing the torsion of the body when the visceral commissure passes above the intestine; A, normal condition; B, quarter rotation; C, complete half rotation; *a*, anus; *l*, left, *r*, right renal organ.

The twisted or straight character of the visceral nervous loop gives a foundation for a division of the Anisopleura into two groups, to which the names Streptoneura and Euthyneura have been applied. To the former belong the great majority of the aquatic and some of the terrestrial species, while the latter contains only the Opisthobranchs and Pulmonifera.

ORDER I.—OPISTHOBRANCHIATA.

This group is exclusively marine, and is composed of forms with a large foot, while the visceral hump, so characteristic of most gasteropods, is very small, or wanting. The name has reference to the fact that the gills are placed behind the heart, which is but another statement of the fact that the torsion of the body has not been carried to its full extent. In the adult stage some are provided with a shell, while in others this is lacking; but all, without exception, have a shell in their earlier stages.

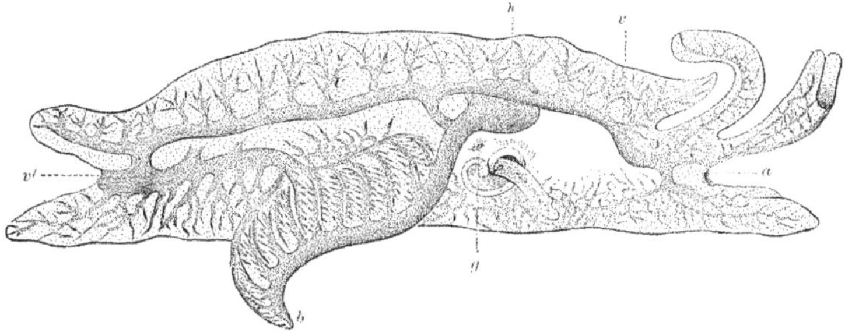

FIG. 332.—Circulation in *Pleurobranchus auriantiacus*, showing posterior position of the gills; *a*, mouth; b, gill; *g*, renal opening; *h*, heart; *v*, veins.

In some, gills are present, and may either project freely into the water or be concealed in the mantle cavity, while others have no specialized respiratory organs. The position of the gills, mentioned above, has also its effect upon the circulatory organs, and hence the auricle is here behind the ventricle of the heart. The vent is upon the side of the body, and the sexes are united in the same individual. Two well-marked sub-orders may be distinguished.

SUB-ORDER I.—NUDIBRANCHIATA.

Possibly no group of molluscs possesses more beautiful forms, or affords more instances of protective resemblances, than does that which has received the name Nudibranchiata. This term, which means naked gills, is very appropriate, for these organs, when present, are not enclosed in a special respiratory cavity, but project freely into the surrounding medium, and are borne either on the back or on the sides of the animal. From the fact that in the adult stage no shell is present, these forms are frequently termed, in more common parlance, naked molluscs.

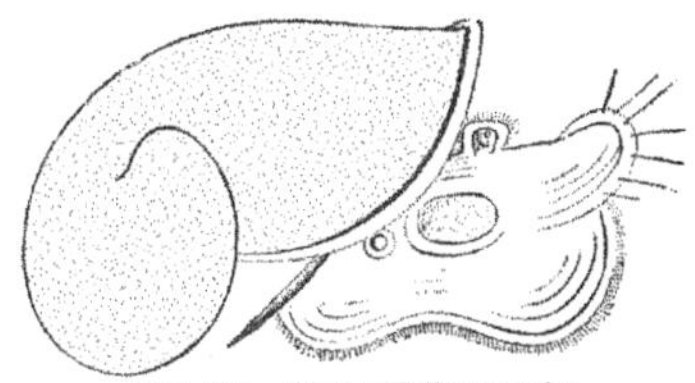

FIG. 333.—Larva of *Entoconcha*.

In the young stage a shell is present. This embryonic shell, which is formed and acquires a spiral or nautiloidal form before the young leaves the egg, disappears with growth. It is transparent, and the young animal can close the aperture with an operculum, while at other times it projects from the opening a ciliated velum, with which it turns and swims as actively as any other gasteropod which retains its calcareous armor throughout life.

The adult, however, is not always without protection other than that afforded by its resemblance to the objects which it frequents, for in some forms the mantle secretes calcareous spicules of various shapes, which sometimes are so numerous and so intertwined that, when the fleshy parts are dissolved in caustic potash, they retain the positions they occupied when in life. When these spicules are numerous, they cause the dorsal surface to be roughened and hardened, forming a protective dorsal shield.

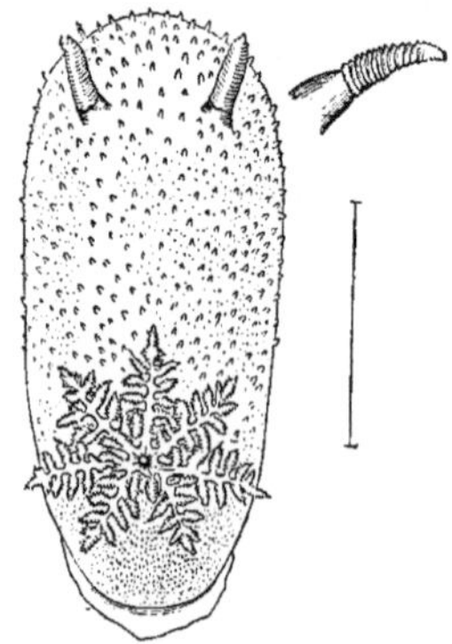

FIG. 334. — *Doris bifida*, showing the gills and a tentacle enlarged.

As is implied in the foregoing, a mantle is sometimes present, but in others this structure is not differentiated. When present it is perforated, and through the openings project the tentacles and the gills. In the young, well-developed eyes are present, but in the adults they appear as minute black dots, just behind the tentacles, or are obsolete. The tentacles are prominent, and seem to serve as olfactory organs, and not as organs of touch. They are frequently made up of a series of plates, presenting an appearance which recalls the antennæ of many insects; at other times they are plaited or simple, and not infrequently they may be retracted into trumpet-shaped sheaths near the base. These variations are of much importance in systematic work.

The gills, as we have said, are typically not inclosed in a cavity of the mantle, but, when present, they project freely into the sea. They vary greatly in form and disposition, furnishing, in these respects, important systematic characters. Sometimes they are in a more or less complete circle, surrounding the posterior opening of the alimentary canal, or they may be arranged in longitudinal series along the sides of the back or body; in the Phyllididæ alone do we find any approach to the formation of a branchial sac. In form they may resemble bushes, or they may be reduced to simple papillæ; all variations between these extremes being found. These gills perform but a part of the respiratory economy, for, in all, the general surface of the body serves for the aëration of the blood, and in the forms without gills it is the sole agent in this process. The forms with gills are said to flourish when deprived by accident of these organs, the skin performing their functions.

In the internal anatomy we find some points which deserve a brief mention. The lingual ribbon varies in the number and arrangement of the teeth, according to the family. The alimentary canal usually terminates on the right side of the body, though in forms like *Doris* it may end medially. The stomach is surrounded by a large, much-branched liver, portions of which extend into the elongated papillæ, which are found on the back. In the apices of these papillæ are found thread cells, recalling the similar defensive organs of the Hydrozoa.

The nudibranchs are mostly littoral forms, and spend their lives creeping among the rocks and seaweeds near the shores. They can, however, swim, and, when employing this mode of locomotion, they usually progress with the back downward and the foot uppermost. The food may be either vegetable or animal. Some forms feed on the more minute algæ, while others create sad havoc among the hydroids. The eggs are laid in bunches, upon stones, hydroids, or sea-weeds, almost every species having its peculiar mode of oviposition. The eggs are imbedded in a transparent gelatinous matrix, allowing the earlier stages of development to be readily seen.

Nearly one thousand species of nudibranchs have been described, from all seas; but as these forms have not been studied to the same extent as their shelled relatives, this number will doubtless be greatly increased by subsequent researches.

The first form requiring notice is the peculiar *Entoconcha mirabilis*, which leads a parasitic life inside the body of *Synapta*, one of the holothurians. So greatly has parasitism altered the form of the body, and all of the organs, that the proper position of this form among the gasteropods is far from certain, some placing it near *Natica*. Indeed, were it not for the characters afforded by the young, its position among the mollusca would not be suspected. Some thirty years ago Johannes Müller found in some specimens of *Synapta digitata* an internal worm-like parasite, attached by one extremity to the alimentary canal, while the other end hung free in the perivisceral cavity. Other observers, notably Baur, have investigated this strange form, but there are many facts concerning it yet to be ascertained.

In about one specimen of *Synapta* out of one or two hundred this strange form occurs. It is a sac, the upper part bearing the female and the lower the male reproductive organs, while the centre of the body serves for a while as a brood pouch, the embryos later passing out from an opening at the free end of the body of the parent. The eggs undergo a tolerably regular development, producing a velum, shell, and operculum, the later stages being found free in the body-cavity of the host. After the stage shown in Fig. 333, nothing more is known of their history. It would appear, from the little that is known of the development, and from the characters of the embryo, that *Entoconcha* should possibly be assigned a place among the nudibranchs. A second species of *Entoconcha* (*E. mülleri*) is found in *Holothuria edulis*, in the Philippines.

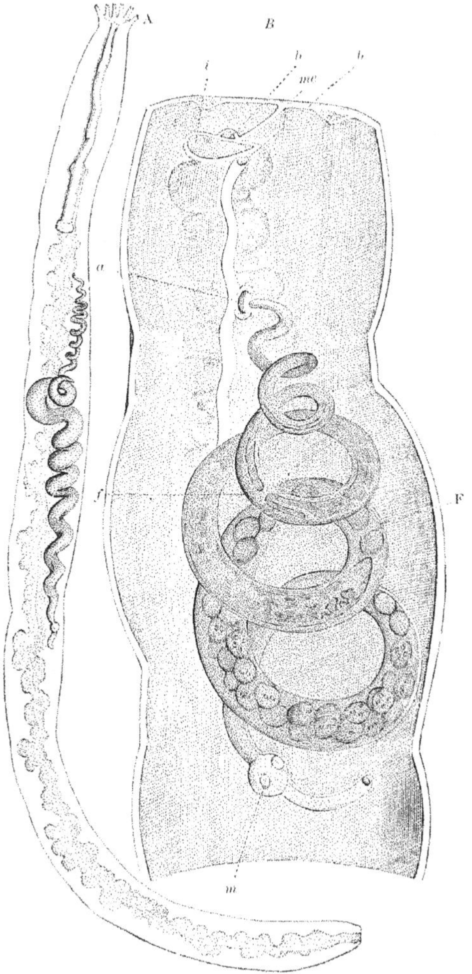

FIG. 335.—A, *Synapta digitata*, with parasitic *Entoconcha*, natural size; B, a portion of *Synapta* with *Entoconcha* (F) enlarged; *a*, point of attachment; *b*, blood vessels; *f*, female portion; *i*, intestine; *m*, male portion; *me*, mesentery.

Leaving these curious parasites, which, so far as known, are unrepresented on our shores, we come to forms which undoubtedly belong to the Nudibranchiata, and which lead free lives in the seas of all parts of the globe.

Passing by the PHYLLIDIDÆ, a small family of tropical and semi-tropical forms, in which the gills are either absent or enclosed between the mantles and the foot, we come to the ELYSIIDÆ, in which the body is shaped much like a common garden slug,

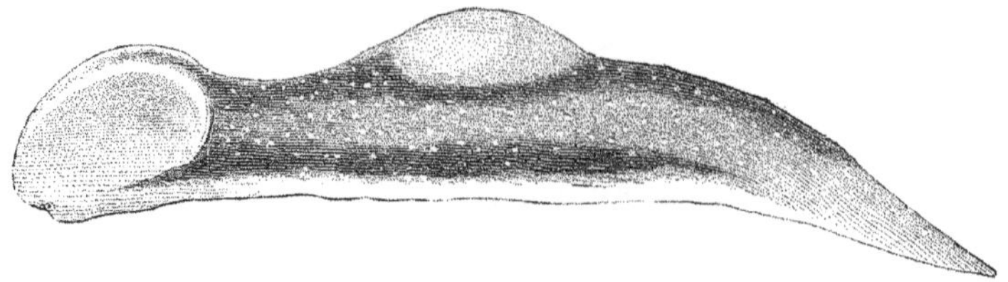

FIG. 336.—*Pontolimax capitans.*

the gills have disappeared, and the tentacles are simple or absent. This family is represented in our figures by *Pontolimax capitans*, a form only a third of an inch in length, found on the coasts of northern Europe. It lives between tides, feeding on minute algæ, and lays its eggs in small, pear-shaped capsules, each containing on the average about one hundred eggs. *Pontolimax zonata* occurs on the New England coast. In *Elysia*, the typical genus of the family, the tentacles are well developed and the sides of the body are expanded into a pair of wings, which stop just behind the neck. *Elysia viridis* of the European seas is of a green color, as is also our New England *E. chlorotica*, and the closely allied *Elysiella catulus*. These forms are not uncommon, creeping about on the eel grass (*Zostera*) of our northern coasts.

FIG. 337.—*Elysia viridis.*

In the EOLIDÆ, a much larger family than the last, the gills, which may be laminated, papillose, or like plumes, are arranged along the sides of the back, while the tentacles are capable of being retracted into sheaths. The genus *Tergipes*, which is represented by a little species common upon the stems of hydroids, received its name from old Forskål, from a belief that it walked upon its back, using its gills as locomotory organs. The branchiæ are eight in number, arranged in a single row of four on each side. In *Hermæa* the gills are more numerous, and the tentacles, which are broad and flattened, are usually folded. Our New England species, *H. cruciata*, has received its specific name from the cross-like marking of each gill. The species of *Montagua*, of which we have over half a dozen forms, and the closely allied *Eolis*, have a large number of gills arranged in transverse rows upon the sides of the back. They are very active animals (for nudibranchs) and are common on piles of bridges, among the roots of sea-weeds, and on rocky bottoms. They lay their eggs in gelatinous spirals with wavy margins, resembling a lady's frill in general appearance. Frequently bright colors are present, making these among the most attractive of marine objects.

FIG. 338. — *Eolis pilata.*

In *Doto* and allied forms the tentacles are retractile into cup-shaped sheaths, while the branchiæ are most curious bodies covered with minute papillæ. *Doto coronata*, which extends from our shores to those of northern Europe, is a handsome object. It is scarcely more than half an inch in length, but, small as it is, there is room for spots of orange, pink, yellow, carmine, purple, and white. Possibly one of our most striking forms is *Dendronotus arborescens*, with its curiously branching gills, which, from their thin, bushy appearance, have given rise to both its generic and specific names. This branching feature is also seen in the tentacular sheaths which are split up like the calyx of a flower. The general color is flesh-red or brown. This is one of the most active of the naked molluscs, and, when confined in an aquarium, is scarcely ever quiet. It lies on hydroids and sea-weeds, being especially adapted for creeping around upon them by its long and slender foot.

FIG. 339. — *Dendronotus arborescens*, bushy sea-slug.

The genus *Scyllea*, which has the body expanded into two long lobes, bearing the gills on either side, is interesting from the fine instance of mimicry it affords. It lives upon the gulf weed (*Sargassum*) of the Atlantic and other seas, and with it is occasionally drifted upon our shores. The large fields of this sea-weed which exist in the tropical Atlantic have a fauna of its own, and among other forms are numbers of fishes, crabs, shrimps, and the slugs now under discussion. Were it not for its protective resemblance to the sea-weed on which it dwells, a resemblance embracing both form and color, *Scyllea pelagica* would furnish many a fine mouthful for its voracious associates, and the species would soon become extinct.

In *Tethys* we have another peculiar form embracing some of the largest of the naked molluscs, *T. fimbriata*, occasionally reaching a foot in length. Its general appearance can be seen from our illustration, which, however, fails to convey any impression of the coloration of the animal. It is nearly transparent, and covered with dots and spots of red of different shades, some so dark as to be almost black. The curious gills on the upper surface were once described as parasites. It is a native of the Mediterranean, and, though often captured, it lives but a short time in aquaria, even in the large ones of the Naples Zoological Station. It is a rapacious animal, feeding upon other molluscs and small crustaceans.

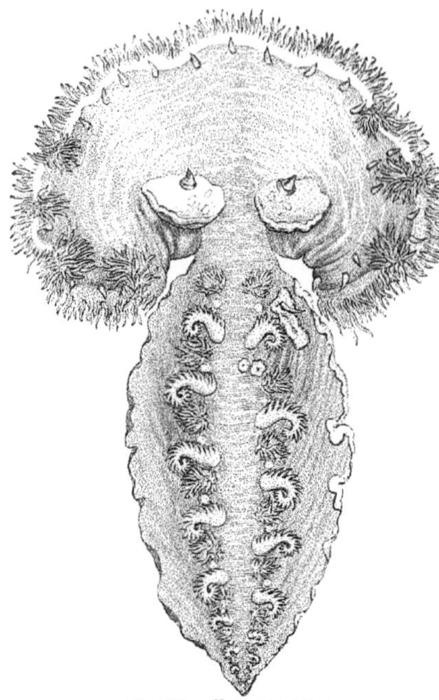

FIG. 340. — *Tethys fimbriata.*

In the remaining forms the branchiæ are arranged upon the back in a more or less complete circle which surrounds the anus. As an example of the POLYCERIDÆ we may mention the beautiful *Polycera lessonii* of our coast, with a pale, flesh-colored body, flecked with bright green, while the tentacles, gills, and tubercles on the back are variously spotted with white or yellow, and occasionally green. There are several other American forms in this family.

The PHYLLIRHOIDÆ is a very peculiar family, whose position among the molluscs would not be certain were it not for the fact that it possesses a lingual ribbon. *Phyllirhoe bucephalus*, the best known species, is a thin, compressed, translucent ani-

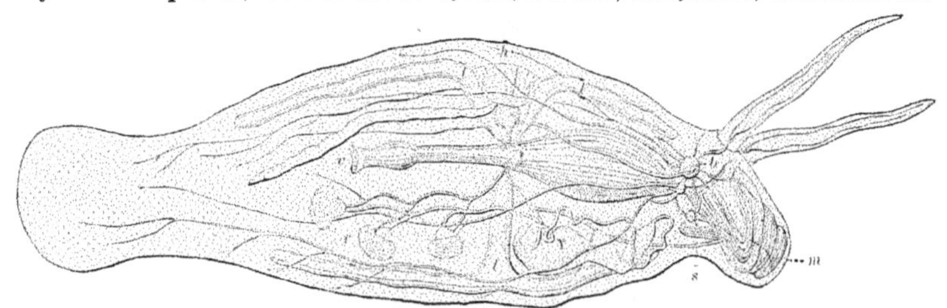

FIG. 341. — *Phyllirhoe bucephalus;* b, brain; *h*, heart; *i*, intestine; *l*, liver; *m*, mouth; *r*, renal organs; *s*, salivary gland; *v*, vent.

mal with a rounded, fin-like tail, which swims freely through the water in much the same manner as a fish. The head is furnished with two long tentacles, gills are absent, and the intestine terminates on the right side of the body. Most of the specimens bear a parasitic medusa, *Mnestra parasitica.* The most interesting fact connected this animal is its phosphorescence. At night, when swimming in the sea or in an aquarium, when disturbed, its whole body is instantly illuminated by points and dots of light.

The Doridopsidæ is noticeable from the fact that the species, which in general appearance resemble those of the next family, have a sucking mouth, and are destitute of an odontophore and jaw, thus presenting a marked exception to all other gasteropods.

The last and largest family of the nudibranchs is the Dorididæ, in which the tentacles are laminate and retractile within sheaths, the shape of which varies according to the genus. There are about four hundred known species distributed in all the seas of the world. The branchiæ vary considerably in shape, but are usually branched, and when expanded, the circle presents a close resemblance to a flower, the effect of which is strengthened by the brilliant colors which are frequently present.

Species are most numerous north of Cape Cod. In their habits they resemble the forms previously described. In *Onchidoris* the lower pair of tentacles are replaced by

Fig. 342. — *Doris pilosa.*

a broad membrane. In *Doris* the oral tentacles are distinct, and the branchiæ, the character of which is well shown in our illustrations, are capable of being retracted into a cavity. Our species which are somewhat numerous, appear in favorable localities in large numbers; but, owing to the protective coloration, which may be similar to the dull sea-weeds or the bright hydroids among which they dwell, they readily escape the collector's eye. Other dark-colored forms are frequent under stones at or near low-water mark.

Sub-Order II. — Tectibranchiata.

The name for this group is the antithesis of that employed for the last, and is used to indicate the fact that the gills are covered and concealed by a flap of the mantle. The gills, it should be said in passing, are not homologous throughout the group. The shell, which is usually present, is thin and delicate, and is not unfrequently concealed by a flap of the mantle which is bent back over it. Another fact of importance is the great development of the epipodia found in most members of the group. The eggs are laid in long ribbons.

The first family we have to mention is the Tornatellidæ, which possesses an ovoid spiral shell, which is usually marked with one or more spiral rows of punctures. The body is large, but usually can be completely retracted into the shell. The cephalic tentacles are large and broad, and united at the base, while the eyes are situated on the outside of the tentacles near their junction with the head. The shells are mostly small, and possess but little interest; a large proportion are fossil, ranging from the carboniferous to the present time.

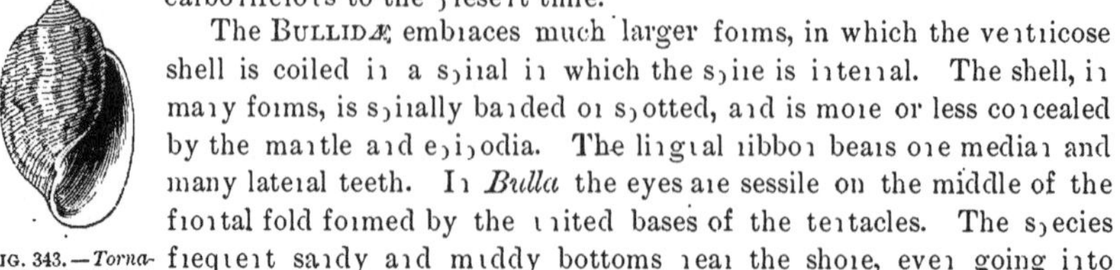

FIG. 343.—*Tornatella flammea.*

The BULLIDÆ embraces much larger forms, in which the ventricose shell is coiled in a spiral in which the spire is internal. The shell, in many forms, is spirally banded or spotted, and is more or less concealed by the mantle and epipodia. The lingual ribbon bears one median and many lateral teeth. In *Bulla* the eyes are sessile on the middle of the frontal fold formed by the united bases of the tentacles. The species frequent sandy and muddy bottoms near the shore, even going into brackish water. At the retreat of the tide they burrow into the mud or hide themselves beneath masses of seaweed. On our east coast is found *B. solitaria*, a brownish spotted form. *Cylichna*, which possibly deserves family rank, is represented on our shores by several small cylindrical shells which frequent slightly deeper water than the Bullas. They move very slowly. *Haminea* may be readily separated from *Bulla* by the lack of color in the shell.

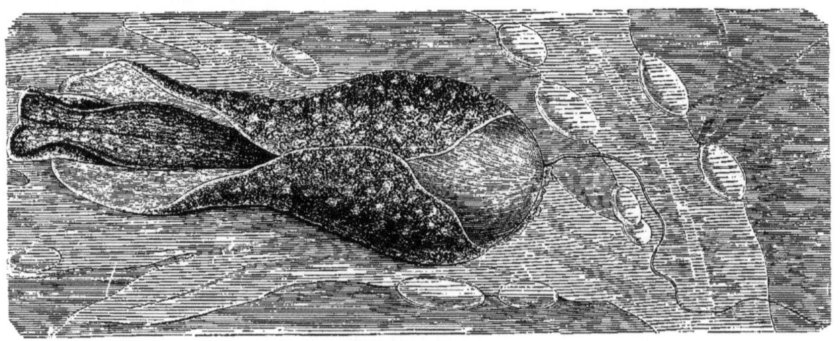

FIG. 344.—*Acera bullata.*

In the PHILINIDÆ the bases of the tentacles are united to form a broad cephalic disc. The shell, which is covered by the mantle and epipodia, is shaped like that of *Bulla*, but scarcely forms a single whorl; in some it is internal and in others external. Eyes may be present or absent. The species are found in water of moderate depth, many species of *Philine* frequenting the shallow water along the shores.

The APLYSIIDÆ embraces slug-like forms known in popular parlance as 'sea-hares.' The shell is small or wanting, and when present is covered by the mantle. The stomach is armed with hardened teeth which play no unimportant part in preparing the food for digestion. The animals feed principally on other molluscs, especially on species of *Acera* (one of the Bullidæ). *Aplysia*, the principal genus, has a pointed oval shell, and the epipodia are extended in swimming. In one species (*A. camelus*) numerous small glands are found beneath the free edge of the mantle which secrete the purple for which these animals were celebrated among the ancients. Near the base of the gill is the outlet of a gland, the secretion of which is said to be poisonous, but whether any of the sea-hares have the toxic effects attributed to them, or even have any poisonous qualities, is yet to be determined. Certain it is that all of the group are not poisonous, for one species forms an article of diet among the South Sea Islands. Some of the European species have a very nauseous smell.

About sixty species of *Aplysia* are known from the whole world, though none are found on our northern coasts. On the Portuguese shores they exist in large numbers,

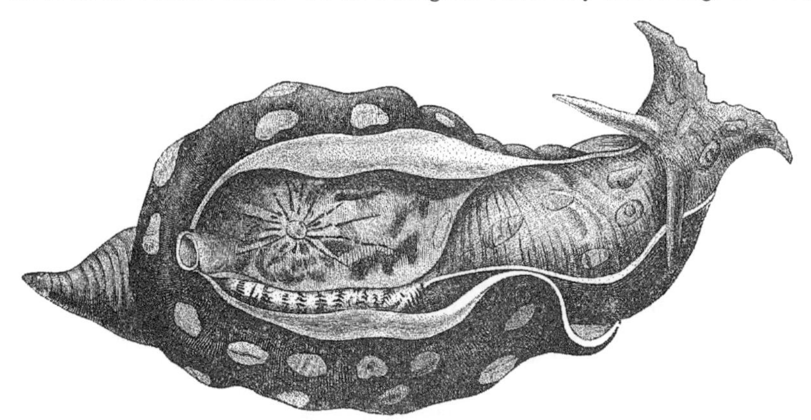

Fig. 345. — *Aplysia depilans*, sea-hare.

and occasionally an easterly storm will throw them up in such quantities on the beach as to cause an epidemic of sickness as well as to render the extraction of the purple a matter of economic importance.

The last family of the Opisthobranchs to be mentioned is the PLEUROBRANCHIDÆ, represented on our coasts by the recently discovered *Koonsia obesa*. In all the members of the family the upper jaw is wanting, the stomach very complicated, and divided into several compartments. The shell, which is usually present, is either borne on the back like that of a limpet, or it is concealed as in the typical genus *Pleurobranchus*. These forms, when creeping slowly through the water, remind one of turtles, and in some the resemblance is strengthened by the distribution of color. In their living state most of the forms are very handsome. *Umbrella* is an aptly named genus, for the shell which covers the back bears no little resemblance to the familiar object bearing the same name.

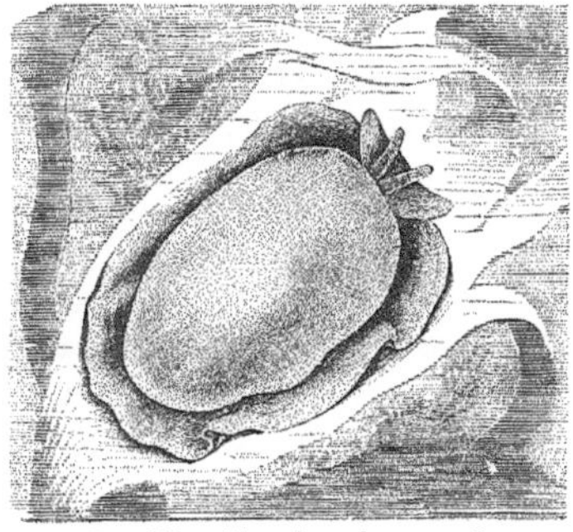

Fig. 346. — *Pleurobranchus peronii.*

Order II. — PULMONATA.

The Pulmonata or Pulmonifera is a group of terrestrial or fresh-water molluscs in which respiration is effected by means of a lung or pulmonary sac, no gills being developed. All the members are hermaphroditic, and an operculum is never formed. Not all the land and fresh-water gasteropods are here included, for, as we shall see further on, many families which have the same habits are entirely at variance with the Pulmonata in the essentials of structure — most prominent being, that, in the one, the visceral nervous loop is straight; in the other, a twisted condition is found.

The pulmonary sac is formed by the union of the edge of the mantle to the body, leaving a small round or oval entrance to a large sac, richly supplied with blood-vessels. In most of the order this lung serves for breathing air, even in the aquatic forms. The operation can readily be witnessed in such a form as *Limnæa*, when kept in confinement. At nearly regular intervals the snail will creep to the surface of the water, and force a bubble of air out of the respiratory orifice, then more air is taken in, and the snail descends again to its pastures. Recent investigations on the Limneans from the profound depths of some of the Swiss lakes have shown some interesting features in connection with this lung. Of course, snails living at these great depths could not ascend to the surface for a supply of air, and it was found that they had acquired a capacity of breathing the oxygen contained in the water, although no gills were present. It is also interesting to observe that all the fresh-water pulmonates fill their lung sac with water in the younger stages, and only later do they adopt the aerial respiration. The pulmonary sac also subserves another function; it forms avery effective hydrostatic apparatus, as, by variations in its size, and consequently in the amount of air, these animals are enabled to rise or sink in the water at will.

The opening to the pulmonary chamber is, of course, at the edge of the mantle and hence, as this part of the body is subject to considerable variation in size, the respiratory opening is far from constant in position. In the common snails (*Helix*), for instance, it is on the right side of the body, just within the shell as the animal is crawling along, while in *Testacella*, where the mantle is very small, it is found near the posterior end of the body. Even the side of the body on which the respiratory orifice is placed is not constant; in most it is found on the right side of the body, but occasionally it occupies the median line behind.

Even greater variations are found in the shell. In some it is large enough to contain the whole body when retracted, some have it reduced to a scale-like plate on the surface, others have it small and internal, while in still others it is entirely absent. Usually it is coiled in a spiral the whorls of which are dextral (revolving from left to right) but in some genera the reverse is the case and the shell is wound from right to left. Even in those which are normally dextral, sinistral monstrosities occur. It is a common idea, but an utterly erroneous one, that the shells north of the equator are coiled in one direction, and south of that line in the other, the coil of the shell following the sun. A little investigation of our fresh-water shells will produce evidence utterly contradicting this. Only in the family Amphibolidæ is an operculum formed, but the other members of the order secrete a mucus which hardens and tightly seals the aperture of the shell. This is known as the epiphragm and is formed when the animal retires in winter or in a season of drought. On the return of moist weather, it is broken down and the snail resumes its wanderings. In *Clausilia* this epiphragm is a permanent structure and is fastened to the mouth of the shell by an elastic stalk, so that it works as a spring door.

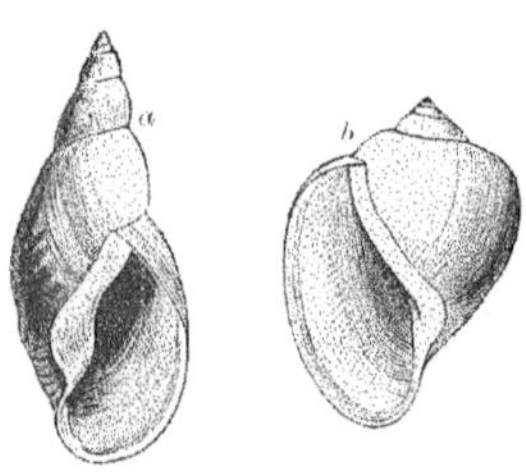

FIG. 347.—*a*, *Limnæa elodes*, a dextral; *b*, *Physa lordii*, a sinistral shell.

The lingual ribbon is short and broad, and is armed with rows of very numerous teeth, there being sometimes over two hundred in a single row. Each tooth has a broad base and acute or denticulated recurved tips. This ribbon is opposed in some forms by an upper jaw composed of either one or three pieces; in others, no upper jaw is present.

The generative apparatus is rather complex and affords good systematic characters. The most noticeable feature is that from the ovotestis a single duct proceeds, which afterward divides into two tubes, one connecting with the male, the other with the female copulatory organs. In the Helicidæ peculiar crystalline, fluted, chitinous, or calcified rods or darts (*spicula amoris*) are formed, the functions of which are still problematical.

The eggs are laid in moist places, in damp earth, under dead leaves, etc.; or, by the aquatic species, in the water. Those of *Limnæa* and *Planorbis* are easily studied during their development. In *Limax*, — the eggs of which are laid separately, each one resembling a drop of dew — when the embryo is far along in its development, a peculiar pulsating sac is formed in the middle of the foot, the function of which is as yet unknown.

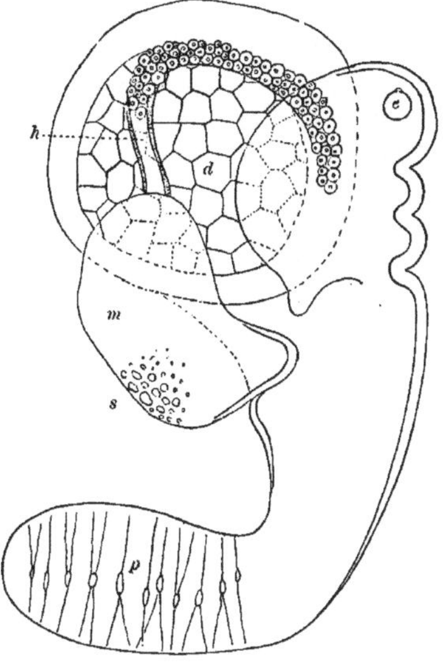

FIG. 348. — Embryo of *Limax*; *d*, yolk; *e*, eye; *f*, foot; *h*, heart; *m*, mantle; *p*, pulsating vesicle; *s*, shell, arising as a number of calcareous granules.

SUB-ORDER I. — BASOMMATOPHORA.

The position of the eyes affords a good character for a division of the Pulmonata into two sub-orders. In the present group the visual organs are seated at the base of the solid, contractile feelers; the velum of the larva is retained in the adult; and the male and female generative apertures are separate and placed on the right side of the neck. Most of the members of the sub-order are aquatic in habits, though some lead more or less terrestrial lives.

The family AMPHIBOLIDÆ, which occurs only in the New Zealand seas, serves to connect the pulmonates with the opisthobranchs. They live in the salt marshes, where the water is at least brackish, but are partially aerial in their respiration, although rudimentary gills are present. The shells are closed by a horny operculum. In these two features the Amphibolidæ differ from all other pulmonates. The shell is spiral and thick, the spire short and the whorls shouldered. The native New Zealanders eat the animal.

The families GADINIDÆ and SIPHONARIDÆ embrace together about a hundred species of limpet-like pulmonates, with shell and habits nearly like *Acmæa* and *Patella.* No species are found on the east coast of the United States.

Concerning the AURICULIDÆ more can be said. The animals are mostly tropical; still several small species are found even in the northern states. The spiral shell is usually thick and solid, and covered with an epidermis. The spire is short, the body whorl large, and the outer lip is thick and frequently armed interiorly with teeth which considerably contract the aperture. Similar teeth are found on the columellar lip. The respiratory pore is posterior, and the male and female reproductive organs are widely separated. The mouth is armed with a horny jaw.

The Auriculidæ are mostly found in the neighborhood of the sea, especially in salt marshes. The genus *Auricula*, from which the family derives its name, has but a very remote resemblance to an ear; the species are all inhabitants of brackish-water swamps in tropical regions, and are characterized by an absence of teeth on the outer lip of the long and narrow aperture. In *Scarabus* the shell is laterally compressed, so that the edges are angular; the aperture would be large were it not for the teeth which arise from both lips, and the spire of the shell is acute. The species all come from the tropical parts of the eastern hemisphere, where they live in the woods near the shore.

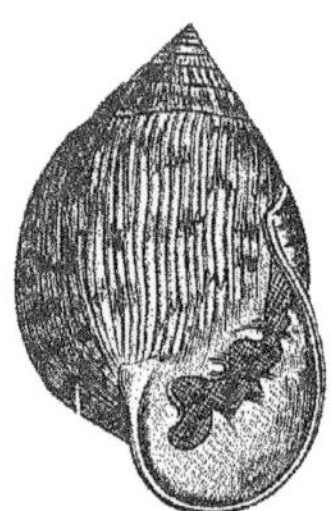

FIG. 349. — *Scarabus fasciatus.*

Alexia is represented in the United States by a single species, *A. myosotis*, which does not extend farther south than New Jersey. In Europe it is found on the shores of the Atlantic and the Mediterranean. It frequents places where it is covered by the tide for several hours each day, moving in a very sluggish manner. Fresh water kills it. Other species are found in Europe and the West Indies. The shell is of a general dark horn color and bears two tooth-like folds on the columella. *Carychium* is much like *Pupa* in shape, and our single species, *C. exiguum*, is widely distributed under stones and moss in damp places, and is the only member of the family which in the United States is found far from the sea.

FIG. 350. — *Scarabus imbrium.*

FIG. 351. — *Alexia myosotis.*

In the species of *Melampus* the shell is ovate in outline, the spire short, and the outer lip acute. Four species are found in the United States, one on the Pacific and three on the Atlantic coasts. One of the southern forms has received the specific name *coffea*, from its resemblance in size, shape, and color to a kernel of coffee. This species and another, *M. flavus*, occur in the United States only in Florida, except when introduced elsewhere by means of vessels trading with southern and West Indian ports. The remaining species, *M. bidentatus*, is common in the grass of every salt marsh from Massachusetts to Texas. When young, this is a very pretty species, being brownish in color, marked with revolving reddish bands, and the whole highly polished; but the adults are dirty and eroded. The length of a large specimen is about half an inch.

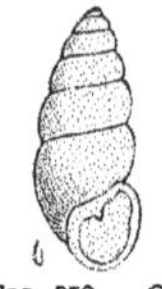

FIG. 352. — *Carychium exiguum.*

FIG. 353. — *Melampus bidentatus.*

The species of *Cassidula* have a subquadrate body whorl and very short spire; and the outer and inner lips are toothed. The species all belong to the Indo-Pacific region, frequenting mangrove swamps and rocky shores. The species of *Pedipes* are all tropical and sub-tropical. They have a looping gait like that of a measuring worm, and with this peculiarity in locomotion is correllated a transverse groove on the foot. The members of the genus are among the most active of molluscs. No species occur in the United States, the nearest approach being Lower California, where *P. lirata* occurs. *Otina* is not represented in the United States. The shells are ear-shaped, and colored, and the animals have the same method of locomotion as *Pedipes*, living among rocks between tide-marks.

All of members of the **Limnæidæ** or pond-snail family are inhabitants of fresh water, and, so far as investigations show, are most numerous in temperate regions. Over six hundred nominal species exist, which belong largely to the genera *Planorbis*, *Physa*, and *Limnæa*. The shells are very variable in shape, but, with but little experience, one readily recognizes the members of the family. In some, the shell is a long

Fig. 354.—Different forms of *Limnæa elodes*.

spiral, others have it coiled like a bit of tape, while still others have a shell without trace of spiral, but flattened and limpet-like. Each of these three types of shell is regarded as affording characters of sub-family rank, and we have the **Limnæinæ**, Planorbinæ, and Ancylinæ respectively.

The pond-snails are exclusively vegetable feeders. They frequent the still waters of sluggish streams and ponds, their thin shells being poorly adapted for a life in rapidly running creeks. They usually require to go to the surface to breathe, but, as noticed on a previous page, this is not always the case. The eggs are laid in clusters attached to sub-aquatic objects and imbedded in a gelatinous matrix. Frequently specimens may be seen progressing at the surface of the water shell downwards, the bottom of

Fig. 355.—*Limnæa stagnalis*, pond-snail.

the foot being just above the liquid. An interesting fact, first pointed out by Professor E. Ray Lankester, is that in this family the velum persists in the adult, nearly as large relatively as in the veliger stage. Though these forms have been much studied, this fact escaped observation until 1883.

First in order comes the genus *Limnæa* with a thin, horn-colored, slender, spiral shell and a large aperture. Important in separating this from the next genus to be mentioned is the fact that the shell is always dextral, that is, the whorls revolve from left to right. In times of drought the **Limneans** burrow into the mud and close the aperture of the shell with an epiphragm like that of the Helicidæ, thus preventing any desiccation of the fluids of the body. When the rains again fill the ponds, they come out of their burrows and lead a free life. Of the species but little can be said, and we will let our figures speak for themselves. All figured are from the United States, though several are found in Europe as well. Doubtless, when these forms are studied in the proper manner, by rearing in confinement all of the progeny of a single pair, and continuing the operation for several generations, it will be found that many of the so-called species are but vagaries. Indeed, Mr. P. R. Whitfield has done this with specimens of *Limnæa megasoma* and has found that, in this way, variations were produced, which conchologists, not knowing the history of the group, would describe as distinct species. Mr. Whitfield's experiments however, were not conclusive, as there was apparently a lack of nutrition and a higher temperature than the normal, which doubtless had an effect on the forms produced.

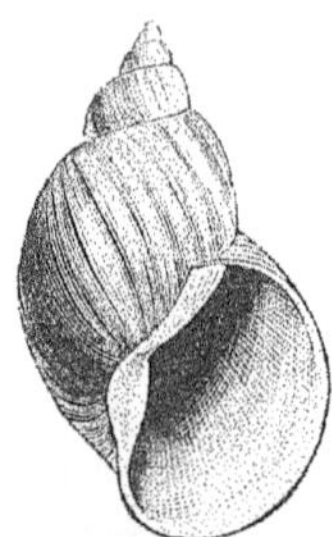
FIG. 356. — *Limnæa megasoma.*

Limnæa is essentially a northern genus, reaching its highest development in North America, in the British possessions; *Physa* on the other hand, is more southerly. It has a thin, amber-colored shell, the whorls of which revolve from right to left. The species are much more active than those of *Limnæa.* The tentacles are long and slender, and the jaw is formed of a single piece. Twenty-three so-called species occur in North America.

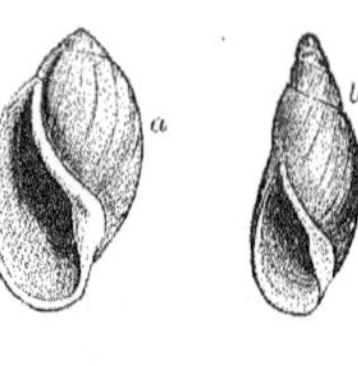

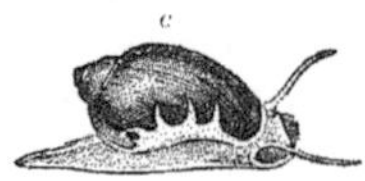

FIG. 357. — *a, Physa ampullacea; b, P. elongata; c, P. heterostropha.*

The genus *Pompholyx* is noticeable, from the fact that, while the shell is dextral, the genitalia open on the left side of the body. The shell is short and broad. Three species are known, all from the Pacific region of the United States. They were formerly supposed to have two pairs of eyes.

In *Planorbis* the shell is wound in a flat spiral, like a roll of tape, showing the whorls on either side. The animals prefer still water, where they move about in a sluggish manner. About a hundred and fifty species have been described, of which about twenty-five occur in the United States. Any description of the various forms would prove tiresome reading for any but the systematic student.

FIG. 358. — *Planorbis.*

The species of *Ancylus* and *Gundlachia* are shaped nearly like the limpets *Acmæa*. and *Crepidula*, and, judging from the shell alone, one would not realize that the animals were so distinct. These fresh-water limpets have the same habits as their marine equivalents. They live attached to the under sides of stones below the surface of the water, feeding on confervæ and other plants. In both, the body is sinistral, the genital openings being on the left side of the body. In *Ancylus*, of which we have about twenty species in the United States, the shell is not at all spiral, but in *Gundlachia* it resembles *Crepidula* in this respect. Two species of the latter genus are found within our boundaries, the three remaining species being from the West Indies and Tasmania.

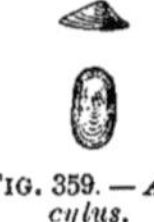
FIG. 359. — *Ancylus.*

The fresh-water pulmonates are badly infested with parasites, most of which are stages of worms which reach their complete development in some of the vertebrates.

On a preceding page has been detailed the history of the liver-fluke, *Distoma*, which passes a portion of its life in a species of *Limnæa*, and this is far from a solitary example. These snails are eaten by fishes and birds, and in the stomach of the eater, the larvæ are set free, and enter into a new stage of existence. In some few cases the history has been thoroughly worked out, but in the majority there is a field for investigation, which will give the careful student wonderful results. The subject is difficult to study, but, with time and patience its problems may be solved.

SUB-ORDER II. — STYLOMMATOPHORA.

The great majority of the Stylommatophora are terrestrial, and are readily distinguished from the other sub-order by having two pairs of tentacles, the superior pair bearing the eyes at the extremity. These tentacles may be simply retracted, or, as in the common snails the tip can be turned in like the finger of a glove, a condition described as invaginate. In most of the sub-order the genital orifices are united.

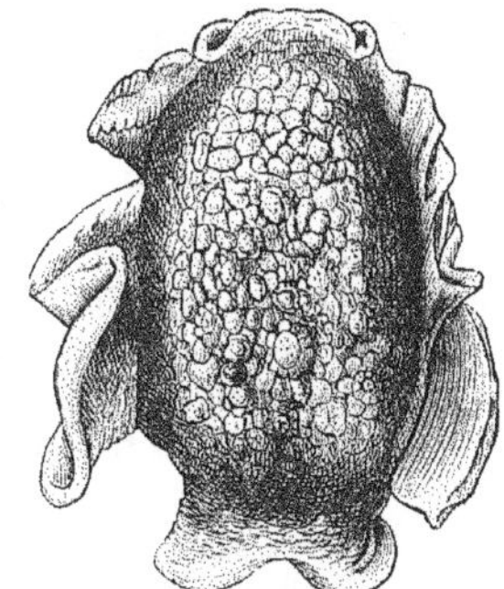

FIG. 360. — *Onchidium tonganum.*

The first family to be considered is the ONCHIDIIDÆ, which embraces a few terrestrial and aquatic forms from warm latitudes. They have the male and female orifices widely separated, and the eye-bearing tentacles simply contractile and not capable of being invaginated. No shell occurs in the family. *Onchidium* has been rendered prominent by the researches of Semper. This naturalist studied one of the species found in the eastern seas, and found that, besides the eyes borne on the end of the tentacles, the whole dorsal surface was covered with visual organs, a fact which will recall the more recent discoveries of Moseley with regard to the chitons. These eyes are different from those borne on the tentacles in the fact that, in structure and development, they are like those of vertebrates, the nervous fibres penetrating the layer of rods and cones, and being distributed over their inner surface.

Whether these eyes exist in all species of *Onchidium* is not known, as all have not yet been made the subject of proper histological investigation. Why these eyes should be developed here is uncertain. The only explanation as yet advanced is that of Semper. The *Onchidia* live on the shores of the ocean, where they creep about in a slug-like manner. They play an important part in the diet of the jumping fish, *Periophthalmus*, which leaves the water and travels about on the beach left bare by the retreating tide, looking for food. Semper supposes that these eyes are of considerable use in avoiding this enemy of the race.

The genera *Onchidella* and *Peronia* are also marine and live on algæ. They are amphibious, and if kept moist, they can live for a long time removed from the water. *Veronicella* is a terrestrial genus represented by a single species in Florida, and several others in the tropics of the old and new worlds. They live in families under trees and stones, whence they come forth at night, ascending trees, etc., in their search for food. Unlike the slugs, which they resemble in general appearance, they leave no slimy tracks behind them. They lay their eggs in long gelatinous threads, fifteen or twenty being contained in a single string.

In all the remaining families the generative orifices are united, but the question of what constitutes a family here is not yet settled. We have endeavored to take a conservative course in this respect, and hence but few families represent the many divisions which exist in some systems. The first is the TESTACELLIDÆ, in which the animal is like the familiar garden slug, but bears a small shell on the hinder end of the body, and the mouth has no upper jaw. The genus *Testacella* is European and is noticeable from the fact that it forms an exception to the other pulmonates in being predatory and carnivorous. Its principal diet is earth-worms. It lives beneath the surface of the earth, and follows the worms down their burrows. Other articles of food are snails and slugs, and it will even eat its own species. It, however, wants its prey alive, and even refuses pieces of a fresh worm which has been chopped up to feed it. Many tales are told of its ferocity and cunning. They are said to live for five or six years; at the approach of cold weather they burrow deep into the earth, and, with the mucus they secrete, they form a cocoon in which they spend the winter.

FIG. 361. — *Testacella haliotidea.*

Allied to *Testacella*, but still entitled to family rank, is a group which is called OLEACINIDÆ. As in the last family, an upper jaw is wanting, but the shell is much better developed and capable of containing the body when retracted. Most prominent is the genus *Glandina*, with about a hundred and twenty-five species. It has a fusiform shell with a thin, sharp, outer lip. *Glandina truncata*, our best known species, extends from South Carolina to Texas, and possibly further south. It prefers moist situations, and thrives in the Everglades of Florida, living in humps of coarse grass. It is partially if not wholly carnivorous, but, unlike *Testacella*, it is not averse to dead animal matter, and will eat that which is partially decayed. It is even cannibalistic. Its tongue is armed with numerous long, sharp teeth, with which it rasps off large mouthfuls of its prey. The shell is usually ashy fawn color, more or less tinged with pink, which soon fades after death. In a Central American species, *G. rosea*, the color persists to a much greater extent. Some of the South American species are much more predacious than our forms, and do not hesitate to attack snails as large or larger than themselves.

In *Streptaxis*, a South American genus, the shell is more like that of the normal species of *Helix* (to be described below), but there is a curious distortion. The axis of the shell is bent so that the lower whorls are not parallel to the earlier ones.

The American family CYLINDRELLIDÆ needs but a passing mention. The shell, as the name indicates, is shaped like a cylinder, composed of many whorls, the last being usually more or less detached from the others, and terminated by a circular mouth. The animals have sluggish motions, and drag their shell horizontally behind them. A few species are found in Florida, but the family reaches its highest development in the tropics, especially in the West India Islands.

The HELICIDÆ is by far the largest family of pulmonates; indeed, it contains more species than all the other families together. Over sixty-five hundred species have been described. A concise definition of the group is impossible, yet all of its members are readily recognized by the tyro as belonging to the family. There is an indescribable something which at once tells the student that the specimen before him belongs to the family Helicidæ. Still, notwithstanding the fact that we cannot frame a satisfactory definition, it will be well to review a few of the characters found in the group.

In all, the upper jaw is present and opposable to the lingual ribbon; the tentacles which bear the eyes are longest and can be invaginated. The shell is spiral, usually well developed, and capable of containing the whole animal; the reproductive orifice is near the base of the right ocular tentacle. An immense number of genera and subgenera have been made in order to render the identification and classification of the numerous species an easier task. Even the family Helicidæ has been broken up into divisions, each of which have been accorded family rank, but which here are regarded as sub-families. Space and the patience of our readers will allow but the mention of but a few forms, while our illustrations will show the general appearance of many of the species found in the United States, as well as a few from foreign countries.

The Helicidæ are all terrestrial, herbivorous animals, which delight in woods, especially in limestone regions. In Europe, some species have proved themselves nuisances to the agriculturist, but with us they have not yet done much damage. Our American forms seem to avoid cultivated places, and the little damage done the farmer or gardener by the molluscs is occasioned by the slugs (*Limax*) and a few imported snails. Why there should be this difference between the snails of Europe and America is not easy to say; possibly it is because our native species have not yet had time to adapt themselves to the changed conditions which accompany civilization; and they still adhere to the traditions of their fathers.

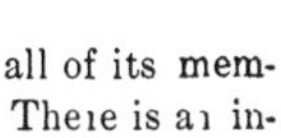

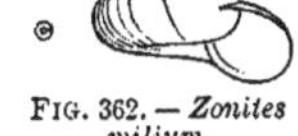

FIG. 362. — *Zonites milium.*

The land snails possess great vitality, and as an illustration we cannot refrain from quoting the wonderful history of a specimen of *Helix desertorum*, which has figured in many a work on the subject of the Mollusca. This specimen was brought from Egypt to England. It "was fixed to a tablet in the British museum on the 25th of March, 1846; and on the 7th of March, 1850, it was observed that he must have come out of his shell in the interval (as the paper had been discolored, apparently in his attempt to get away); but, finding escape impossible, had again retired, closing his shell with the usual glistening film; this led to his immersion in tepid water and marvellous recovery." Even longer was the life of a specimen of *Helix veatchii*, from Lower California, detailed by Dr. R. E. C. Stearns. This individual lived six years, from 1859 to 1865, in confinement, without food.

The time of oviposition is from April to June. The number of eggs varies from thirty to fifty or more. They are laid in the light, moist mould, each one separate, or united by the slightly adhesive exterior. There is no gelatinous matrix like that found in the aquatic forms. In laying the eggs, the snail usually burrows its head into the soil, stretching the body to the utmost extent. Since the reproductive orifice, as has been said, is beneath the upper tentacles, this places the eggs at a distance beneath the surface about equal to the length of the body in front of the shell. Other species actually burrow beneath the surface to the depth of three or four inches before laying their eggs, so as to insure a moist condition.

It is related that the eggs possess great vitality, and that they are capable of withstanding desiccation. When so dry that they had lost all form, and were reduced to a friable condition, an exposure of but a single hour to moisture restored their former form and elasticity, and the egg developed in the normal manner. The writer, in his studies of the development of *Limax*, did not have such results. The eggs, after drying, were readily swollen by a moist atmosphere; but if the desiccation had been too long continued (even without heat), the eggs failed to develop farther.

Like most of the shelled pulmonates, the Helicidæ in temperate climates form an epiphragm to close the shell during the winter hibernation, and in the hotter portions of the globe during the dry season. The method of forming this has thus been described. "The animal being withdrawn into the shell, the collar is brought to a level with the aperture, and a quantity of mucus is poured out from it and covers it. A small quantity of air is then emitted from the respiratory foramen, which detaches the mucus from the collar, and projects it in a convex form like a bubble. At the same time the animal retreats farther into the shell, leaving a vacuum between itself and the membrane, which is consequently pressed back by the external air to a level with the aperture, or even farther, so as to form a concave surface, where, having become desiccated and hard, it remains fixed. These operations are nearly simultaneous, and occupy but an instant. As the weather becomes colder, the animal retires farther into the shell and makes another septum, and so on, until sometimes there are as many as six of these partitions; the circulation becomes slow; the pulsations of the heart, which in the season of activity vary from forty to sixty in a minute, according to the temperature of the air, decrease in frequency and strength, until they at length become imperceptible; the other functions of the body cease, and a state of torpidity succeeds, which is interrupted only by the heat of the next spring's sun." With the snails which occupy a constantly warm, moist climate like that of Florida, there is no period of hibernation; they are active throughout the year.

First in order comes the Vitrininæ, of which the genus *Vitrina* is the type. Here the thin spiral shell is too small to contain the entire animal, and is composed of a few rapidly enlarging whorls. The species are very active and live in moist situations, usually feeding on vegetable substances, but not in all cases being averse to an animal diet. Three species of this genus are found in the United States, while there are about a hundred in the entire world. The most common form in our territory is that figured, *V. pellucida*.

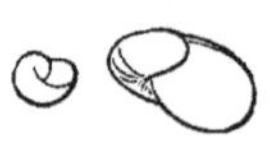

FIG. 363. — *Vitrina pellucida.*

In the next sub-family, the Zonitinæ, but two genera need our attention. In the genus *Macrocyclis*, of which only one species is found east of the Rocky Mountains, the thin shell has a wide umbilicus and a sharp outer lip. *M. concava* is comparatively common and leads an active life. It is very voracious, and feeds upon other species of the family. Its body is narrow and very extensible, and it thrusts it into the shell of other species and feeds on the soft parts at its leisure. *Zonites* contains many more species than the genus just mentioned, in which the shell is much like that in *Macrocyclis*, the differences being found almost entirely in the dentition and in the soft parts. *Z. cellaria* is an European species which has been introduced into America, where it is now common in the seaport towns along the Atlantic coasts. It lives in cellars and in hothouses and gardens. The way in which it has been introduced is uncertain. From its habits it would

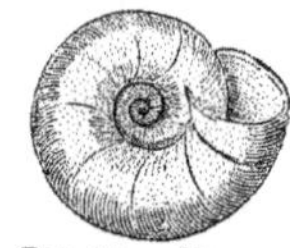

FIG. 364. — *Macrocyclis concava.*

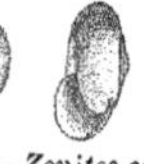

FIG. 365. — *Zonites cellaria*, cellar snail.

appear probable that it came along with hot-house plants, or that its eggs may have adhered to some wine cask and found suitable conditions for development in the cellars of the new world. Many other species are found in the United States, most of them being small and inconspicuous, *Z. milium* being one of our smallest shells.

The genus *Helix* has been divided into innumerable sub-genera and tribes, the details of which should be sought in special works. This genus is the first of the sub-family Helicinæ, in which the spiral shell is thicker and stouter than in the preceding divisions, and capable of containing the entire animal when retracted. Most of the species have the outer lip thickened and reflected, and not infrequently the aperture is greatly reduced by tooth-like processes which may arise from the columella or the outer lip, or from both. The species are usually much larger than those in the sub-families just passed.

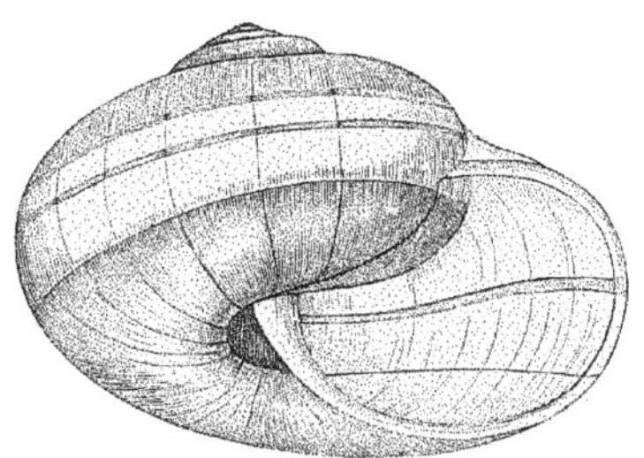
FIG. 366. — *Helix sudanensis.*

The characters of the genus *Helix* are very poorly defined, and the shape of the shell varies between very wide extremes. In some the spiral is high, in others it is nearly flat. In most of our northern species the shell is horn-colored without ornamentation, but in the tropics brightly colored species are the rule. The color may be laid on in blotches, or more frequently in stripes, which follow the spiral of the shell as in the adjacent figure of *Helix sudanensis*, which, as its name indicates, comes from Africa.

With such a wealth of species to choose from (about thirty-five hundred being known) it is a difficult task to select the few which our space will admit. Our most common species is possibly *Helix albolabris*, which, when adult, reaches a diameter of about one inch. In the young the outer lip is thin and sharp; but when the full size has been reached, the lip becomes thickened and reflected, or turned outwards, and covered with a white porcellanous deposit which gives the specific name. Usually the columella is smooth, but occasionally specimens are found in which a tooth is developed. This species is found most abundantly in forests of hard wood. In the southern states its place is taken by a similar but much larger species, *Helix major*.

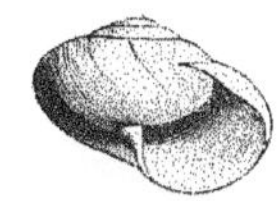
FIG. 367. — *Helix albolabris*, young.

The garden-snail of Europe, *H. hortensis*, has been introduced into several places along our eastern coasts. It is very common on the islands in and near the harbor of Salem, Mass., where, together with *Helix alternata*, it lives in the long grass and among the juniper-trees. This species has a white lip, and is usually ornamented with a varying number of reddish lines which follow the spiral of the shell. Each of the islands mentioned has its own peculiar pattern of ornamentation, which seems to have been derived from the first animals introduced. The method in which this species obtained a foothold on these islands (several of which are small and uninhabited, and separated by a mile or more of salt water from the shore) is even less easily decided than in the case of the *Zonites cellaria*.

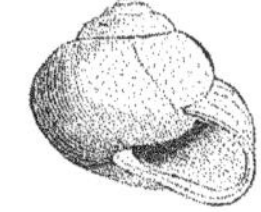
FIG. 368. — *Helix hortensis*, garden snail.

Several of the European species are used as food, and one, *Helix pomatia*, the Roman snail, has long occupied a place in the economy of the Latin races. This and *Helix aspersa* are to-day extensively eaten by the French, and the latter species was introduced into Charleston, S. C., by the French inhabitants for the purposes of food. The writer several years ago, tried the experiment of introducing it into New England; but although the places where specimens were distributed have since been carefully searched, none have been found. It may be that the east winds and the cold of winter prove too much for it. Some if not all of our American species are edible; *H. thyroides*, when treated with vinegar, has a very peculiar but pleasant taste, excelling, in this respect, *Helix albolabris*.

FIG. 369. — *Helix thyroides.*

Another common species in the United States is *Helix alternata*, in which the outer lip is sharp and the horn-colored shell is ornamented with blotches of dark brown. In New York and New England it is even more common than *H. albolabris*, occurring not only in the woods, but in the open fields as well, although it seems far more dependent on moisture than some of the other species. It is not so palatable as *H. thyroides*. Allied to *H. alternata* is the pretty, but small species, *H. asteriscus*, in which the whorls are ornamented by a number of transverse ribs. It is only found in the northern states. *Helix harpa*, which has a boreal distribution, is found on both continents. The shell is high, and ornamented on the two lower whorls by transverse ribs. "The body is so translucent, that, when extended, the ganglionic centres can be plainly seen. In motion it is exceedingly graceful, at times poising its beautiful shell high above its body, and twirling it around, not unlike a *Physa*, again hugging its pretty harp close to its body; the shell when in this last position, continually oscillates as if the animal [which is very small in proportion to its shell] could not balance it; it rarely ever moves in a straight line but is always turning and whisking about, and this is done at times very quickly and abruptly."

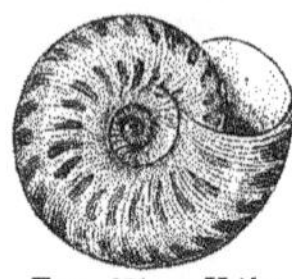
FIG. 370. — *Helix alternata.*

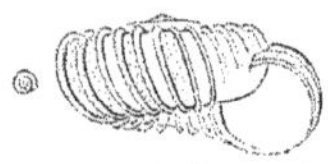
FIG. 371. — *Helix asteriscus.*

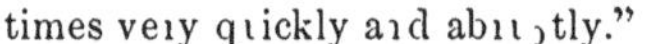

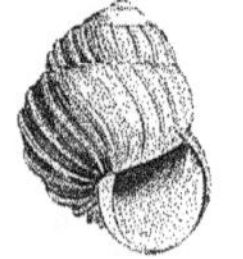
FIG. 372. — *Helix harpa.*

In a large number of species like *Helix sayi*, *dentifera*, etc., a tooth is always developed on the columella, like that occasionally found in *Helix albolabris*, while in another series, including *tridentata*, *palliata*, etc., the aperture is still farther contracted by the development of one or more teeth from the inside of the outer lip. In one of this latter group, *H. hirsuta*, the aperture is very narrow, and the outside of the shell is covered with numerous short, stiff hairs.

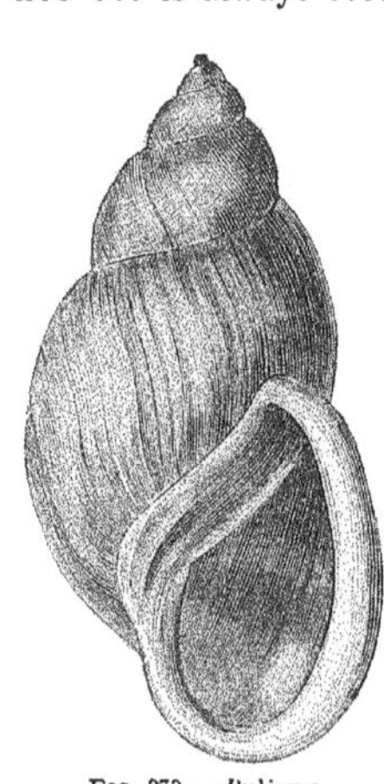
FIG. 373. — *Bulimus.*

The species of *Bulimus* are largely tropical, and the majority of the three hundred and odd species come from South America. The animal is much like that of *Helix*, but the shell is longer and has but a few whorls, while the lip is thickened, reflected, and continuous with a callus layer on the columella. Most of the species are large, some being among the giants of the pulmonates, only exceeded by the *Achatinæ* to be mentioned in a moment. The largest species is *Bulimus ovatus* which is common in the forests of southeastern Brazil; the shell reaches a length of six inches.

This species is an article of food and is sold in the markets of Rio Janeiro. Its eggs are also very large; they have a white calcareous shell, and equal in size those of a pigeon.

FIG. 374. — *Helix palliata.*

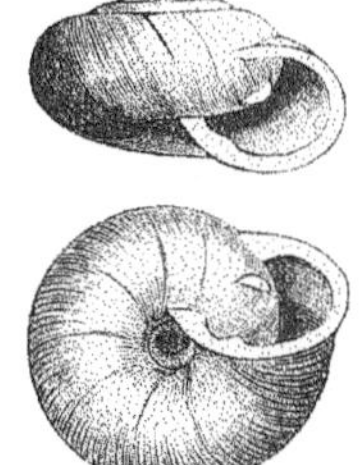

FIG. 375. — *Helix sayi.*

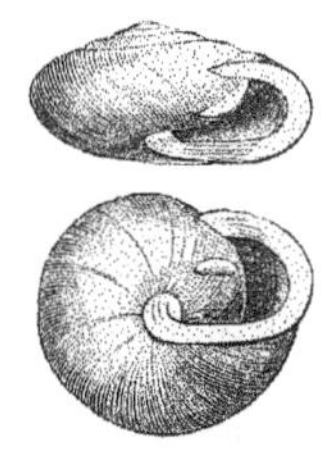

FIG. 376. — *Helix dentifera.*

In the older works several species of *Bulimus* were credited to the United States, but more recent studies show that these forms belong elsewhere in schemes of classification.

In some of the South Sea Islands, especially in the Society group, occur a number of land shells united under the generic name *Partula.* These are brightly colored and much like *Bulimus* in shape. Formerly they were very abundant; but a few years ago a great storm utterly destroyed the groves in which they were found and almost extinguished the genus. Unlike most of the pulmonates they bring forth their young alive,

FIG. 377. — *Achatina mauritanica.*

and the shells are more frequently sinistral than in the other genera of the Helicidæ. In the genus *Binneya* occurs a peculiarity first noticed by Dr. J. G. Cooper. The species are all inhabitants of Mexico and Southern California. At the approach of the dry season they retreat as do the *Helices* of more northern climes in the winter. Still, as the shell is too small to contain the whole body, the epiphragm is greatly enlarged so that it covers all the parts which would otherwise be exposed. This epiphragmal envelope is white and parchment-like.

The sub-family Achatininæ embraces forms much like the Helicinæ but distinguished by lingual dentition and by the fact that the lip is usually sharp, the columella truncated, the shell with an elongate spire, the body whorl being swollen. The genus *Achatina*, the agate shells, derives its name from the usually banded species. It embraces the largest species of pulmonates known, even exceeding the genus *Bulimus* in this respect, as some of the shells measure ten inches in length. The eggs are of proportionate size and have a calcareous shell. Most of the species are found in Africa, where they live in trees, descending to the ground to lay their eggs.

In the genus *Achatinella*, the dextral or sinistral shell is much like that of *Bulimus* in outline, but is distinguished among other characters by the spiral fold which accompanies the columella. The species are confined to the Hawaiian Islands, but their number has been multiplied to an utterly unwarranted extent, no less than three hundred having been described. All are very pretty shells, with a polished exterior, and striped and spotted with bright colors, red, green, and brown predominating. We well know

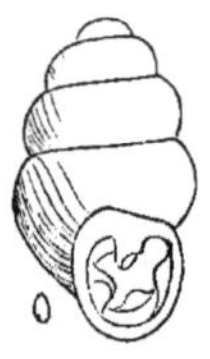

FIG. 378. —*Pupa contracta.*

FIG. 379. — *Pupa armifera.*

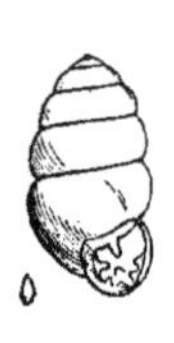

FIG. 380. — *Pupa pentodon.*

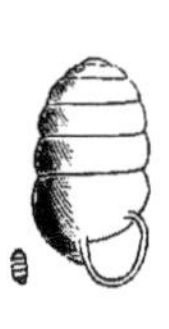

FIG. 381. — *Pupa badia.*

FIG. 382. — *Pupa fallax.*

how inconstant is the number of bands in the land shells of the United States, where the same species may be plain or ornamented with one or several spiral bands, but these *Achatinellæ* have been divided up mostly on similar characters. They live largely on the low shrubbery near the sea, but since the introduction of cattle on the islands they have become much less common than formerly, on account of the destruction of their food plants; and their ultimate extinction is but a question of time.

In the PUPIDÆ, we have a large number of generally small, many whorled, more or less cylindrical shells, in which the aperture is frequently contracted by tooth-like processes, like those previously described in some of the *Helices*. Our American species of *Pupa* are almost all very minute, so that it requires good eyes to collect

FIG. 383. — *Vertigo ovata.*

FIG. 384. — *Vertigo milium.*

FIG. 385. — *Vertigo bollesianus.*

FIG. 386. — *Vertigo ventricosa.*

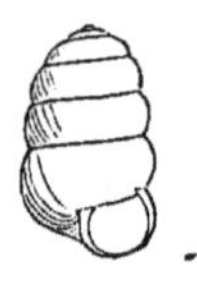

FIG. 387. — *Vertigo simplex.*

them. They seem to be even more dependent on moisture than most other land shells. The species are largely based on the number and form of the teeth of the aperture, the variations in which may be seen in our figures of some of the more common species from the United States. One of the most important distinctions between *Vertigo* and *Pupa* lies in the fact that in the latter genus the cephalic tentacles are present, though small, while in the former they are absent.

The genus *Clausilia* occurs in the regions surrounding the Mediterranean, its seven hundred nominal species being distributed in Europe, Asia, and Africa. The shell is long and cylindrical or fusiform, and is usually coiled from right to left, although dextral forms occur. The animal is also sinistral, the genital and respiratory orifices being on the left side of the body. The aperture is usually distinct from the rest of the shell, being separated by a neck or constriction. We have already alluded to the peculiar permanent epiphragm with which these forms close the aperture.

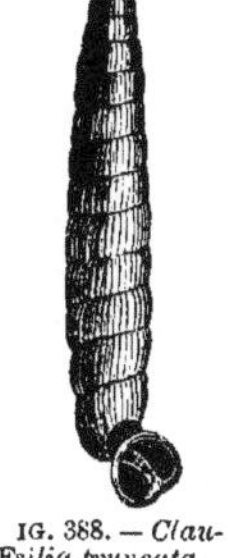

FIG. 388. — *Clausilia truncata.*

The members of the family SUCCINIDÆ have a world-wide distribution, and are usually found near the margins of ponds and streams. The shell much resembles that of the Limneans, though the two families are widely different. The family is distinguished from all others by the upper jaw, which consists of the usual arcuate portion backed up by a quadrate plate. The shell is very thin and transparent, and made up of a few rapidly enlarging whorls. The principal genus is *Succinea*, of which about two hundred species are known. These forms have an oval aperture and a sharp outer lip, and are usually regarded as amphibious or even as preferring a subaquatic life. This belief does not appear to be well founded, for, although they are found near the margins of streams, they live exclusively in the air, and some of them are found far from any body of water.

FIG. 389. — *a, Succinea totteniana; b, S. ovalis; c, S. avara.*

At the time of drought, and at the approach of winter, they draw the body completely within the shell, and form an epiphragm like that of the *Helices*. The shell is amber-colored or whitish. Our most common species are *Succinea avara*, and *S. obliqua*.

The terrestrial pulmonates in which the shell is internal or absent are known in popular parlance as slugs, while in scientific works they are united into a family to which the name LIMACIDÆ is applied. Their general appearance is too well known to

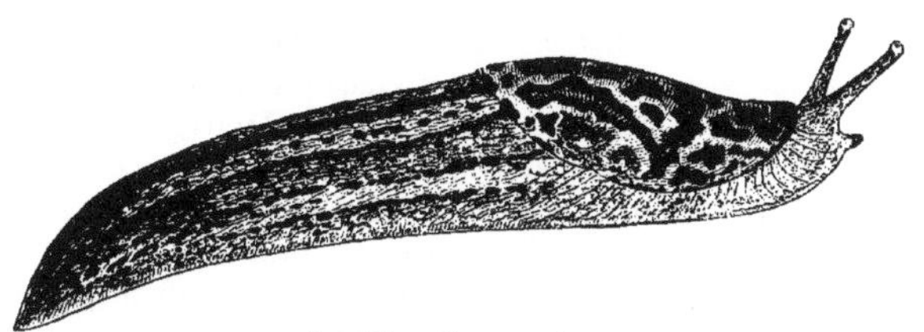

FIG. 390. — *Limax maximus.*

call for any detailed description, yet there are certain features which have a morphological significance to be mentioned. On the dorsal surface of the body, near the anterior end, is a fleshy plate, the mantle. At or near the right margin of this, is the opening of the respiratory pore. The head is well defined and provided with tentacles.

The slugs are chiefly nocturnal, and this fact accounts for the few ordinarily seen, although there may be thousands about. In the daytime they secrete themselves under boards, fallen trees, etc., where there is at least partial darkness, but at night they come out to feed. They do a great amount of damage in gardens, as they feed largely on vegetation, although they are not averse to an animal diet. Since they hide themselves during the day, the damage they occasion is usually attributed to birds, and the larvæ of insects, but the presence of slugs can usually be recognized by the presence of streaks of glistening slime in the neighborhood. Most of the terrestrial pulmonates are able to secrete a mucus from their body, and in some there are special pores for its emission. In the slugs this capacity reaches a great development, and as they crawl along they leave a streak behind them, which, on drying, produces the glistening marks referred to. This secretion of mucus is to a certain extent defensive, and when the animals are irritated the amount is greatly increased. This fact gives us a simple method of checking their ravages, which is to sprinkle coal-ashes around the plants which it is desired to protect. The fine grit of the ashes irritates them, and they pour out the mucus to such an extent that they are soon exhausted, and besides, since it rapidly hardens on exposure to the air, they are soon rendered prisoners.

This secretion of mucus is used in another way. Slugs frequently climb trees in search of fruit, and when through feeding they take a quicker method of descending than their ordinary snail's pace. The foot pours out a lot of mucus, which is passed along to the posterior end. This mucus is then attached to the limb on which the animal is, and then the slug casts itself loose. Its weight draws the mucus out into a fine thread, and, more being secreted, the slug lets itself down after the fashion of a spider, with this exception; it has not the power of returning to the point of support. This power of forming a thread has been observed in almost all the American species, at least when young; but some of the larger forms, when adult, are too heavy to trust their weight to such a slender support.

We have spoken of their ravages in gardens, but in America they have not yet become such a pest as in Europe. There they are classed along with caterpillars, locusts, and rats, and a war of extermination is waged against them. In older times the power of the church was invoked against them, but prayers and anathemas failed to cause their extinction, or in fact any appreciable diminution of their numbers. There is another aspect which should not be passed by without mention. Slugs have long been supposed to have medicinal qualities, the rudimentary shell being regarded as especially efficacious. This belief can hardly be regarded as extinct, as Mr. Binney says that "during the year 1863, a syrup of snails was prescribed to members of my family, by two regular French physicians in Paris." During the middle ages, when superstition ran riot, of course they were much more highly esteemed. The shell was regarded as an amulet, protecting the wearer against certain diseases and witchcraft, while the liquid obtained by their distillation was used to improve the complexion. In Europe they are eaten, but in America neither dietetic nor magic qualities have been assigned to these loathsome appearing animals.

The Limacidæ are divisible into three sub-families. In the first, the Tebenophorinæ, the mantle covers the entire back, and no shell is present. Our only species is *Tebenophorus carolinensis*, a sluggish, inactive form found in the woods, usually under the bark, or in the interior of decaying logs. It varies considerably in color, from nearly white without spots to white with brown blotches or black spots, and to blackish gray. It reaches a length of about four inches when fully extended, though at such times the head is not projected beyond the mantle.

In the Arioninæ the shell may be present, though concealed by the mantle, or it may be represented by a number of calcareous grains scattered through the corresponding portion of the mantle, a condition which recalls the embryonic condition of

Limax, as shown in Fig. 348 on a preceding page. In the principal genus, *Arion*, there is a triangular pore at the upper posterior part of the body, which readily separates it from *Limax*. The only species in the United States which undoubtedly belongs to this genus is *Arion fuscus*, which has been introduced from Europe into Boston, where a colony has existed for many years. It lives in gardens, and occasionally strays into cellars and other dark places. It is not known elsewhere in America. In Europe it is a common species, and its eggs are said to be phosphorescent, shining in the dark for several days after being laid. In color this species is whitish or grayish, sometimes tinged with brown. It reaches a length of about two inches.

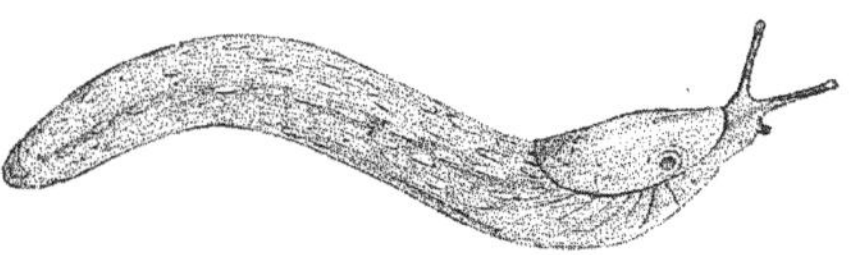

FIG. 391. — *Arion fuscus*.

Three other genera of Arioninæ, *Ariolimax*, *Prophysaon*, and *Hemphillia*, are found on the Pacific coast.

The last sub-family, the Limacinæ, embraces the largest proportion of the slugs, the typical genus, *Limax*, containing about one hundred species. This is the only genus represented in the United States, where, besides our native species, we have several introduced from Europe. Our largest species, *Limax maximus*, is one of these immigrants, which has been found in several places in America. Its rich brown or black spots and stripes upon an ashy or light brown groundwork make it a conspicuous form.

Another imported species is *L. flavus*, brown or brownish in color, with lighter spots. This is more common than *L. maximus*, and is found in various Atlantic cities from Boston to Charleston. It lives in cellars and in gardens, preferring the former. Still more common is the smaller *L. agrestis*, which is also an introduced form. It is smaller than the others, and is extremely variable in color. It lays more eggs than the two species mentioned, and the period of reproduction appears to last through the warmer months of the year. Our native *Limax campestris* is very common, and is found in the woods and the open fields, along the sides of the roads and in gardens. It is brownish gray or amber colored, and is smaller than the other species mentioned. The eggs are rather numerous and transparent, and are laid under leaves or in moist earth. Dr. E. L. Mark has studied the earlier stages of the development of this species; a later stage is shown in Fig. 348.

FIG. 392. — *Limax flavus*.

Another genus, *Phosphorax*, which is very imperfectly known, comes from the Cape Verdes. The only species is said to be phosphorescent, as is indicated by both its generic and specific names (*P. noctilucens*).

ORDER III. — ZYGOBRANCHIA.

All of the gasteropods which follow belong to the Streptoneurous group, the characters of which were detailed on a preceding page. In the first division, the Zygobranchia, the torsion of the body has not been accompanied by an atrophy or disappearance of the organs of the primitive left side, and we thus have the gills and openings of the renal glands of both sides remaining, thus showing that the group is more primitive than those which follow it. Another fact that also emphasizes this inferiority is the absence of distinct genital ducts, the products of the reproductive organs escaping by the larger renal opening. The lingual ribbon is well developed.

The family HALIOTIDÆ embraces the forms which are familiarly known as ear-shells, and to which the local terms ormer and abalone are applied in the Channel Islands and southern California respectively. The shell is spiral, the body whorl being flattened and very large. The dorsal surface of the shell is perforated by a line of openings through which pass a series of tentacular processes from the mantle. As growth proceeds, these are closed up posteriorly. The eyes are on short stalks. The forms are mostly tropical and semi-tropical in their distribution, and are extensively collected for their beautiful shells, which are an article of commerce. The shells furnish a large proportion of the mother-of-pearl, especially that used in inlaying *papier maché* ornaments. In France and the Channel Islands, ormers are used as an article of food, but on account of their toughness they require pounding and mashing before cooking.

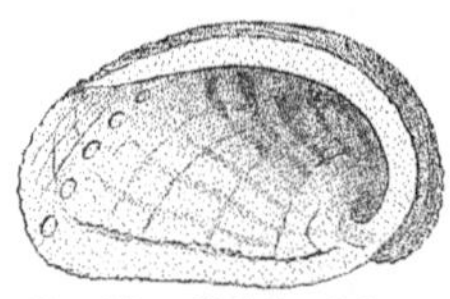

FIG. 393. — *Haliotis*, abalone.

The FISSURELLIDÆ, or key-hole limpets, are structurally closely allied to the last family, but in external appearance they seem far different. The shell is conical and shows but very slightly any spiral. The series of openings of the *Haliotis* are replaced by a hole at or near the apex of the shell, or by a notch in the front margin. On the inside of the shell is a horseshoe-shaped impression, indicating the surface of attachment of the muscles of the foot. The eyes, instead of being placed on stalks, are scarcely elevated above the surrounding surface. Like the members of the last family, the species are largely inhabitants of the warmer seas of the globe, although some forms are boreal in their range. They are mostly found near the shores, where they feed upon the smaller seaweeds. In their habits they are not different from the other limpets.

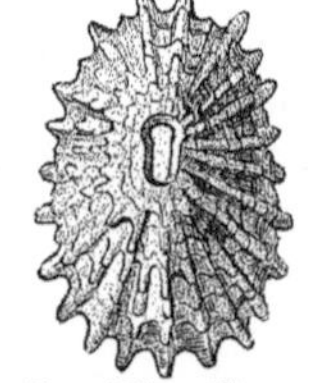

FIG. 394. — *Fissurella nodosa*, key-hole limpet.

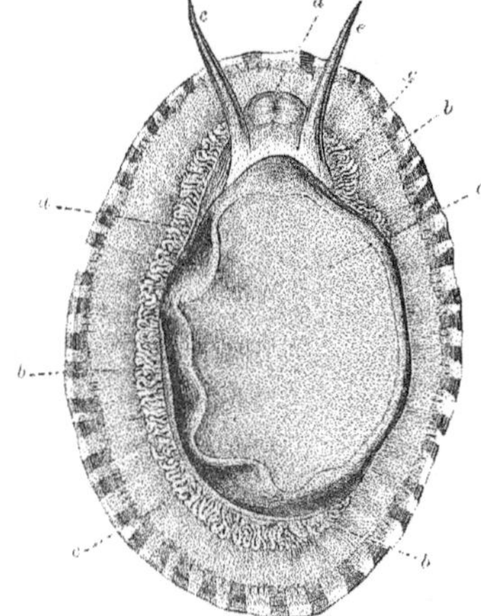

FIG. 395. — Under surface of *Patella algira*; *a*, foot; *b*, edge of mantle; *c*, gill; *d*, head; *e*, tentacles.

The third family of the Zygobranchia, the PATELLIDÆ, is apparently far different from the other two in the structure of the gills, and the fact that it really should have a place here is shown by one of the neatest bits of morphological logic with which we are acquainted. On the first examination of a *Patella* we find a respiratory organ in the form of a circle just beneath the mantle, while the branchiæ above the neck, comparable to those of *Haliotis* and *Fissurella*, are absent. Spengel, however, found in this region two little prominences, the homologies and functions of which were obscure. The thought, however, suggested itself that these might be the rudiments of the true gills and an anatomical examination showed that this view of the homology was correct. It will be remembered that the typical gill is innervated from the visceral loop of the nervous system, and that near its base is a patch of olfactory epithelium (see p. 250). Dissection of a *Patella* showed that these prominences received their nerve supply from the visceral loop, and near each was found an olfactory organ, thus making the homology complete and indisputable. The circular functional gill is therefore a superadded structure which has arisen in a manner and from some cause not yet explained.

Not all limpets have been shown to belong in this place, although future investigation may demonstrate that they should be classed here, a view which is strengthened by the similarity in lingual dentition. Still, for convenience, it will be well to consider the ACMÆIDÆ together with this family, as in general appearance and in many important points of structure they are closely similar. In these forms there is a single cervical gill, while the circular marginal gill may be either present or absent. The prominent genera are *Acmæa*, *Lottia*, and *Scurria*. In *Lepeta* no gills are found.

FIG. 396. — *Acmæa testudinalis*, limpet.

The shell in the limpets is conical, usually considerably depressed, and is so characteristic as to have given rise to the adjectives patelliform and limpet-like. The apex points forward, and the internal horseshoe-shaped muscular impression, like that of the Fissurellidæ, is open in front. A large number of genera and sub-genera have been made, the characters resting upon the respiratory organs, shape and ornamentation of the shell, etc.

As ordinarily found, the limpets are attached to some rock or other object by their broad foot, and the strength with which they hold is astonishing. If the collector approach the animals suddenly, and with a quick motion slides them from their attachment, he can get them easily, but if an incautious touch gives them warning of danger, the shell will not infrequently break before the animal loosens its hold. The strange story is told that each limpet has its own abiding place. At the time of high tide he wanders off to find pastures of algæ suitable for his palate, but as it ebbs he returns to his chosen spot, and at low tide clings fast to the same spot on the rock which he left a few hours before.

On the European shores, limpets play an important part in the diet of the people living near the shores, but on our coasts, except a very few used as bait, they have no economic importance. Why it is that our people neglect so many articles of food it is impossible to say. In the northern states, shrimps, limpets, periwinkles, mussels, etc., are scarcely touched; yet the sea teems with them, and everyone who has tried them bears witness to their palatability. In the case of the limpets it may be that the comparative scarcity on our coasts may be the cause of their neglect.

FIG. 397.—*Acmæa alveus.*

Acmæa testudinalis, the most common limpet on our northern coasts, belongs to the family Acmæidæ mentioned above. Its position here is extremely doubtful, and aside from the fact that its branchial system has never been studied, our only excuse for mentioning it here is to treat of all the limpets together. This species is very variable in color, but is usually variously mottled with brown, pale green, and white. Like most of the limpets it lives between tide-marks. *Acmæa alveus* is a variety of the foregoing, but, from the habit that it has of living on eel-grass, it has acquired a narrower shell than the typical form.

ORDER IV. — SCUTIBRANCHIA.

All the remaining Gasteropoda contrast with the Zygobranchiata in the fact that the torsion of the body has caused the obsolescence or abortion of one of the true gills, and for this reason Dr. Lankester has arranged them under one ordinal head Azygobranchia. When, however, we take other characters into consideration, it becomes necessary to divide up this large group, and in the following pages the Scutibranchia, Ctenobranchia, and Heteropoda together equal the Azygobranchia of that able English morphologist.

The first family, the **TROCHIDÆ**, are commonly known as top-shells, the shell of the typical forms, when inverted, being strikingly similar to the plaything of our youth. The shell is spiral, and is either pyramidal or turbinated, and has a nacreous interior. Many of these shells are sold as ornaments, after the epidermis and external layers of the shell have been cut away, leaving the whole a mass of mother-of-pearl. The animal has long and slender tentacles, and at the bases of these arise the peduncles which support the eyes. The head, and sides of the body, are ornamented with fringed lobes and longer tentacles. When the animal withdraws into its shell, it closes the aperture with an operculum, which may be either horny, or calcareous with a horny base. As the animal increases in size, the operculum also grows by additions which are arranged in a spiral. When crawling about, the animal carries its operculum on the dorsal surface of the foot, as do all operculated gasteropods. Some of the opercula of the smaller species are in great repute as eye-stones. Their whole value in this respect is due to the fact that they have no irregularities which would injure the cornea or the inner surface of the eyelid. Otherwise they are no better than any other hard substance of similar shape and size. The physiology of their action is readily understood. The species live in shallow water near the shore and are herbivorous in their diet.

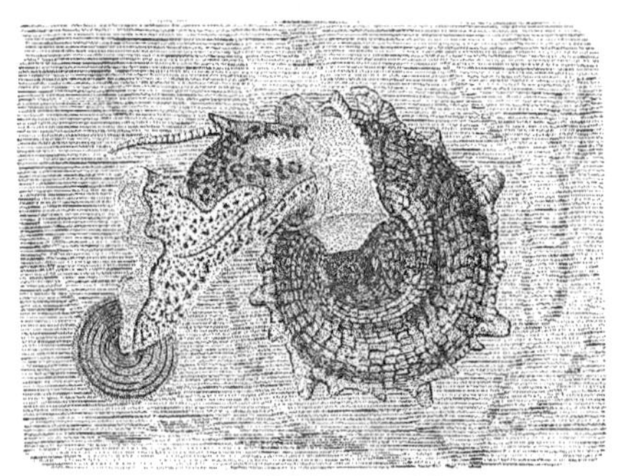

FIG. 398. — *Delphinula laciniata.*

In *Trochus*, the shell forms a regular pyramid, and the base is flattened. The whorls are flattened, the aperture is oblique, and the operculum is horny and multispiral. Of this genus and its various sub-divisions, over two hundred and fifty species have been described. In our northern waters, these forms are represented by several species of the genus *Margarita*, which in many respects is intermediate between *Trochus* and *Turbo*. The whorls of the shell are more ventricose, or swollen, than in *Trochus*, and the thin epidermis allows the pearly shell to be readily seen. The species are found from extreme low-water mark to a depth of one hundred fathoms and over.

FIG. 399. — *Trochus zizyphinus.*

FIG. 400. — *Margarita.*

In *Turbo* the whorls are ventricose, the aperture large and rounded, and the operculum calcareous; the base of the shell is never flattened. The species are mostly tropical and littoral, delighting in rocky coasts where they are exposed to the force of the waves. In the Orient the larger species are eaten. The largest species known is *Turbo marmoratus* of the Chinese Seas. In *Delphinula* the shell is depressed, the aperture round and pearly, the umbilicus open, the operculum horny, and the whorls of the shell are usually spiny. The genus is found on the coral reefs of the Indo-Pacific Seas, near low-water mark. Our figure shows the under surface of the shell, with the body extended.

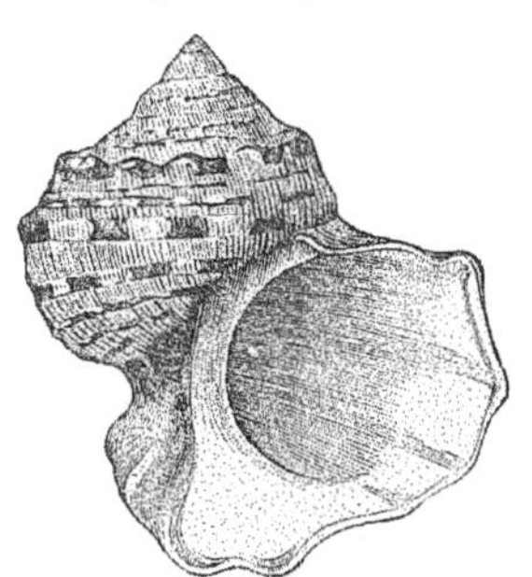

FIG. 401.—*Turbo marmoratus.*

Phasianella contains species which have somewhat the shape of the genus *Bulimus* among the pulmonates. The shell is not pearly but is richly colored; whence the name pheasant shells. About forty species are known, all from tropical seas. Those from Australian and New Zealand seas are large, reaching occasionally a length of about two inches, but those from other parts of the world are smaller, our West Indian forms being very small. *Rotella* contains a number of brightly-colored depressed species from the eastern seas. In *Monodonta*, which is much like *Turbo* in general appearance, the outer lip is much thickened and grooved, while the columella is toothed. It has about the same distribution as the last species. In the Malay Archipelago one of the species is eaten, notwithstanding its peppery taste.

The NERITIDÆ contains thick hemispherical shells with a very small spire, a sharp outer lip, and a calcareous operculum which is frequently irregular in shape. The eyes are placed at the extremity of the slender eye-stalks, which arise from the head outside the long and slender tentacles. The foot is broad and triangular, the apex being behind. As the animal grows, it absorbs the inner part of the whorls of the shell, so that the resulting cavity is simple instead of spiral. The typical genus is *Nerita*, which has a thick or spirally grooved shell. The columella is much thickened and toothed, and in one species, *N. peloronta*, this columellar thickening is ornamented with a blotch of red, giving the shell the common name of bleeding tooth. Most of the species are marine, but many ascend the streams entering the ocean to such a distance that the water in which they live is brackish.

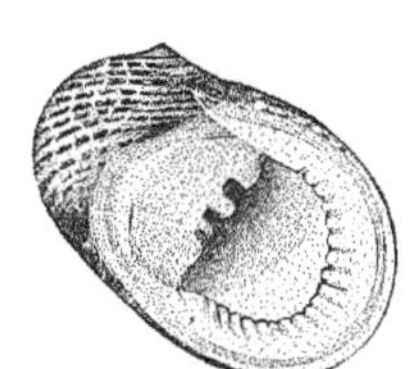

FIG. 402.—*Nerita histrio.*

Neritina is much like *Nerita*, but is more globular. The shells are variously ornamented with spots or bands of black and purple laid upon the polished exterior. The species are mostly confined to the fresh waters of the warmer regions of the earth, but some species are found in the sea. *Navicella* is more like the slipper limpets (*Crepidula*) in appearance, the aperture embracing nearly the entire shell. They are fresh-water forms, and the resemblance to the limpets is strengthened by their mode of life as they attach themselves by their foot to submerged stones and plants.

The family PLEUROTOMARIDÆ shows resemblances to both the Trochidæ and the Haliotidæ. The shell is much like that found in the latter family, except that the outer lip of the aperture is notched, or there is a series of perforations in the upper part of the whorl. The species are largely fossil, the living forms being few in number and comparatively rare. In most of the species the notch in the aperture of the shell is closed behind as growth progresses, and the result is that each whorl receives a band, which is quite distinct from the rest of the shell. In some of the fossil species this is very marked, and stands up elevated to a considerable distance beyond the surrounding surface; in others the closing up is not complete, and the result is that there remains a series of holes like those of the abalone (*Haliotis*). Recent deep-sea explorations have largely increased the number of known species of this family, most of which are apparently inhabitants of water from four hundred to a thousand fathoms in depth.

The HELICINIDÆ is a family of terrestrial gasteropods variously placed by different naturalists. Most commonly the members are placed among the true land shells (pulmonates), but with these they have little or no affinity. We follow Claus in assigning them their present position. In appearance of the shell, in the structure of the lingual ribbon, as well as in their habits, they are much like *Helix*, living as they do upon the land, either concealed under the dead leaves on the ground, or among the branches or foliage of the trees. They differ however, from the Helicidæ, among other important points, in the possession of an operculum. The aperture of the shell is semilunar, and the umbilicus is covered by a callus. The species are all tropical, and are mostly confined to the American continent and the West India Islands. Only a few species out of the five hundred known are found within the limits of the United States. The prominent genera are *Helicina*, *Stoastoma*, and *Proserpina*.

FIG. 403. — *Helicina variegata.*

ORDER V. — CTENOBRANCHIA.

Most of the members of this group despise a vegetable diet and prefer to live on animal matter whether living or dead. Still, some exceptions occur which will be noticed in the proper places. In all, the shell is spiral, and the gill of the normally right side is alone present. What has previously been considered as the rudimentary gill of the left side has been shown by Spengel to be the highly developed olfactory organ. The gills have a comb-like shape and the axis is frequently attached to the roof of the branchial chamber, which by that torsion of the body described on a preceding page, is brought above the neck of the animal. In many, a well-developed copulatory organ is found on the right side of the neck, and the proboscis may or may not be retractile. The former condition of affairs has given rise to a group called Proboscidifera, containing the families Tritonidæ, Doliidæ, and part of the Muricidæ as here limited, while another group, containing the Cypræidæ, Velutinidæ, and Naticidæ, have the rostrum invertible only at the tip.

The Ctenobranchia is divided into four sub-orders, the distinctions being largely founded upon the arrangement of the teeth upon the lingual ribbon, although other characters are of course employed.

SUB-ORDER I. — PTENOGLOSSA.

In this group the shell has the aperture entire; that is, the whorls are complete and the edges of the aperture are not notched or prolonged into canals. No respiratory siphon is formed, and copulatory organs are absent. The tongue is armed with numerous small teeth on either side, but lacks the normal middle row.

The IANTHINIDÆ are remarkable for the beautiful purple color of their thin shells. They are pelagic, oceanic snails, which lead a predaceous life. At times on the high seas the navigator encounters vast numbers of them, forming immense schools, and feeding upon the other forms of life; medusæ, Crustacea, etc., with which they are surrounded. The animal has a large head furnished with an extensible proboscis. The eyes are minute and situated on the extremities of the ocular peduncles, while the foot is small and divided. The shells are thin and delicate, the whorls of the spiral being few in number. At the base they are of a deep violet color, but the apex is nearly or entirely white.

One of the most interesting features connected with these shells is the enormous float which they form to support the eggs. The foot secretes a glutinous secretion which hardens to a slight extent when brought in contact with the water. During the reproductive season the formation of this egg float is continuous, and, as it is formed, eggs are fastened to its lower surface. From this mode of formation, that part of the float farthest from the animal contains the most advanced eggs; and, in fact, the eggs in this portion may have hatched and the embryos have begun their free life ere those

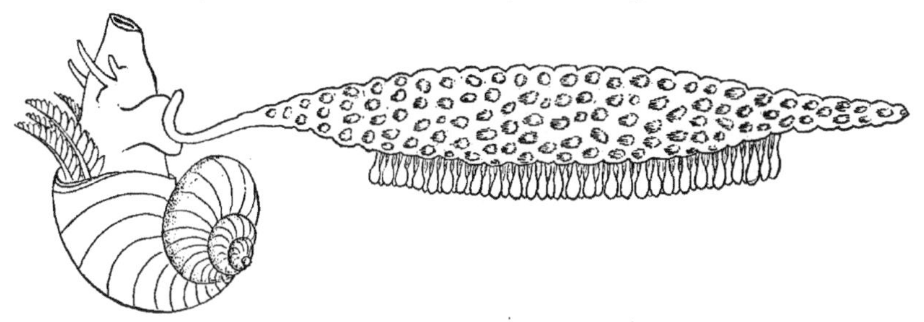

FIG. 404. — *Ianthina*, purple shell, with the float supporting the eggs.

nearest the body have passed through the earlier stages of development. Although the parent usually carries the float attached to the body, still it has apparently the power to cast it off at will, while the action of storms usually separates the mother from the egg. When thus cast adrift, the float still sustains the eggs, and they pursue their development as usual. Each egg is fastened to the float by a short peduncle, while the float itself is composed of numerous little bubbles, thus securing great buoyant powers.

The *Ianthinæ* do not appear to have the power of sinking in the water unless the float is detached, and so at the time of storms they are frequently cast upon the shores in large numbers. At such times they are utterly helpless and make no attempts to crawl. They, however, frequently adhere to each other by means of the foot, and, when handled, secrete a violet-colored fluid.

Ianthina, the most prominent genus, contains about ten species, one of which (*I. fragilis*) is occasionally thrown up on the southern New England coasts by severe southeast storms. It is not properly a member of the American fauna, but like the rest of the genus is an inhabitant of the high seas. The only other living genus of the family (*Recluzia*) is covered with a brownish epidermis. Like *Ianthina*, it forms a float.

The SOLARIDÆ embraces a group of molluscs which, from the shape of the shell, was formerly included in the Trochidæ. The shell is orbicular and forms a more or less flattened cone, usually perforated by a wide and deep umbilicus. The shells are not nacreous, and a horny spiral operculum closes the usually angular aperture. The animal has sessile eyes, long, retractile proboscis, and the gill chamber divided into two parts.

Solarium perspectivum has received the common names of perspective shell and sun-dial shell. It is about two and a half inches in diameter, and is of a yellowish hue, prettily spotted and banded with red. It comes from the Indian Ocean, and specimens are found in most collections.

The only other genus which needs to be mentioned is *Phorus*, which embraces the carrier or mason-shells of the eastern seas. These forms have the habit (if habit it may be called) of covering their shells with all sorts of extraneous objects,—shells, stones, bits of coral, and the like. These foreign bodies are fastened by the substance of the shell and doubtless are protective; for, viewed from above, a shell thus tricked out has but slight resemblance to a properly conducted mollusc, and thus runs a better chance of escaping the maw of the bottom-feeding fishes. Now that this peculiar habit exists, we can readily see how it is retained, but the way in which it was first acquired is not so readily explained.

The mode of progression of the mason shells is rather peculiar. Most of the gasteropodous molluscs have a gliding motion, the various parts of the foot acting in a manner best described by comparing it to the locomotion of a thousand-legged worm. The *Phori*, on the other hand, have a gait like that of a measuring worm. They extend the small cylindrical foot, attach the anterior portion, and then draw the hind portion forward. This latter now affords a foothold; the anterior portion is again extended, and the operation is repeated. This gives rise to an interrupted, almost jumping movement, well adapted to the banks of broken coral and dead shells inhabited by these animals.

The SCALARIDÆ, or wentle-trap family, embraces but a single genus and about a hundred and fifty species, distributed through all the seas of the world. The common name is a corruption of the German word for a spiral stairway, and would be eminently appropriate, were it not for the fact that the tread of the stairs goes the wrong way. The shells are usually pure white, and composed of several rounded whorls ornamented with transverse ribs, roughly corresponding to the steps of a flight of stairs. The aperture of the shell is round and the edge continuous, the inner lip not being formed by the columella. The active, predaceous animal has a retractile proboscis, and in the existence of a rudimentary siphonal fold shows an approach to the members of the next sub-order. The eyes are near the outer bases of the slender, pointed tentacles. Several species of *Scalaria* are found on our New England coasts. Mr. Couthouy kept a specimen of *S. grönlandica* in confinement; it was rather sluggish in its movements, and fed eagerly on fresh beef, especially if somewhat macerated. Some of the species are said to secrete a purple fluid.

FIG. 405.—*Scalaria pretiosa*, precious wentle-trap.

The most noted member of the family is *Scalaria pretiosa*, the precious wentle trap. This species, which comes from the Chinese Seas, has always been highly valued by collectors on account of its rarity, and a single specimen has in times past been sold for about two hundred dollars. Now they are much more common, and are sold by dealers for an average price of one or two dollars.

SUB-ORDER II. — RHACHIGLOSSA.

The Rhachiglossa are all predaceous marine snails with well-developed proboscis and a respiratory siphon, which, when the animal is extended, lies in a notch in the aperture or in a long, more or less tubular canal of the shell. The tongue is long and small, and bears at the most but three teeth in a transverse row, one rhachidian and one lateral on either side, the latter being occasionally reduced to mere hooks, or, as in the case of the Volutidæ, they have entirely disappeared, and the central or rhachidian teeth alone remain. All are predacious and carnivorous.

We have just referred to one of the characters of the VOLUTIDÆ, but now we may give some others derived from other parts. The shell is thick and heavy, and the spire is short, rising but little above the body whorl. The anterior margin or base of the aperture is deeply notched for the respiratory siphon, while the columella bears strong spiral plaits or folds. The eyes are placed at the base of the tentacles, and the foot is large and broad. The typical genus is *Voluta*, in which the spire is short, the mouth wide, and the first fold on the columella is the largest. The species are largely from the Indo-Pacific region, although some are found in other seas. One of the most interesting species is that figured. It derives its specific name, *musica*, from a number of fine dark lines interspersed with blotches, which follow the whorls of the shell and bear no distant resemblance to written music. In some the similarity is more marked than in the specimen figured. This species presents an exception to most of the Volutes, in having a small operculum developed. It comes from the West Indies.

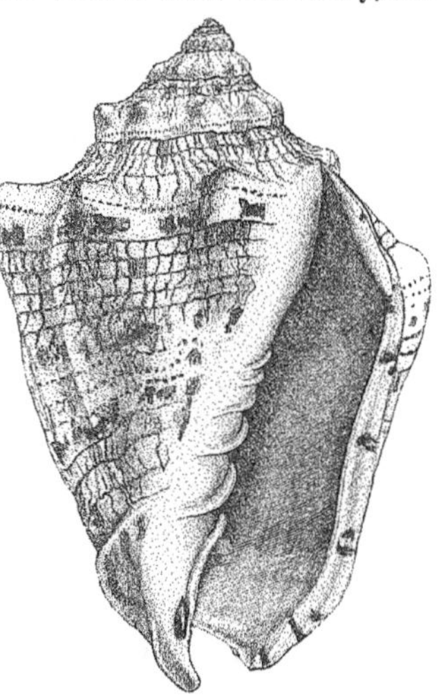

FIG. 406. — *Voluta musica.*

As an example of the forms from the eastern seas we may mention the beautifully shaped species from the Philippine Islands, which, from its diadem of spines and its size, well deserves the name, *Voluta imperialis*, which science has given it. It is common in collections. The last species which we can mention is the rare *Voluta junonia*, or, as it is called by dealers, the peacock-tail volute. The figure represents the shell of the natural size; it is white, spotted with orange. For many years it was considered among the rarest of shells, specimens having been sold for about two hundred dollars, and no later than 1876 a specimen brought fifty dollars. Recently quite a number have been brought from the West Indies, and now they can be bought of the dealers for eight or ten dollars.

Most of the Volutidæ are ovo-viviparous; that is, they bring forth living young: but some, if not all, of the genus *Voluta* lay eggs, which are enveloped in a perfectly transparent, corneous corpuscle half as large as the parent. *Cymbium* and *Melo* bring forth their young alive, a brood containing four or more individuals.

In *Marginella* we have some two hundred species of small, polished oval shells with the respiratory notch small. Most of them are brightly colored, and in life the markings of the body are even more beautiful than those of the shell, as is shown by the following description of the colors of a Javanese species, — "a pale, semi-transparent, pinkish-yellow mantle, with a range of semi-elliptic crimson spots around the

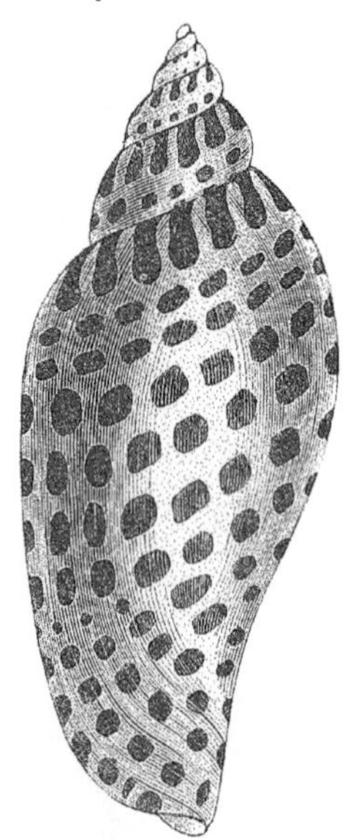

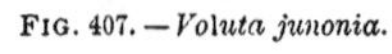

FIG. 407. — *Voluta junonia.*

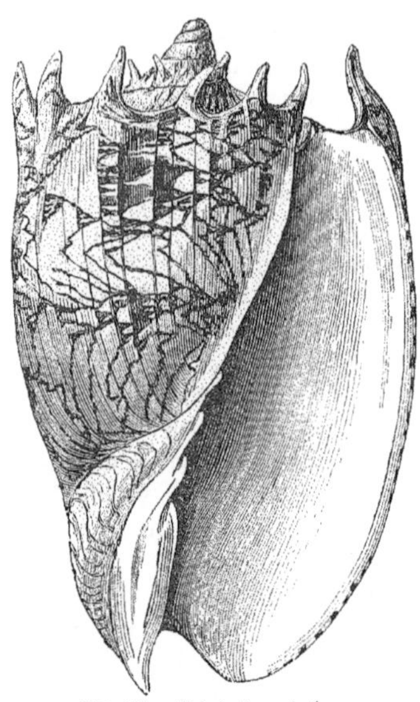

FIG. 408. — *Voluta imperialis.*

thin free edge, and the remainder covered with vertically-radiating linear spots, and short waved lines of the same color; the foot, also of a yellowish, delicate pink, is marbled all over with the deepest and richest crimson, and the same with the siphon. The tentacles are yellowish, with a row of marbled crimson spots." Species of *Marginella* are found in all the warmer seas of the world, some being found in the West Indies and on the coasts of Georgia and Florida.

The olive shells, belonging to the family OLIVIDÆ, have always been favorites with collectors on account of the beauty of their smooth and polished porcellanous shells. In these forms the spire is short, the aperture deeply notched, the columellar lip is covered with a callous deposit and usually ornamented with oblique folds. In the genus *Oliva*, which receives its name from a shape somewhat like that of an olive, the aperture is long and narrow, and the columellar lip is plicate. The foot is very large and is laterally extended into two lobes, which, when the animal

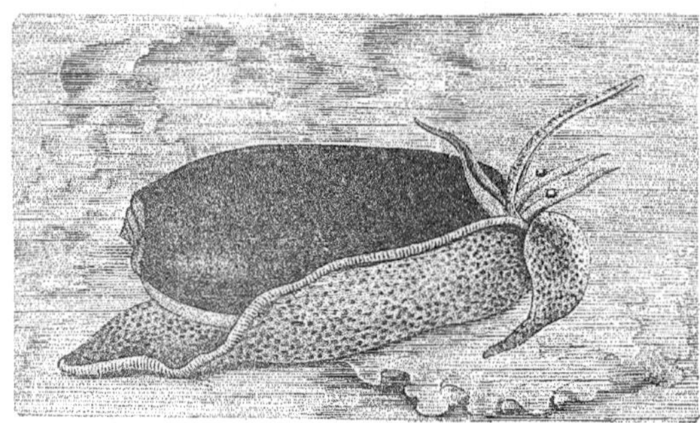

FIG. 409. — *Oliva maura.*

is in motion, are folded up over the shell. The proboscis is short, the siphon long, and the eyes are placed at about the middle of the tentacles. The eighty and odd known species all come from the tropics, a few only extending their range outside. They are all active, predaceous forms, and in some localities are caught by lowering a net with a piece of meat inside as a bait. In some places the number of specimens is almost infinite; at low tide miles and miles of flats are covered with them.

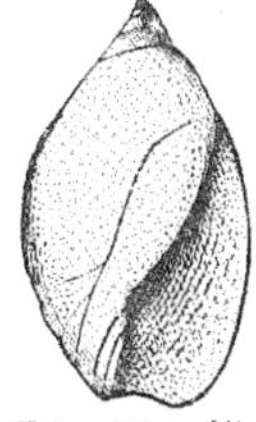

FIG. 410. — *Olivella biplicata.*

One of the most common species in collections is a little white form belonging to the sub-genus *Olivella*; from its resemblance to a grain of rice it has received the specific name *oryza*. It comes from the West Indies, in some parts of which it occurs in vast numbers. The sub-genus *Olivella* is distinguished from *Oliva* proper by the longer spire of the shell and the absence of tentacles and eyes. *Olivella biplicata*, which is figured, comes from the Pacific coast.

Of the true *Olivas*, the most common species in the southern United States is *O. litterata*, marked with angular markings, which by a stretch of the imagination might be regarded as resembling writing. The general color is a yellowish white, the markings brownish. Another lot of shells are named, from their resemblance to certain rocks, *jaspidea*, *porphyria*, etc. The latter species comes from Panama.

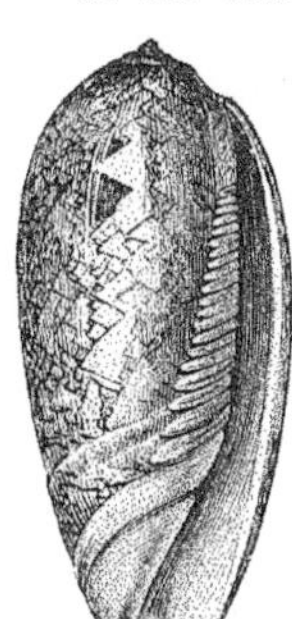

FIG. 411. — *Oliva porphyria.*

The harp-shells (genus *Harpa*) differ markedly from the other members of the family by their broad aperture, and swollen, transversely-ribbed whorls. Although only nine species are known, every collection contains several specimens, those of *Harpa ventricosa* being possibly the most common. Large as is the shell, it is not sufficient to contain the whole animal, and Semper, as well as the older naturalists, record a peculiar habit of self-mutilation with some of the species from the eastern seas. When captured, a part of the foot remains outside of the shell. This the animal brings across the sharp edge of the aperture, thus cutting it off. In time a new portion grows out, replacing that which was amputated. The animals are lively, and bright colored. They are found in all the tropical seas, except the Atlantic. *Harpa ventricòsa* comes from the East Indies, *H. imperialis* from Bourbon, while *H. crenata* and *H. scriba* are found at Panama.

In the MITRIDÆ but a single genus, *Mitra*, needs mention. Here the shell is thick, long, and fusiform, the spire being well developed and the columella plicate; the aperture is narrow, notched anteriorly, and in some species is partly closed by a small horny operculum. *Mitra episcopalis*, possibly the most common member of the genus, is ornamented with spots (usually quadrangular in outline) of red, salmon, or orange. It comes from the Philippine Islands, where it moves rather sluggishly over the flats, especially when the tide has just begun to come in. When the tide recedes, it buries itself just beneath the surface. Some of the smaller species are more lively, and others crawl about on the surface of the sand when the tide is out. Although frequently seen in the day-time, the *Mitras* are essentially nocturnal, and spend most of the hours of daylight hidden under rocks and in holes in the coral reefs. Some species, when irritated, defend themselves by secreting a purple fluid, the odor of which is said to be very nauseating.

Another eastern species, nearly or quite as common as the one mentioned, is *M.*

papalis, which has each whorl of the spire crowned with knobs, and the small colored spots of the shell much more irregular in shape and distribution than in *M. episcopalis*. The genus *Mitra* contains over two hundred species, a large proportion of which come from the Philippines and the neighboring seas. Almost all of the genus are tropical and semitropical; several being found in the West Indies; but an exception to this distribution is found in *M. grönlandica*, as its name indicates, an Arctic form, which, on account of peculiarities of its lingual dentition, has been separated as a sub-genus *Volutomitra*.

The family MURICIDÆ, as at present limited, embraces a heterogeneous assemblage of forms. Several attempts have been made to divide it without doing violence to the affinities of one or more genera, but no scheme has as yet received universal acceptance. If we base the division on the lingual dentition, it does not agree with characters derived from the animal and from the shell; if on the anatomy of the animals, still other features are not in accord, etc. With this uncertainty it is best, at least in a popular work, to leave the classification in its present condition, and to define the family as embracing a group of molluscs, in which the foot is broad and of moderate length, the siphon long, the eyes at the base of the tentacles, while the characters derived from the shell are the presence of a long or short, straight, anterior canal, and an oval operculum with the nucleus at the smaller end. Necessarily where so much confusion exists there will be an inequality in the relative rank of certain of the included types, and in the following remarks some genera named will possibly not be worthy of generic rank, while others, on the other hand, may deserve to be regarded as really of the grade of sub-families. The same trouble also occurs with the next family, the Buccinidæ.

The first sub-family, the Muricinæ, is well marked by characters derived from the shell. The growth is apparently marked by periods of rest, and at each of these the aperture is thickened and marked by ornamentations of various kinds. Then the shell grows again, and shortly another period of rest ensues, when the nodes, spines, or thickenings (varices they are called) of the mouth are repeated. These interruptions occur at varying intervals in different species, and are of some use in defining generic limits. A similar process occurs in some other families.

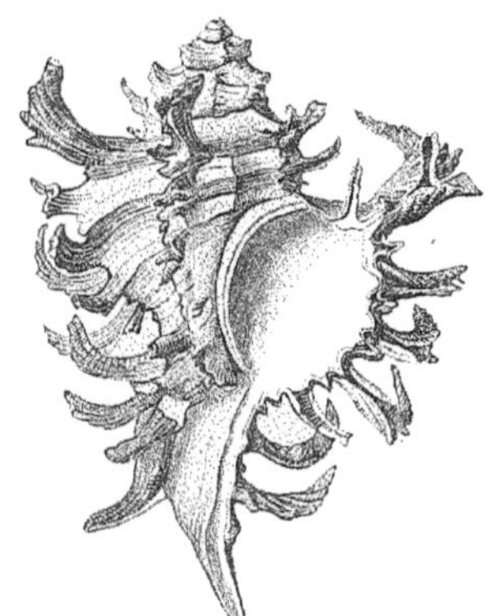

FIG. 412. — *Murex endiva.*

The typical genus is *Murex*, in which the canal is long and straight, the aperture round, and the shell is interrupted by varices and spines at least three times in the course of the growth of the whorl. In the colder waters the colors are subdued, and the shell does not acquire that fantastic form that is frequent in the tropical species. The species are among the most rapacious of molluscs, boring through the shells of other species in the same way as does the *Natica*, to be described on a subsequent page. In Europe, *Murex erinaceus* does great damage to the oyster beds. Allied species (*M. brandaris* and *M. trunculus*) were employed by the ancient inhabitants of Syria and Greece in the preparation of the celebrated Tyrian purple. Of the two hundred and odd species of this genus, we need only mention, in addition to those just referred to, the *Murex tenuispina*, in which the canal is very long and almost converted into a closed tube, while the surface of the shell is armed with very long and slender spines, evidently defensive in their nature. In forms like *M. endiva* and *M. scorpio*, the spines are stouter and broadened at the extremity. None of the species of *Murex* proper extend into the colder waters of our Atlantic coast; indeed the genus belongs largely to the tropical waters of the old world. In their place are found a few small species belonging to allied genera, of which *Eupleura caudata* may be mentioned first. In this species but two prominent varices are formed to a whorl, giving the shell a flattened appearance, a fact which led to its original description under the generic name *Ranella*. The aperture is toothed within, and smaller ridges occur between the well-marked varices. The shell is brown in color, and reaches a length of about an inch. It is rather uncommon on the southern shores of New England, except in certain localities, but farther south it is very abundant.

Urosalpinx cinerea, on our coasts, plays the same destructive part that *Murex brandaris* does in Europe. The fishermen have applied to it the name 'drill,' on account of its settling down on oysters and boring a hole through the shell, through which the soft parts are eaten. The drill is sluggish in its motions. It is about the size of the last species, ashy or brownish in color, and ornamented with ten or twelve undulations on the lower whorl. It lays its eggs in capsules, of about the same size as those of *Purpura lapillus*, to be described in the next family, but differing from them in being flattened and keeled at the edges. Each capsule, on the average, contains ten or twelve eggs. The drill ranges from Massachusetts Bay to Florida; north of Massachusetts it is rare and local; yet a colony exists in the southern part of the Gulf of St. Lawrence, a fact which at once recalls the existence of oysters in the same region.

FIG. 413. — *Urosalpinx cinerea*, drill.

In the Fusinæ the shell is spindle-shaped, and the edge is never thickened so that no varices are formed. *Fusus* contains a number of tropical and sub-tropical forms in which the spindle shape is especially well marked. The northern species, formerly referred to this genus, are now referred to the next family. In the typical forms the columella is smooth, a fact which separates it from *Fasciolaria*, in which it bears oblique folds. *Fasciolaria gigantea*, which occurs on the southern coast of the United States, is the largest known gasteropod, its shell reaching a length of nearly two feet.

The BUCCINIDÆ is closely related to the Muricidæ. The spiral shell, instead of a long siphonal canal, has a notch through which the long siphon is extended. While many of the included forms are very distinct, there are others which can scarcely be separated from the preceding family. This is especially true of a group of boreal shells, represented on our coasts by the genera *Neptunea* and *Sipho*. Here the shell is much like that in *Fusus*, and indeed, the species were formerly included in that genus. *N. decemcostatus* is marked with ten large revolving ribs on the body whorl. *Sipho islandicus* is even more like a *Fusus*. Both are large shells occurring in the cold deep water north of Cape Cod, reaching a length of nearly or quite three inches.

Another problematical genus is *Pyrula*, with its various sub-divisions, *Fulgur*, *Sycotypus*, etc., some of which probably belong here, while others should be transferred to the Muricidæ, the Doliidæ, etc. In all, the shell is somewhat pear-shaped in outline, the spire being short, while the anterior end is greatly prolonged to correspond to the stem. Our two best-known species are *Fulgur carica* and *Sycotypus canaliculatus.* The former is a heavy shell with a short spire ornamented by a row of tubercles. The latter is much more delicate, and is covered with a hairy epidermis, the sutures of the spire being marked with a deep revolving channel. These shells are both inhabitants of water of moderate depth, coming to the shore only for the purpose of oviposition. Their egg-cases, which are very peculiar, are frequently cast on the shore and attract the attention of the most casual observer. They consist of a series of flattened membranous capsules attached by one edge to a cord, and having opposite the point of attachment a more transparent spot, indicating the place where the young are subsequently to make their exit. When laying these long strings, the snail goes beneath the surface and as the ribbon begins to be formed, it appears above the sand, slowly

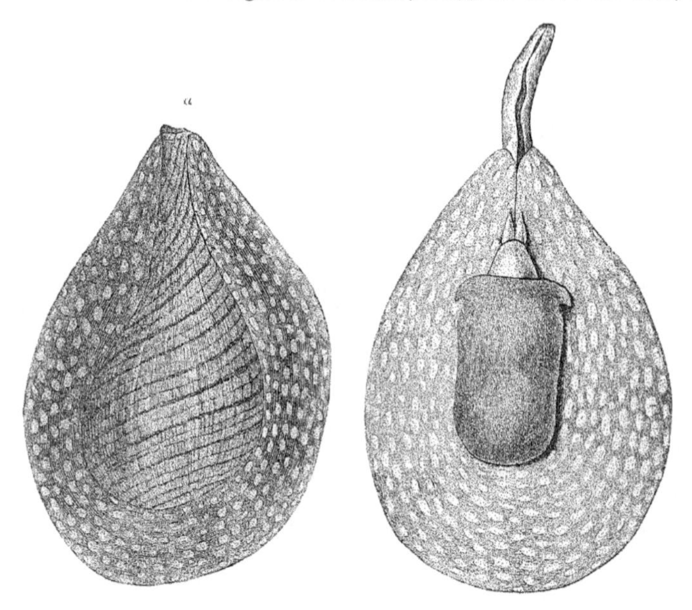

FIG. 414.—*Pyrula decussata*, pear-snail; *a*, dorsal, *b*, lower surface.

increasing in length, until the whole of its two or three feet of extent are formed. The first part of the cord is without capsules for about three inches, then come a few cases imperfectly formed. Each capsule contains a number of eggs, and specimens taken some time after oviposition show the shell formed, repeating in miniature the essential features of the parent.

The name *Fulgur* (lightning) was applied on account of the zigzag brown streaks with which young specimens (and older ones in warmer waters) are marked. *Fulgur carica* is used extensively by fishermen as bait. Another species, living further south, is noticeable from the fact that it is reversed or sinistral, receiving on this account the specific name *perversa*.

With the genus *Buccinum* we take up a series of forms, whose position is less doubtful than some of those just mentioned. *Buccinum* is a northern genus of shells covered with a horny epidermis, having a large aperture, a siphonal notch rather than a canal, a smooth columella, and an untoothed outer lip. The most common species is the whelk, *B. undatum*, common to the northern Atlantic shores of Europe and America. It burrows in the sand below low-water mark. Its eggs are laid in hemispherical capsules, yellow in color, piled up in a heap, and presenting an appearance well described by the name 'sea-corn' applied to them by the New England fishermen. In England they are called 'sea wash-balls' from the fact that they are employed in washing the hands, their parchment-like texture tending to scour away the dirt. Each capsule, when first laid, contains a number of eggs, but of these but few develop, the others being swallowed by the young which have got a little start in development. In this respect, they are like the young spiders.

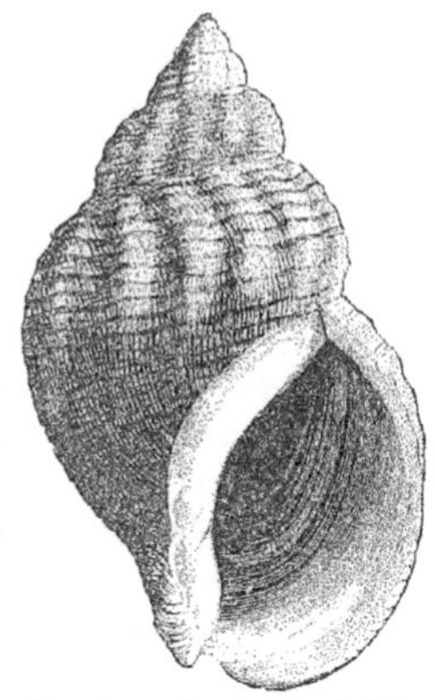

FIG. 415. — *Buccinum undatum*, whelk.

In America, the whelk is not used as food, but in England large numbers are brought to the market, the annual catch at Whitstable (a small village at the mouth of the Thames) being worth, in 1866, £12,000. Although the shell of the whelk is stout and strong, it is eaten in great numbers by the larger bottom-feeding fishes, some of which are furnished with teeth strong enough to crush the shell like a stone breaker, while others bolt shell and all whole, leaving the gastric juices the labor of dissolving the nutritious portions.

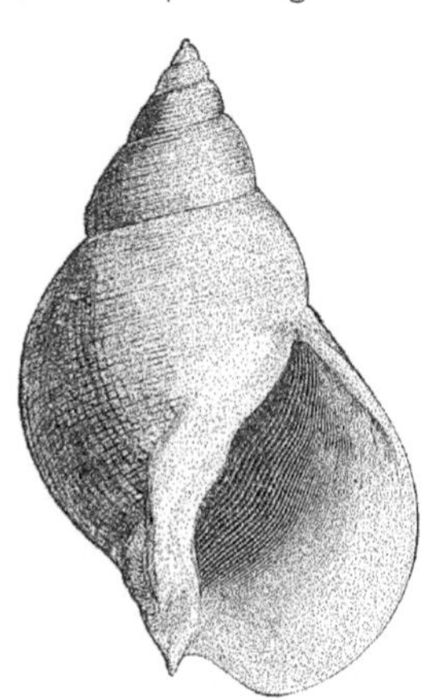

FIG. 416. — *Buccinum ciliatum*.

Besides the common *Buccinum undatum*, several other species are found in the north Atlantic, north of the New England shores. One of these boreal forms, *B. ciliatum*, is figured; the differences between this and the common whelk are evident. Most of the specimens in collections are obtained from the stomachs of fishes caught on the Grand Bank.

The genus *Eburna* embraces the ivory shells, so called from the color and texture of some of the forms. In the dozen oriental species comprised in the genus, the shell is thick, deeply umbilicate, the columella and outer lip without folds or teeth, and the suture between the whorls channelled. The surface of the shell is ivory-white, spotted with an orange red. The animals usually move along at a leisurely pace, but when alarmed they are capable of much quicker motions. They frequent muddy bottoms where the water is ten or twelve fathoms in depth, and are caught in considerable numbers in the nets of the Chinese fishermen, who use them as food.

Nassa contains a large number of species divided up into the sub-genera, *Ilyanassa*, *Tritia*, etc. The general shape and appearance may be seen from our figures, a common character being the tooth or plait at the upper part of the columella, much more marked in some species than in others, and the extensive deposition of enamel on the columellar lip, which not infrequently extends to a considerable distance outside the aperture. Most of the species are littoral, and at low tide our New England flats are covered with myriads of *Nassa trivittata* and *N. obsoleta*. Farther south a third species, *N. vibex*, becomes prominent. The trails of these species are common on the soft mud, and frequently at the end will be found a little pellet of mud beneath which the animal is hidden. All the specimens, however, do not bury themselves at the retreat of the tide, as they are able to live for a considerable time out of water. Possibly *N. obsoleta* is the more common form. In this, the shell is dark brown, and ornamented by a net work of reticulating lines. It does not thrive well where exposed to the ocean surf, but prefers sheltered inlets, extending in large numbers into inlets where the water is decidedly brackish. In size it reaches a length of about an inch. *N. trivittata* is slightly smaller, and white or greenish white in color. The third species, *N. vibex*, is still smaller, reaching a length of half an inch, and banded with ashy white and pale red, the colors being brightest in the southern forms. All of the *Nassæ* are carnivorous, drilling holes through the shell of other mollusks and then feeding on the flesh. They are, however, not confined to living objects, for they will accumulate in large numbers around any decaying crab or fish, and, together with the amphipods, soon devour all the fleshy portions. In Europe an allied species, *N. reticulata* is an enemy of the oyster beds, drilling through the shell in a short time. They usually select the young oysters, but will destroy one three years old in about eight hours.

FIG. 417.—*Nassa trivittata.*

FIG. 418.—*Nassa obsoleta.*

The egg-cases of *Nassa obsoleta* are among the most common of marine objects. They are placed on any solid object that is handy, dead shells and the 'sand-saucer' egg masses of *Natica* being most frequently used. The capsules are curiously fluted and ridged, and are crowded together without order, each attached by its own pedicel.

FIG. 419.—*Purpura lapillus.*

Purpura and its allies are by some placed in the Muricidæ, by others in the Buccinidæ. In *Purpura* the aperture is wide, the spire short, the whorls enlarging rapidly; the columella is flattened, and the outer lip is toothed. *Purpura lapillus*, a dirty white or ashen species, is common to the shores of Europe and North America, thriving better and growing to a larger size in the old world, where specimens are frequently zoned with brown. On our own coast there is much variation in appearance, individuals from the rocky coasts, where they are exposed to the surf, having the ribs of the shell nearly smooth, while those from sheltered localities have them roughened by scale-like projections. This species does not range much south of Cape Cod, but north of that barrier it is very common. It feeds on other animals, being especially fond of the acorn barnacles (*Balanus balanoides*) which flourish between tides. The eggs are laid in small oval capsules supported on slender stalks. Each capsule contains numbers of eggs, only a few of which eventually hatch, the others furnishing food for those that develop.

FIG. 420.—Egg capsules of *Purpura lapillus*, enlarged.

Purpura patula was one of the forms which furnished the famous Tyrian purple, the others belonging to the Muricidæ. The animals were gathered in large numbers and crushed, shells and all, in mortar-shaped holes in the rocks, two or three feet in depth. From the bruised mass a liquid was obtained which was mixed with a small amount of soda and diluted with several times its weight of sea-water. At first the fluid was yellow, but, after exposure for a time to the rays of the sun, it changed to purple. Then the wool was dipped into the dye for a few hours and taken out colored. During the change in color a fetid odor like that of assafœtida is given off. To obtain the finest color a mixture of two species of *Purpura* or *Murex* in certain proportions was employed. For a long time the art of coloring with the secretions of molluscs was entirely lost, but at the close of the middle ages it was rediscovered. Modern chemistry has, however, replaced it, and now the use of molluscan dyes is nearly or quite extinct.

In *Concholepas peruviana*, the only species of the genus, the body whorl increases so rapidly in size as to render the shell much like that of a limpet. This species occurs along nearly the whole of the western coast of South America, and is extensively eaten by the Chilians and Peruvians. The flesh is tough, and is beaten to make it more tender. The species of *Rapana* live upon coral reefs, feeding upon the polyps. The genus *Rhizochilus* is noticeable from the fact that the young of one species has a well-formed shell much like that of *Rapana*; but as the adult condition is reached it cements to the shell, branches of the coral *Antipathes* or other shells, or both, until at length all means of communication with the exterior is by means of the siphonal canal, the aperture being completely closed. What is the cause of this peculiar self-immurement no one has yet been able to decide. Other species of the genus are not known to possess such habits as those just described of *R. antipatharum*. In *R. madreporarium* the animal attaches itself to the larger reef-building corals by means of the foot.

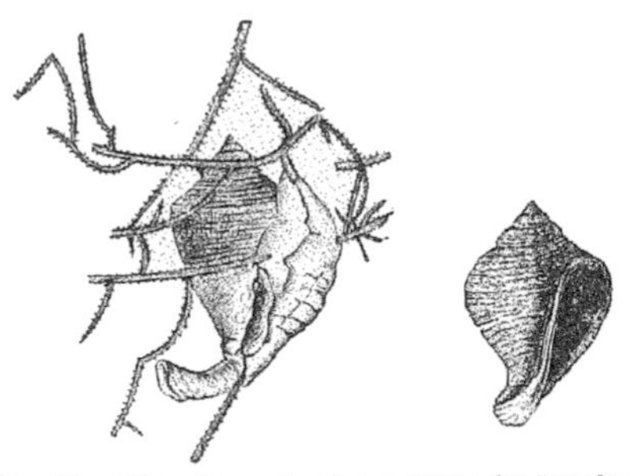

FIG. 421.—*Rhizochilus antipatharum*, fastened to branches of *Antipathes*; on the right a young specimen.

Another interesting genus is *Magilus*, the species of which all belong to the eastern seas. It is a fine example of that degeneracy which occurs in certain molluscs. The young *Magilus* begins life as a well-behaved mollusc with a regular spiral shell; but shortly it settles down on some growing coral, and then a race begins between the two slow-growing forms. If *Magilus* kept quiet and grew no further, a short time would suffice to completely envelop him in the stony coral; but, as soon as he is partially covered, the whorls of the shell leave their spiral course, and grow out as an irregular tube. As the coral grows, new additions are made to the shell, and the neck-and-neck race is kept up until the mollusc or the coral dies. Soon the tube becomes too long for the mollusc, and he leaves the spiral portion and comes out to live in the outer straight tube, filling up the deserted whorls with a solid deposit of lime.

In the COLUMBELLIDÆ the shell is oval, the spire moderately short, the aperture narrow, and terminated by a very short anterior canal; the outer lip is thick and internally crenulated, while the columellar lip is toothed. The species are mostly small, and many are brightly colored. On our eastern coasts several species occur, among them *Columbella avara, lunata, ornata*, etc., while on the west shores the genus is represented by four species. In the tropics the number is much larger, some three hundred being known from the whole world. All are littoral, carnivorous forms, abundant on seaweeds and hydroids, and in pools left by the retreating tides.

SUB-ORDER III. — TOXIGLOSSA.

These animals are all predaceous and carnivorous, for which they are well adapted. They have a strong proboscis, which can be extended some distance from the shell. The lingual ribbon is armed with two rows of teeth, the middle or rhachidian series being absent. The teeth are long and hollow, and it would appear that the animals have the power to poison their prey.

The largest family, and the one best known to collectors, is the CONIDÆ, which receives its name from the conical shape of the shell. The members are almost all tropical or sub-tropical, the number of species and the brightness of the colors increasing as we approach the equatorial regions. Notwithstanding their carnivorous propensities they are apparently timorous, preferring to live in holes in the rocks and coral reefs, and retiring within the shell at the approach of danger. They crawl in a slow, sluggish manner, with their tentacles stretched straight out before them. The only genus is *Conus*, of which about three hundred species are known, most of them being inhabitants of the eastern seas, only about fifty being found in the tropical waters of America. The general appearance of the animal may be seen from our figure of one of an oriental species, *Conus textilis*. The eyes are near the base of the tentacles, the foot is narrow and long, and furnished in the middle with a large opening, the object of which is frequently asserted to be the admission of water to the circulatory system. This connection of the blood vessels with the external world has been lately denied in any and all mollusos, and apparently with reason. Usually a small operculum is present, but not infrequently it is absent. The shell is thick, cone-shaped, the spire short, aperture narrow, the outer lip sharp and neither toothed.

FIG. 422. — *Conus textilis.*

We have just referred to the fact that some, if not all, of the Toxiglossa are poisonous, and the reader will doubtless pardon the following quotation from the pages of Mr. Arthur Adams. Speaking of *Conus aulicus* he says, — "Its bite produces a venomed wound accompanied by acute pain, and making a small, deep, triangular mark, which is succeeded by a watery vescicle. At the little island of Meyo, one of the Moluccas, near Ternate, Sir Edward Belcher was bitten by one of these cones, which suddenly exserted its proboscis as he took it out of the water with his hand, and he compares the sensation he experienced to that produced by burning phosphorus under the skin." In the South Sea Islands, *Conus textilis* and *C. marmoreus* are also considered as poisonous, though cases where their bite is fatal are rare and not well authenticated. It is supposed that the teeth break off and are left in the wound.

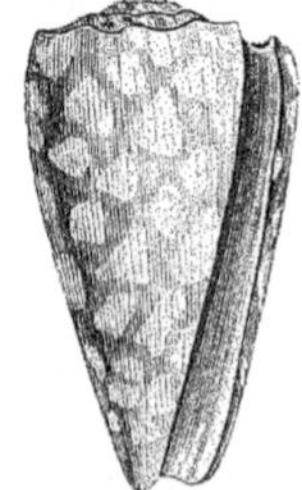

FIG. 423. — *Conus marmoreus.*

A description of the more common species of cones would prove dull reading, and so we merely mention the colors of the two species figured. *Conus marmoreus* is dark or even black, marked with triangles of white, while *C. textile* is very variable, the general ground color being golden or orange, on which are laid brown reticulating lines and white spots. The cones are favorites with collectors, and rightly so, for they are among the most handsome of shells. Some of the species are very rare. *Conus gloria-maris*, a white species with orange spots and triangular lines, has been sold for two hundred dollars, while some of the rarer varieties of *C. cedo-nulli* (a very variable species) have brought over one hundred dollars. The former comes from the eastern seas, while the latter is West Indian. Of course these prices do not indicate any intrinsic value in the shell, but are merely indices of the comparative rarity, and of the prices which rich collectors are willing to pay for certain noted species. Other species equally rare would not command a small fraction of these prices, merely for the reason that they are not so well known, and dealers have not yet attempted to speculate upon them.

FIG. 424.—*Terebra oculata.*

The TEREBRIDÆ contains about two hundred species of long, slender, many-whorled shells from the tropical seas. They are readily distinguished from other similar forms by the small aperture with an anterior siphonal notch, and by the absence of true plaits on the columella. The tentacles are short, and the eyes, when present, are near or at the tips. The *Terebras* are known among the sailors as auger-shells.

About as little need be said of the PLEUROTOMIDÆ, in which the shell is spindle-shaped, the aperture prolonged anteriorly while near the suture there is a notch. An operculum is not always present. Although some five hundred species are known, but little of popular interest can be detailed concerning them. The genus *Pleurotoma* is represented on our eastern shores by a few small and inconspicuous species usually assigned to the sub-genera *Bela* and *Mangelia*, while on the Pacific coast the species are about equally numerous. In the West Indies and at Panama many more forms are found.

FIG. 425.—*Pleurotoma babylonia.*

The CANCELLARIDÆ differ from the other Toxiglossa in being vegetarians, and they differ further in having the proboscis rudimentary. The shells may be recognized by the folds on the columella and on the outer lip, and the fact that the shell is almost always marked off into squares by transverse ribs and revolving lines which gives rise to the name of the principal genus *Cancellaria*. The species live in comparatively shallow water, though they are but very rarely found above low-water mark. In the northern Atlantic the family is represented by a small white shell about half an inch in length, known as *Admete viridula*. No specimens are known to have come from south of Cape Cod.

SUB-ORDER IV. — TÆNIOGLOSSA.

The Tænioglossate Mollusca are largely marine, though one or two families are found in fresh water. The shell is spiral, though in a few forms this appearance is obscured. The lingual ribbon in most forms has the shape of a band, and is armed with seven teeth in a transverse row, though in a few forms here admitted there are nine, while in others the number is reduced to three, and occasionally all are absent. Two tentacles are always present. In some the aperture is entire, and in others it is notched or produced into a canal for the respiratory siphon. We will first consider the holostomate (entire mouthed) forms.

Our first family contains the periwinkles, the LITTORINIDÆ of scientific nomenclature. This latter name is very appropriate, for they are all shore-living forms. The shell is ovate, with a short, sharp spire, and a round mouth which is closed, on the retreat of the animal, by a horny operculum. They have a thick foot, large snout, and the eyes are placed at the base of the antennæ. They live on the shore between tide marks and feed upon the smaller algæ. The principal genus is *Littorina.* The first species which we will mention is the periwinkle proper, *L. litorea*, the mollusc that is eaten after being extracted from its shell by a bent pin. Our figure represents the species (which is one of the largest of the genus) natural size. The shell is solid and very variable in color. Some are banded and some a uniform tint of red, brown, or black, the darker colors predominating. The advance of this species on our shores is very remarkable. It is a native of Europe and was first noticed at Halifax several years ago. In 1870 it had appeared on the coast of Maine. In 1872 it had reached Massachusetts, but it did not appear south of Cape Cod until a year or two later. Now on both the northern and the southern shores of New England it is one of the most common molluscs. In England it is extensively used as food, the annual catch amounting to many hundreds of tons. It is prepared by boiling, then the operculum is pulled off and the meat extracted from the shell. In taste it is much like the clam, only far more delicate. Though very abundant it has not yet acquired any economic importance here, as the American people seem greatly averse to trying experiments in the gastronomic line. Were it better known, it would be appreciated.

FIG. 426. — *Littorina litorea*, periwinkle.

FIG. 427. — *Littorina rudis.*

Another species which is common to both continents, *L. rudis*, has received nearly twenty specific names on account of its variations and its extensive range. The general appearance of the shell may be seen from our figure, but the color is variable. Usually it is yellow or olive-green, and without markings, but occasionally specimens are found banded or blotched with some lighter color. Until the advent of *L. litorea*, *Littorina palliata* was the most abundant species on the New England coast. The shell may be either plain or variously ornamented with bands and blotches of color, — white, green, or brown. It is common on rocky shores, and is especially fond of creeping over the rock-weed (*Fucus*) or eel-grass (*Zostera*). Further south the common species is *L. irrorata*, which is a little larger than *L. litorea*, but is longer and has a more acute spire. It is comparatively rare in southern New England, its metropolis being further south. Its introduction into the waters of Long Island Sound may have been with oysters transplanted from the Chesapeake.

FIG. 428. — *Littorina palliata.*

Another member of the family which should be mentioned is *Lacuna vincta*. It has a thinner and more slender shell than any of our *Littorinas*, and is reddish or horn-colored usually, with two or more darker reddish bands, which follow the spiral of the shell. Like the rest of the family, it is a vegetarian, feeding on algæ.

FIG. 429. — *Lacuna vincta.*

Closely allied to the Littorinidæ is the RISSOIDÆ, which needs but a passing mention. The family, and its principal genus, *Rissoa*, derive their name from Risso, a naturalist who in the early part of this century studied the fauna of the Mediterranean. Some of the members inhabit the sea, while others live in brackish water, or even in that which is entirely fresh. In all, the spire is long, the lip thickened, and the aperture rounded. *Rissoella*, which is found in Europe and Japan, lives between tide-marks. The shell is very thin, and the eyes, which are placed upon the surface of the head, are "so far behind the tentacles that the transparency of the shell seems to be essential to the vision of the animal." The species of *Rissoa* are numerous, several being found in American waters. *Bythinia* is one of the most prominent of the fresh-water genera. Its fifty species belong to the eastern hemisphere. In the United States, *Amnicola* is distributed through all parts of the country, the small species living in fresh water. They were formerly included in the Paludinidæ. *Pomatiopsis* should be mentioned, from the fact that the species are air-breathers.

The CYCLOSTOMIDÆ are exclusively terrestrial, and were formerly included among the Pulmonata. Like the members of that order, they breathe the air by an essentially similar pulmonary organ, but in the rest of their anatomy they deserve a place where we have put them. This is one of the many instances where a too close attention to one organ or one physiological operation would lead to erroneous ideas of relationship. Still the other process is next to impossible, and on this point we can do no better than quote the words of Fritz Müller: "Of a hundred who feel themselves compelled to give their systematic confession of faith as the introduction to a manual or monographic memoir, ninety-nine will commence by saying that a natural system cannot be founded on a single character, but has to take into account all characters, and the general structure of the animal, but that we must not sum up these characters as equivalent magnitudes, that we must not count, but weigh them, and determine the importance to be ascribed to each of them according to its physiological significance. This is probably followed by a little jingle of words in general terms on the comparative importance of animal and vegetative organs, circulation, respiration, and the like. But when we come to the work itself, to the discrimination and arrangement of the species, genera, families, etc., in all probability not one of the ninety-and-nine will pay the least attention to these fine rules or undertake the hopeless attempt to carry them out in detail." Notwithstanding this melancholy picture, science is constantly striving to arrange the groups of animals and plants; facts of structure are weighed, and it is by just this process that the Cyclostomidæ, a family without gills and with a pulmonary respiration, are accorded a position here among the branchiate molluscs.

FIG. 430. — *Cyclostoma inca.*

In this family the whorls of the spiral shell are rounded and the aperture is circular, hence the family name. As in the branchiate forms with which they are associated, the members of the family close the shell by an operculum which is round and increases in size with the growth of the animal, the new additions being placed in a spiral manner, the nucleus being central. The animal is much like that of *Littorina*;

it has a long proboscis and two contractile feelers, with the eyes at their bases. The lingual teeth are seven in a transverse band.

The Cyclostomes are largely tropical, but very few species straying into temperate regions. They live in damp places, some on the ground, some in trees, while others are found far from the sea. Over a thousand species have been described. Some have a peculiar gait; the foot is divided into halves by a longitudinal furrow, and in walking the animal advances one side and then puts it down, then the other side is moved forward in the same way; the two sides corresponding to the two feet of man. The principal genera are *Cyclostoma*, *Cyclophorus*, *Cyclotus*, and *Chondropoma*.

In the ACICULIDÆ the shell is nearly cylindrical, the margins of the aperture being nearly parallel. From the wave-like motion with which they progress, they have been termed 'looping-snails.' The species are amphibious, and live among the sea-weeds thrown up on the shore, or in shallow water. The species are small. The only genus which needs notice at our hands is *Truncatella*, in which, as the animal approaches maturity, the upper parts of the spire are broken away and the animal repairs the damage by closing up the broken whorls by a calcareous deposit. On account of this truncation of the shell, the genus has received its name. The family contains about a hundred species, mostly from the tropics.

The PALUDINIDÆ shares with the Limnæidæ, already mentioned, the common name, 'pond-snails.' Its numbers are widely distributed through the temperate zone of the northern hemisphere, but few being found within the tropics. They live in muddy ponds or streams, where they crawl slowly over the bottom or even burrow in the soft mud. The foot is large and broad, a well-developed proboscis is present, and the long cylindrical tentacles bear the eyes on little projections near the base. The water required for respiratory purposes is conveyed to the gills by an interesting contrivance. On either side of the neck are developed little fleshy outgrowths which, together with the mantle, form little tubes. In the adjacent figure these are seen on either side, projecting a little outside the shell. The water goes in through the right tube, passes over the gills, and then is forced out through the tube on the left.

FIG. 431. — *Paludina intertexta*, showing respiratory tubes.

The sexes are distinct, and the young are brought forth alive. The eggs undergo their development inside of the mother, and in the species with thin shells they may occasionally be seen inside the body. The development requires about two months, and at the end of that period the young are sent forth, three or four at a time, to begin life for themselves. The shells are thin in some forms and more solid in others; the prevailing color is some shade of green or greenish brown, banded with darker, though sometimes the bands lacking.

The principal genus, *Paludina*, has been divided up into several sub-genera, of which *Melantho* and *Tulotoma* are strictly North American. A curious instance of the fact that a low temperature, which affects so greatly the land shells, has but little influence on the size of fresh-water forms, is found in the case of *Paludina ussuriensis*, the largest Siberian species, which occurs in a latitude but little south of the line of perpetual frost, and where the mean annual temperature is the same as in Iceland. On the other hand, recent explorers have brought home large species found in the lakes of Central Africa. No species have as yet been found in Australia, New Zealand, Polynesia, or the West Indies.

The MELANIIDÆ are also fresh-water inhabitants, in which the shell is covered with a thick epidermis, which, in some species, is so dark that the family name (*melas*, black) is very appropriate. Others are brown or dark green. The shell is usually long, turreted, or conical, with a small mouth. The foot is large and triangular, the proboscis short but stout, and the eyes are near the bases of the tentacles. The family is readily separable into two sub-families, both on structural characters, and on geographical distribution. The first, the Melaniinæ, are oriental, only a few being found on the North American continent. In these, the aperture is usually broadly rounded and

FIG. 432. — *Paludina vivipara.*

not produced in front, though often channelled or notched, while the margin of the mantle is fringed, and many of the species are ovo-viviparous; that is, the eggs undergo their development and are hatched outside the parent. In the other sub-family, the Strepomatinæ, the margin of the mantle is plain, and the eggs are laid and attached to stones and plants. With three or four exceptions, all the Strepomatinæ are confined to the United States, a few extending to the West Indies.

Of the Melaniinæ, the most prominent genus is *Melania*, which contains about four hundred species, mostly distributed through Asia and Polynesia, though a few are found in tropical America and southern Europe. In many of the specimens the apex of the shell has disappeared, owing to the erosive action of the water which they inhabit. *Melanopsis costata*, a common Syrian species of a genus allied to *Melania*, is said to inhabit the Dead Sea, the only exception, so far as I am aware, to an exclusively fresh-water habitat in the family.

FIG. 433. — *Pleurocera pallidum.*

Of the Strepomatinæ, about five hundred nominal species have been described, but until we know more of the anatomy, the life history, and the variations of these forms, nothing definite can be known of their classification. Here is a wide field for study open to those students of the central portions of the United States, which will

FIG. 434. — *Anculotus prærosa.*

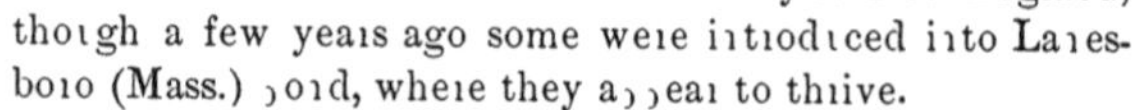

FIG. 435. — *Anculotus plicata.* FIG. 436. — *Lithasia fuliginosa.* FIG. 437. — *Goniobasis depygis.* FIG. 438. — *Goniobasis impressa.*

be productive of great results. Of the development we know absolutely nothing, not even if a veliger is formed. Ten genera and sub-genera are recognized, and the great bulk of our species come from the Ohio River and its tributaries. Species of *Anculotus* are found in the Potomac and Susquehanna, while others of the family have been introduced into the Erie Canal. None are known to occur naturally in New England, though a few years ago some were introduced into Lanesboro (Mass.) pond, where they appear to thrive.

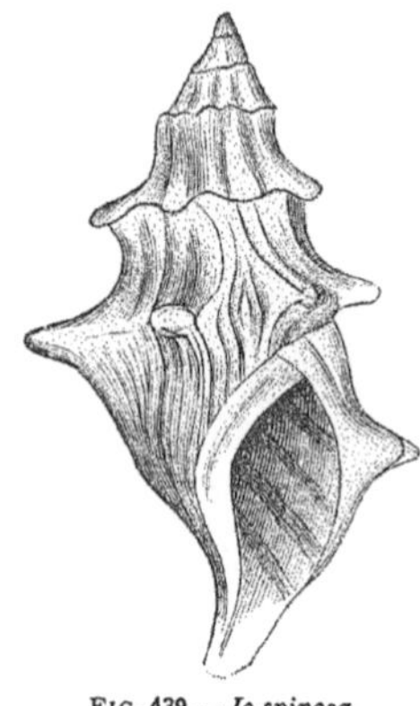

FIG. 439. — *Io spinosa.*

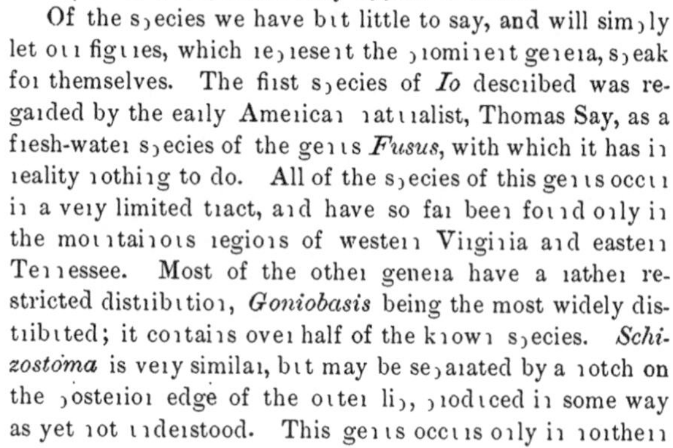

Of the species we have but little to say, and will simply let our figures, which represent the prominent genera, speak for themselves. The first species of *Io* described was regarded by the early American naturalist, Thomas Say, as a fresh-water species of the genus *Fusus*, with which it has in reality nothing to do. All of the species of this genus occur in a very limited tract, and have so far been found only in the mountainous regions of western Virginia and eastern Tennessee. Most of the other genera have a rather restricted distribution, *Goniobasis* being the most widely distributed; it contains over half of the known species. *Schizostoma* is very similar, but may be separated by a notch on the posterior edge of the outer lip, produced in some way as yet not understood. This genus occurs only in northern Alabama, in the streams which flow into the Tennessee River.

The PYRAMIDELLIDÆ is a small family of marine molluscs, with a long, slender shell, in which the columella has frequently one or more prominent folds; the eyes are sessile, the proboscis retractile, and the tentacles either broad or long and slender. In many the lingual teeth have entirely disappeared, owing to their parasitic habits.

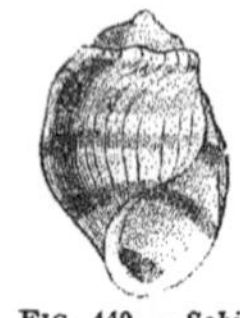

FIG. 440. — *Schizostoma salibrosa.*

Pyramidella is a tropical genus of littoral molluscs, the members of which burrow along just beneath the surface of the mud. The members of the allied fossil genus *Nerinæa* were much larger than any of the existing members of the family. The species of *Odostomia* are numerous and widely distributed, the genus being represented by several species on our coasts. Some of these usually live beneath stones, moving in a very slow manner, while others are usually found on the shells of scallops (*Pecten*) and but rarely anywhere else. Whether they live a semi-parasitic life or are commensals has not been settled. *Stylifer* is much more of a parasite, and it is a curious fact that all the parasitic Mollusca affect only the echinoderms and cœlenterates. On our New England coast *S. stimpsoni* occurs on the common sea-urchin, *Strongylocentrotus;* in England *Stylina turtoni* also affects sea-urchins; a Mediterranean species lives firmly attached to the anal tube of the feather star, *Comatula;* while, to mention but one more species, *S. astericola*, as its name implies, dwells on starfish. Some of these forms are true parasites. With their slender foot they obtain an entrance between the calcareous plates of the starfish or sea-urchin, and there they live on the juices of their host, only the apex of the shell being visible from the exterior.

In *Eulima* the parasitic habit is carried still farther. Most of these live on holothurians, although some attack starfishes. They even enter the alimentary tract of some of the sea-cucumbers, where they feed on the food of the host. One species has become so modified by parasitism that, though it still retains a shell, it has lost many of its distinctive molluscan characters. This species, as described by Semper, lives on the outside of a species of holothurian, and its proboscis is enormously developed so that it pierces the tissues of its host and enters the cœlomatic cavity. Feeding thus on the fluids of the sea-cucumber, it has no need for teeth, and hence its lingual ribbon has disappeared, and from the same cause, disuse, the foot and eyes have followed a like course. Our American *Eulima oleacea* does not seem to have carried its parasitic habits so far. It occurs on the skin of *Thyone briareus*, a common holothurian south of Cape Cod.

It is doubtless the fact that it lives a parasitic life that has led some naturalists to place that most degraded of molluscs, *Entoconcha*, in this family. Not enough is known of it to warrant such a position on any other ground. It is referred to on a previous page (p. 297).

The TURRITELLIDÆ embraces forms with a shell much like that of the Pyramidellidæ; long and turreted, with a round mouth which is not notched or produced into a canal. The operculum is horny and many-whorled. The animal has a moderate, but short foot, a short muzzle, and the eyes are placed at the outer base of the tentacles. The margin of the mantle is fringed. Nearly a hundred species are known, all from salt water. Most of them are covered with a brownish epidermis.

Closely allied to the last family is the VERMETIDÆ, some of the members of which recall the *Serpulæ* among the worms (p. 227) on account of the irregularity and vermian shape of the shells. The animal has an elongate head bearing two long tentacles. The shell begins as a regular spiral, but after a few turns the regularity is lost, and the tubular shell grows in any direction, the spiral character almost entirely disappearing. The shell grows more rapidly than the animal, and, as a result, the posterior portions of the tube are partitioned

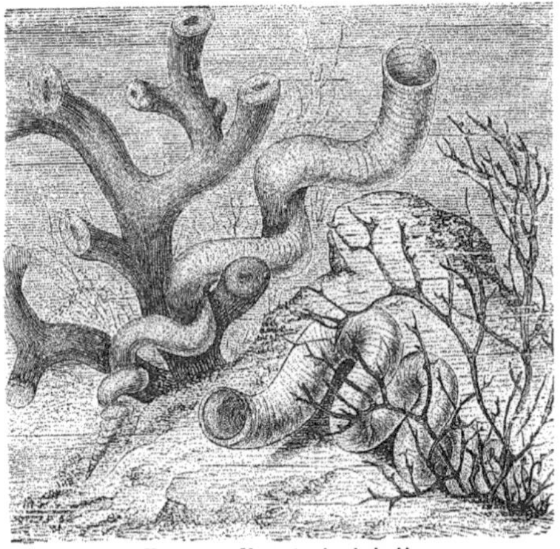

FIG. 441.—*Vermetus lumbricalis.*

off by a calcareous deposit, thus giving a ready means of separating the shell of the mollusc from that of the worm. In *Vermetus* the shell is entire, but in *Siliquaria* the tube has a longitudinal slit. *Cæcum* is usually placed here, but it differs greatly in its lingual dentition, only two lateral teeth on either side being present, and the rhachidians are absent. The shell begins as a spiral, but with growth the older portions are lost and the shell becomes simply a curved tube embracing an arc of from sixty to ninety degrees. Both *Vermetus* and *Cæcum* are represented on the shores of the United States; the species of the latter genus are very minute, and in assorting the contents of the dredge they readily escape notice. Some of the Vermetidæ become attached to sub-marine objects, while others are always free. Specific limits are not very well defined, owing to the irregularity of the shells.

The apple-shells of dealers belong to the family AMPULLARIIDÆ. These live in tropical countries, where they replace in the ponds and marshes the Paludinidæ of more temperate climes. Africa and South America seem to be especially favorable to their growth, and from these regions we have the finest specimens. They are an amphibious group, living apparently equally well either in or out of the water, yet they require a certain amount of moisture for their well being, and when their native marshes dry up they burrow deep into the mud. They are well fitted by nature for their amphibious life, for they have both lungs and gills. The lung cavity is placed above that which contains the gills, but the two are connected by means of an opening through the fleshy partition. This structure enables them to breathe either air or water, and Professor Carl Semper, of Würzburg, who spent a long time in the Philippines, frequently observed "that the *Ampullariæ* breathe not only with both gills and lungs, but they do so in regular alternation; for a certain time they inhale air at the surface of the water, forming a hollow elongated tube by incurving the margin of the mantle, so that the hollow surface is closed against the water and open only at the top. When they have thus sucked in a sufficient quantity of air, they reverse the margin of the mantle, opening the tube, into which the water streams. The changes are tolerably frequent, once or twice in a few minutes, depending probably on the temperature."

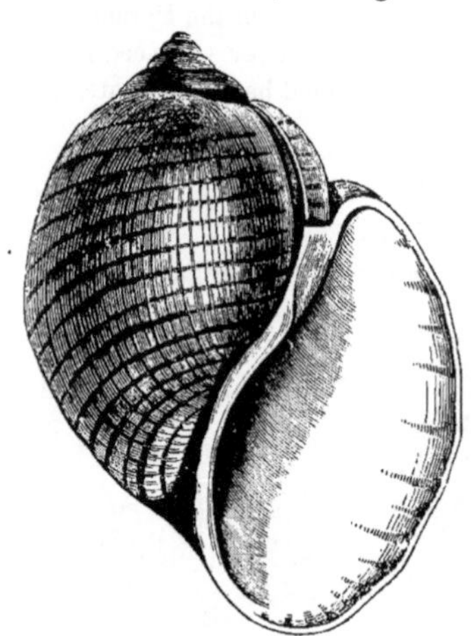

FIG. 442.—*Ampullaria canaliculas*, apple snail.

The apple snails are long-lived, and specimens have been brought to northern climes hidden in the hollows of logs of mahogany from Honduras, but no one knows how long before. Mr. Laidly, who lived for many years at Calcutta, tried many experiments with them, and some specimens which were confined for five years in a drawer were alive at the end of that period.

The shells are large, thin, and globular, covered with a glossy epidermis, and closed by a horny operculum. The eyes are good-sized, and are placed on short stalks, and the tentacles are very long and slender. The eggs are large and spherical and are laid upon the stems of water-plants in large bunches, furnishing many a meal for the marsh birds. The most prominent genus is *Ampullaria*, in which the epidermis is usually green, and the respiratory tube long. The South

Americans call them idol shells. A few species are found in the southern portion of North America, living in the rice swamps of Georgia, etc.

The family Valvatidæ embraces a few small fresh-water shells, usually associated with the Paludinidæ. The shells are discoidal or conical and are usually covered with a greenish epidermis. The animal has a small foot, and when in motion the delicate branchial plume is extended outside the gill cavity, and may be seen, an object of beauty, above the neck. The species are distributed through the temperate regions of the globe, frequenting slow-running rivers and ponds. They are hermaphroditic and lay their eggs in a single capsule attached to stones or to aquatic plants. *Valvata*, the only genus, has been divided into three or four sub-genera. Our figure represents *Valvata tricarinata*, a common species in some parts of the United States. Other forms have the whorls rounded instead of angular.

FIG. 443. — *Valvata.*

We have already referred to some of the limpets on a preceding page, but it is extremely doubtful if all should be included in the Zygobranchiate Mollusca and if any value is to be placed on lingual dentition, some of them should be assigned a position here. We have already mentioned the Acmæidæ, and would merely state that usually they are considered as closely allied to the Calyptræidæ which will receive brief mention. In this family the shell is limpet-like and with the beginning somewhat spiral. In some the spiral is so far developed as to form a partial partition (columellar lip) so that the common name 'slipper-limpet' is very appropriate. In others the interior of the shell is simple. Unlike the other limpets previously mentioned, these forms are sedentary, and most of them never seem to leave the spot where they settle down in early life. That this stationary habit occurs is shown by the shape of the shell, which is always moulded to fit any irregularities of the surface to which they are attached. Thus, when on an oyster shell, the edge of the slipper-limpet reproduces (of course reversed) all the elevations and depressions of the other. With this stationary life it is not readily seen what they live upon, unless it be the microscopic life so abundant in the sea water, for they cannot move far to obtain the attached algæ. In confinement some have been known to devour animal food.

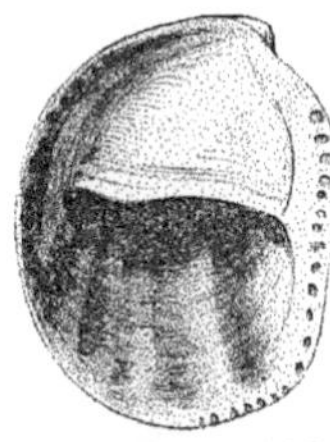

FIG. 444. — *Crepidula fornicata*, slipper limpet.

The genus *Crepidula* departs but little in conchological characters from the type usual in the univalve molluscs. The general appearance may be seen from our figures, which represent two of the common American forms. The aperture is divided by a posterior lamina (the columellar lip) so that the resemblance to a slipper is striking, whence the common name, slipper-limpet. These forms are almost always found attached to the shells of other molluscs, the outer part of the aperture of *Natica* and *Nassa*, when inhabited by hermit crabs, being a favorite place. Why this association of forms should exist has not been explained; possibly the crumbs dropped from the hermit's dinner may afford abundant food for the limpet. Frequently several individuals are placed one above another, the lowest one adhering to some living or dead shell or to the king-crab (*Limulus*).

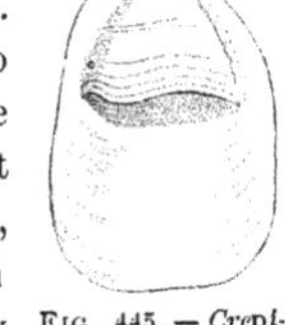

FIG. 445. — *Crepidula plana.*

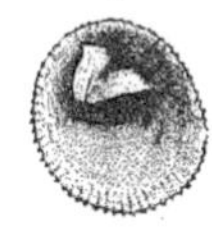

FIG. 446. — *Crucibulum striatum*, cup-and-saucer limpet.

The genera *Calyptræa* and *Crucibulum* are known as cup-and-saucer limpets. In the centre of the inner surface of the shell is a calcareous cup, the shell proper corresponding to the saucer. In habits the members of these genera are much like the species of *Crepidula*. Among the other genera of the family may be mentioned *Infundibulum*, *Capulus*, and *Hipponyx*.

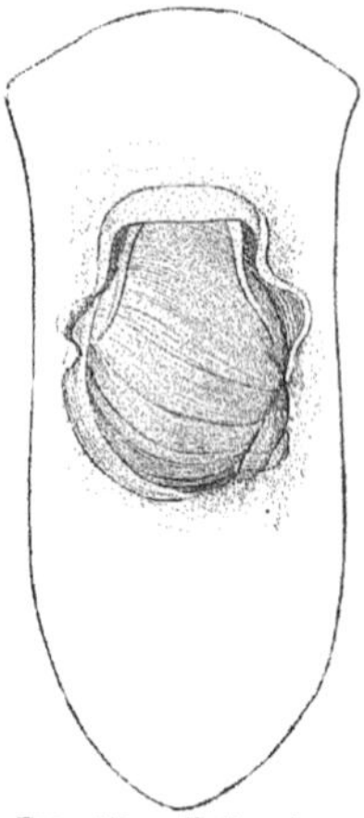

FIG. 447. — *Natica heros* crawling, showing the Very large foot.

The NATICIDÆ are carnivorous marine snails with globular shells and an entire half-moon shaped aperture, the outer lip of which is sharp, while the inner one is usually ornamented with a callus deposit which is frequently thick and extensive. The animal has a very large foot, and, when extended, broad folds nearly or quite envelop the whole shell. The tentacles are small, slender, and widely separated, though connected for a portion of their length by a fleshy membrane arising from the top of the head. Eyes may be present or absent; when present they are placed at the base of the tentacles.

The genus *Natica*, with its two hundred species, is distributed all over the world. It has been divided into the subgenera *Neverita*, *Lunatia*, *Mamma*, *Amaura*, etc., which need not be characterized here. All are active predaceous forms, living by preference on sandy bottoms and shores, often straying above low-water mark so that they are left exposed by the receding tide. At such times they usually burrow just beneath the surface, creating a little hummock of sand. In northern New England the most common species is *Natica (Lunatia) heros*. It is there the largest littoral univalve, the shell in large specimens attaining a diameter of three and a half and a length of five inches. In color the adult shell is ashy gray or brownish, the inside of the aperture being reddish; but the young are frequently marked with three rows of dark, usually reddish spots. This was formerly regarded as a distinct species under the name *triseriata*. South of Cape Cod *Natica (Neverita) duplicata* is fully as abundant as the species just named. It may readily be distinguished by the callus which in this species completely covers the umbilicus, as shown in the cut. *Natica heros* extends south to Georgia, and *N. duplicata* has been reported from Mexico and Yucatan. Several other species of the genus are found on our eastern coasts.

FIG. 448. — *Natica heros.*

The food of *Natica* is mostly bivalve molluscs. Burrowing through the mud, the snail runs across some unfortunate clam, and immediately the siege begins. Close its valves as tightly as possible, the clam is not secure; for the snail brings its well-armed lingual ribbon into play and by drawing it across the shell, slightly rotating the body meanwhile, it rasps away a portion of the hard calcareous armor, and in time it makes a hole through to the soft parts, which are then devoured. The hole through the shell is almost perfectly circular, and is bevelled like a countersunk hole made by the artisan.

The egg masses of the *Naticas* bear the common name sand-saucers; they are among the most abundant objects on the sea-shore during the spawning season. The general shape is shown in the figure; it is somewhat like a saucer without a bottom, but the two ends of the ribbon are not united. On holding one up to the light, the numerous eggs are seen as light spots arranged in quincunx. For a long time these sand-saucers were a puzzle to naturalists, and under several different names they were regarded as Polyzoa. The peculiar shape of the ribbon is due to the method of its

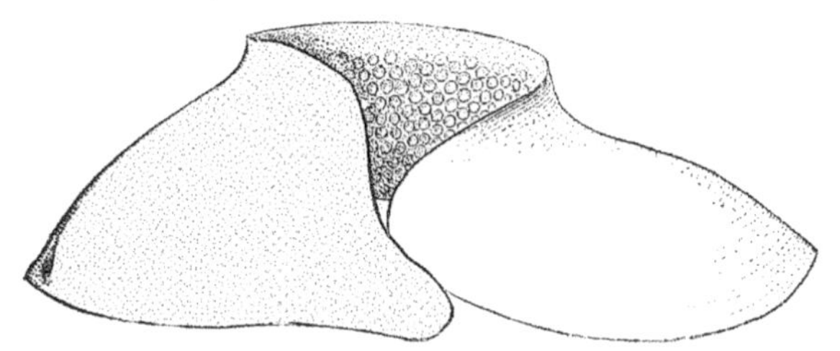

FIG. 449. — Sand-saucer, egg mass of *Natica heros.*

formation. As it passes out of the parent it passes over the foot and is pressed against the body of the shell. At first it is soft, gelatinous, and adhesive, but soon it hardens and unites firmly to the surface some of the surrounding sand. Mr. A. H. Tuttle has recently studied the development of these eggs.

The shell of *Sigaretus* is much like that of *Haliotis*, but lacks the series of perforations characteristic of that form. The species frequent sand flats in the warmer waters of the world, where they crawl in a sluggish manner searching for food. They are very timorous and retract themselves on the slightest alarm; but, owing to the large size of the aperture and the minute operculum, they are not much better off after having withdrawn themselves as much as possible into the shell. *Lamellaria*, in shape of shell, holds a position about midway between *Natica* and *Sigaretus*, but has no operculum. Only ten species are known, none of them from America. The following observations apply to *L. perspicua*, a European species. The time of spawning is during February and March. At that time the female eats a round hole in the colonies of one or more species of compound ascidians, this it lines with a pot-shaped capsule furnished with a transparent lid. In this the eggs are placed, and, as they increase in size, the whole nest rises up beyond the surrounding portion of the ascidian. At last the embryos are ready to begin their free life, and then the lid bursts open and out come the young. By some the genus *Entoconcha* is placed in or near this family.

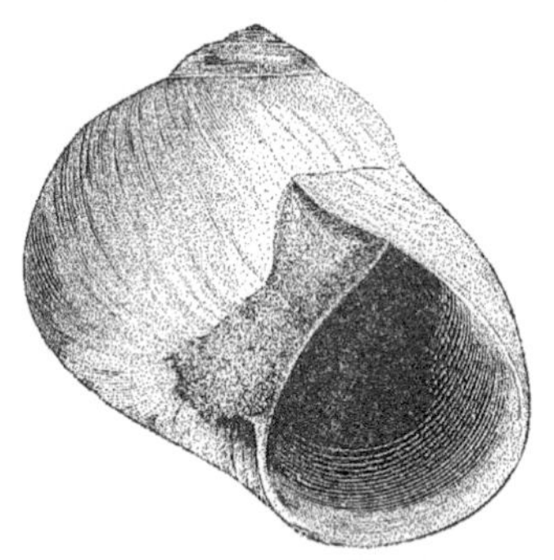

FIG. 450. — *Natica duplicata.*

With the family CERITHIIDÆ we return to a group of molluscs with long and slender shells, resembling in many respects the Melanians, but in others departing considerably from them. The shell is long and many-whorled, the aperture with an anterior canal and a second less distinct posteriorly. The proper habitat of the family is in the tropics, though numerous members are found in colder waters. Some of the species are marine, some live in brackish water, while others inhabit streams and ponds. The species of *Cerithium* are numerous and variable, and it is said that it was the difficulty of classifying the species of this genus that first led Lamarck to speculate upon the origin of species. Our species are mostly placed among the subgenera *Cerithiopsis*, *Bittium*, and *Triforis*. They are all marine. The species of the latter have reversed shells, and one (*T. nigrocinctus*) is abundant on bottoms where the algæ are numerous.

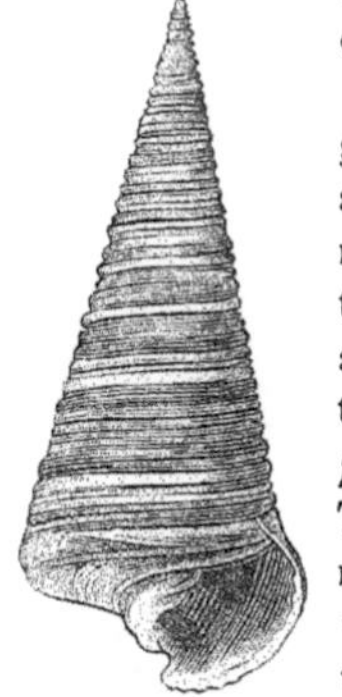

FIG. 451. — *Potamides telescopium.*

The fresh and brackish water species are mostly placed in the genus *Potamides*, of which *Pyrazus*, *Telescopium*, and *Cerithidea* are to be regarded as sub-genera. The fifty known species are mostly from the eastern hemisphere. The brackish-water species of the East Indies frequently leave the water and climb on the roots and branches of the mangrove trees, suspending themselves by glutinous threads. The Malays are fond of *P. telescopium* and *P. palustre*, which they cook by throwing them on their wood fires. The bent pin, so useful (according to Dickens) in extracting the meat of the periwinkle, would here avail but little, on account of the numerous whorls; so the natives break off the apex of the shell and then suck the animal through the opening thus made. The individuals of this genus are very numerous in suitable places, and near Calcutta *Potamides telescopium* is so abundant that the shells are gathered and burned for lime, the animals being previously killed by exposure to the sun.

The cowries, or porcelain shells, are known in scientific nomenclature as the CYPRÆIDÆ. Like the cones, they are favorites with collectors, and a drawer filled with their enamelled and brightly-colored shells is a beautiful sight. By characters derived from the shell they are readily distinguished from all other molluscs, for the spire is nearly or quite concealed by the body whorl. The aperture is long and narrow, terminating at each end in a notch, while the lips are usually toothed, the outer one being thickened and rolled inwards in the adult, though in the young it is thin and sharp, and the spire is visible. The animal has a broad foot, while the mantle is expanded on either side so as to form broad lobes (usually fringed on the margin), which turn up over the shell. When full grown these lobes secrete the shining enamel which covers the entire exterior of the shell and obscures the spire. The line where the lobes of the mantle meet upon the back is usually marked with a lighter line.

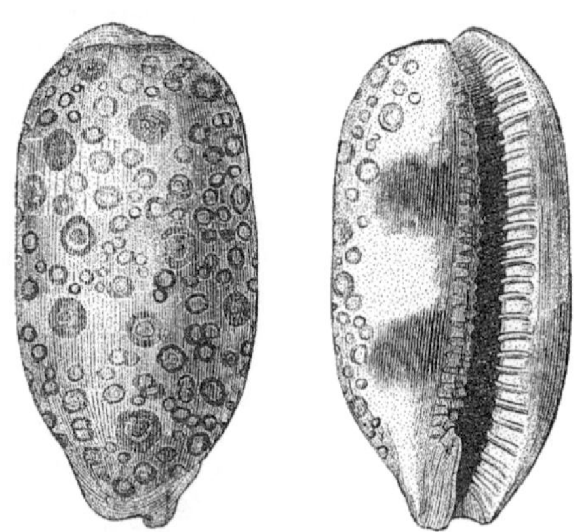

FIG. 452. — *Cypræa argus*, eyed cowry.

The principal genus, *Cypræa*, contains about two hundred tropical and semitropical species, only a few of which occur in the Atlantic, the great majority being from the East Indies and the islands of the Pacific. In habit some are retiring, hiding themselves under stones and among the branches of coral, while others crawl about, exposed to the full rays of the sun. They are littoral animals, occurring most numerously near or just below low-water mark. The colors of the shells are bright, or are laid on in pretty and striking patterns, while the animals are, if possible, even more beautiful than the shells. Some of the species are among the conchological rarities, and for them, in times past, enormous prices have been paid. The rarest species (or at least the highest priced ones) are *C. guttata* and *C. princeps*. Certain species have always been highly prized by savage races, and *Cypræa moneta*, the money cowry, passes, or used to pass, as the current medium of exchange among some of the African tribes. It is a native of the Pacific Ocean, and is brought in large numbers to England and thence shipped to Africa to be used in barter.

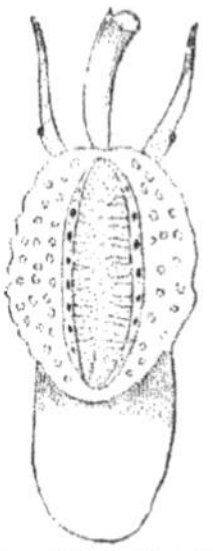

FIG. 453.—*Trivia californica.*

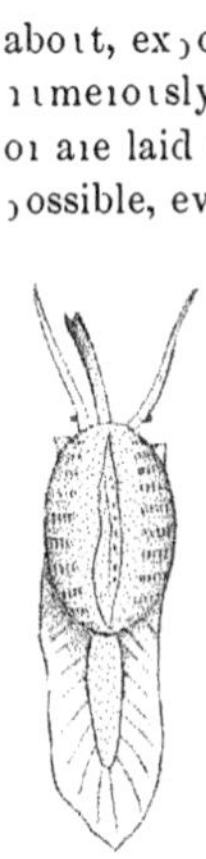

Among the sub-genera may be mentioned *Trivia*, *Luponia*, and *Aricia*, *C. moneta* being a representative of the latter. *Trivia* strays into northern waters, one species being found in European seas, while another is found in California. The sub-genus *Luponia* is also represented by two species in the latter region. All the species of *Cypræa* are carnivorous, and they eat living or dead food.

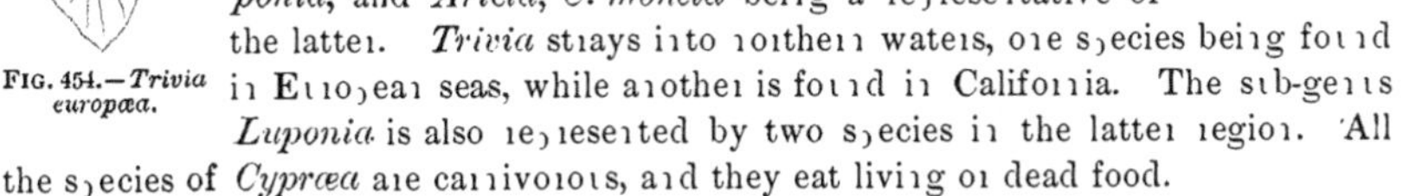

FIG. 454.—*Trivia europæa.*

The species of *Ovulum* are to be recognized by their elongate shell, which, in some forms, is greatly drawn out at the extremities, as in the weaver's-shuttle shell, *O. volva*, from the Philippines. These shells are not so prettily marked as the cowries are, but

FIG. 455. — *Ovulum volva*, weaver's shuttle shell.

are usually more or less unicolored, and the color sometimes corresponds to that of the surroundings, and the Floridian species, *Ovulum uniplicatum*, is said to be yellow or purple according to the color of the gorgonid corals on which it dwells, and on which it is supposed to feed.

The STROMBIDÆ receive the common name, wing shells, on account of the broad wing-like expansion of the outer lip in some of the species. Other common names are applied to some of the members of the family. The animals are among the most active of the molluscs. The long and muscular foot is divided into two halves, adapting it for progression by a series of leaps instead of the ordinary creeping motion. The operculum is also an aid in locomotion; it is claw-shaped and toothed on the outer edge, and these serrations are used to obtain a foothold. Then the foot is straightened by a violent muscular action, and the animal jumps forward. The other parts are equally well developed, and seem to demand for these animals a place near the top of the gasteropod series. The eyes are large and placed at the extremities of the thick stalks, while the tentacles are long and slender, the tentacles and eye-stalks being united for a portion of their length. The muzzle is long and extensible. The strombs are carnivorous and

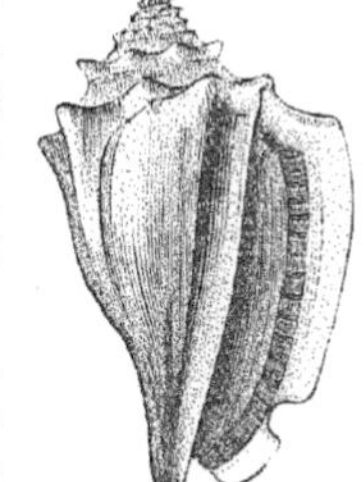

FIG. 456. — *Strombus pugilis.*

scavengers, feeding almost entirely on dead and decaying animal matter. The shell has a sharp conical spire, the outer lip is expanded, and usually is notched near the anterior canal.

The genus *Strombus* has the outer lip entire, with the exception of the notch mentioned above, and the aperture is long and narrow. In the young the outer lip is small, and the shell looks like that of a cone, but as the period of maturity is approached it becomes flared out. Species of the larger wing-shells are among the most common ornaments, large numbers being brought from the West Indies. In the United States, the species most commonly seen is *Strombus gigas*, a West Indian form, with a delicate red or pink interior. As the name indicates, this is one of the largest members of the genus. Another species common on the Florida reefs is *S. pugilis*, the general appearance of which may be seen from our figure.

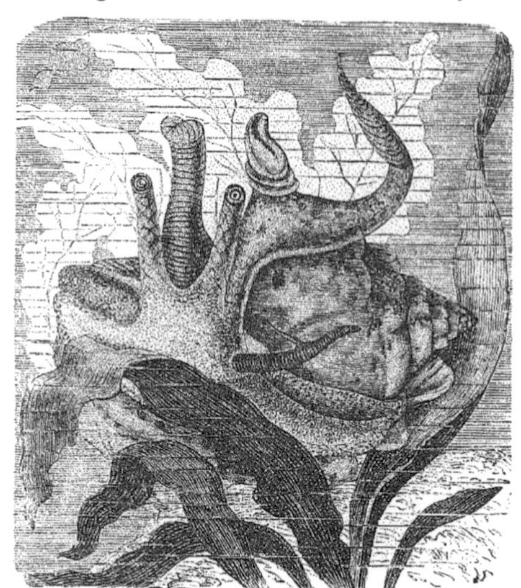

FIG. 457. — *Strombus lentiginosus.*

Allied to *Strombus* are the scorpion shells forming the genus *Pterocera*. In early life the shells are much like those of the stroms, but when approaching the adult condition the long finger-like processes of the mantle begin the secretion of shelly matter, so that the outer lip is armed with a number of strong horns. At first these horns are channeled, but later in life they become solid. In some they are short and straight, but in others they are considerably bent. All the species come from the Indo-Pacific regions. In *Rostellaria*, which has much the same distribution, the shell is much like that of *Fusus*, the spire and the anterior canal being longer than in *Pteroceras*, while the horns of the aperture are very much smaller. *Rostellaria* moves about in the same manner as do the stroms, but some species are more timid. It is an inhabitant of deeper water than the other. The only other oriental genus which we will mention is *Terebellum* the species of which live in deep water. They are very active and at the same time very timorous. "It will remain stationary for a long time when, suddenly, it will roll over with its shell, and continue again perfectly quiet." At other times they will leap several inches from the ground, like species of *Strombus*. In some species one of the ocular peduncles is longer than the other, and this one is used almost exclusively. As in the stroms, the iris is colored and the pupil black. *Aporrhais* con-

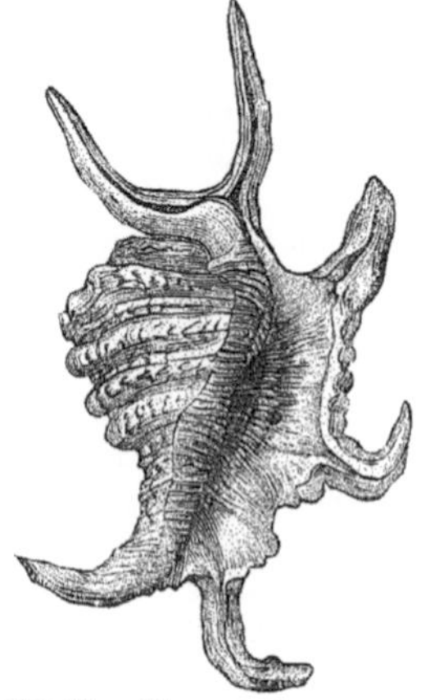

FIG. 458. — *Pterocera chiragra*, scorpion shell.

tains but four living species, one of which, *A. occidentalis*, is occasionally found off our northern coasts. Of its habits almost nothing is known. In the pelican's-foot shell (*A. pes-pelecani*) of Europe the outer lip is produced into two or three long horns, like those of *Pterocera*.

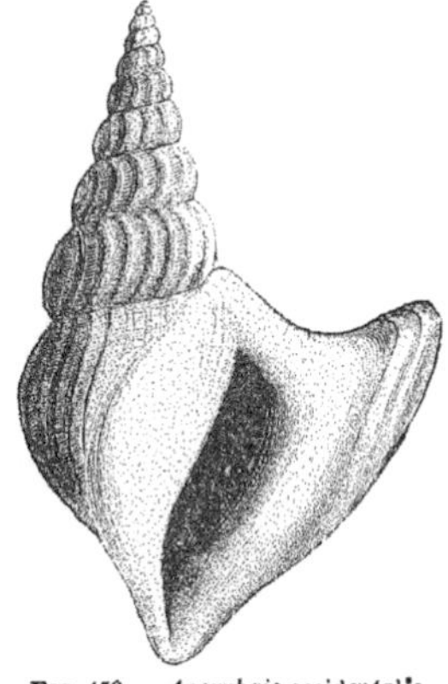

FIG. 459. — *Aporrhais occidentalis*.

The helmet shells, CASSIDÆ, are thick and heavy, the spire is short, and the aperture terminates in front in a short recurved canal, the columellar lip is covered with a thick deposit of callus, and, like the other, is toothed or ribbed. The animal has a large head, a long extensible proboscis, and a large foot. These species are mostly large and active carnivorous animals, feeding chiefly on bivalve molluscs. Owing to the fact that the shell is made up of differently colored layers, the helmet shells are extensively used in the manufacture of shell cameos. "Genoa and Rome are the seats of the best work, although many common ones are cut in France. In Rome there are about eighty shell-cameo cutters, and in Genoa thirty, some of whom also carve in coral. The art of cameo cutting was confined to Rome for upwards of forty years, and to Italy until the last twenty-six years, at which time an Italian began cutting cameos in Paris, and now over three hundred persons are employed in that city. . . . The shell is first cut into pieces the size of the required cameos, by means of diamond dust and the slitting mill, or by a blade of steel [soft iron ?] fed with emery and water. It is then carefully shaped into a square, oval, or other form on the grindstone, and the edge finished with the oilstone. It is next cemented to a block of wood, which

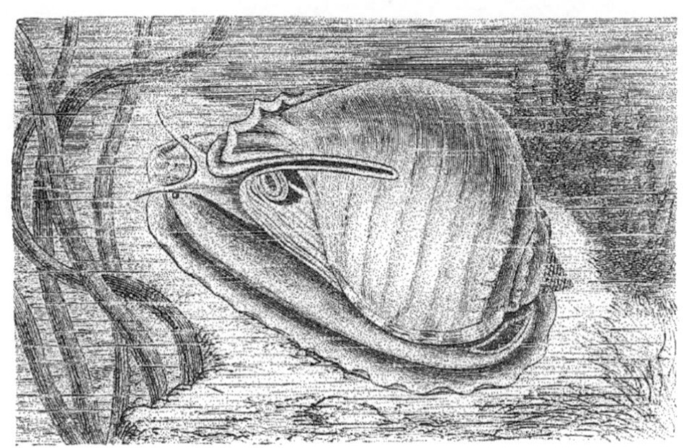

FIG. 460. — *Cassis glauca*, helmet shell.

serves as a handle to be grasped by the artist while tracing out with a pencil the figure to be cut on the shell. The pencil mark is followed by a sharp point, which scratches the desired outline, and this again by delicate tools of steel wire, flattened at the end and hardened, and by files and gravers, for the removal of the superfluous portions of the white enamel. A common darning-needle, fixed in a wooden handle, forms a useful tool in this very minute and delicate species of carving. The careful manipulation necessary in this work can only be acquired by experience; the general shape must first be wrought out, care being taken to leave every projection rather in excess, to be gradually reduced as the details and finish of the work are approached."

About fifty species of helmet shells are known, all from the warmer seas. *Cassis*, the largest genus, has its metropolis in the eastern seas, a few species being found in the West Indies and in the Mediterranean. The species most used for cameos is the black helmet, *C. madagascarensis*, in which the white outer layer covers a darker, almost black, second layer. The external surface is not, as the common name would apparently imply, wholly black. The whorls are a dirty white, and the outer part of the lips are rosy, the black being at and near the edges of the aperture. It reaches a length of nearly a foot. *C. glauca*, the species figured, is smaller; it comes from the East Indies. The other genera of the family are *Cassidaria*, *Oniscia*, and *Pachybatron*.

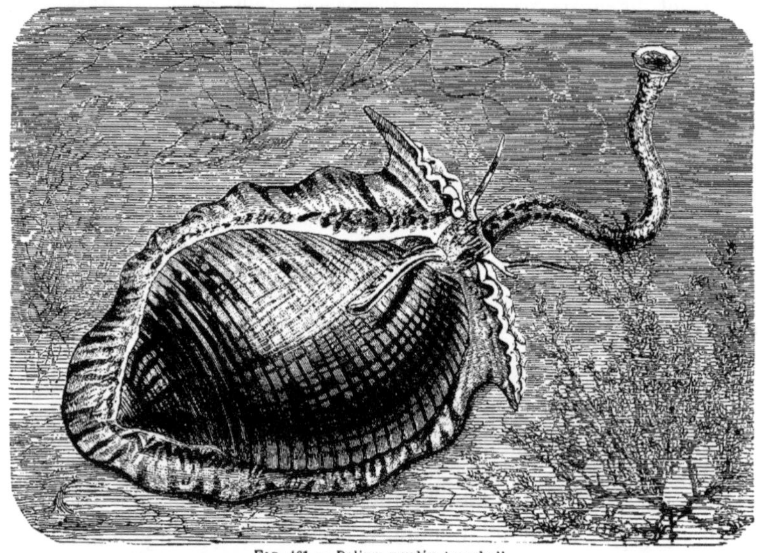

FIG. 461. — *Dolium perdix*, tun shell.

The members of the family DOLIIDÆ are large, with thin shells, the whorls of which are ventricose, the body whorl being very large, and ornamented with revolving ribs. The animal is large, and is provided with a remarkably developed proboscis, which is long, cylindrical and flexible. The foot is very large and lobed in front; the tentacles arise from a distinct head, and the eyes are placed on small pedicels growing out from the bases of the tentacles. In the young the shell is closed by an operculum, but it disappears with growth.

The species of *Dolium* are more or less globular, with a very large aperture, and have received the common name, tun shells, while the species of *Ficula* are known from their shape as fig or pear shells. Both genera are tropical or sub-tropical, active molluscs, living on animal food. The shells are not usually brightly colored, but the animals themselves are handsomely marked.

Last in the order comes the TRITONIDÆ. In conchological characters it belongs near the Muricidæ, but in dentition, development etc., its place is here, the lingual ribbon bearing seven rows of teeth, one rachidian, flanked on either side by three laterals. The shells are ornamented with varices, the character of which serves to distinguish the two principal genera. In *Ranella* a varix is formed at each half turn, so that the shell bears a flattened appearance, due to the prominent and continuous varices of the whorls; in *Tritonium* the varices are not continuous, but are found at irregular intervals. Almost all of the hundred and fifty species live between the tropics, where they feed upon decaying animal matter, the species of *Ranella* being among the most prominent of the molluscan scavengers. The species of both genera are handsomely colored, and some of those of *Tritonium* are among the largest of gasteropods. The shells are heavy, but the animals are, nevertheless, very active; some are littoral, while others live in deeper water.

FIG. 462. — *Ranella bituberculavis.*

ORDER VI. — HETEROPODA.

The naturalist, when skimming the surface of the ocean far from land, usually captures numbers of pelagic forms with bodies as transparent as glass. When carefully studied, these forms are of the greatest interest; there will be the curious worms known as *Sagitta*, eel-like fishes (*Leptocephalus*) whose position is far from certain, shrimps and larvæ of Crustacea, *Salpæ*, and jelly-fishes. Among the number will usually be found specimens of the groups of molluscs known as heteropods and pteropods. With the former of these two we have now to deal. The Heteropods are streptoneurous gasteropods specially organized for a pelagic life, and the modifications which they have undergone serve to place them as highest in the series. Yet small as is the group, the variations existing in it are by no means inconsiderable. The foot has become modified for swimming, and may have three typical divisions well marked, or it may be reduced to a thin vertical fin, small in proportion to the body. The body may be developed in the normal manner, and the visceral hump coiled in a spiral, bearing a shell, or it may be simply cylindrical and the visceral hump completely atrophied.

The nervous system is constructed on the usual gasteropod type, and, together with the sense organs, is highly developed. The eyes, two in number, are enclosed in separate capsules, and are moved by appropriate muscles. They are highly organized, with cornea, iris, lens, and retina, and, like those of the cephalopods to be described farther on, they are much like those of vertebrates, the principal difference being in the relative positions of the fibres of the optic nerve and the layer of rods and cones. The large ears are placed at or near the sides of the cerebral ganglion, and receive their nervous supply from it. The organ of smell is also highly developed, and is placed near the gills when these latter are present, or in a corresponding position when the branchiæ have disappeared. The organ consists of a groove of ciliated sensory epithelium placed just above an olfactory ganglion and innervated in the normal manner.

Owing to the great transparency of the living animals, the study of the internal structure is an easy task; no dissection is necessary to ascertain the principal features, all that is required being to place the animal under the microscope. The adjoining figure of the structure of *Atlanta* was originally made in this way. The head extends forward, more or less like a proboscis, and bears the eyes and usually a pair of tentaeles. At its extremity is the large buccal mass, which can be protruded so as to bring the well-developed lingual ribbon, with its sharp and movable teeth, against its

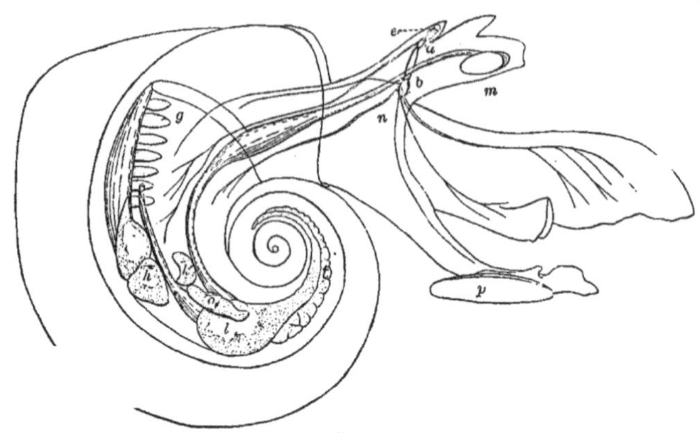

FIG. 463. — *Atlanta peronii*; *a*, brain; b, infra-œsophageal ganglia; *e*, eye; *g*, gills; *h*, heart; *k*, kidney; *l*, liver; *m*, mouth; *o*, ovary; *p*, operculum; *t*, male reproductive gland.

prey. The general positions and relations of the viscera can be seen from our figures, and need not be described at length. Gills are usually present, but not constantly so, occasionally individuals of the same species varying in this respect.

The sexes are separate in the heteropods. The eggs are usually laid in long cylindrical cords, which in the case of *Firoloides* soon break up into short pieces. In the Atlantidæ, on the other hand, the eggs are deposited separately. The young, in their development, pass through a trochosphere and then a veliger stage.

All of the heteropods are predatory animals, feeding on the numerous forms around them. In seizing their prey, the buccal mass and lingual ribbon are extended, and the lateral teeth are bent outwards; then by a muscular movement the curved

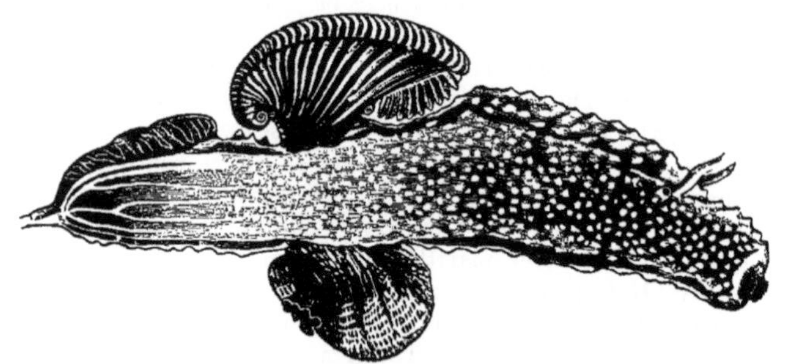

FIG. 464. — *Carinaria*.

laterals are closed against each other, forceps-like, and the prey is secured. The animals are very ravenous, and in their world they play no inconspicuous part in the reduction of the numbers of other animals. Comparatively few naturalists have studied the habits of these beautiful animals, except those found in the Mediterranean, and we therefore copy, sometimes word for word, the account given by Mr. Arthur Adams:—Among the pelagic heteropodous molluscs which we found in crossing the South Atlantic Ocean were vast numbers of *Atlantæ* and numerous *Carinariæ*. They are crepuscular animals like the pteropods, and are furnished with hyaline shells of the greatest delicacy and beauty. The *Atlanta*, with an elegant, glassy, spiral, carinated shell, globose in one species and flattened in another, is quite a sprightly little mollusc, probing every object within its reach by means of its elongated trunk, twisting its body about, and swimming in every direction by the lateral movements of its vertical dilated foot. I have frequently seen them descend to the bottom of the glass vessel in which they were kept, fix themselves there in the manner of a leech by their sucking disc, and carefully examine the nature of their prison by protruding the front portion of their foot in every direction. Almost all pelagic molluscs usually swim on their backs in a reversed position, and although I have seen them commonly swim in this way after capture, they frequently progress feebly with the shell uppermost. When fresh and just taken, I have seen both *Atlanta* and *Carinaria* swim with their bodies in every position; on their sides, on their backs, and with the foot downwards. The *Carinariæ* are swift and rapid in their movements, and dart for-

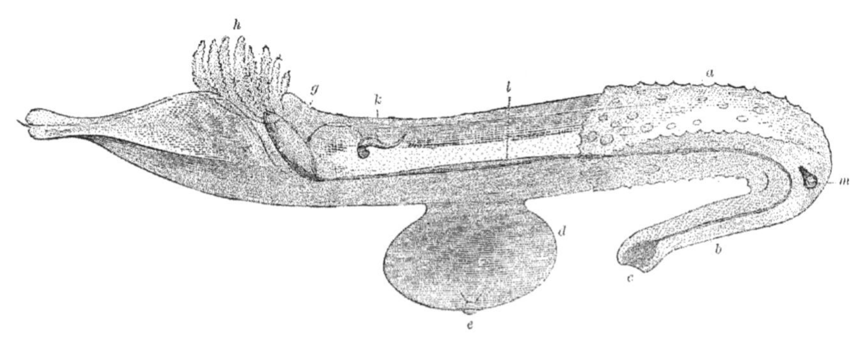

FIG. 465.—*Pterotrachea scutata*; *a*, shield; b, proboscis; *c*, mouth; *d*, float; *e*, sucker; *g*, cloacal sac; *h*, gills; *k*, genital gland; *l*, alimentary canal; *m*, eye.

wards by a continuous effort, moving their foot and caudal appendage from side to side as a powerful natatory organ; they do not progress by sudden jerks like *Atlanta*.

Three well-marked families of Heteropoda are recognized by naturalists. In the ATLANTIDÆ the visceral hump is well developed, and is enclosed in a spiral, glassy shell, large enough to contain the animal when retracted. The shell starts in a dextral spiral, but soon becomes flattened and bilaterally symmetrical. Its edge is provided with a sharp keel, and the aperture is closed on the retraction of the animal by a thin ovate operculum. But two genera are known, *Atlanta* and *Oxygyrus*, containing a total of about twenty-five species, all from the warmer seas. The fossil genus *Bellerophon* possibly belongs here.

In the CARINARIDÆ the visceral hump and the shell have undergone considerable reduction in size, as shown by our figure. The shell is hyaline, and from beneath its margin project the gills. The middle lobe of the foot is alone prominent, and extends from the lower surface of the body like a vertical fin. The tentacles are well developed. The two genera *Carinaria* and *Cardiapoda* contain a baker's dozen of species.

In the PTEROTRACHEIDÆ the reduction of the visceral hump is carried to its fullest extent. The animal forms a long, cylindrical body, relieved only by the vertical foot and the small and inconspicuous gills, which are unprotected in some forms, project but a short distance from the body. The shell is absent, and the foot is provided only in the male with a sucking disc. Two genera are recognized. In *Pterotrachea* (= *Firola*) no tentacles are present in either sex, and the tail is pinnate. In the other genus, *Firoloides*, tentacles are present in the male, but the tail fin is simple and slender, and the gills may be present or absent even in the same species. The egg string of the Pterotracheans is much like that of *Carinaria;* it is long and subcylindrical, and the eggs are imbedded in a single row in the glassy matrix. Oviposition appears to take place the whole year round, but most abundantly between September and March. The *Pterotracheæ* seem to possess great vitality and to withstand injuries which would be certain death to other forms. The loss of the gills is of no especial consequence, and indeed they are so small in proportion to the body that they are of no great use, and are to be regarded as rudiments of respiratory organs which have been retained during a retrograde development. This vitality goes even farther, as the following quotation will show: — "*Anops peronii*, described and figured as having no head, was probably a mutilated *Firola*. 'Such specimens are very common, and seem just as lively as the rest.'"

Sub-Class III. — Pteropoda.

The wing-footed Mollusca is another group, the position of which is far from certain. By some it is made a class equivalent to the acephals and cephalopods, some place it near the base of the cephalophorous series, and one recent author has regarded it as a member of the Cephalopoda. The best authorities are, however, inclined for the present to regard it as a member of the Cephalophora, assigning it a position at the top of the group, a course which we will adopt.

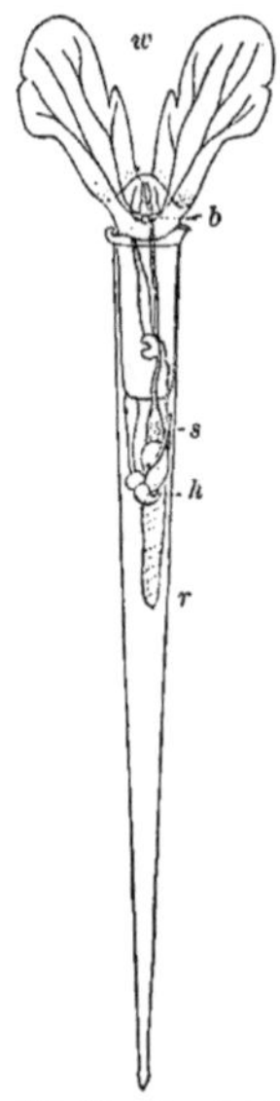

The pteropods are pelagic molluscs which derive their name from the fact that portions of the foot (epipodia) are expanded and fitted for aquatic flight. The body is sometimes long and stretched out in a straight line; at others it is coiled in a spiral. The anterior part, or head, is usually not plainly differentiated from the foot. In some the mantle is not well marked, and the body is naked; in others the mantle is well developed and secretes a glassy, horny, cartilaginous, or even calcareous shell, the two sides of which are almost invariably alike, and into which the whole body can be retracted. At the anterior end is the mouth, which is either surrounded by two tentacles, or six protrusible processes armed with minute sucking discs, or by (as in *Pneumodermon*) two long arms bearing suckers like those of a cuttle-fish. The mouth is provided with jaws and a lingual ribbon, the teeth of which vary between moderate limits, some having twenty-five, and others as few as three in a transverse row. Salivary glands are present, and just behind the entrance of their ducts the alimentary canal widens into a stomach, and then contracts to form the intestine, which, after a number of convolutions, bends on itself, terminating usually on the right side near the edge of the mantle.

Fig. 466.—*Creseis acicula*; *b*, brain; *h*, heart; *r*, reproductive organ; *s*, stomach; *w*, wings.

The circulatory organs are but poorly developed, and a closed system does not exist. The arteries end with wide openings in the body cavity. Here the blood passes about between the tissues through spaces without proper walls, going to the respiratory organs, and thence to the heart. In some forms no respiratory organs are present, the skin serving the purpose, but in others gills occur, none of which (except possibly those of *Pneumodermon*) are the homologues of those of most molluscs. In *Pneumodermon* the gills are borne on the hinder end of the body, but in the thecosomatous forms there are folds in the mantle cavity which subserve respiratory purposes. The kidneys are long, contractile tubes, communicating internally with the pericardial cavity, and externally either with the mantle cavity, or with the exterior. In nervous system the pteropods are very similar to the opisthobranchs. The cerebral ganglia in one group are nearly united and placed above the throat, but in others they occupy widely separated, lateral positions. Two auditory capsules are present, placed near the pedal ganglia, but eyes are either absent or remain in a very rudimentary condition, a fact which accords with their nocturnal habits.

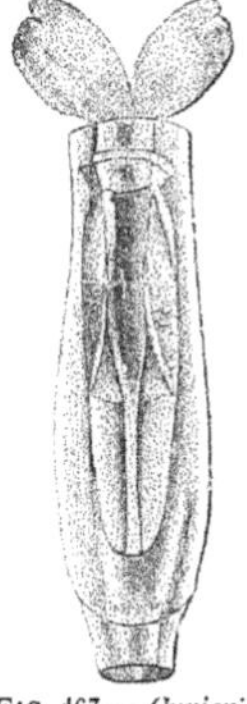

FIG. 467. — *Cuvieria columella.*

The pteropods are hermaphroditic, the genital glands being placed beneath the heart in the visceral sac, and commonly possessing a common duct. The eggs are laid in long cylindrical strings, which float freely about in the sea. The development of some of the species has been followed by the eminent Swiss embryologist, Herman Fol. The larva is provided with a velum and a shell, the latter capable of being closed by an operculum. With growth, the embryonic shell is cast aside, and in some is replaced by a second permanent shell, while others remain naked through life. This presence or absence of a shell in the adult is correllated with other structural peculiarities which served, in the hands of De Blainville, to divide the sub-class into two orders.

The pteropods are mostly small, but few acquiring a length of an inch or more; yet, notwithstanding their small size, they are objects of beauty. They live upon the high seas in all quarters of the globe, at times appearing in vast multitudes, especially at the approach of darkness and during the later moments of twilight. They swim about by the violent flappings of their wings, or they retract these organs and sink at will into the depths of the sea, out of the reach of storms.

ORDER I. — THECOSOMATA.

As the name implies, this division of the pteropods has the body enclosed in a case or shell, and with this protection of the body are associated other structural features, which need to be only briefly alluded to. The head is but feebly developed; indeed, is frequently not distinct from the rudimentary foot which remains in connection with the wings or greatly developed epipodia. Rudimentary tentacles are present.

In the HYALEIDÆ the shell is horny or calcareous, never spiral, but either nearly globular or needle-shaped, the two sides being symmetrical, terminating with one or three sharp points. The cavity of the mantle opens ventrally and contains a horse-shoe-shaped plaited and ciliated gill. Best known in the family is the genus *Hyalea*, or, as it is frequently termed, *Cavolina*, the species of which belong to temperate or tropical seas. The shell is nearly globular, somewhat flattened above and rounded below, while the posterior extremity terminates in three points. The aperture is narrow and is continued by a slit on either side, through which are protruded folds of the mantle (shown in our figure), the function of which is uncertain. The *Hyaleæ* swim in a very erratic manner, darting hither and thither with great rapidity, turning suddenly, and being always on the alert for food. They, like all the rest of the group, are carnivorous, devouring immense quantities of microscopic animals. The species of *Cleodora* swim in a much more leisurely manner, not turning as abruptly as those of the former genus. They have a straight, triangular shell, terminating in a sharp point behind, and in front with a triangular mouth. In *Styliola* (or *Creseis*) the shell is round and needle-shaped, running to an acute tip. The species average about half an inch in length and have perfectly transparent, glassy shells, through which all parts of their internal structure can be distinctly seen. *S. vitrea* occurs on the New England coast. *Cuvieria* is another genus which should receive mention. In early life it is much like *Styliola*, but with growth it partitions off the hinder parts of the shell, and the body moves forward and builds a swollen and more nearly cylindrical shell. The deserted portions of the shell soon become broken off, and the result is the truncated appearance shown in our figure, which illustrates a species common to the Mediterranean and the tropical Atlantic. I have seen numerous specimens, each about half an inch in length, collected between the Bermudas and Florida.

FIG. 468. — *Hyalea tridentata.*

Here, in all probability, belong the problematical fossils known as *Conularia* or 'cone-in-cone.' They first appear in the Silurian, and some reach, for pteropods, an enormous size, an Australian species being estimated to have had a length of about sixteen inches. They have a four-sided shell, with the apex partitioned off by narrow, closely placed septa, so that the whole resembles a series of cones, or rather of pyramids, nested within one another. Those other long and slender fossils, ornamented with rings, and known Tentaculites, should also be placed here. They occur in Silurian and Devonian rocks.

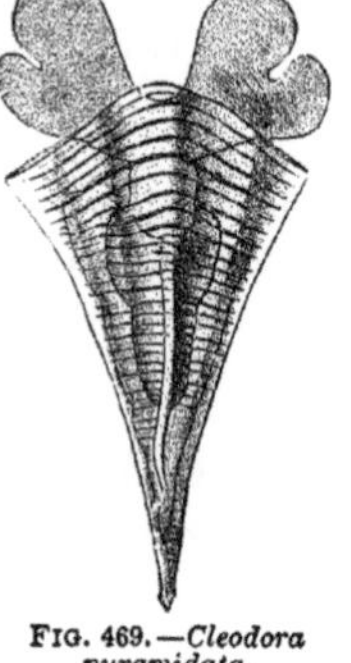

FIG. 469. — *Cleodora pyramidata.*

The LIMACINIDÆ are readily recognized by the spirally-coiled shell, the whorls of which are sinistral. The mantle is large and opens dorsally. The species are largely inhabitants of the colder waters, some being found in the Arctic and Antarctic seas. On our coast a member of the family, *Spirialis gouldii*, is not very uncommon in the evening, both north and south of Cape Cod. Mr. Alexander Agassiz kept several in confinement for a while, and observed that during the day they were quiet, and rarely left the bottom of the jars in which they were confined, but after dark they became very active and arose to the surface.

The CYMBULIDÆ are noticeable for their comparatively large size and the very peculiar shell which they secrete. In early life, like the rest of the group, they have a small, spiral, horny shell, but this becomes lost and in its place the animal secretes a cartilaginous slipper-shaped shell, apparently possessing no more consistency than ordinary gelatine jelly. In this thick, transparent, flexible shell sits the mollusc, like the old woman in her shoe, paddling about by the large oval wings; but, unlike the other shelled pteropods, it has not the power to retract itself within the aperture. Two genera are known; *Cymbulia*, with a slipper-shaped shell pointed in front and square behind, and *Tiedemannia*, with a body much like that of *Cymbulia*, but with the shell internal. *Tiedemannia* can change its colors at will by means of chromatophores much like those of the cephalopods soon to be described.

Order II. — GYMNOSOMATA.

The Gymnosomata are naked pteropods, in which the head is distinct and well separated from the body and the foot, and in which well-developed tentacles are present. The wings are distinct from the foot, and external gills are present in one family. The young at first are provided with a shell and swim by means of a velum, but soon both these embryonic structures are lost, and, pending the development of the wings, the larva swims by means of bands of cilia which surround the body, as shown in our figure of the larva of *Pneumodermon*.

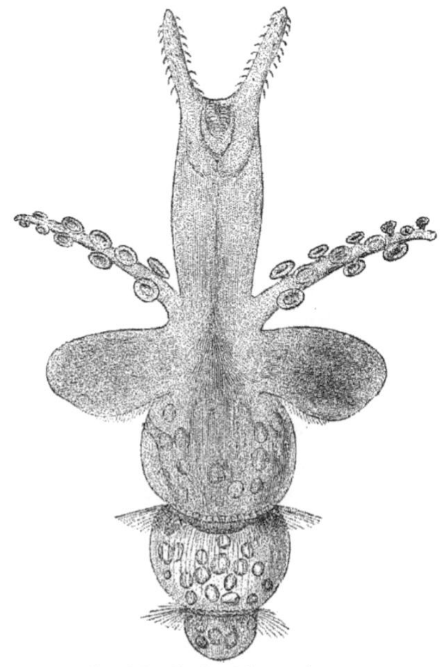

FIG. 470. — Larva of *Pneumodermon*.

In the CLIONIDÆ the body is spindle-shaped, and the head bears tentacles which are without suckers. In *Clione* there are two of these tentacles, and, besides, from either side of the mouth arise three pairs of extensible processes armed with minute suckers. The number of these suckers is enormous. In the fresh specimen the appendages are reddish in color, but under a low power of the microscope the color is seen to be due to a number (about three thousand on each tentacle) of red cylindrical sheaths, and upon the use of a higher amplification each of these cylinders is seen to contain about twenty suckers placed on slender stems, and capable of being protruded from the mouth of the sheath. Now, since a sheath contains twenty suckers, and since there are about three thousand sheaths on each appendage, a simple mathematical operation shows us that each *Clione* is furnished with about 360,000 suckers.

Clione borealis, as its name implies, is an Arctic form which is frequently seen in vast patches. Together with *Limacina borealis*, a species belonging to the last order, it furnishes the principal food supply of the whalebone whales. The whalers call it 'brit,' and the presence of one of these schools of brit is considered as indicative of a good whaling ground. *Clione papilionacea* occasionally occurs on the eastern coasts of the United States, and has been reported as far south as New York. It may be that its apparent rarity is due to the facts that it is a northern form, and that naturalists are not in the habit of trailing the surface net in the winter. This species reaches a length of about an inch and a half.

In the PNEUMODERMONIDÆ the body has much the same shape as in the other, but it is distinguished by the presence of posterior external gills, and two extensible arms bearing suckers much like those of the cephalopods. Specimens of *Pneumodermon* are common in the Mediterranean and in the warmer Atlantic. They swim strongly, but when touched by a foreign object they roll themselves up like an armadillo, and, feigning death, sink until out of the reach of apparent danger.

CLASS III.—CEPHALOPODA.

We have now reached the highest group of the molluscs, the one which embraces the squids and cuttle-fishes, and which has given rise to many of the tales and legends of the sea-serpent and other marine monsters. But before indulging in these stories, we must first look at some of the structural and embryological features which mark off the cephalopods from the rest of the molluscs, a distinction which has always been recognized, though the reasons therefor have not until recently been scientifically formulated. This confusion has been due to the fact that the cephalopods are a highly organized type, which, in course of a long residence on the earth, have eliminated all traces of a metamorphosis from their development, and which consequently progress in a straight line from the egg to the adult.

The body of the cephalopods is enveloped in a mantle, open only at the anterior end, from whence proceeds the head, separated from the body by a slight constriction or neck. On either side of the head is a large eye, the structure of which will be described farther on. Beyond the eyes arises a circle of usually eight or ten lobes, or 'arms,' in the centre of which is the mouth. Part of the difficulty in homologizing the cephalopods with the other gasteropods has arisen in connection with these arms. They were early recognized as corresponding to the foot, and, as they arose from the head, the name Cephalopoda (head-footed) was applied to the group. At a later date, the siphon (a structure to be described immediately) was studied, and again the foot was recognized. Which was right? Further investigation showed that both were correct. The arms correspond to the anterior division (propodium), the siphon to the middle one (mesopodium), while the metapodium is absent, or represented by a rudiment, forming a valve in the siphon of some species.

The arms, or tentacular lobes, as we have just said, surround the mouth. In the Nautilus they bear peculiar tentacles, capable of being retracted into tubular sheaths. In the rest of the cephalopods they are longer, and are armed on one side with spherical sucking organs,—acetabula, they are called. Each one of these sucking organs has a rim, which can be closely applied to any foreign body, and the hold increased by means of numerous fine hooks around the edge. In the centre is a moveable portion, which, from its physiological resemblance to the plunger of a pump, has received the name of piston. The animal applies these suckers to any foreign object, and then pulls out the piston. Besides these suckers, certain arms of some species bear numbers of sharp, recurved hooks. Of the manner of capturing the prey we shall speak later.

Morphology shows (in a way which need not here be discussed) that these arms and tentacles correspond to the first division of the foot (propodium) of other molluscs. Next comes the siphon, the homological mesopodium. This is a tubular structure, arising from what is usually called the ventral surface of the body and projecting a short distance outside the mantle. In *Nautilus* the two lobes of the siphon are separate, being merely folded one upon the other in order to form the tube, but in all other living forms this condition is characteristic of an embryonic stage, and, long before the adult is reached, the two halves unite.

The shape of the mantle depends on that of the body. In all it is a cup-shaped or conical envelope, attached to the rest of the animal by a narrow line on the dorsal surface. Its anterior margin is free, but is provided with little depressions which fit over corresponding cartilages on the siphon, so that the only communication between the cavity of the mantle and the exterior may be through the siphon. The mantle is highly muscular and in life is constantly expanded and contracted, alternately filling and emptying the cavity. The water thus taken in subserves the purposes of respiration, and also plays an important part in locomotion. The animal takes the water in through the free space between the neck and the mantle, and then, if using it only for respiratory purposes, passes it out again in the same way; but if it desires to change its position, the mantle is hooked to the siphonal cartilages, and then the water is forced out through the siphon. Changes in direction are provided for by the flexibility of the end of the siphon. When the animal wishes to go in a backward course, the siphon is left in its normal condition; when in a forward direction, the end of the siphon is bent over the edge of the mantle, so that the current is directed toward the end of the body, and by the reaction locomotion is effected.

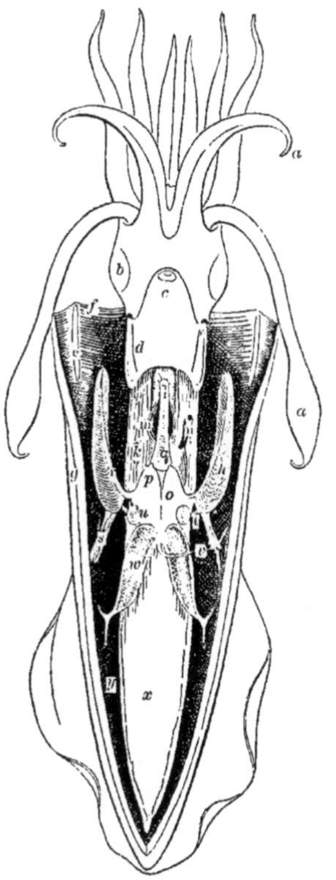

FIG. 471. — External anatomy of *Loligo pealei*, as shown by cutting open the mantle; *a*, arms; *b*, eye; *c*, siphon; d, cartilages of siphon, over which hook those of the mantle, *e*; *f*, free edge of mantle; *g*, mantle; *h*, gills; *i*, rectum; *m*, ink-bag; *n*, penis; *q*, intestine; *r*, Veins from gills; *s*, gill muscles; *t*, branchial artery; *u*, branchial heart; *v*, mantle artery; *w*, posterior Venæ caVæ; *x*, Visceral sac; *y*, caVity of mantle.

In all but a very few of the cephalopods, the mantle secretes a shell, which may be either horny or calcareous. In the young of all, a shell gland appears exactly as in all other molluscs, but in the great majority of the living forms it soon becomes enclosed by the ingrowth and coalescence of the edges, and hence the shell is internal. In this case it is usually much longer than broad, and serves to give strength and stiffness to the otherwise soft body. In *Spirula* an exception is found. The shell is enclosed in the body, but the gland which forms it does not become completely closed. The shell, also, is not a long flat object, but is tubular and coiled in a spiral, and, moreover, is partitioned off and traversed by a slender tube, much like that in the shell of *Nautilus* now to be described. In this genus the shell is external, but the early stages

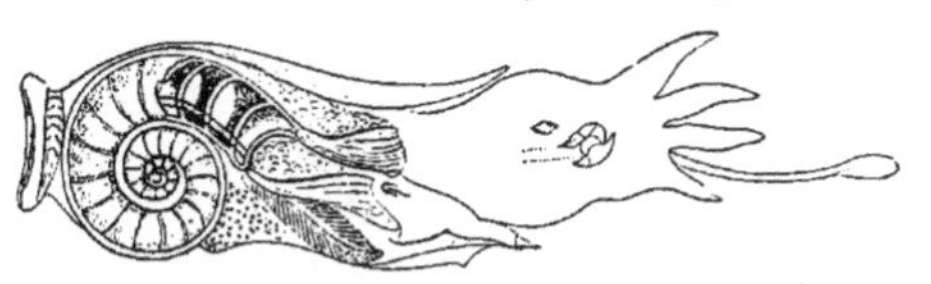

FIG. 472. — Section of *Spirula australis*, showing the relations of the shell to the animal.

of its growth have not yet been studied. In the adult it is coiled in a flattened spiral, like that of a *Planorbis*. A very important distinction is, however, to be noticed in the relations of the shell to the animals in these two forms. In *Planorbis* the foot is turned toward the inside of the coils; in *Nautilus* the reverse is the case. The shell, to exactly correspond, should be unrolled and coiled in the opposite direction. Besides this peculiarity the shell of the *Nautilus* is chambered. At regular periods of growth the animal moves forward in its shell, and partitions off a chamber by means of a calcareous partition; again and again is the process repeated until the result is the series of cells so familiar to all in the sawed shell of the pearly nautilus.

A similar partitioning off of the shell is somewhat common among the gasteropods, and several instances have been mentioned in the preceding pages, but in *Nautilus* and its allies a feature is introduced unknown elsewhere in the Mollusca, and, indeed, in the animal kingdom. In these forms the partitions of the chambered shell are traversed by a cord-like pedicle arising from the central portion of the dorsal region of the body. This pedicle, in passing through the partitions, forms a more or less tubular series of openings known as the siphuncle, and by this tube any recent or fossil shell may at once be recognized as belonging to the Cephalopoda. The deserted chambers are filled in the living specimens with a gas, the purpose of which seems to be to lessen the specific gravity of the animal. The gas is said to be a mixture of oxygen and nitrogen, the latter in greater proportion than in the atmosphere. How it obtains entrance to the chambers is unknown; possibly it is secreted by the animal, as is the case in many other animals. *Arcella* (one of the Protozoa allied to *Difflugia*) fills its shell with gas; the float of the siphonophores acquires its buoyant qualities from the contained air, while in the vertebrates the gaseous contents of the air-bladder in the bony fishes will at once suggest itself in this connection.

The mode of the formation of the chambers and the forward movement of the animal are also obscure. It would, however, appear probable that there is a periodicity in the operations, and it has been suggested that "each septum shutting off an air-containing chamber is formed during a period of quiescence, probably after the reproductive act, when the visceral mass of the Nautilus may be slightly shrunk, and gas is secreted from the dorsal integument so as to fill up the space previously occupied by the animal."

Between the shell of the Nautilus and that of other forms a considerable range occurs, the study of which requires a reference to both living and fossil forms. A series can be arranged from large shells like that of the Nautilus, leading down through small and internal but still chambered forms, like that of *Spirula*, and others in which the chambers still persist, but are strengthened by the addition of external layers, as in the fossil Belemnites; next come forms in which the shell proper disappears, and the super-added lamina alone appear (*Loligo*), and, lastly, all trace of a shell is lost in the adult *Octopus*. Besides this, another series may be traced from the closely coiled shell of the Nautilus and the Ammonites through fossil forms like *Crioceras*, *Ancyloceras*, *Anisoceras*, etc., until we reach the perfectly straight shells of the palæozoic Baculites and Orthoceratites. In the septa of the shells a similar gradation may be seen, reaching from the simple ones of Nautilus and the Orthoceratites to the highly complicated ones found in the Ammonites. One more shell is all that our

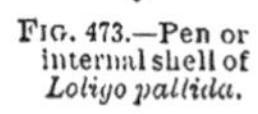

FIG. 473.—Pen or internal shell of *Loligo pallida*.

space will at present allow us to allude, that of the paper sailor, *Argonauta*. In this genus the shell is without connection with the body, but is secreted by two of the arms, which are greatly expanded for the purpose, and serves as a nest for the eggs.

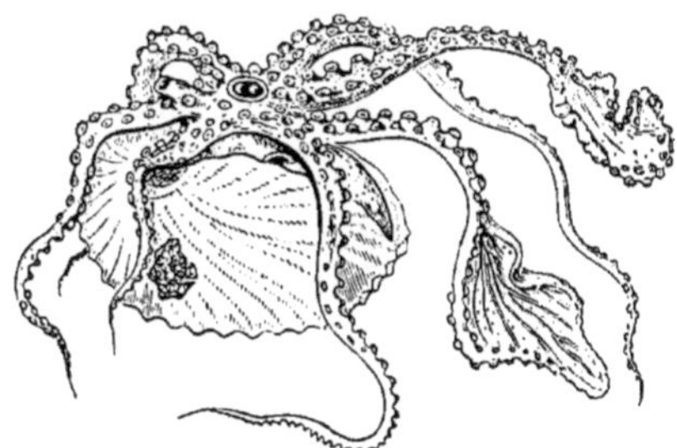

FIG. 474. — *Argonauta* with shell broken to show the eggs.

The mouth of the cephalopods is armed with a pair of beaks like those of a parrot, with sharp cutting edges. In the Nautilus these beaks are calcified, but in all other living forms they are horny. Between the jaws is the tongue, and below them is a lingual ribbon, to all intents and purposes like that of the cephalophorous Mollusca. Behind the lingual ribbon begins the usually narrow œsophagus, into which one or two pairs of salivary glands pour their secretions. The œsophagus empties into the stomach, which is of moderate size, but strong and muscular, and provided with a blind appendage (cæcum), which frequently is larger than the stomach itself. From the stomach the intestine goes forward in a course nearly parallel to that followed by the œsophagus, and terminates in a median vent on the lower surface of the body just within the siphon. The liver is large, and opens into the cæcum. In the dibranchiate forms there is an ink-bag, borne near the intestine, and pouring its inky secretions into the rectum, or into the mantle cavity through an opening near the anus. The ink is brownish or black, and is used to create a cloud in the water when the animal wishes to escape from some danger. The ink is not readily decomposed; on the contrary, it is occasionally found fossil in the rocks along with the remains of the animal which produced it. So well has it been preserved that in one celebrated instance a naturalist drew the portrait of a fossil squid with the sepia derived from its fossil, but not fossilized, ink-bag.

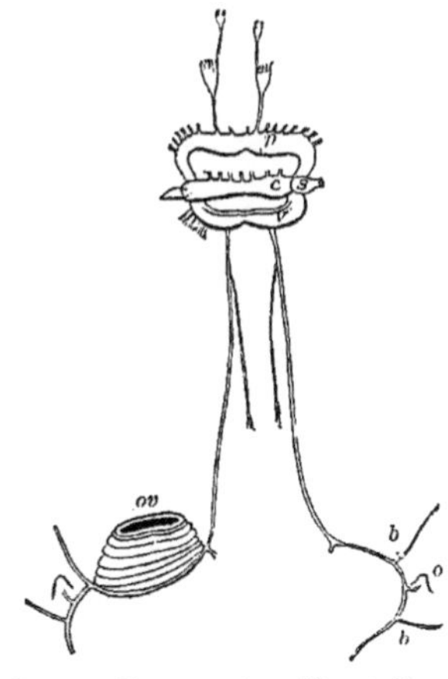

FIG. 475.—Nervous system of female *Nautilus*; *b*, branchial nerve; *c*, cerebral ganglia; *o*, olfactory organ; *ov*, oviduct, *p*, pedal ganglion; *v*, visceral ganglion; *s*, optic nerve.

The circulatory system is well developed in all the cephalopods, especially in the higher groups. The details would prove dry reading, and so we will merely mention a few facts. There is only one ventricle, but the auricles may vary from two to four, in accordance with the number of gills. On each vein going to the gills is a pulsating vesicle, the branchial heart. It is in the problematical appendages to these that the curious worms, Dicyemida, referred to on a preceding page, have been found. In some true capillaries are developed, while in others the arteries terminate in the lacunæ of the body. We have just referred to the existence of one or two pairs of gills. This character is associated with others, and is used in the naming and definition of the two primary groups, Dibranchiata and Tetrabranchiata. Correllated with this variation in the number of gills and hearts is one in the kidneys, which are also two or four in number. It is in these points only that any metamerism of the cephalopods is noticeable.

The nervous system is greatly concentrated, the cerebral, pleural, and pedal ganglia being placed close together around the throat. The organs of smell have been found only in the Nautilus, where they occupy the normal position at the bases of the gills. In this genus, as well as in all others, close by the eyes are the openings to a pair of pits, which have been supposed to have olfactory functions. The ears are two in number, and are small pits or vesicles which in the adult become entirely shut off from the exterior. They are placed in the sides of the head below the eyes, and each contains a single otolith.

In the Nautilus the eye has not reached that high development found in the squids and cuttle-fish, but in fact it is extremely simple. On either side of the head are two prominences which contain the eyes. In the centre of each is a minute hole communicating with a large internal chamber filled with sea-water. The inner wall of this is lined by the retina, in which the nerves terminate. As will be seen, this eye is extremely simple. There is no lens, no cornea; nothing but the features enumerated. All are familiar with the fact that a lens is not necessary for the production of an image. If we admit the light to a darkened room, by means of a very small hole, so as to prevent any superimposition of images, a picture of external objects can be thrown on a screen. This is the physiology of the eye of the Nautilus. In the Dibranchiata the visual organ is very complex, and it is to be noticed that in its development it passes through a stage closely similar to that which persists through life in the Nautilus. With development it goes farther,

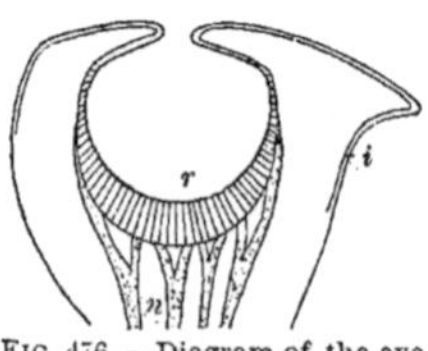

FIG. 476.—Diagram of the eye of *Nautilus*; *c*, cavity; *i*, integument; *n*, nerves; *r*, retina.

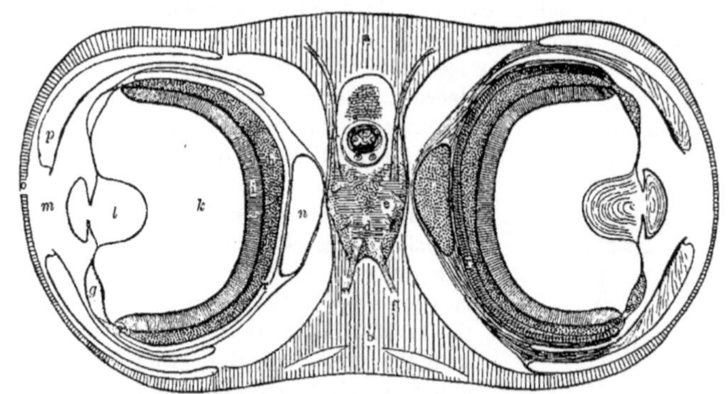

FIG. 477.—Section through the head of a young *Loligo* to show the structure of the eyes; *c*, œsophagus; *d*, *e*, pedal ganglion; *g*, optic ganglion; *h*, ganglionic layer of retina; *i*, layer of rods; *k*, posterior chamber of the eye; *l*, lens; *m*, anterior chamber; *o*, cornea; *p*, iris.

and the result is an optical organ which presents some startling analogies to that found in the vertebrates. When carefully studied it is seen that this resemblance is superficial, and that fundamentally the two are entirely different. In both the eye is provided with a cornea, then comes an iris with a circular pupil, and next a lens dividing the cavity into two chambers. Here, however, the correspondence stops. In the squid the posterior lining of the inner chamber is the retina, with its rods and cones, and then comes the ganglionic layer. In the vertebrates the relations of these two are reversed. Embryology reinforces these differences, and shows that the eye of the cephalopod is greatly different in its development from that of the vertebrate.

Among the peculiar features of the cephalopods should be mentioned the chromatophores, by which the animal is enabled to change its color quickly. Such structures are found elsewhere in the animal kingdom, but here they attain the greatest development. They are best studied in the young just after hatching, as then the whole animal can be placed under the microscope and examined with comparatively high powers. Each chromatophore is in reality a cell filled with pigment, and expanded by suitable muscles attached to its wall. These cells are placed just beneath the epidermis, and, when expanded, the contiguous ones nearly touch, so that the animal appears of a uniform tint. When the chromatophore contracts, as it does by its own elasticity, the color is condensed so that it appears as a very minute black dot, and then the translucent, almost transparent, character of the tissues is seen. The expansion and contractions of the chromatophores take place with considerable rapidity, so that the changes in color are remarkable. There are several sets of these chromatophores, each distinguished by its color, yellow, red, blue, or brown; and each set expands independently of the others, so that an animal almost transparent will suddenly turn bright red, instantly change to a blue, and then a yellow or a brown, or back to its previous colorless condition. The chromatophores in different parts of the body are also independent, and sometimes the fins will be bright blue and the anterior part of the mantle an equally vivid red. When examined under the microscope the whole of these changes can be carefully studied, and then it is found that the chromatophores are suddenly expanded, but that their contraction is accomplished in a more leisurely manner. When expanded they have a diameter of about a fiftieth of an inch, but when reduced to their smallest size they measure but one two hundred and fiftieth of an inch. The purpose of this change of color is protective; and as by it the animal can assimilate its color to any surroundings, it can readily escape observation by any fishes on which it may wish to feed or which might otherwise feed on it. In the Nautilus these ebromatophores do not appear to be present, though they exist in all other forms. With this capacity for change it will readily be seen that color is a character of no importance in discriminating the species of cephalopods.

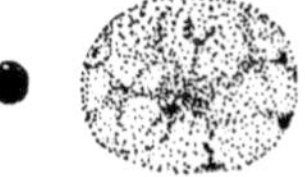

FIG. 478. — Chromatophore of squid, expanded and contracted; enlarged.

The sexes of the cephalopods are separate. Of the internal reproductive organs we need say nothing, but in connection with the reproductive act there are some very interesting features. Many years ago Cuvier in dissecting a female *Parasira*, found attached to it a peculiar worm-like object furnished with numerous suckers. He regarded it as a parasite, and gave to it the name *Hectocotylus*. At about the same time the Italian naturalist Delle Chiaje found a similar structure in *Argonauta*, which he called *Trichocephalus*. Several years elapsed before the true nature of these structures was recognized. It was then discovered that this supposed parasitic worm was in reality one of the arms of the male, which separates and serves to convey the male reproductive element to the female. This 'hectocotylization' reaches its highest development in the two genera named above. In these, one of the third pair of arms (right or left, according to the genus) becomes thus modified. Just before the period of reproduction, this arm appears as a nearly globular sac. This then bursts, and from it issues an arm

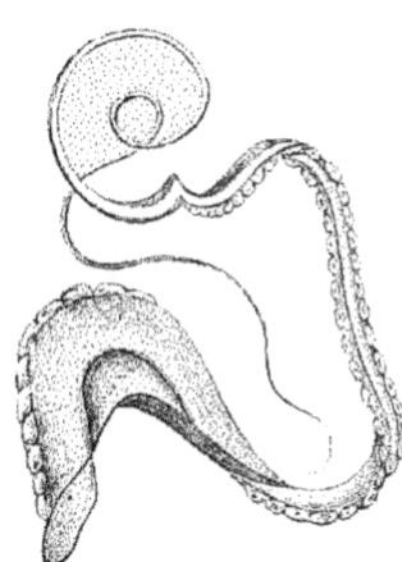

FIG. 479.—Hectocotylized arm of *Argonauta* after separation from the male.

larger than the rest, which the male then charges with one of the packets of spermatozoa soon to be described. During the coitus this arm becomes detached and

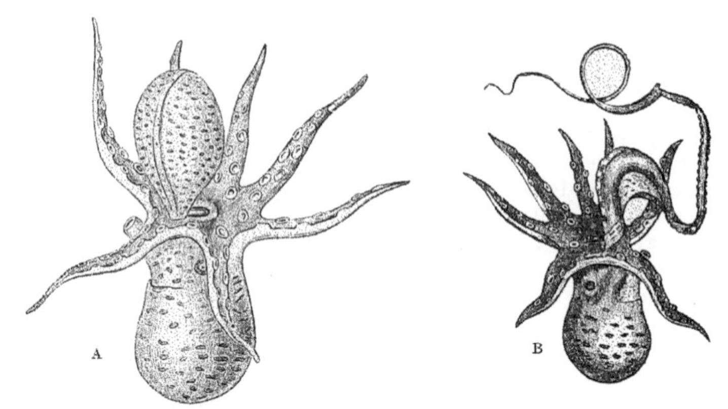

FIG. 480. — Hectocotylization in the male *Argonauta;* in A is shown the sac which in B has burst open, allowing the hectocotylized arm to extend itself.

remains adherent to the female, which then swims away. A new arm is formed in the male before the return of the breeding season.

This subject of hectocotylization has been studied by Steenstrup, a Danish naturalist, who finds that the arm or extent of modification is variable in the different genera, but in some (among which is *Ommastrephes*, so common on our coast) this process has not been observed. We have just alluded to the packets of spermatozoa. These are called spermatophores and are found variously modified in different invertebrates. In *Loligo*, one of the squids which we select as a type, it consists of a slender capsule about half an inch in length, two thirds of which is occupied by the spermatozoa, while the remainder contains a complicated discharging apparatus. On contact with the salt water the sheath ruptures at one end, and forth issues the discharging apparatus, dragging the spermatozoa after it.

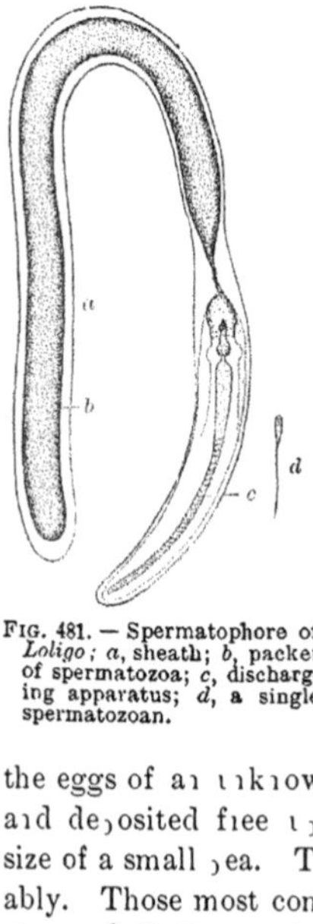

FIG. 481. — Spermatophore of *Loligo; a*, sheath; *b*, packet of spermatozoa; *c*, discharging apparatus; *d*, a single spermatozoan.

The eggs of the cephalopods are usually enveloped in capsules, some of which contain as many as two hundred eggs. This is not the case with all, for the writer once dredged the eggs of an unknown cephalopod (? *Rossia*) which were separate and deposited free upon the ocean's bottom. They were about the size of a small pea. The shape and size of the capsules vary considerably. Those most commonly found on the New England coasts are those of *Loligo pealei*. They are finger-shaped and gelatinous, and so transparent that the embryos can be readily seen. Usually from twenty-five to a hundred of these

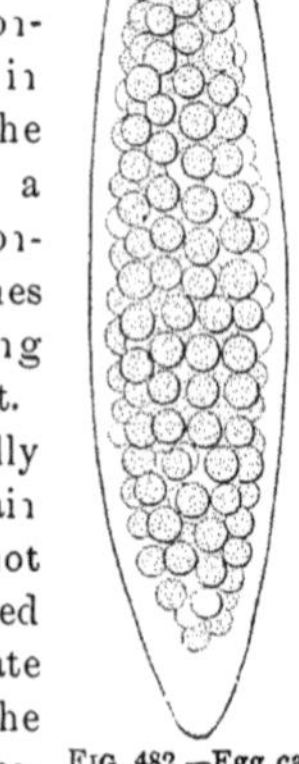

FIG. 482.—Egg capsule of *Loligo*.

capsules are united in a bunch, but I have seen one mass of the eggs of this species which would more than fill a bushel basket. Owing to the great amount of food-yolk present, the eggs of all cephalopods are very large, and the development is extensively modified from the same reason.

The development of but few of the cephalopods has been studied, and all that is known throws but little light on the relationships of the group. A knowledge of the development of *Nautilus* is a great desideratum. The following account relates largely to *Loligo pealei*, the only species studied by the writer. Owing to the great quantity of food yolk and its distribution, the segmentation of the egg is at first confined to one pole, and the result is that the first stages of the blastoderm are very like those in the chick. The blastoderm gradually increases until it envelopes the whole yolk, but before that stage is reached some of the organs begin to be outlined. First to appear are a shallow pit, the shell gland, at the extremity of the body, and two others closely similar, the rudiments of the eyes, on the sides of the body. Then a fold arises, the first traces of the mantle. The two siphonal folds next appear, and as development progresses they take the form of two distinct flaps, a condition which is permanent in *Nautilus*, and then the edges fuse together. The first appearance of the arms is in the shape of simple prominences. At this time the yolk extends as a huge mass from between the arms, but, as growth continues, it is gradually absorbed and transformed into food for the rapidly increasing tissues. Thus the development is direct, and not a trace of a metamorphosis appears. Even the veliger and trochosphere conditions are lost. The reasons for this are to be found in the very long history of the group, together with the high point to which they have attained. In their structure they are, as we have seen, the highest of the Mollusca, and hence the transition from the egg to the adult is very extensive. On this account there is a tendency to drop all useless and larval features, and to pursue the shortest course. This tendency exists everywhere in the animal kingdom, but of course it requires time to eliminate them all. Time the cephalopods have had. In the lower Silurian rocks, fossil forms not generically distinguishable from living *Nautili* are found, and the tetrabranchs flourished in palæozoic time, while forms assigned to the dibranchs first appeared in the triassic strata.

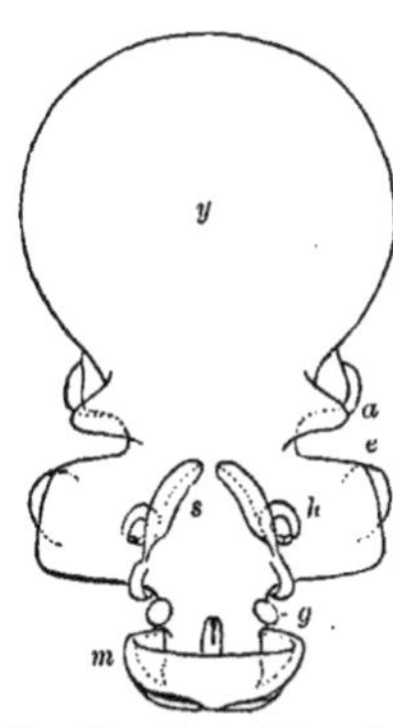

FIG. 483. — Ventral view of a rather early embryo of *Loligo*; *a*, arms; *e*, eye; *g*, gill; *h*, ear; *m*, mantle; *s*, siphonal folds not yet united; *v*, vent; *y*, yolk sac.

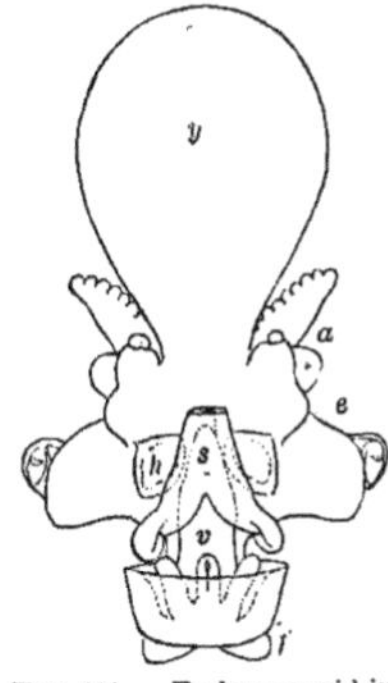

FIG. 484. — Embryo squid in which the siphonal folds (*s*) have united, and the suckers or acetabula are appearing upon two pairs of the arms; *a*, arms; *e*, eye; *f*, fin; *h*, ear; *v*, vent.

SUB-CLASS I. — TETRABRANCHIATA.

We have incidentally alluded in the preceding pages to some of the characters which serve to separate the cephalopods into two groups, the names of which have reference to the number of gills. The lower of the two, the Tetrabranchiata, are represented by but a single living genus, *Nautilus*, and from that we have to derive all our knowledge of the soft and perishable portions. These characters may be briefly enumerated as follows: The two siphonal folds are never united; the tentacles are numerous and never bear suckers; the gills are four in number, and, corresponding to them, four auricles are present; the renal organs are also in two pairs, and branchial hearts do not exist. The character of the eye has already been described. The shell is either straight or curved, never internal, and is divided into a series of chambers. Two well-marked families exist, but lately they have been broken up into a large number of groups, to which family rank has been accorded, but with these we need not concern ourselves.

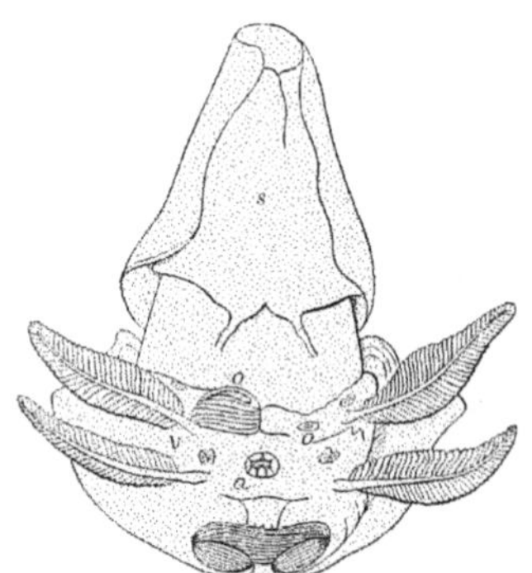

FIG. 485. — Under surface of a part of a *Nautilus*, showing the siphon (*s*) and gills; *a*, vent; *o*, oviducts.

In the NAUTILIDÆ, the partitions forming the chambers of the shell are simple, and concave on the outer surface; the siphuncle occupies a nearly central position; the aperture is simple, and the external surface is nearly or entirely smooth. Over two thousand species have been described, of which only six are living at the present time, all belonging to the genus *Nautilus*, in which the whorls of shell are few, but closely coiled in a flat spiral. The shells are very common, but the animals are rarely seen; this may partly be due to the habits, but more from the fact that the natives of the South Seas eat the animal when obtained, but use the shell for barter. These shells are composed of two layers, the outer one being dull and opaque, but the inner one is pearly, whence all the species derive the common name pearly nautilus. Specimens converted into cameos by removing parts of the outer layer by means of acid are very common, some being really artistic; this point, of course, depending entirely on the skill of the artist. The *Nautili* live in comparatively shallow water, where they creep slowly over the bottom, or swim by means of the water forced out of the siphon. They live on animal matter, and the traps in which they are caught are baited with crabs, sea-urchins, and the like.

Of the many fossil genera we need say but little. They occur in all the rocks from the lower Silurian to the present time. In shape, all transitions may be found, from the perfectly straight *Orthoceras* to flat spirals like *Nautilus*, and conical spirals like *Trochoceras*. Some of the species of the former genus grew to an enormous size, and Prof. J. S. Newberry estimates that the fossil *Orthoceras titan* weighed some tons.

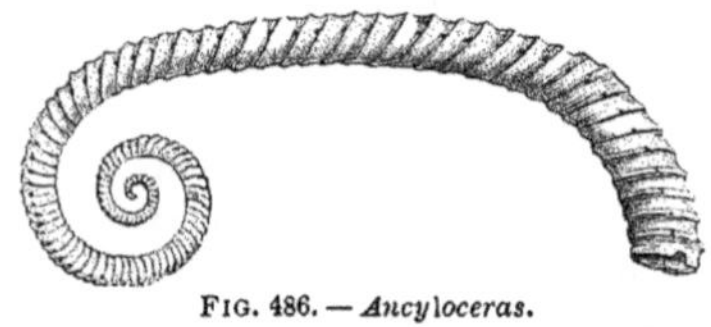

FIG. 486. — *Ancyloceras.*

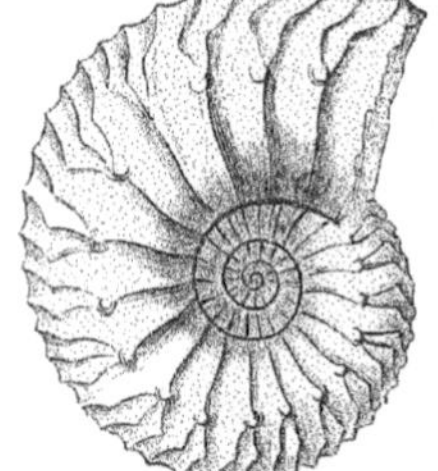

FIG. 487. — *Ammonite.*

In the AMMONITIDÆ the septa are lobed or folded at their margins, so that the line of their insertion into the shell frequently has a bushy or dendritic appearance. Associated with the *Ammonites* frequently occur horny or shelly plates called aptychi or anaptychi, the relations and functions of which are entirely problematical. None of the family are now living, and but a few extended into the tertiaries. The bulk of the species belonged to the mesozoic rocks. In form they vary almost exactly as do the members of the last family, loosely coiled and straight forms being common. One of the most interesting of modern investigations was that conducted by Professor Alpheus Hyatt on the embryology of these extinct forms. Every specimen contains in its interior the shell possessed by the embryo, and by breaking open the fossil this can be examined and studied. In this way, by studying large collections, Professor Hyatt arrived at a clear idea of the development of the group, and consequently of the manner in which the numerous genera should be arranged in order to exhibit their affinities.

Sub-Class II. — Dibranchiata.

In its characters this group differs considerably from the last. As the name implies, but one pair of gills are present, and as a consequence but one pair of auricles exists. The circulation, however, is reinforced by a pair of contractile organs (branchial hearts they are called) on the vessels which convey the blood from all parts of the body to the heart. A pair of kidneys are present. The folds of the siphon are united to form a complete tube; and the arms which surround the head are furnished with sucking cups. The shell, when present, is internal, and may be either calcareous or horny; among those with horny shells, it takes the shape of a flat plate or a long style, while in others it may be chambered or flat. An ink bag is always present. This is the group which contains the forms commonly known as squid, cuttle-fish, kraken, poulpe, and the like, and embraces some of the largest animals now living, as well as numerous smaller forms. Thanks to the labors of Prof. A. E. Verrill, the species found on the northeastern coasts of America are as well known as those from any part of the world. The number of arms is made the basis of division into two orders.

Order I. — OCTOPODA.

In this division the arms which surround the mouth are eight in number, and are covered with sessile suckers in which the horny lip is absent; the eyes are small, and can be covered by the skin, which closes over from all sides; the body is short and nearly spherical, and lacks either internal or external shell, and usually also the fins so characteristic of most of the decapodous forms; the mantle is without cartilage, and on the dorsal surface is united to the head by a broad band. The siphon is without valves, and the oviducts are paired in all except *Cirrhoteuthis*, where the right one is aborted.

In the Cirrhoteuthidæ a peculiar modification of the arms occurs; they are united nearly to their tips by a thin membrane, so that the whole corresponds to an umbrella, the eight arms answering to the ribs, while the rather long body bears a fin on either side. The genera are *Cirrhoteuthis* and *Stauroteuthis*, and the known species come from the northern seas. Of the latter genus but a single specimen is known; it is called *S. syrtensis*, and was taken about thirty miles east of Sable Island, Nova Scotia. *Cirrhoteuthis mülleri* is found in the Greenland seas.

The next family, the Philonexidæ, has no fins; the mantle has an apparatus wherewith it may be locked to the siphon; the arms are united to a greater or less extent by a membrane, and the dorsal ones are most developed; there are several pores in the surface of the head which communicate with aquiferous cavities. In this family hectocotylization reaches its greatest extent, the third arm of the right or left (*Argonauta*) side being thus modified. The animals swim well.

The species known as *Parasira catenulata* occurs on both sides of the Atlantic, specimens being most numerous in the West Indies and the Mediterranean. A few years ago a single specimen was found on the southern New England coast. Two species of the genus have been described, but, according to Steenstrup, the differences are sexual, the form known as *P. carena* being the male. The flesh of this species is tough and unwholesome. The ventral surface of the body is ornamented by tubercles and reticulating ridges. Closely allied is the genus to which Professor Verrill has applied the name *Alloposus*, but which at the same time presents resemblances to the last family in the membrane which unites the arms for about two thirds their length. Large female specimens of the only species (*A. mollis*) weigh over twenty pounds, and have a total length of thirty-two inches.

The only other genus of the family which we need to mention is *Argonauta*, the one which embraces the paper sailors, so well known for the delicate and beautiful shell which they secrete. This shell, however, is not homologous with that of other molluscs, as is seen by the method of its formation. In the female the two dorsal arms are expanded into two broad membranes at their extremities, and it is these two membranes that secrete the shell. The purpose of the shell is merely to protect the eggs, and, although the female sits in it, she has no organic connection with it. This fact, taken with the finding of female paper sailors without shells, led to many a dispute among the older naturalists, and for a long time it was maintained that *Argonauta* took the shell of some other animal. These delicate egg-nests, which are produced by the females alone, are frequently washed ashore in tropical countries in large quantities. On our shores they are rare, only two with the animal having been reported, though about a dozen dead shells have been dredged by the U. S. Fish Commission a hundred miles south of the New England coast.

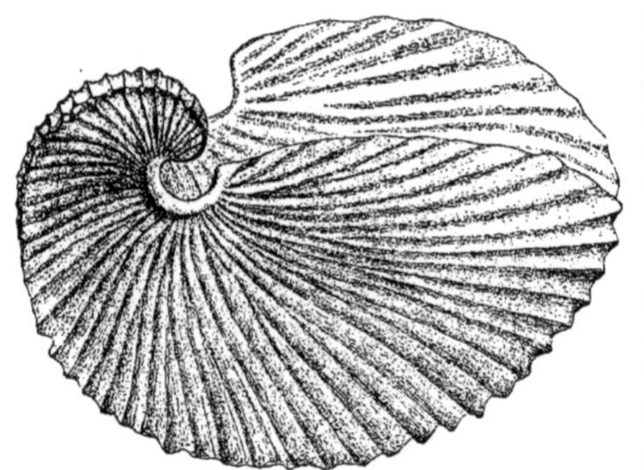
FIG. 488. — Shell or egg-nest of *Argonauta argo*, the paper sailor.

The paper sailor is often described at the present day as taking advantage of fine weather and coming to the surface, when it lifts its broad arms and sails away before the breeze. It does nothing of the kind; it swims, as do all other cephalopods, by forcing water through its siphon, its broad arms clasping the shell and the others trailing behind. The *Argonautæ*, of which nine species have been described, are all pelagic, coming to the surface at the spawning season, and sinking to the bottom at other times. The male is but about an inch in length, being sometimes scarce a tenth of the size of the female.

The OCTOPODIDÆ are more littoral than the last family, and have the mantle connected to the visceral sac by muscular bands, while there are no aquiferous pores in the head; the arms are long and more or less webbed, and are furnished with one, two, or three rows of acetabula, and lateral fins may be present or absent.

First in order comes the *Octopus* of science, the subject of many a weird and blood-curdling tale. Most prominent among these is that description of the devil-fish given by Victor Hugo in his "Toilers of the Sea"; but, after so many have shown the falsity of this creation, it is only necessary to briefly refer to the wonderful composite. The French name for these animals (poulpe) is but little different from the term polyp; an encyclopedia was consulted for both words, and the information obtained combined, and then, for a name, the term *Cephaloptera* (the true devil-fish, one of the rays) was stolen, and the whole set forth in the most vigorous language; a description of the appearance and habits of an animal which never existed, and which never will. The *Octopi* sometimes reach a large size, *O. vuglaris* of the West Indies and the Mediterranean reaching a length of nine feet and a weight of sixty-eight pounds, while the *O. punctatus* of our Pacific coast "reaches a length of sixteen feet, or a radial spread of nearly twenty-eight feet," the body in such a specimen being about six inches in diameter and a foot in length. There is no satisfactory evidence that any species of *Octopus* has ever intentionally attacked a human being, or that any one has been seriously injured by them. In habits they are timorous, hiding among rocks and feeding on crabs and molluscs. In their nests will frequently be found bushels of clam shells, which tell the tale of their feasts.

Some fifty species of this genus have been described from the whole world, and within the past few years several have been ascertained to exist on or near the New England shores; but of these, with the exception of *O. bairdii*, a widely distributed form, specimens are as yet rare. The characters which separate *Octopus* from the other genera of the family are a small rounded body without fins, two rows of sessile suckers on each of the long and slender arms; the right arm of the third pair is hectocotylized in the male.

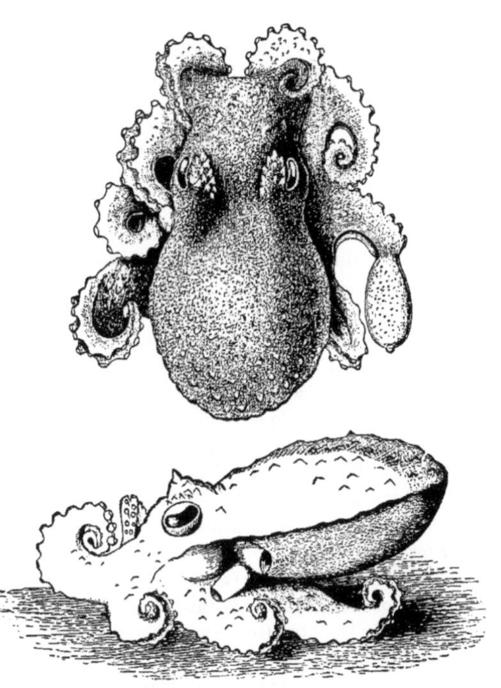

FIG. 489.—Dorsal and side Views of *Octopus bairdii*; the upper figure shows on the right side an hectocotylized arm.

The best known species is *O. vulgaris*, which has often been studied by European naturalists, and which plays an important part in the food supply of the countries around the Mediterranean. They are eaten especially during Lent, the traditions and decisions of the mother church not recognizing their meat as flesh, but rather as fish.

Professor Verrill thus describes the habits of our *Octopus bairdii*, specimens of which he kept in confinement for several days. "When at rest it remained at the bottom of the vessel, adhering firmly by some of the basal suckers of its arms, while the outer portions of the arms were curled back in various positions; the body was held in a nearly horizontal position, and the eyes were usually half closed and had a sleepy look. . . . When disturbed, or in any way excited, the eyes opened more widely, especially at night; the small tubercles over its surface, and the larger ones above the eyes, were erected, giving it a decided appearance of excitement and watchfulness.

"It was rarely, if ever, observed actually to creep about by means of its arms and suckers, but it would swim readily and actively, circling around the pan or jar, in which it was kept, many times before resting again. In swimming backward, the partial web connecting the arms together was used as an organ of locomotion, as well as the siphon; the web and the arms were alternately spread and closed, the closing being done energetically and coincidently with the ejection of the water from the siphon; and the arms, after each contraction, were all held pointing forward in a compact bundle, so as to afford the least resistance to the motion. As soon as the motion resulting from each impulse began to diminish sensibly the arms were again spread and the same actions repeated. This action of the web and arms recalled that of the disc of the jelly-fishes but it was more energetic.

"The siphon was bent in different directions to alter the direction of the motions, and, by bending it to the right or left side, backward motions in oblique or circular directions were given, but it was often bent directly downward and curved backward, so that the jet of water from it served to propel the animal directly forward. This, so far as observed, was its only mode of moving forward. The same mode of swimming forward has previously been observed in cuttle-fishes (*Sepia*) and in squids (*Loligo*)." It was much more active in the night, and in freedom is probably nocturnal. In captivity they could not be induced to touch food.

One species of *Eledone* (a genus in which the arms bear but a single row of suckers) is found in our waters, but of this but two specimens have as yet been found. Professor Verrill has called it *E. verrucosa*, in allusion to its warty appearance. In the European seas three species are found, one of which is named *E. moschata*, in allusion to its musky odor, which, however, is shared by some other cephalopods. In confinement, this species, when tranquil, is yellowish in color, and holds itself in much the same position as that described for *Octopus bairdii*, but a slight touch is all that is necessary to excite and enrage it. Then the color becomes a fine maroon, the tubercles on the body become prominent and the expansions and contractions of the mantle are very irregular; in short, the animal shows every appearance of anger. When swimming, which it accomplishes in much the same way as does *Octopus*, the body takes on another shade of yellow ornamented by bright spots of red; and when walking about by means of its arms and suckers, still another hue is seen.

Eledone moschata, as well as other species of the genus, is used as food, notwithstanding its musky taste and odor, which persists after death. Its flesh is more tender than that of *Octopus*, and is eaten boiled, fried, or in salads. This species frequents water usually from ten to twenty fathoms in depth, and in its capture the same methods are employed as in taking *Octopi*; earthen jars are lowered to the bottom, and, for some reason as yet unexplained, these animals soon esconce themselves inside. In a few hours the jars are pulled up and the prey taken.

ORDER II.—DECACERA.

The name usually applied to this division (Decapoda) is rather unfortunate, since it has also been used to indicate the group of Crustacea which contains the shrimps, lobsters, and crabs. For this reason the name Decacera has been substituted. Both terms have reference to the number of divisions of the foot which surround the head, the ten arms of this order contrasting strongly with the eight found in the last. Besides this, there are other characters to be enumerated. The body is usually long, and comes to an acute point behind, and bears on either side a fin and is strengthened internally by a horny 'pen' or a calcareous 'bone.' The two extra arms arise between the third and fourth pairs of the Octopods, and usually differ considerably from the others. They are usually longer and more slender, and have the basal portion narrower than the apex. This slender portion is without suckers, but the distal portions bear these organs, which, like those of the other arms, are supported on short stalks or peduncles. These long arms are usually called tentacles or tentacular arms.

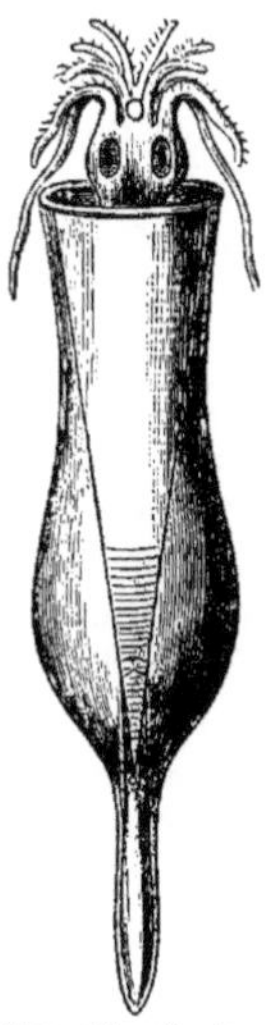

FIG. 490. — Restoration of Belemnite.

Among the fossil forms referred to this order, is the family known as BELEMNITIDÆ, the exact relationships of which are even yet uncertain. Different authorities have attempted to restore the animal, and one of these restorations, that of Quenstedt, is shown in the adjacent figure. Between these various restorations considerable differences exist. From certain well-preserved fossils it is known that the arms were furnished with hooks. The shell, however, is better preserved. It consists of a pen much like that of existing forms, but to this is added a calcareous chambered shell, the phragmocone, the partitions of which are traversed by a siphuncle. The apex of this phragmocone is enveloped in a second calcareous shell, the rostrum or guard. These fossils, which are very abundant in some parts of the Jurassic and cretaceous rocks, have received a large number of popular names, indicative of their general shape or of the superstitions and myths which have been associated with them. Among these may be mentioned arrowheads, thunder-bolts, petrified fingers, spectre candles, etc. In the outer chamber of the phragmocone of some specimens the remains of the ink sac have been found. Some of these fossils are very large, indicating an animal several feet in length.

As we have just said, the position and structure of the Belemnites are very uncertain. In some respects they seem to be related to *Spirula*; in others, to forms like *Ommastrephes*. Several attempts have been made to divide the living species of the order into sub-ordinal groups, the divisions being based upon the calcareous or horny nature of the shell, or upon the perforate or entire condition of the cornea, but with all considerable fault can be found, and the result is a highly artificial arrangement.

Of all the families the SPIRULIDÆ seems the most nearly related to the tetrabranchiates as well as to the Belemnites, and hence it is well to consider it first. The chambered shell of *Spirula* at once suggests that of *Nautilus* (the only other living genus in which it is present) but, as was first pointed out by Owen, the relationship of the body to the shell is diametrically opposite in the two; in *Nautilus* the ventral wall of the shell describes a convex curve; in *Spirula* the reverse is the case. Another important distinction exists in the fact that in the former the shell is external, in the latter internal; indeed, it seems to represent the phragmocone of the Belemnite, the guard and pen being absent, and the resemblance is strengthened by an examination of the remains of the genus *Spirulirostra*. In *Spirula*

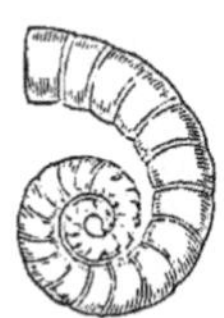

FIG. 491. — *Spirula peronii.*

the eye has an imperforate cornea, ten arms (one pair tentacular) which in the female bear six rows of minute suckers, but none in the male. The body is oblong, with minute terminal fins; the female has an oviduct on the right side, and two nidamental glands; in the male no hectocotylus is formed. *Spirula* is the only genus.

Although on some shores the dead shells are washed up by thousands, the animals are among the greatest rarities in the animal kingdom. Only three perfect, and a few mutilated specimens have as yet been found. Our knowledge of the anatomy is almost wholly due to Owen, whose papers on the subject are models in the line of research with very limited material. The shells of *Spirula fragilis* are occasionally cast up on the outer shores of Nantucket, while a living specimen was dredged by the U. S. coast survey in the West Indies in 1878. It came from a depth of nine hundred and fifty fathoms. *Spirula peronii* and *S. australis* come from the Indo-Pacific and Australian seas.

The family SEPIADÆ embraces the true cuttle-fishes, in which the internal shell is calcareous, the cornea imperforate, the body oval, with long fins, the tentacular arms very long, and capable of being completely retracted into pouches at the base; in the male, the left arm of the fourth pair is hectocotylized. The only genus, *Sepia*, is the one which furnishes the cuttle-bone which is found thrust between the bars of every canary cage, and whose ink is the basis of the pigment sepia. This is still manufactured in Rome. The cuttle-fishes are essentially littoral animals and are extensively caught, not only for the ink and the bone, but for the sake of the flesh, which is used as food.

The animals, when undisturbed, swim rather slowly and gracefully by means of undulations of their fins, but when startled or alarmed, the siphon and the arms are worked as in the case of *Octopus*, and then the fins are wrapped tightly around the body. They lay their eggs in black egg-shaped or pyriform capsules attached to submerged objects. They feed upon fishes, crabs, and molluscs. As in all cephalopods, the colors are changeable, but in *Sepia officinalis* zebra-like bands of blackish brown cross the body.

With the family LOLIGINIDÆ we take up some of the squids in which the body is more or less conical, tapering to a point behind. The fins are large, sometimes extending the whole length of the body. The cornea is entire, the eyes without lids, the tentacular arms but partially retractile and distally furnished with four rows of suckers, while on the other arms there are but two rows. The fourth arm on the left side is hectocotylized at the tip. The pen is long, slender, and flattened. Three living genera are known, only one of which, *Loligo*, is represented on our coasts; *Sepioteuthis* occurs in the West Indies, and *Loliolus* in the Pacific. *Loligo pealei* is the common squid south of Cape Cod; it extends south to the Carolinas, but north of the Cape it is not so common as *Ommastrephes illicebrosa*. A second species, *L. brevis*, extends from Virginia to Brazil, while *L. galei* occurs in the Gulf of Mexico. The anatomy, habits, and external development of the former species are pretty well known. They lay their eggs during the warmer months of the year, in large bunches of gelatinous capsules which are usually attached to some algæ. The young develop rather rapidly, and at certain times are taken in large numbers in the surface net. They are eagerly eaten, not only by fishes, but by the larger individuals of their own species. None of the squid are used in America as food, but immense numbers are caught and used as bait by the fishermen.

The SEPIOLIDÆ are closely related to the last family, but differ in having a short, thick body, rounded behind, the eyes with a lower and sometimes an upper lid, the tentacles more or less completely retractile, and the arms furnished with two rows of acetabula. The fins are separate, and attached near the middle of the body. This family is represented on our coasts by the genera *Sepiola*, *Rossia*, and *Heteroteuthis*, the species of which are far from common.

The species of *Cranchia*, the only genus of the family CRANCHIIDÆ, have a short, round body, with two small fins on the hinder end a small head, with large eyes, the cornea of which is perforated so that the sea-water penetrates to the lens, two rows of suckers on the arms, the tentacles long and armed with eight rows of acetabula. Closely allied is the family DESMOTEUTHIDÆ, with its two genera, *Desmoteuthis* and *Taonius*, in which the body is longer and pointed posteriorly.

In the LOLIGOPSIDÆ the form is longer and the fins are large, the head very small, the arms with two rows of suckers, the tentacles not retractile, the siphon without valves. Of *Histioteuthis* three species are known, two of which are found in the Mediterranean, while of the third, *H. collinsii*, found off Nova Scotia, one imperfect specimen and the beaks of two others are all that have been found. The other prominent genera are *Loligopsis*, *Mastigoteuthis*, *Chiroteuthis*, and *Thysanoteuthis*.

In the family TEUTHIDÆ the tentacular arms are distally armed with sharp, horny and recurved hooks, which, to a greater or less extent replace the sucking discs, and in some of the genera the other arms bear hooks as well as suckers, while others have hooks alone. The tentacular arms, which are used, as in all decacerous cephalopods, for the capture of the prey, can be fastened together by the sucking discs for a greater or less proportion of their length, leaving the extremities to form a living forceps, from which escape is next to impossible. The eyes are provided with lids, and the cornea is perforated as in the last family.

Possibly to be placed here is the large species *Moroteuthis robusta*, found by Mr. Dall in Alaska. Three mutilated specimens were seen, the largest of which, when found, had a total length of fourteen feet, but the terminal portions of the tentacular arms were gone, and so the actual size was somewhat greater. The length of the mantle was seven feet seven and a half inches, and the diameter of the body eighteen inches. A peculiarity exists in the pen, the posterior end of which "is one-sided, funnel-shaped close to the tip," and "inserted into a long, round, thick, firm, cartilaginous cone, which tapers to a point posteriorly," which thus, as pointed out by Professor Verrill, corresponds to the guard of the extinct Belemnites, both in position and in its relation to the pen. Nothing like this is known in any other recent genus.

Among the undoubted members of the family may be mentioned the genus *Onychoteuthis*, in which the forceps of the tentacular arms is well developed. The club which terminates each of these arms bears on its inner surface two rows of recurved hooks, while just below are a number of suckers which serve to unite the two arms. The sessile arms have suckers only. *O. banksii*, one of the ten known species, ranges from the Arctic seas to the Cape of Good Hope and the Indian Ocean. It is solitary in its habits, frequenting the open seas, and is most numerous round the banks of gulf weed. In *Enoploteuthis* and three or four other genera, all the sessile arms bear hooks but no suckers, the tentacular arms being much as in the last genus.

The last family to be mentioned is the OMMASTREPHIDÆ, in which the body is long and tapers to a point behind; the arms are short and without hooks, but furnished with two rows of suckers; the tentacular arms are not retractile, but terminate in an expanded club, armed with four rows of suckers. The eyes are provided with lids, and the cornea is perforated so that the salt water bathes the lens; the siphon is held by four bands and contains an internal valve (probably the homologue of the mesopodium).

The typical genus is *Ommastrephes*, of which one species, *O. illecibrosus*, is the most common squid north of Cape Cod, though further south it is less common, and replaced by *Loligo pealei.* From an economic point of view it is a very important species, for its distribution is coincident with that of the cod-fishery, in which it is used as bait. They swim in large schools, and are frequently found following schools of young mackerel and herring, on which they feed. They usually swim backwards, with the ease and grace of a trout, and a school of them furnishes a beautiful sight, their color changing instantly to accord with the bottom over which they pass. They are largely nocturnal, though they are seen in large numbers in the daytime. In the morning, on Cape Ann, the flats will sometimes be found covered by specimens left by the retreating tide during the night. When stranded they can do but little to help themselves. Their usual recourse is to begin to pump water through the siphon, but this forces them farther on the beach as frequently as it aids them in their escape to deeper water. Left on the shore, and exposed to the air, they quickly die. In confinement they usually live but a short time, as they are very timorous and dart backward with great velocity on the slightest alarm. In doing this they almost always strike the hinder end of the body against the walls of the tank in which they are kept, inflicting injuries which soon prove fatal. At times I have known them to leap from the tank in which they were kept, clearing the sides, which extended eight inches above the water.

In confinement I have watched them capture small fishes. They advance stealthily towards the proposed victim by undulations of the fins, when they suddenly seize it by means of the tentacular arms, and kill it by biting a piece out of the back of the neck with their powerful jaws. In their natural freedom, according to Professor S. I. Smith, they dart suddenly backward amongst the school of young fish, and then suddenly turn obliquely to the right or left to seize their victim.

In some places they are caught for bait by driving the school on shore; at other times by the use of nets and pounds. This latter method is the one most adopted by the fishermen on Cape Ann. In Newfoundland 'jigging' is the process employed. A jig is a bit of lead armed with hooks radially arranged, which is let down from the boat and kept constantly moving up and down. This in some way exerts a fatal fascinating power upon the squid, which seizes it, and, becoming entangled in the hooks, they are quickly drawn to the surface. In Japan an allied species is caught in a similar manner. The number of squid used annually in the cod fishery is almost beyond computation; a small vessel in six weeks will sometimes use eighty thousand squid. For the purposes of bait some are kept fresh, and others are salted or pickled in brine. As would be supposed from their being used as bait, many fish are fond of squid, and not unfrequently schools of these molluscs which are pursuing young mackerel are in turn pursued by older fish of the same species.

In the older works many allusions are made to huge marine monsters, the description and figure of Pontoppidian being most frequently copied. In all of these accounts, the animal partakes of the nature of a squid or poulpe. One of these early accounts, which relates to a specimen stranded on the coast of Ireland, is more accurate than most, and is withal so quaint that it deserves to be quoted here. It was published in 1673, and is quoted from Prof. Verrill.

" *The Monster Described.*

"This Monster was taken at Dingle-I-cosh in the county of Kerry, being driven up by a great storm in the Month of October last 1673; having two heads, one great head (out of which spring a little head two foot, or a yard from the great head) with two great eyes, each as big as a pewter dish, the length of it being about nineteen foot, bigger in the body than any horse, of the shape represented by this figure, having upon the great head ten horns, some of six some of eight or ten one of eleven foot long, the biggest horns as big as a man's Leg, the least as big as his wrist, which horns it threw from it on both sides; And to it again to defend itself having two of the ten horns plain, and smooth that were the middle and biggest horns, the other eight had one hundred Crowns a peece, placed by two and two on each of them, in all 800 crowns, each Crown having teeth, that tore anything that touched them, by shutting together the sharp teeth, being like the wheels of a watch, the Crowns were as big as a mans thumb or something bigger, that a man might put his finger in the hollow part of them, and had in them something like a pearl or eye in the middle; over this Monster's back was a mantle of a bright Red Color, with a fringe round it, it hung down on both sides like a Carpet or a table, falling back on each side, and faced with white; the crowns and mantle were glorious to behold: This monster had not one bone about him, nor skin nor scales, or feet but had a smooth skin like a man's belly. It swoom by the lappits of the mantle; The little head it could dart forth a yard from the great, and draw it in again at plesure, being like a hawks beak and having in the little head two tongues by which it is thought it received all its nourishment; when it was dead and opened the liver wayed 30 pounds. The man that took it came to Clonmel the 4th of this instant December, with two of the horns in a long box with the little head, and the figure of the fish drawn on a painted-cloth, which was the full proportion of it, and he went up to Dublin, with an intent to shew it to the Lord Lieutenant."

With our present knowledge it is easy to recognize this monster as one of the giant squids, belonging, doubtless, to the genus *Architeuthis*. The whalers have long had accounts of the sperm whale eating giant squid, portions of the arms being vomited by these animals in their death flurry, but science has recognized the existence of these huge monsters for only a few years, and for the greatest portion of our knowledge we are indebted to Prof. Verrill. On our shores the first reliable account was published in 1873, and merely described the jaws of a large individual found on the Grand Banks. Since that time several specimens or parts of specimens have been found, the number at present amounting to nearly thirty. These are at present referred to the species *Architeuthis princeps*, *A. harveyi*, and *A. megaptera*. Some five or six other species have been described from other parts of the world, but on our coasts all the specimens as yet received have come from the Grand Banks, or Newfoundland. In shape, general appearance, and almost everything except size, they are closely allied to the little species of *Ommastrephes*, but in size the difference is enormous. The Irish specimen of two hundred years ago had a length of thirty-one feet, but apparently the tentacular arms were damaged. All of the specimens which have as yet been examined by scientific men have been more or less imperfect. Yet some of the measurements may prove interesting. One caught on the coast of Labrador, and used as dog's meat, was said to have a total length of fifty-two feet, of which thirty-seven belonged to the tentacular arms; another, cast on shore at Catalina, Newfoundland, in 1877, had a head and body nine and a half feet long, and tentacular arms thirty feet in length, the circumference of the body being seven feet. This was the specimen exhibited at the New York Aquarium, and afterwards in other parts of the country. It was the best specimen ever obtained. These large specimens all belong to the species called *A. princeps*. The largest specimen thus far seen may also possibly belong to this species. It was described by the Rev. Mr. Harvey of Newfoundland, who has been the source of a large amount of our information about these monsters, and who has supplied Prof. Verrill with some of his specimens and much of his data. His account, which was published in the Boston *Traveller* of Jan. 30, 1879, runs as follows: —

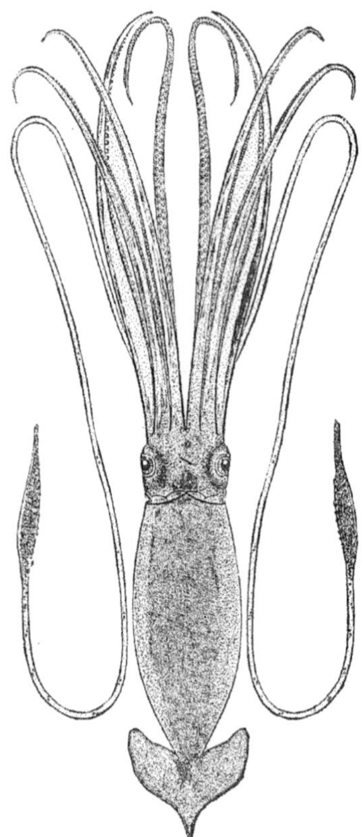

FIG. 492. — *Architeuthis princeps*, giant squid, one fiftieth the natural size.

"On the second day of November last, Stephen Sherring, a fisherman residing in Thimble Tickle (Notre Dame Bay), not far from the locality where the other devil-fish was cast ashore, was out in a boat with two other men; not far from the shore they observed some bulky object, and supposing it might be part of a wreck, they rowed toward it, and, to their horror, found themselves close to a huge fish, having large glassy eyes, which was making desperate efforts to escape, and churning the water into foam by the motion of its immense arms and tail. It was aground, and the tide was ebbing. From the funnel at the back of its head it was ejecting large volumes of water, this being its method of moving backward, the force of the stream by the reaction of the surrounding medium, driving it in the required direction. At times the water from the siphon was black as ink.

"Finding the monster partially disabled, the fishermen plucked up courage and ventured near enough to throw the grapnel of their boat, the sharp flukes of which, having barbed points, sunk into the soft body. To the grapnel they attached a stout rope which they carried ashore and tied to a tree, so as to prevent the fish from going out with the tide. It was a happy thought for the devil-fish found himself effectually moored to the shore. His struggles were terrific as he flung his ten arms about in dying agony. The fishermen took care to keep a respectful distance from the long tentacles which ever and anon darted out like great tongues from the central mass. At length it became exhausted, and as the water receded it expired.

"The fishermen, alas! knowing no better, proceeded to convert it into dog's meat. It was a splendid specimen — the largest yet taken — the body measuring twenty feet from the beak to the extremity of the tail. It was thus exactly double the size of the New York [Aquarium] specimen, and five feet longer than the one taken by Budgell.

The circumference of the body was not stated, but one of the arms measured thirty-five feet. This must have been a tentacle."

In order that the reader may obtain some idea of the size of these giants of the Mollusca, we introduce side by side figures of the jaws of one of the common squid (*Loligo*) and of the giant squid (*Architeuthis*), both natural size.

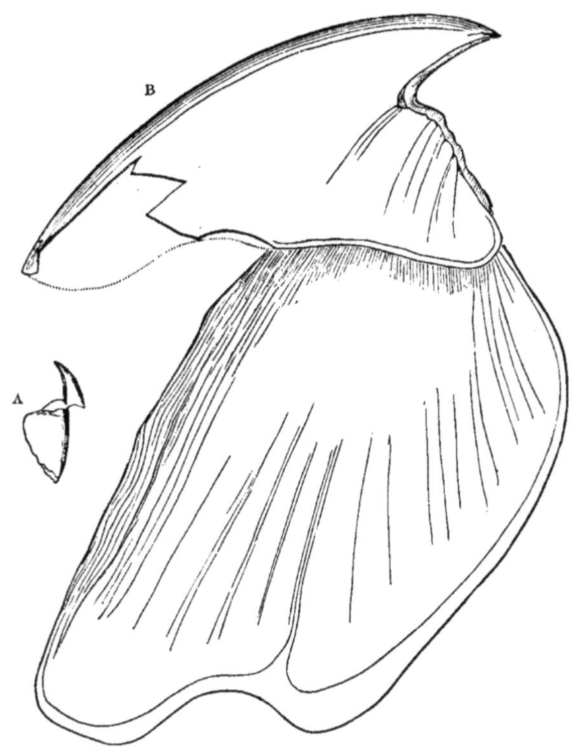

FIG. 493.—Jaws of squid; A, *Loligo*; B, *Architeuthis*.

Other giant squid have been taken in various parts of the world, Iceland, Sweden, Ireland, Japan, New Zealand, etc., but none equal in size the specimen just mentioned. Of the habits of these monsters we know nothing; they are apparently nocturnal, as the specimens have almost always been cast ashore in the night. It is thought that our species frequent the deep fiords which cut the Newfoundland coasts, hiding in the dark places during the day.

J. S. KINGSLEY.

SPECIAL NOTICE.

The dates of issuance of the parts composing this volume are as follows: pages 1 to 48, October 15, 1883; pages 49 to 144, June 12, 1884; pages 145 to 192, July 27, 1884; pages 193 to 272, December 2, 1884; pages 273 to 390, December 27, 1884; Introduction, January 2, 1885.

INDEX.

Lightning Source UK Ltd.
Milton Keynes UK
UKHW052348190219
337529UK00021B/976/P

9 781331 982494